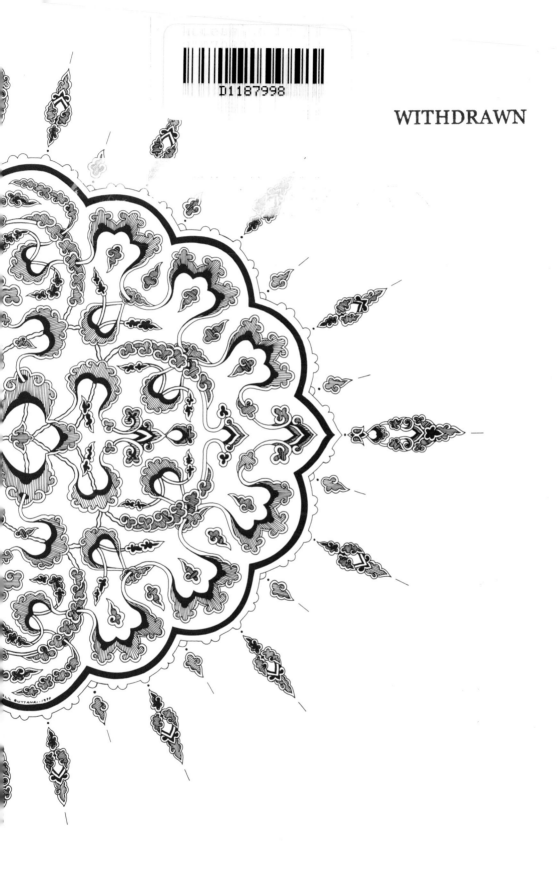

Texts in Applied Mathematics 3

Springer
New York
Berlin
Heidelberg
Barcelona
Hong Kong
London
Milan
Paris
Singapore
Tokyo

Texts in Applied Mathematics

(continued after index)

Jack K. Hale Hüseyin Koçak

Dynamics and Bifurcations

With 314 Illustrations

Springer

Jack K. Hale
School of Mathematics
Georgia Institute of Technology
Atlanta, GA 30332
USA
hale@math.gatech.edu

Hüseyin Koçak
Department of Mathematics and
    Computer Science
University of Miami
Coral Gables, FL 33124
USA
hk@math.miami.edu

Editors

J.E. Marsden
Control and Dynamical Systems 107-81
California Institute of Technology
Pasadena, CA 91125
USA

L. Sirovich
Division of Applied Mathematics
Brown University
Providence, RI 02912
USA

M. Golubitsky
Department of Mathematics
University of Houston
Houston, TX 77204-3476
USA

W. Jäger
Department of Applied Mathematics
Universität Heidelberg
Im Neuenheimer Feld 294
69120 Heidelberg, Germany

Cover and text art by Halil Buttanri.

Mathematics Subject Classifications: 58 Fxx, 34 xx, 58 F 14

Library of Congress Catalog Card Number: 92-10512

Printed on acid-free paper.

Photocomposed copy prepared from the authors' T_EX file.
Printed and bound by Edwards Brothers, Inc., Ann Arbor, MI.
Printed in the United States of America.

9 8 7 6 5 4

ISBN 0-387-97141-6 Springer-Verlag New York Berlin Heidelberg
ISBN 3-540-97141-6 Springer-Verlag Berlin Heidelberg New York   SPIN 10747387

To Students:

Who are the primary reason
for the existence of
our profession
and
this book

Greeting

Thank you for opening our book. Inside you will find ideas and examples about the geometry of dynamics and bifurcations of ordinary differential and difference equations. As it is an unusual book in both content and style, let us explain how it evolved from our courses in the Division of Applied Mathematics at Brown University during a three-year period, and came into being.

The subject of differential and difference equations, alias *dynamical systems*, is an old and much-honored chapter in science, one which germinated in applied fields such as celestial mechanics, nonlinear oscillations, and fluid dynamics. Over the centuries, as a result of the efforts of scientists and mathematicians alike, an attractive and far-reaching theory has emerged. In recent years, due primarily to the proliferation of computers, dynamical systems has once more turned to its roots in applications with perhaps a more mature outlook. Currently, the level of excitement and activity, not only on the mathematical front but in almost all allied fields of learning, is unique. It is the aim of our book to provide a modest foundation for taking part in certain theoretical and practical facets of these exciting developments.

The subject of dynamical systems is a vast one not easily accessible to undergraduate and beginning graduate students in mathematics or science and engineering. Many of the available books and expository narratives either require extensive mathematical preparation, or are not designed to be used as textbooks. It is with the desire to fill this void that we have written the present book.

It is both our conviction and our experience that many of the fundamental ideas of dynamics and bifurcations can be explained in a simple setting, one that is mathematically insightful yet devoid of extensive formalism. Accordingly, we have opted in the present book to proceed by low-dimensional dynamical systems. We will momentarily give a brief summary of some of the central topics of our book, one which necessarily contains some technical terms. If you are a beginning student of dynamics, however, rest assured that precise mathematical definitions of all these terms, as well as ample realizations of the dynamical phenomena in specific equations, will unfold as you turn the pages.

Equations in dimensions one, "one and one half," and two constitute the majority of the text. Indeed, nearly one hundred pages are devoted to scalar equations where, despite their simplicity and apparent triviality, many of the contemporary ideas of our subject are already visible. We demonstrate, in particular, that the basic notions of stability and bifurcations of vector fields are easily explained for scalar autonomous equations—dimension one—because their flows are determined from the equilibrium points. We also explore how numerical solutions of such equations lead to scalar maps, and show some of the "anomalies," albeit profound and exciting, that may arise when numerical approximation is poor—period-doubling bifurcation, chaos, etc. We then turn to the dynamics and bifurcations of periodic solutions of nonautonomous equations with periodic coefficients—dimension one and one half—where scalar maps reappear naturally as Poincaré maps. In our discussion of the stability of periodic solutions of such equations, we demonstrate how one naturally encounters elementary but essential ideas from the transformation theory of differential equations—normal form theory. These ideas, presented in the context of scalar equations, and more importantly, the philosophical outlook of the subject that these ideas convey, recur frequently in later chapters, with a few technical embellishments.

We next proceed to investigate the dynamics of planar autonomous equations—dimension two—where, in addition to equilibria, new dynamical behavior, such as periodic and homoclinic orbits, appears. In studying the stability of an equilibrium point, we touch upon certain subtle topological aspects of linear systems as well as the standard theory of Liapunov functions. The bifurcation theory of equilibrium points of planar equations gives rise to a number of new ideas. When, for example, the bifurcation is to other equilibria, one is led naturally to introduce center manifolds and the method of Liapunov–Schmidt to make a reduction to a scalar autonomous equation. The other important bifurcation from an equilibrium point is to a periodic orbit—Poincaré–Andronov–Hopf bifurcation—and its analysis can be reduced to that of a nonautonomous periodic equation. There are, of course, other properties of planar differential equations that are more global in character and hence cannot be investigated in terms of scalar equations. Among these interesting topics, we have chosen to include the Poincaré–Bendixson theory of planar limit sets, geometry and bifurcations of conservative and gradient systems, and a discussion of structural stability—with, of course, an emphasis on the ideas rather than on extensive technical details.

We subsequently include an abbreviated discussion of certain aspects of the theory of planar maps. As in the case of scalar equations, we explore some of the difficulties associated with numerical solutions of differential equations in light of the dynamics and bifurcations of such maps. To indicate not only the richness but also the bewildering complexity of this topic, we include computer simulations of some of the famous maps.

The final part of the book consists of several substantial examples in dimensions "two and one half," three, and four. This section is more discursive than the previous ones; it is more like a preview designed to provide a smooth entry into certain areas of current research—forced oscillations, strange attractors, chaos, completely integrable Hamiltonian systems, etc.

For a more detailed list of the topics covered in the book, you are, of course, invited to browse through the Table of Contents. As you peruse the entries, however, bear in mind that in dynamical systems, as in most parts of mathematics, while general theorems certainly occupy a central place, one must ultimately face the task of analyzing the dynamics of specific equations, especially in applications. Moreover, even for the most abstractly inclined, grappling with specific examples usually proves to be an irreplaceable source of general theoretical observations. With this philosophy in mind, the text and the exercises alike are interwoven with numerous specific differential and difference equations of theoretical and practical interest. Unfortunately, unraveling the dynamics of specific equations often turns out to be analytically insurmountable. The computer, in all its present versatility, is, however, beginning to prove its utility in this pursuit. On this new front, our favorite computer program is, of course, *PHASER: An Animator/Simulator for Dynamical Systems,* accompanying one of our earlier books *Differential and Difference Equations through Computer Experiments.* Our students have found *PHASER* to be an ideal medium to see the "dynamics" in dynamical systems and to do some of their assignments; we, too, used it to produce many of the illustrations for our book.

Dynamical systems is a vast and vibrant area. We hope that *Dynamics & Bifurcations* will arouse your interest in bifurcation theory sufficiently that you will be inclined to explore this exciting subject further using, of course, our other favorite book—*Methods of Bifurcation Theory.*

We would like to record in closing our gratitude to those who have contributed unselfishly to the realization of our book. In particular, the enthusiastic participation of our students—a lively group consisting of undergraduate and graduate students of pure and applied mathematics, and of science and engineering—helped considerably in fixing our ideas and setting realistic bounds for our own enthusiasm. Critical readings of the text and insightful suggestions by Nathaniel Chafee, Brian Coomes, Philip

Davis, Robert Griffiths, Henry Hamman, Albert Harum–Alvarez, Arnoldo Horta, Şahin Koçak, Nancy Lawther, Alan Lazer, Konstantin Mischaikow, Kenneth Palmer, Jack Pipkin, Plácido Táboas, Natalia Sternberg, Michael Wolfson, Gaetano Zampieri, and Lee Zia have been invaluable in shaping the manuscript. We are equally grateful to Halil Buttanrı for his classical art work, and Fred Bisshopp, Kim Foster–Cosner, Sam Fulcomer, Stuart Geman, Donald McClure, Andrew Mossberg, and Jim Yorke for their assistance on matters *de modus vivendi*.

Jack Hale and Hüseyin Koçak
August 1991, Druid Hills

Contents

PART I: Dimension One

PART II: Dimension One and One Half

PART III: Dimension Two

PART IV: Higher Dimensions

1D

HALIL BUTTANRI - 1991

HALIL BUTTANRI - 1991

1

Scalar Autonomous Equations

 In this opening chapter, we present selected basic concepts about the geometry of solutions of ordinary differential equations. To keep the ideas free from technical complications, the setting is one-dimensional—the scalar autonomous differential equations. Despite their simplicity, these concepts are central to our subject and reappear in various incarnations throughout the book. Following a collection of examples, we first state a theorem on the existence and uniqueness of solutions. Then we explain what a differential equation is geometrically. To facilitate qualitative analysis, geometric concepts such as vector field, orbit, equilibrium point, and limit set are included in this discussion. The next topic is the notion of stability of an equilibrium point and the role of linear approximation in determining stability. We conclude the chapter with an example of a scalar differential equation defined on a one-dimensional space other than the real line—a circle.

1.1. Existence and Uniqueness

In this introductory section we establish our notation for differential equations and their solutions. Then, after several motivational examples, we state a basic existence and uniqueness theorem.

Let I be an open interval of the real line $\mathbb{R}$ and let

$$x : I \to \mathbb{R}; \quad t \mapsto x(t)$$

be a real-valued differentiable function of a real variable t. We will use the notation $\dot{x}$ to denote the derivative dx/dt, and refer to t as *time* or the *independent variable*. Also, let

$$f : \mathbb{R} \to \mathbb{R}; \quad x \mapsto f(x)$$

be a given real-valued function. In Chapter 1, we will consider differential equations of the form

$$\dot{x} = f(x), \tag{1.1}$$

where x is an unknown function of t and f is a given function of x. Equation (1.1) is called a *scalar autonomous* differential equation; scalar because x is one dimensional (real-valued) and autonomous because the function f does not depend on t.

We say that a function x is a solution of Eq. (1.1) on the interval I if $\dot{x}(t) = f(x(t))$ for all $t \in I$. We will often be interested in a *specific solution* of Eq. (1.1) which at some initial time $t_0 \in I$ has the value x_0. Thus we will study x satisfying

$$\dot{x} = f(x), \qquad x(t_0) = x_0. \tag{1.2}$$

Equation (1.2) is referred to as an *initial-value problem* and any of its solutions is called a *solution through* x_0 *at* t_0. A useful consequence of the autonomous character of the differential equation in Eq. (1.2) is that there is no loss of generality in assuming that the initial value-problem is specified with $t_0 = 0$, and we will often tacitly do so. To wit, let $x(t)$ be a solution of Eq. (1.2) through x_0 at t_0 and define $y(t) \equiv x(t + t_0)$. Now, observe that $y(t)$ is a solution of Eq. (1.2) through x_0 at zero since

$$\dot{y}(t) = \dot{x}(t + t_0) = f(x(t + t_0)) = f(y(t)) \quad \text{and} \quad y(0) = x_0.$$

As you may recall from your previous studies, a solution of Eq. (1.2) through x_0 at t_0 is given implicitly, using the method of "separation of variables," by the formula

$$\int_{x_0}^{x} \frac{1}{f(s)}\,ds = t - t_0, \tag{1.3}$$

when the integral is defined. One obtains $x(t)$ by finding the inverse of the function on the left-hand side of this equation. Occasionally, we will use this formula to exhibit solutions of special differential equations for the purposes of illustrations. However, in general, it is impossible to perform these integrations and one should not expect to obtain explicit formulas for solutions. It is important to realize this fact from the beginning. In fact, our objective in this book is to understand as much as possible about the behavior of solutions of differential equations without the knowledge of an explicit formula for the solutions.

Let us now give several examples of differential equations and their solutions in order to realize some of the difficulties that arise in laying the foundations for the theory, that is, the existence and the uniqueness of solutions of Eq. (1.2).

Example 1.1. *The first example:* Consider the differential equation

$$\dot{x} = -x. \tag{1.4}$$

It can be seen by simple differentiation that $x(t) = e^{-t}x_0$ is a solution through x_0 at $t_0 = 0$, and it is defined for all $t \in \mathbb{R}$. Question: is this the only solution of Eq. (1.4) satisfying the initial value $x(0) = x_0$? ◇

Example 1.2. *Finite time:* Consider the initial-value problem

$$\dot{x} = x^2, \qquad x(0) = x_0. \tag{1.5}$$

It is easy to verify by direct substitution, or using formula (1.3), that the function

$$x(t) = \frac{x_0}{1 - x_0 t}$$

is a solution. Notice that, although the function $f(x) = x^2$ is remarkably "nice," the solution $x(t)$ is defined on the interval $(-\infty, 1/x_0)$ for $x_0 > 0$, on $(-\infty, +\infty)$ for $x_0 = 0$, and $(1/x_0, +\infty)$ for $x_0 < 0$. The importance of this example is that the solution is not always defined on all of $\mathbb{R}$ and the interval of definition of the solution varies with the initial condition. Furthermore, the solution becomes unbounded as t approaches $1/x_0$, the boundary of the interval of definition. ◇

Example 1.3. *Multiple solutions:* Consider the initial-value problem

$$\dot{x} = \sqrt{x}, \qquad x(0) = x_0, \qquad \text{with} \quad x \geq 0.$$

A solution is given by $x(t) = \left(t + 2\sqrt{x_0}\right)^2/4$. If $x_0 = 0$, then there is also the solution which is identically zero for all t. Therefore, the initial-value problem above *does not have a unique solution* through x_0 at zero.

In this example, the domain of $f(x) = \sqrt{x}$ is naturally restricted to a subset of $\mathbb{R}$. In applications, this situation arises often, for instance, a population of insects cannot grow to be negative. $\diamond$

The examples above show the necessity of certain conditions on the function f in order to guarantee the existence and the uniqueness of solutions to the initial-value problem Eq. (1.2). We will state such a theorem below, and also present a more general result in the Appendix. First, however, we need to introduce a small piece of notation.

We will denote the set of all continuous functions $f : \mathbb{R} \to \mathbb{R}$ by $C^0(\mathbb{R}, \mathbb{R})$, and the set of all differentiable functions with continuous first derivatives by $C^1(\mathbb{R}, \mathbb{R})$. Analogously, we will use $C^n(\mathbb{R}, \mathbb{R})$ to indicate the functions with continuous derivatives up through order n. If the domain of functions is a subset U of $\mathbb{R}$, then we will use the notation $C^0(U, \mathbb{R})$, etc. If there is no ambiguity, we will usually omit the dependence on the domain and simply refer to a member of one of these sets as a C^0, C^1, or C^n function, etc. In the case of a real-valued continuous function of several variables, $f : \mathbb{R}^k \to \mathbb{R}$ is said to be a C^1 function if all the first partial derivatives are continuous.

To emphasize the dependence of a solution $x(t)$ of Eq. (1.2) through x_0 at $t_0 = 0$ on the initial condition, we will often use the notation $\varphi(t, x_0)$ for this solution. In other words, $\varphi(t, x_0) = x(t)$ and $\varphi(0, x_0) = x_0$.

Theorem 1.4. (Existence and Uniqueness of Solutions)
(i) *If $f \in C^0(\mathbb{R}, \mathbb{R})$, then, for any $x_0 \in \mathbb{R}$, there is an interval (possibly infinite) $I_{x_0} \equiv (\alpha_{x_0}, \beta_{x_0})$ containing $t_0 = 0$ and a solution $\varphi(t, x_0)$ of the initial-value problem*

$$\dot{x} = f(x), \qquad x(0) = x_0,$$

defined for all $t \in I_{x_0}$, satisfying the initial condition $\varphi(0, x_0) = x_0$. Also, if α_{x_0} is finite, then

$$\lim_{t \to \alpha_{x_0}^+} |\varphi(t, x_0)| = +\infty,$$

or, if β_{x_0} is finite, then

$$\lim_{t \to \beta_{x_0}^-} |\varphi(t, x_0)| = +\infty.$$

(ii) *If, in addition, $f \in C^1(\mathbb{R}, \mathbb{R})$, then $\varphi(t, x_0)$ is unique on I_{x_0} and $\varphi(t, x_0)$ is continuous in (t, x_0) together with its first partial derivatives, that is, $\varphi(t, x_0)$ is a C^1 function.* $\diamond$

The largest possible interval I_{x_0} in part (i) of the theorem above is called the *maximal interval of existence* of the solution $\varphi(t, x_0)$. The maximal interval of existence of a solution of Example 1.2 is shown in Figure 1.0.

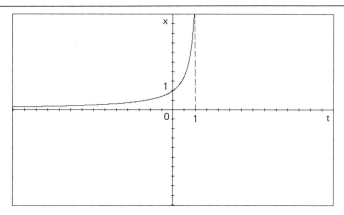

Figure 1.0. *Maximal interval of existence of the solution of $\dot{x} = x^2$ with initial value $x(0) = 1$ is $(-\infty, 1)$.*

In applications, the function f may not be defined on all of $\mathbb{R}$. One common situation is that $f \in C^n(U, \mathbb{R})$, where U is an open and bounded subset of $\mathbb{R}$. In this case, the conclusions of Theorem 1.4 are the same except that all of the limit points of $\varphi(t, x_0)$ as $t \to \alpha_{x_0}^+$ (or $t \to \beta_{x_0}^-$) must belong to the boundary of U.

Let us now return briefly to the notation $\varphi(t, x_0)$ for the solution of an initial-value problem and reexamine it in light of our foregoing discussions. For a given C^1 function f, Theorem 1.4 implies that the family of all specific solutions of $\dot{x} = f(x)$ can be represented by $\varphi(t, x_0)$ viewed as a function of two variables, where $t \in I_{x_0}$ and $x_0 \in \mathbb{R}$. As such, $\varphi(t, x_0)$ is called the *flow* of $\dot{x} = f(x)$. The domain of this function of two variables could be somewhat complicated because the domain of t may depend on x_0, as seen in Example 1.2.

The fine structures of flows will be one of our main concerns in the following chapters. For the moment, we will be content to introduce a common name for our subject—dynamical systems. If f is a C^1 function, then, for each t, the flow $\varphi(t, x_0)$ gives rise to a map of $\mathbb{R}$ into itself (with possibly restricted domain) given by $x_0 \mapsto \varphi(t, x_0)$. Here are some of the important properties of this map:

(i) $\varphi(0, x_0) = x_0$,

(ii) $\varphi(t + s, x_0) = \varphi(t, \varphi(s, x_0))$ for each t and s when the map on either side is defined,

(iii) $\varphi(t, x_0)$ is a C^1 map for each t and it has a C^1 inverse given by $\varphi(-t, x_0)$.

A map of $\mathbb{R}$ into itself satisfying these three properties is called a C^1 *dynamical system* on $\mathbb{R}$. So, we can say in conclusion that the flow of a scalar autonomous differential equation gives rise to a dynamical system on $\mathbb{R}$. There are also other ways of obtaining dynamical systems and we will see one such important case in Chapter 3.

Exercises ———————————————————— ♣ ♡ ♠ ◇

1.1. *Verifying hypotheses:* Show that the initial-value problem $\dot{x} = 1+x^2$, $x(0) = 0$ has a unique solution by verifying the hypotheses of Theorem 1.4. Find the solution. What is the maximal interval of existence of the solution?

1.2. *Multiple solutions and numerics:* Reexamine Example 1.3, $\dot{x} = \sqrt{x}$, with the initial value $x(0) = 0$ in light of Theorem 1.4. Which hypothesis of the theorem is not met by the example to prevent the uniqueness of solutions? Solve this initial value problem numerically on the computer. If you have not studied numerical solutions of differential equations before, you may want to return to this task after reading Chapter 3 or consult the reference below. Which solution do you obtain? Try several different numerical algorithms. Do you succeed in obtaining the nonzero solution?
   Help: In the sequel, some of the exercises will include numerical experiments with differential equations. To eliminate the burden of programming, we suggest the computer program *PHASER: An Animator/Simulator for Dynamical Systems* which accompanies one of our earlier books, Koçak [1989].

1.3. *Infinitely many solutions:* Show that the differential equation $\dot{x} = x^{2/3}$ has infinitely many solutions satisfying $x(0) = 0$ on every interval $[0, a]$.

1.4. *No solution:* Consider the function

$$f(x) = \begin{cases} 1 & \text{if } x \leq 0 \\ 2 & \text{if } x > 0. \end{cases}$$

Show that the differential equation $\dot{x} = f(x)$ has no solution satisfying $x(0) = 0$ on any open interval about $t_0 = 0$.

1.2. Geometry of Flows

In preparation for the qualitative study of differential equations we now reconsider the scalar autonomous differential equation (1.1) and its flow $\varphi(t, x_0)$ from a geometric point of view.

At each point on the (t, x)-plane where $f(x)$ is defined, the right-hand side of Eq. (1.1) gives a value of the derivative dx/dt which can be thought of as the slope of a line segment passing through that point. The collection of all such line segments is called the *direction field* of the differential equation (1.1).

The graph of a solution of Eq. (1.2) through x_0, that is, the subset of the (t, x)-plane defined by $\{ (t, \varphi(t, x_0)) : t \in I_{x_0} \}$ is called the *trajectory* through x_0. A trajectory is tangent to the line segments of the direction field at each point on the plane it passes through. The direction fields and several representative trajectories of Examples 1.1 and 1.2 are shown in Figures 1.1a and 1.2a, respectively.

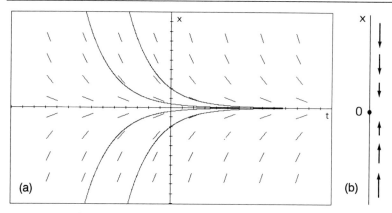

Figure 1.1. (a) *Direction field along with several trajectories, and* (b) *vector field of* $\dot{x} = -x$.

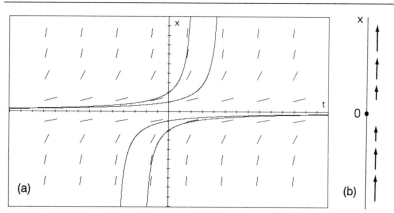

Figure 1.2. (a) *Direction field along with several trajectories, and* (b) *vector field of* $\dot{x} = x^2$.

Since $f(x)$ is independent of t, on any line parallel to the t-axis the segments of the direction field all have the same slope. Therefore, it is natural to consider the projection onto the x-axis of the direction field and the trajectories of Eq. (1.1).

To each point x on the x-axis we can associate the directed line segment from x to $x + f(x)$. We can view this directed line segment as a vector based at x. The collection of all such vectors is called the *vector field* generated by Eq. (1.1), or simply the vector field f; see Figures 1.1b and 1.2b.

Definition 1.5. *The positive orbit* $\gamma^+(x_0)$, *negative orbit* $\gamma^-(x_0)$, *and orbit*

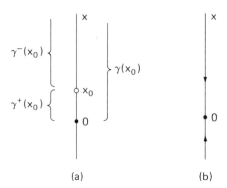

Figure 1.3. (a) *Positive orbit, negative orbit, orbit through x_0, and* (b) *phase portrait of $\dot{x} = -x$.*

$\gamma(x_0)$ of x_0 are defined, respectively, as the following subsets of the x-axis:

$$\gamma^+(x_0) = \bigcup_{t \in [0, \beta_{x_0})} \varphi(t, x_0),$$

$$\gamma^-(x_0) = \bigcup_{t \in (\alpha_{x_0}, 0]} \varphi(t, x_0),$$

$$\gamma(x_0) = \bigcup_{t \in (\alpha_{x_0}, \beta_{x_0})} \varphi(t, x_0).$$

The "velocity" of an orbit at a point x is given by the element of the vector field at that point; see, for example, Figures 1.1b and 1.2b. As shown in Figures 1.3 and 1.4, on the orbit $\gamma(x_0)$ we insert arrows to indicate the direction $\varphi(t, x_0)$ is changing as t increases. The flow of a differential equation is then drawn as the collection of all its orbits together with the direction arrows and the resulting picture is called the *phase portrait* of the differential equation; see Figures 1.3b and 1.4b.

It is clear from the definitions above that the orbit $\gamma(x_0)$ is the projection onto the x-axis of the trajectory through x_0. Consequently, in our investigation of qualitative properties of solutions of differential equations, it will be advantageous to focus our attention on orbits (see Lemma 1.7 below). In the process, however, we will lose the information about the time-parametrization, the speed, of solutions. For instance, while it takes an infinite amount of time to trace the negative orbit $\gamma^-(x_0) = [x_0, +\infty)$, with $x_0 > 0$, of Example 1.1, it takes only finite time to trace the positive orbit $\gamma^+(x_0) = [x_0, +\infty)$, with $x_0 > 0$, of Example 1.2; see Figures 1.3a and 1.4a.

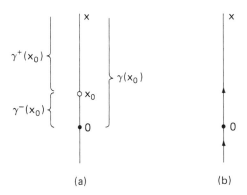

Figure 1.4. (a) *Positive orbit, negative orbit, orbit through* x_0, *and* (b) *phase portrait of* $\dot{x} = x^2$.

There are some orbits which are especially simple, but they play a central role in the qualitative study of differential equations, as well as in applications.

Definition 1.6. *A point* $\bar{x} \in \mathbb{R}$ *is called an* equilibrium point *(also* critical point, steady state solution, *etc.) of* $\dot{x} = f(x)$, *if* $f(\bar{x}) = 0$.

When $\bar{x}$ is an equilibrium point, the constant function $x(t) = \bar{x}$ for all t is a solution, and thus the orbit $\gamma(\bar{x})$ is $\bar{x}$ itself.

It is very easy to draw orbits of Eq. (1.1) from the graph of $f(x)$. In fact, the sign of f determines the direction of the motion along an orbit. If $f(x_0) < 0$, then the solution is decreasing in t, and $\varphi(t, x_0)$ either approaches an equilibrium point or tends to $-\infty$ as $t \to \beta_{x_0}$. Similarly, if $f(x_0) > 0$, then the solution is increasing in t, and $\varphi(t, x_0)$ either approaches an equilibrium point or tends to $+\infty$ as $t \to \beta_{x_0}$; see Figures 1.5a and 1.6a. Furthermore, if solutions of the initial-value problem for Eq. (1.1) are unique, then the solutions through two different initial conditions with $x_0 < y_0$ satisfy $\varphi(t, x_0) < \varphi(t, y_0)$. Thus we have the following lemma:

Lemma 1.7. *Suppose that the solution* $\varphi(t, x_0)$ *of the initial-value problem is unique for every* x_0. *Then*
  (i) $\varphi(t, x_0)$ *is a monotone function in* t;
 (ii) $\varphi(t, x_0) < \varphi(t, y_0)$ *for all* t *if* $x_0 < y_0$;
(iii) *if* $\gamma^+(x_0)$ *[respectively,* $\gamma^-(x_0)$] *is bounded, then* $\beta_{x_0} = +\infty$ *[respectively,* $\alpha_{x_0} = -\infty$] *and* $\varphi(t, x_0) \to \bar{x}$ *as* $t \to +\infty$ *[respectively,* $t \to -\infty$], *where* $\bar{x}$ *is an equilibrium point.* $\Diamond$

Let us now illustrate the approach to drawing phase portraits discussed above. We note, however, that unlike the following example, it may not

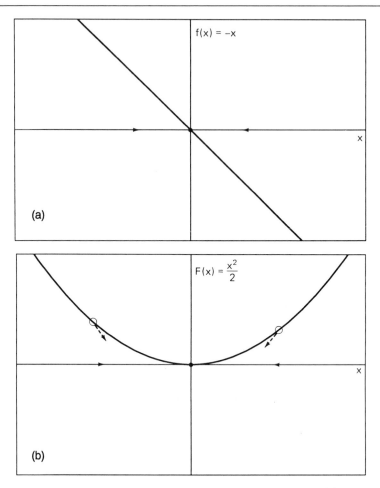

Figure 1.5. *Determining the phase portrait of* $\dot{x} = -x$ *from (a) the function* $f(x) = -x$, *and (b) from the corresponding potential function* $F(x) = x^2/2$. *Notice that, unlike in the previous figures, the variable* x *is assigned to the horizontal axis. In the subsequent figures we will use whichever assignment is more convenient.*

be so easy to locate the equilibria of a given differential equation. We will address this difficulty later.

Example 1.8. Consider the differential equation

$$\dot{x} = x - x^3. \tag{1.6}$$

The equilibrium points of this equation are -1, 0, and 1, and the function $f(x) = x - x^3$ is positive on the interval $(-\infty, -1)$, negative on $(-1, 0)$,

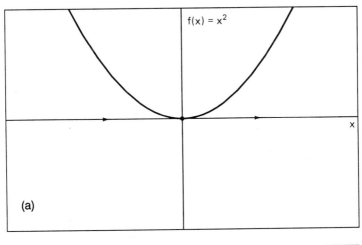

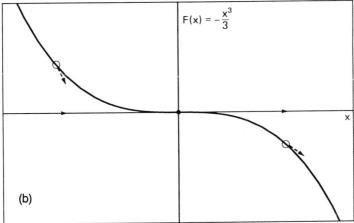

Figure 1.6. *Determining the phase portrait of $\dot{x} = x^2$ from (a) the function $f(x) = x^2$, and (b) from the potential function $F(x) = -x^3/3$.*

positive on $(0, 1)$, and negative on $(1, +\infty)$. Therefore, its phase portrait can easily be drawn as in Figure 1.7a. The orbits are the open intervals $(-\infty, -1)$, $(-1, 0)$, $(0, 1)$, $(1, +\infty)$, and the points $\{-1\}$, $\{0\}$, and $\{1\}$. $\Diamond$

Determining the origins and ultimate destinations of orbits will be one of our primary concerns. Therefore, we introduce the following important concepts:

Definition 1.9. *If $\gamma^-(x_0)$ is bounded, then the set*

$$\alpha(x_0) = \lim_{t \to \alpha_{x_0}^+} \varphi(t, x_0)$$

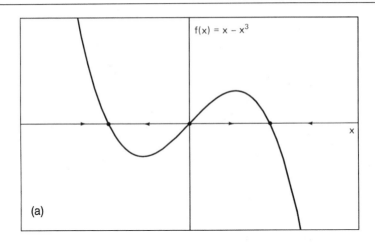

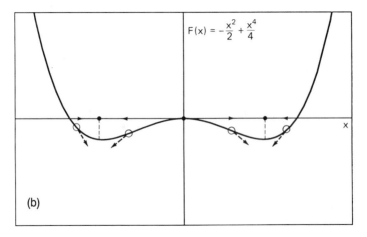

Figure 1.7. *Determining the phase portrait of $\dot{x} = x - x^3$ from (a) the function $f(x) = x - x^3$, and (b) from the corresponding potential function $F(x) = -x^2/2 + x^4/4$.*

is called the *α-limit set* of x_0. Similarly, if $\gamma^+(x_0)$ is bounded, then the set

$$\omega(x_0) = \lim_{t \to \beta_{x_0}^-} \varphi(t, x_0)$$

is called the *ω-limit set* of x_0.

The last part of Lemma 1.7 can now be restated as follows: the limit sets $\alpha(x_0)$ and $\omega(x_0)$ are equilibrium points, if they exist. In case you are wondering, the reason for this somewhat strange choice of terminology is that α and ω are, respectively, the first and the last letters of the Greek alphabet.

We now present another method that is particularly useful in determining the flows of certain specific differential equations. Equation (1.1) can be rewritten in the form

$$\dot{x} = f(x) = -\frac{d}{dx}F(x), \qquad (1.7)$$

where

$$F(x) \equiv -\int_0^x f(s)\,ds.$$

Equation (1.7) in this form is a special case of *gradient systems* which we will study later in a more general setting. At this time, it is sufficient to note that, if $x(t)$ is a solution of Eq. (1.7), then

$$\frac{d}{dt}F\left(x(t)\right) = \frac{d}{dx}F\left(x(t)\right) \cdot \frac{d}{dt}x(t) = -\left[f\left(x(t)\right)\right]^2 \leq 0.$$

Thus F is always decreasing along the solution curves, and hence can be thought of as a *"potential" function* of Eq. (1.1). It is evident that the equilibrium points of the differential equation (1.7) are extreme points of the potential function F.

Let us now reconsider the previous examples from this new viewpoint.

Example 1.1 revisited. Equation (1.4) can be written as a gradient system (1.7) with the potential function $F(x) = x^2/2$. The orbits of

$$\dot{x} = -x = -\frac{d}{dx}\left(\frac{x^2}{2}\right)$$

can be drawn by thinking of the motion of a particle on the graph of the potential function $F(x)$. As shown in Figure 1.5b, a particle at any point x_0 goes downhill with (ever decreasing) velocity $f(x)$ towards the equilibrium point 0. Therefore, the orbits for this equation are the intervals $(-\infty, 0)$, $(0, +\infty)$, and the equilibrium point 0. $\Diamond$

Example 1.2 revisited. The differential equation (1.5) can be written as

$$\dot{x} = x^2 = -\frac{d}{dx}\left(-\frac{x^3}{3}\right)$$

with the potential function $F(x) = -x^3/3$. The graph of $F(x)$ and the flow are shown in Figure 1.6b. Here the orbits are again the same intervals as in the previous example, however, the two flows are different. $\Diamond$

Example 1.8 revisited. The differential equation (1.6) can be written as

$$\dot{x} = x - x^3 = -\frac{d}{dx}\left(-\frac{x^2}{2} + \frac{x^4}{4}\right)$$

with the potential function $F(x) = -x^2/2 + x^4/4$. The graph of $F(x)$ and the flow are shown in Figure 1.7b. Here the orbits are the intervals $(-\infty, -1)$, $(-1, 0)$, $(0, 1)$, and $(1, +\infty)$, and the equilibrium points $-1, 0$, and 1. $\diamondsuit$

Exercises _____

1.5. *Many examples:* Describe all of the orbits and sketch the phase portrait of each of the following scalar differential equations in two ways: first, by determining the intervals on which the vector field is of constant sign; second, by using a potential function:
(a) $\dot{x} = -2x$;   (b) $\dot{x} = 1 + x$;   (c) $\dot{x} = x(1 - x)$;
(d) $\dot{x} = x - x^3 + 1$;   (e) $\dot{x} = x - x^3 + 0.2$;
(f) $\dot{x} = -x - x^3 + 1$;   (g) $\dot{x} = -x - x^3 + \lambda$, where λ is a constant.
(h) $\dot{x} = 2\sin x$;   (i) $\dot{x} = 1 - 2\sin x$;   (j) $\dot{x} = 1 - \sin x$;
(k) $\dot{x} = 2 - \sin x$;   (l) $\dot{x} = \tanh x$;   (m) $\dot{x} = \begin{cases} 0 & \text{if } x = 0 \\ x\ln|x| & \text{if } x \neq 0. \end{cases}$
Suggestion: You may wish to determine the phase portraits of some of these examples numerically using PHASER. In this case, to compute a negative orbit you should use a negative step size and a negative "end time."

1.6. Show that if $f \in C^1(\mathbb{R}, \mathbb{R})$ and the positive orbit $\gamma^+(x_0)$ is bounded, then $\beta_{x_0} = +\infty$. State and prove a similar fact for negative orbits.
Hint: Use uniqueness.

1.7. Show that if $f \in C^1(\mathbb{R}, \mathbb{R})$ and the positive orbit $\gamma^+(x_0)$ is bounded, then $\omega(x_0)$ is an equilibrium point. State and prove a similar fact for negative orbits.
Hint: Use the Mean Value Theorem to show that $\dot{\varphi}(t, x_0) \to 0$ as $t \to +\infty$.

1.3. Stability of Equilibria

In this section we introduce the concept of stability of an equilibrium point and present several theorems for determining the flow near an equilibrium point. The notion of stability is of considerable theoretical and practical importance; we will return to this topic in later chapters and explore other forms of stability.

Roughly speaking, an equilibrium point $\bar{x}$ is stable if all solutions starting near $\bar{x}$ stay nearby. If, in addition, nearby solutions tend to $\bar{x}$ as $t \to +\infty$, then $\bar{x}$ is asymptotically stable. Precise definitions are given below.

Definition 1.10. *An equilibrium point $\bar{x}$ of Eq. (1.1) is said to be stable if, for any given $\epsilon > 0$, there is a $\delta > 0$, depending on ϵ, such that, for every x_0 for which $|x_0 - \bar{x}| < \delta$, the solution $\varphi(t, x_0)$ of Eq. (1.1) through x_0 at 0 satisfies the inequality $|\varphi(t, x_0) - \bar{x}| < \epsilon$ for all $t \geq 0$. The equilibrium point $\bar{x}$ is said to be unstable if it is not stable.*

Definition 1.11. *An equilibrium point $\bar{x}$ is said to be asymptotically stable if it is stable and, in addition, there is an $r > 0$ such that $|\varphi(t, x_0) - \bar{x}| \to 0$ as $t \to +\infty$ for all x_0 satisfying $|x_0 - \bar{x}| < r$.*

We should point out that we have purposely introduced a redundancy in this definition of asymptotic stability. In the case of a scalar differential equation, if every solution with initial value close to $\bar{x}$ approaches $\bar{x}$ as $t \to +\infty$, then it follows that $\bar{x}$ is stable. However, this is not so in higher dimensions and the definition of asymptotic stability as given will be necessary.

The following lemma, whose proof is left as an exercise, is helpful in determining the stability of an equilibrium point of Eq. (1.1) from the function $f(x)$.

Lemma 1.12. *An equilibrium point $\bar{x}$ of $\dot{x} = f(x)$ is stable if there is a $\delta > 0$ such that $(x-\bar{x})f(x) \leq 0$ for $|x-\bar{x}| < \delta$. Similarly, $\bar{x}$ is asymptotically stable if and only if there is a $\delta > 0$ such that $(x - \bar{x})f(x) < 0$ for $0 < |x-\bar{x}| < \delta$. An equilibrium point $\bar{x}$ of $\dot{x} = f(x)$ is unstable if there is a $\delta > 0$ such that $(x - \bar{x})f(x) > 0$ for either $0 < x - \bar{x} < \delta$ or $-\delta < x - \bar{x} < 0$.* ◊

Using the lemma above, it is easy to verify that, in Example 1.1, $\dot{x} = -x$, the equilibrium point at 0 is asymptotically stable, while, in Example 1.2, $\dot{x} = x^2$, the equilibrium point at 0 is unstable; see Figures 1.5a and 1.6a. In Example 1.8, $\dot{x} = x - x^3$, the equilibrium points at -1 and 1 are asymptotically stable, and 0 is unstable; see Figure 1.7a.

Let us now examine a somewhat intricate example concerning the first sentence of Lemma 1.12.

Example 1.13. Consider the differential equation

$$\dot{x} = f(x) \equiv \begin{cases} 0 & \text{if } x = 0; \\ -x^3 \sin \frac{1}{x} & \text{otherwise.} \end{cases}$$

The function $f(x)$ above is continuous together with its first derivative. The graph of f and the phase portrait of the differential equation above are depicted in Figure 1.8. The equilibrium points are $\bar{x} = 0$, and $\bar{x} = (k\pi)^{-1}$ where k is any integer. Notice that the equilibrium points $[(2k + 1)\pi]^{-1}$ and $-[(2k + 2)\pi]^{-1}$ are asymptotically stable while $-[(2k + 1)\pi]^{-1}$ and $[(2k + 2)\pi]^{-1}$ are unstable for $k = 0, 1, 2, \ldots$. The equilibrium point 0 is stable, but not asymptotically; there is no $\delta > 0$ such that $xf(x) < 0$ for $0 < |x| < \delta$. So, it is not possible to improve Lemma 1.12 by changing the first statement to say "if and only if." ◊

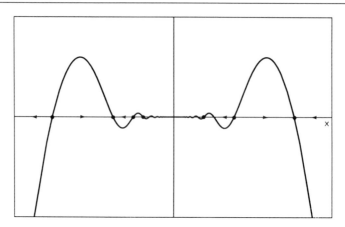

Figure 1.8. *Phase portrait of $\dot{x} = -x^3 \sin x^{-1}$. The origin is stable, but not asymptotically stable.*

It is instructive to contrast stability of an equilibrium point with continuous dependence of $\varphi(t, x_0)$ in t, x_0 (see Theorem 1.4). For instance, Example 1.2 has $\varphi(t, x_0)$ continuous in t, x_0, but the equilibrium point at 0 is not stable. To understand the difference between these two important concepts, a certain amount of technical language is unavoidable. Continuity of $\varphi(t, x_0)$ in t and x_0 near an equilibrium point $\bar{x}$ implies the following type of uniformity. Given any interval $[0, \tau]$, and any $\varepsilon > 0$, there is a $\delta(\varepsilon, \tau) > 0$ such that $|x_0 - \bar{x}| < \delta(\varepsilon, \tau)$ implies $|\varphi(t, x_0) - \bar{x}| < \varepsilon$ for $0 \leq t \leq \tau$, that is, the function $\varphi(t, x_0)$ is continuous in x_0 uniformly with respect to t in closed bounded sets. Notice that the number $\delta(\varepsilon, \tau)$ in Example 1.2 must approach zero as $\tau \to +\infty$. On the other hand, if $\bar{x}$ is stable, then δ depends only on ε and not on τ; that is, the stability of an equilibrium point $\bar{x}$ is equivalent to the statement that $\varphi(t, x_0)$ is continuous in x_0 at $\bar{x}$ *uniformly* with respect to $t \geq 0$.

It is evident from Definitions 1.10 and 1.11 that the stability of an equilibrium point $\bar{x}$ of Eq. (1.1) is a local property of the flow near the equilibrium. Therefore, it is reasonable to expect that under certain conditions the stability properties of $\bar{x}$ can be determined from the linear approximation, that is, the derivative $f'(x) \equiv (d/dx) f(x)$ of the function f near $\bar{x}$. Since the equilibrium point $\bar{x} = 0$ of the linear differential equation $\dot{x} = cx$ is asymptotically stable if $c < 0$ and unstable if $c > 0$, we are naturally led to the following statement:

Theorem 1.14. *Suppose that f is a C^1 function and $\bar{x}$ is an equilibrium point of $\dot{x} = f(x)$, that is, $f(\bar{x}) = 0$. Suppose also that $f'(\bar{x}) \neq 0$. Then the equilibrium point $\bar{x}$ is asymptotically stable if $f'(\bar{x}) < 0$, and unstable if $f'(\bar{x}) > 0$.*

Proof. We first shift the x-axis by introducing the new variable $y \equiv x - \bar{x}$, so that the equilibrium $\bar{x}$ of $\dot{x} = f(x)$ corresponds to the equilibrium of the differential equation $\dot{y} = f(\bar{x} + y)$ at $y = 0$. If we expand the function $f(\bar{x} + y)$ into its Taylor expansion near 0, we obtain

$$\dot{y} = f'(\bar{x})y + g(y),$$

which can be considered as a perturbation of the linear differential equation $\dot{y} = f'(\bar{x})y$. In fact, the function $g(y)$ satisfies $g(0) = 0$ and $g'(0) = 0$. Since $g'(0) = 0$, for any $\epsilon > 0$, there is a $\delta > 0$ such that $|g'(y)| < \epsilon$ if $|y| < \delta$. Using the formula $g(y) = \int_0^y g'(s)\,ds$, it follows that $|g(y)| \leq \epsilon|y|$ if $|y| < \delta$. Now suppose that $f'(\bar{x}) \neq 0$ and $\epsilon < |f'(\bar{x})|$. Then $|y| < \delta$ implies that the sign of the function $f(\bar{x} + y) = f'(\bar{x})y + g(y)$ is determined by the sign of $f'(\bar{x})y$. Therefore, the conclusion of the theorem follows from Lemma 1.12. $\Diamond$

The linear differential equation $\dot{x} = f'(\bar{x})x$ is called the *linear variational equation* or the *linearization* of the vector field $\dot{x} = f(x)$ about its equilibrium point $\bar{x}$. Theorem 1.14 asserts that, when $f'(\bar{x}) \neq 0$, the stability type of the equilibrium point $\bar{x}$ of $\dot{x} = f(x)$ is the same as the stability type of the equilibrium point at the origin of its linearized vector field.

We introduce the following common terminology for an equilibrium point satisfying the hypothesis of the theorem above.

Definition 1.15. *An equilibrium point $\bar{x}$ of $\dot{x} = f(x)$ is called a hyperbolic equilibrium if $f'(\bar{x}) \neq 0$.*

If $f'(\bar{x}) = 0$, then $\bar{x}$ is called a *nonhyperbolic* or *degenerate* equilibrium point. Unlike near a hyperbolic equilibrium where the linear term of the vector field determines the flow locally, the stability properties of a nonhyperbolic equilibrium $\bar{x}$ depends on higher-order terms in the Taylor expansion of the function $f(\bar{x} + y)$. For instance, while $\bar{x} = 0$ is an unstable equilibrium for $\dot{x} = x^2$, it is asymptotically stable for $\dot{x} = -x^3$. There are other complications associated with nonhyperbolic equilibria; infinitely many equilibria are present in any open neighborhood of the nonhyperbolic equilibrium $\bar{x} = 0$ of the differential equation $\dot{x} = -x^3 \sin x^{-1}$, as we saw in Example 1.13. These examples point to the realization that a study of nonhyperbolic equilibria will not be trivial. Despite the difficulties associated with them, however, nonhyperbolic equilibria play a prominent role in our subject, as we shall soon see.

Exercises ⎯⎯⎯⎯⎯⎯⎯⎯⎯⎯⎯⎯⎯⎯⎯⎯⎯⎯⎯⎯ ♣♡♠♢

1.8. *Many examples:* Determine the equilibrium points of the following scalar differential equations and compute the linear variational equations about the equilibria. Identify the hyperbolic equilibria and their stability types. Finally, sketch phase portraits by hand and also by using PHASER.

(a) $\dot{x} = 0$;   (b) $\dot{x} = 2.1x(1 - x)$;   (c) $\dot{x} = 1 + x^2$;
(d) $\dot{x} = 1 - x^2$;   (e) $\dot{x} = 2x^2 - x^3$;   (f) $\dot{x} = x - x^3 + 0.2$;
(g) $\dot{x} = 2\sin x$;   (h) $\dot{x} = 1 - 2\sin x$;   (i) $\dot{x} = 1 - \sin x$;
(j) $\dot{x} = x\sin x$;   (k) $\dot{x} = a - x^3 \sin(x^{-1})$ where $a > 0$ is a constant;
(l) $\dot{x} = x\left[1 - b(e^x - 1)\right]$ for values of $b = -1.1, -1, -0.1, 0,$ and 0.1.

1.9. Give an alternative, and perhaps simpler, proof of Theorem 1.14 using the Mean Value Theorem.

1.10. *Minimum:* Is an equilibrium point corresponding to a local minimum of a potential function always asymptotically stable?

1.11. Consider the differential equation $\dot{x} = (1 - x)x^{-1/2}$ for $x > 0$. What is the α-limit set of a solution? Does a solution reach its α-limit set in finite time? Compare your answer on PHASER. Any problems?

1.12. *Hyperbolic means faster:* As we saw, the phase portraits of the differential equations $\dot{x} = -x$ and $\dot{x} = -x^3$ are qualitatively the same: all orbits eventually approach the unique asymptotically stable equilibrium at the origin. Compare the speeds of approach to the equilibrium. Which one is faster? Pay particular attention to what happens near the origin, say, for $|x| < 1$.

1.13. *Quantitative information is important too:* As we have remarked earlier, phase portraits give no information about the values of solutions along orbits. Such quantitative information, which is of paramount interest in certain applications, is usually obtained by numerical approximations on the computer. In fact, good approximations are the most that should be expected since explicit solutions can be obtained for only very special equations. On the other hand, it is quite remarkable that some famous applications involve the use of linear differential equations, hence reducing numerical tasks to calculating the values of the exponential or logarithm function. Here are several such examples.

Radioactive Decay: It has been observed experimentally, by Rutherford and others, that certain radioactive elements decay at a rate proportional to their mass. By idealizing this complicated natural process, ignoring, for instance, that atoms are discrete entities, it is reasonable to model the phenomenon of radioactive decay using a differential equation: if $N(t)$ is the mass of radioactive substance at time t, then

$$\dot{N} = -\lambda N,$$

where λ is a positive constant which is a characteristic of the radioactive substance.

From the qualitative point of view, all of the matter in such a radioactive substance eventually radiates away. From the quantitative perspective, however, there are some concerns. First, we need to know the value of λ for a given substance. This is done experimentally. The solution of the linear differential equation above satisfying $N(0) = N_0$ is given by $N(t) = N_0 e^{-\lambda t}$. The constant λ can be determined by measuring the amount of remaining mass at some later time, say, τ. It is standard to use the half-life of the substance for τ, that is, the necessary time for the substance to decay to half of its original size. *Show* that $\lambda = (\ln 2)/\tau$, where τ is the half-life. Once λ is determined, $N(t)$ can readily be found for any t by evaluating the exponential function on a pocket calculator. Have you ever wondered how your calculator or computer determines the values of the exponential or logarithm function?

The half-life of the naturally occurring radioactive element ^{14}C, carbon-14, is known to be 5568 years. *Compute* the length of time it takes for a mass of ^{14}C to reduce to 20 percent of its original weight.

Radiocarbon Dating: An effective method of estimating the ages of archeological finds of organic origin is the method of ^{14}C dating discovered by W. Libby in 1949. The key idea of the method is remarkably simple: ^{14}C is in equilibrium in living plants—the amount absorbed from the atmosphere balances the amount that radiates. Once the plant dies, it ceases to absorb any more ^{14}C but the radiation continues. One basic cosmological premise is that the concentration of ^{14}C in the atmosphere has been constant over millennia. Suppose that at $t = 0$ a tree dies. Let $R(t)$ be the rate of disintegration of ^{14}C in the dead wood at time t. *Derive* the formula

$$t = \frac{1}{\lambda} \ln \frac{R(0)}{R(t)}.$$

Now, we can measure $R(t)$ at the present time. $R(0)$ is also measurable using a piece of living plant. So, the age of the dead wood is easy to compute. Here is an example.

Ağrı Dağında: In 1956, a piece of old wood excavated at Mount Ararat gave a count of 5.96 disintegrations per minute per gram of ^{14}C while living wood gave 6.68. Did the piece of old wood come from the Ark? Well, ^{14}C dating is not always reliable over relatively short time spans.

1.4. Equations on a Circle

This brief section may appear to be misplaced but please read on for the future. Here, we examine a special feature in the geometry of the flow of Eq. (1.1) when the vector field $f(x)$ is a periodic function of x with period P. More precisely, we consider differential equations of the form

$$\dot{x} = f(x), \quad \text{with} \quad f(x + P) = f(x). \tag{1.8}$$

It is easy to see that, if $x(t)$ is a solution of Eq. (1.8), then $x(t) + P$ is also a solution. Therefore, all the information about the flow is obtained by studying the flow on an interval of length P. If we identify (glue) the two end points of such an interval, then the resulting geometric object is a circle, which we will denote by S^1. As we shall see in later chapters, it often is advantageous to make such an identification and study the flow of Eq. (1.8) on S^1. For now, we confine our presentation to a simple example.

Example 1.16. Consider the following periodic vector field with period $P = 2\pi$:

$$\dot{x} = \sin x. \tag{1.9}$$

Using the methods of Section 1.2, it is easy to determine the flow of Eq. (1.9) on $\mathbb{R}$. The corresponding flow on S^1 can be obtained by identifying the end points of any interval of length 2π; see Figure 1.9. $\Diamond$

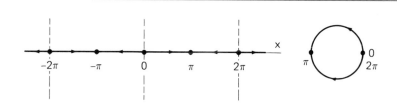

Figure 1.9. *Phase portrait of $\dot{x} = \sin x$ on the line and on the circle.*

Here, we conclude our introduction to the dynamics of scalar autonomous equations and turn to the meaning of the second word in the title of our book—bifurcations.

Exercises _____ ♣♡♠◇

1.14. Sketch the phase portraits on the circle and analyze the stability of equilibria of the following scalar differential equations:
(a) $\dot{x} = 2\sin x$;   (b) $\dot{x} = 1 - 2\sin x$;   (c) $\dot{x} = 1 - \sin x$;
(d) $\dot{x} = 1 - 2\sin(x + 1)$;   (e) $\dot{x} = \cos(2x) - \cos x + 1$.

Bibliographical Notes _____

There is a vast literature on the fundamental theorem of existence and uniqueness of solutions of ordinary differential equations. Variations on the statements and methods of proofs of such theorems abound; see, for example, the Appendix, Coddington and Levinson [1955], Hale [1980], Hartman [1964], and Robbin [1968]. We will have no need to resort to "pathological" functions as vector fields; the dynamics of polynomial vector fields, with a few trigonometric functions thrown in, provide us with more complexity than any one is able to understand.

It is not our intent in this book to dwell on specialized results that are manifestations of low dimensionality. Indeed, many of the concepts and theorems presented in this simple context have counterparts in higher dimensions. However, for simplicity and continuity of exposition, we usually refrain from diversions into higher dimensions until later chapters. For example, we have seen that all scalar autonomous differential equations can be viewed as gradient systems. This is not so in higher dimensions. There are, however, gradient systems in higher dimensions; in fact, because of their relatively simple dynamics, as expounded in Chapter 14, they are of great theoretical interest.

The flow of a differential equation gives rise to a dynamical system. Should certain systems that exhibit "dynamical behavior" be modeled using differential equations? How important is the abstract notion of a dynamical system in the theory of differential equations? On such questions, see the essay by Hirsch [1984].

Vector fields on spaces other than the usual Euclidean space arise naturally in applications—Hamiltonian mechanics, for instance—as we shall see in later chapters. A systematic study of vector fields on compact manifolds (closed and bounded smooth subsets of a Euclidean space) goes under the name of *global analysis*; see, for example, Nitecki [1971] and Palis and de Melo [1982].

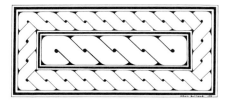

2

Elementary
Bifurcations

 In this chapter, we begin to explore the main theme of our book: *bifurcation theory*, the study of possible changes in the structure of the orbits of a differential equation depending on variable parameters. We first illustrate certain key ideas by way of specific examples. Then we generalize these observations and analyze local bifurcations of an arbitrary scalar differential equation. Since the Implicit Function Theorem is the main ingredient used in these generalizations, we include a precise statement of this celebrated theorem. We subsequently return to a specific example and analyze the bifurcations of a differential equation on the circle. Bifurcation behavior of specific differential equations can be encapsulated in certain pictures called bifurcation diagrams. Next, we give a numerical procedure for determining these diagrams, which are very useful in applications. We conclude the chapter with a discussion of some of the more subtle aspects of the notion of qualitative equivalence of phase portraits.

2.1. Dependence on Parameters – Examples

This section consists of a collection of specific examples designed to illustrate some of the key ideas from bifurcation theory. Despite their simplicity, these examples capture what happens in the general case, as we shall see in the following sections.

Example 2.1. *Hyperbolic equilibrium is insensitive:* Consider the linear differential equation

$$\dot{x} = c - x \equiv F(c,\, x), \tag{2.1}$$

where c is a real parameter. For $c = 0$, we have $F(0,\, x) = -x$, and thus, in this case, Eq. (2.1) becomes Eq. (1.4). Therefore, we refer to Eq. (2.1) as a *perturbation* of Eq. (1.4). The effect of the introduction of the parameter c is that the line $F(0,\, x) = -x$ is vertically translated by a distance c. However, it is more convenient for our purposes to leave the line fixed and vertically translate the x-axis by $-c$. By doing so, we can easily determine the flows for all values of the parameter c from the graph of $F(c,\, x)$ by shifting the x-axis and then using the method in Section 1.2. As shown in Figure 2.1, for all values of the parameter c, there is a single hyperbolic equilibrium which is asymptotically stable. $\diamond$

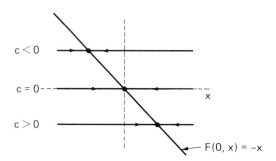

Figure 2.1. *Phase portraits of $\dot{x} = c - x$ for several values of c.*

Example 2.2. *Saddle-node bifurcation:* Consider the quadratic differential equation

$$\dot{x} = c + x^2 \equiv F(c,\, x), \tag{2.2}$$

where c is a real parameter. Notice that Eq. (2.2) is a perturbation of Eq. (1.5), and that the origin is a nonhyperbolic equilibrium point for $c = 0$.

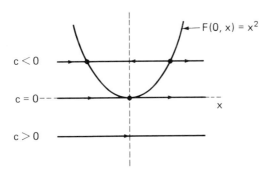

Figure 2.2. *Phase portraits of $\dot{x} = c + x^2$ for several values of c.*

Using the graphical method described in the previous example, we can easily determine the flow of Eq. (2.2) for all values of the parameter c by leaving the original parabola $F(0, x) = x^2$ fixed and vertically translating the x-axis by $-c$. The resulting flows are depicted in Figure 2.2. For all $c < 0$, the orbits are given by the intervals $(-\infty, -\sqrt{-c})$, $(-\sqrt{-c}, \sqrt{-c})$, and $(\sqrt{-c}, +\infty)$, and the equilibrium points $-\sqrt{-c}$ and $\sqrt{-c}$. For $c = 0$, the orbits are $(-\infty, 0)$ and $(0, +\infty)$, and the equilibrium point 0. For all $c > 0$, the only orbit is $(-\infty, +\infty)$, and there is no equilibrium point. We have marked the directions of all these orbits in Figure 2.2.

If the parameter c is varied, as long as $c < 0$, the number and the direction of the orbits remain the same; the only change is the shifting of the location of the equilibrium points $\pm\sqrt{-c}$. Similarly, for all $c > 0$, there is only one orbit and its direction is from left to right. However, if $c = 0$, regardless of how small an amount c is varied, the number of orbits changes: there are two equilibria for any $c < 0$, and none for $c > 0$. $\Diamond$

For a scalar differential equation $\dot{x} = f(x)$, the equilibrium points and the sign of the function $f(x)$ between the equilibria determine the number of orbits and the direction of the flow on the orbits. We refer to the number of orbits and the direction of the flow on the orbits as the *orbit structure* of the differential equation or the *qualitative structure of the flow*.

The study of changes in the qualitative structure of the flow of a differential equation as parameters are varied is called *bifurcation theory*. At a given parameter value, a differential equation is said to have *stable orbit structure* if the qualitative structure of the flow does not change for sufficiently small variations of the parameter. A parameter value for which the flow does not have stable orbit structure is called a *bifurcation value*, and the equation is said to be at a *bifurcation point*. It is evident from the analysis above that Eq. (2.1) has stable orbit structure for all values of

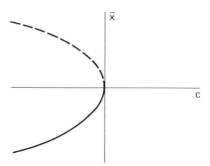

Figure 2.3. *Bifurcation diagram of saddle-node bifurcation. Notice that the parameter is assigned to the horizontal axis; the stable equilibria are drawn in solid lines and the unstable equilibria in dashed lines. We will follow these conventions in bifurcation diagrams.*

c, and that Eq. (2.2) has stable orbit structure for any $c \neq 0$, but is at a bifurcation point for $c = 0$. The particular bifurcation behavior of Eq. (2.2) described above is called *saddle-node bifurcation*. The choice of terminology saddle-node will become apparent when we discuss two-dimensional systems in Part III of our book.

There is another very useful graphical method for depicting some of the important dynamical features in equations $\dot{x} = F(c, x)$ depending on a parameter c. This method consists of drawing curves on the (c, x)-plane, where the curves depict the equilibrium points for each value of the parameter. More specifically, a point (c_0, x_0) lies on one of these curves if and only if $F(c_0, x_0) = 0$. Also, to represent the stability types of these equilibria, we label stable equilibria with solid curves and unstable equilibria with dotted curves. The resulting picture is called a *bifurcation diagram*. For instance, the bifurcation diagram of the saddle-node bifurcation in Example 2.2, $\dot{x} = c + x^2$, is the parabola $c = -x^2$ labeled as in Figure 2.3.

Example 2.3. *Transcritical bifurcation:* Consider the differential equation containing a real parameter c:

$$\dot{x} = cx + x^2, \tag{2.3}$$

which is another perturbation of Eq. (1.5). Unlike the previous example, this perturbation is not a translation of the unperturbed vector field. Nevertheless, it is still easy to determine the phase portrait of Eq. (2.3) from the graph of the function $F(c, x) = cx + x^2$ as shown in Figure 2.4. Notice that the origin is an equilibrium point for all values of the parameter c. For $c < 0$, the origin is asymptotically stable and there is another equilibrium point $\bar{x} = -c$ which is unstable. The parameter value $c = 0$ is a

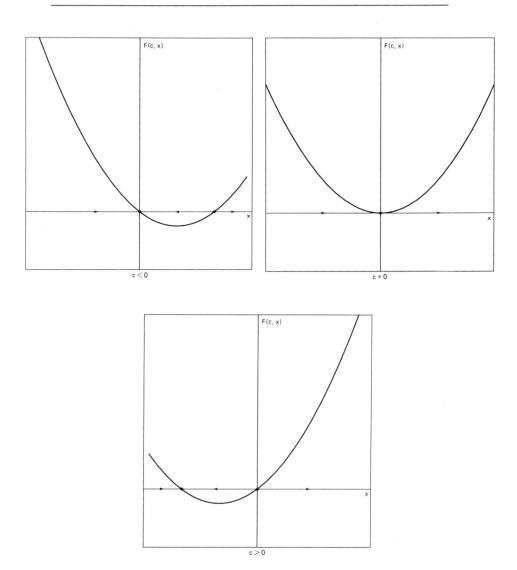

Figure 2.4. *Phase portraits of $\dot{x} = cx + x^2$ for several values of c.*

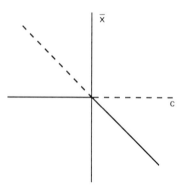

Figure 2.5. *Bifurcation diagram of transcritical bifurcation.*

bifurcation value at which the two equilibria coalesce at the origin, which is a nonhyperbolic unstable equilibrium point. For $c > 0$, the origin becomes unstable by transferring its stability to another equilibrium point, $\bar{x} = -c$. For this reason, the bifurcation that Eq. (2.3) undergoes is called *transcritical* bifurcation; see Figure 2.5. $\Diamond$

Example 2.4. *Hysteresis:* Consider the cubic differential equation containing a real parameter c:

$$\dot{x} = c + x - x^3. \tag{2.4}$$

Varying c corresponds to a vertical shift of the x-axis in the plot of the graph of $F(c, x) = c + x - x^3$; see Figure 2.6 for the flows of Eq. (2.4). For $c = 0$, Eq. (2.4) is Eq. (1.6) and it has stable orbit structure. The flow continues to have stable orbit structure for small values of the parameter, that is, for $-c_1 < c < c_1$, where $c_1 = \frac{2}{3\sqrt{3}}$ is the local maximum value and $-c_1$ is the local minimum value of $F(0, x)$. For $c = -c_1$ or $c = c_1$, the equation is at a bifurcation point. For the parameter values $c < -c_1$ and $c > c_1$, the equation again has stable orbit structure. The bifurcation diagram of Eq. (2.4) is shown in Figure 2.7.

Because of its frequent occurrence in applications, it is worthwhile to explore the dynamics of Eq. (2.4) in a bit more detail. Let us suppose that the differential equation is a model of some physical system and the parameter c is a changeable characteristic of the model. If we start the system with a very large negative value of c, after a long time, regardless of the initial condition x_0, the system will be very near a stable equilibrium state on the left leg of the cubic. Now, let us continuously increase the value of the parameter c. Since the system was near the stable state when we began to vary c, it will stay near this stable state for small variations

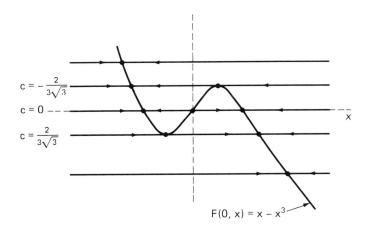

Figure 2.6. *Phase portraits of $\dot{x} = c + x - x^3$ for several values of c.*

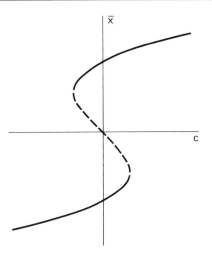

Figure 2.7. *Bifurcation diagram of $\dot{x} = c + x - x^3$.*

in c. In fact, as we increase the parameter c, the system will follow the stable equilibria on the left until $c = c_1$. At this point the system will *jump* to a different stable equilibrium state on the right leg of the cubic. As we continue to increase the parameter c, the system will follow the stable equilibria on the right. The dashed lines in Figure 2.8 show the equilibria the system will follow as c is increased from a very large negative value to a very large positive value. Now, if we start decreasing the parameter c from

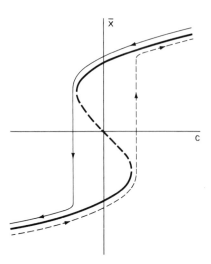

Figure 2.8. *Hysteresis loop.*

a very large positive value, the system will follow the equilibria on the right leg of the cubic until $c = -c_1$, at which point it will jump to the left leg. The solid lines in Figure 2.8 indicate the equilibria the system will follow as c is decreased from a very large positive value to a very large negative value. The important observation about this experiment is that the system experiences a jump at two different values of the parameter; moreover, the parameter value at which the jump takes place is determined by the direction in which the physical parameter is varied! This phenomenon is referred to as *hysteresis* and the part in Figure 2.8 that resembles a parallelogram is called the *hysteresis loop.* ◊

The perturbation of the cubic differential equation given in Eq. (2.4) is, in a way, the simplest one since it is equivalent to a translation of the x-axis. In the next two examples, we will study the effects of other perturbations of Eq. (1.6).

Example 2.5. *Pitchfork bifurcation:* Consider the differential equation

$$\dot{x} = dx - x^3, \tag{2.5}$$

where d is a real parameter. The effect of varying d is equivalent to changing the slope of the cubic at the origin while keeping the x-axis the same. Reasoning as before, it is easy to see that Eq. (2.5) has three equilibria and stable orbit structure for all $d > 0$. At $d = 0$, the equilibria come together at the origin and the system is at a bifurcation point. For all $d < 0$, the equation again has stable orbit structure, with one asymptotically stable equilibrium point; see Figure 2.9.

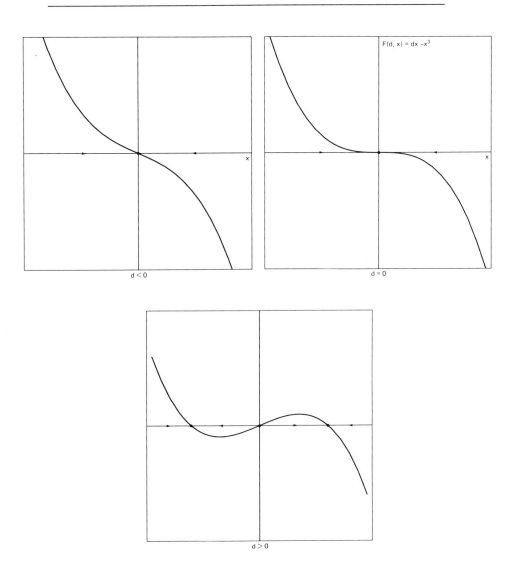

Figure 2.9. *Phase portraits of $\dot{x} = dx - x^3$ for several values of d.*

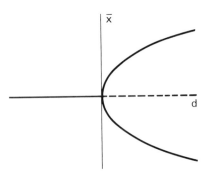

Figure 2.10. *Supercritical pitchfork bifurcation in* $\dot{x} = dx - x^3$.

The bifurcation diagram of Eq. (2.5) is shown in Figure 2.10 and, because of its appearance, that bifurcation is known as the *pitchfork bifurcation*. Notice that $x = 0$ is always an equilibrium point. However, as the parameter d passes through the bifurcation value $d = 0$, the equilibrium at the origin loses its stability by giving it up to two new stable equilibria which bifurcate from the origin.

For this particular example, the pitchfork bifurcation is called *supercritical* because the additional equilibrium points which appear at the bifurcation value occur for the values of the parameter at which the equilibrium point is unstable. When the additional equilibria occur for the values of the parameter at which the original equilibrium point is stable, the bifurcation is called *subcritical*. As illustrated in Figure 2.11, an example of a subcritical pitchfork bifurcation can be seen in the equation $\dot{x} = dx + x^3$. $\diamondsuit$

Example 2.6. *Fold or cusp:* Let us now combine the two different perturbations above and consider the cubic differential equation

$$\dot{x} = c + dx - x^3 \equiv F(c, d, x), \qquad (2.6)$$

depending on two real parameters c and d. The vector field (2.6) is the most general perturbation of the function $-x^3$ with lower order terms because any term involving x^2 can always be eliminated by an appropriate translation of the variable. In fact, for a general cubic $-x^3 + ex^2 + \hat{d}x + \hat{c}$, use the change of variable $x \mapsto x + e/3$ and determine the new coefficients c and d in terms of e, $\hat{c}$, and $\hat{d}$.

We begin the analysis of Example 2.6 by first finding the bifurcation values of the parameters. As we have seen in the previous examples, at bifurcation points, a differential equation must have a nonhyperbolic equi-

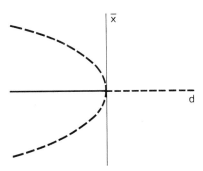

Figure 2.11. *Subcritical pitchfork bifurcation in $\dot{x} = dx + x^3$.*

point, that is,

$$F(c, d, x) = 0 \quad \text{and} \quad \frac{\partial}{\partial x} F(c, d, x) = 0.$$

For this example [Eq. (2.6)], the equations above are equivalent to

$$c + dx - x^3 = 0 \quad \text{and} \quad d - 3x^2 = 0.$$

Our objective is to determine all values of c and d for which these two equations can have some common solution x. Therefore, we can consider these equations as defining c and d parametrically in terms of x. Solving the second equation for d and then substituting the result into the first equation yields

$$d = 3x^2 \quad \text{and} \quad c = -2x^3. \tag{2.7}$$

If we now eliminate x from these two equations, we obtain the following equation for a *cusp*:

$$4d^3 = 27c^2. \tag{2.8}$$

In Figure 2.12 we have drawn the graph of Eq. (2.8) in the (c, d)-plane; this graph is a cusp. In each appropriate region of that (c, d)-plane we have sketched a graph for the function $F(c, d, x)$. In each such sketch we have indicated the flow determined by Eq. (2.6).

There is a wealth of information packed in Figure 2.12, including the dynamics of Eqs. (2.4) and (2.5). To extract the dynamics of Eq. (2.4), we fix d at a positive value, say $d = 1$, and then obtain the bifurcation diagram of hysteresis shown in Figure 2.7. To extract the dynamics of Eq. (2.5), we fix $c = 0$ and obtain the pitchfork bifurcation diagram in Figure 2.10. Equation (2.6) also contains a supercritical *saddle-node* bifurcation in disguise: fix $c \neq 0$, say $c = 1$, and vary d, as illustrated in Figure 2.13.

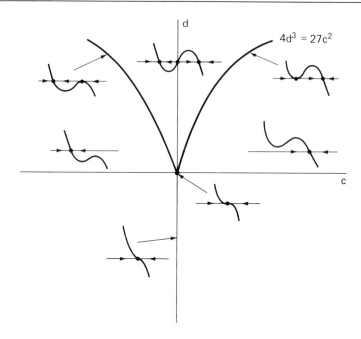

Figure 2.12. *Pictures of the cusp $4d^3 = 27c^2$ in the (c, d)-plane and some representative phase portraits of $\dot{x} = c + dx - x^3$.*

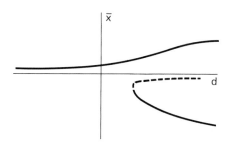

Figure 2.13. *Supercritical saddle-node bifurcation in $\dot{x} = 1 + dx - x^3$.*

Although somewhat difficult to draw, the full bifurcation diagram of Eq. (2.6) in the three-dimensional (c, d, x)-space can be constructed from the equation $c + dx - x^3 = 0$; see Figure 2.14. Of course, Figures 2.7, 2.10, and 2.13 are just various planar slices of the full bifurcation diagram in Figure 2.14. These slices are shown in Figure 2.15. ◊

Example 2.7. As the final example in this section, let us consider the

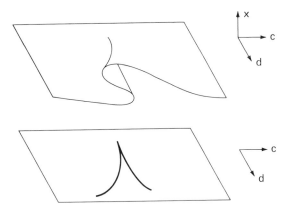

Figure 2.14. *The bifurcation diagram of the cubic differential equation* $\dot{x} = c + dx - x^3$ *in the* (c, d, x)*-space. The cusp below is the set of points in the* (c, d)*-plane for which the folding surface above has vertical tangency; over a point inside the cusp, the surface above has multiple values.*

following differential equation, depending on a scalar parameter λ :

$$\dot{x} = \lambda + (m\lambda + 1)x - x^3, \tag{2.9}$$

where m is a fixed constant. This equation is a special case of Eq. (2.6) with

$$c = \lambda \quad \text{and} \quad d = m\lambda + 1 \tag{2.10}$$

for which Eq. (2.8) of the cusp becomes

$$4(m\lambda + 1)^3 = 27\lambda^2. \tag{2.11}$$

For fixed m, the bifurcation values for Eq. (2.9) correspond to the solutions of the cubic equation (2.11). Geometrically, these bifurcation values are the points of intersection of the cusp [Eq. (2.8)] with the parametrized line [Eq. (2.10)]. Each of the three cases, $m > 1$, $m = 1$, and $m < 1$ are different and are depicted in Figure 2.16, where we have also drawn the bifurcation diagrams and the orbit structures of Eq. (2.9). You are invited to visualize these bifurcation diagrams as appropriate planar slices of the three-dimensional bifurcation diagram given in Figure 2.14.

An important remark for $m < 1$ lies in the observation that the orbit structures in the bifurcation diagram are equivalent when the parameter λ lies in the intervals $(-\infty, \lambda_0)$ and (λ_1, λ_2). However, to go from a flow corresponding to a value of λ in the first interval to an equivalent flow for λ in the second interval one must pass through two saddle-node bifurcations. In other words, the flows of Eq. (2.9) with the same qualitative structure are not "connected." $\diamondsuit$

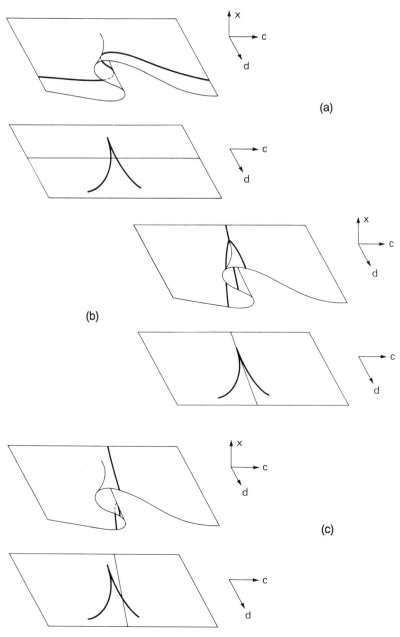

Figure 2.15. *Slices of the bifurcation diagram of* $\dot{x} = c + dx - x^3$: (a) *hysteresis for* $d = 1$, (b) *supercritical pitchfork for* $c = 0$, *and* (c) *supercritical saddle-node for* $c = 1$.

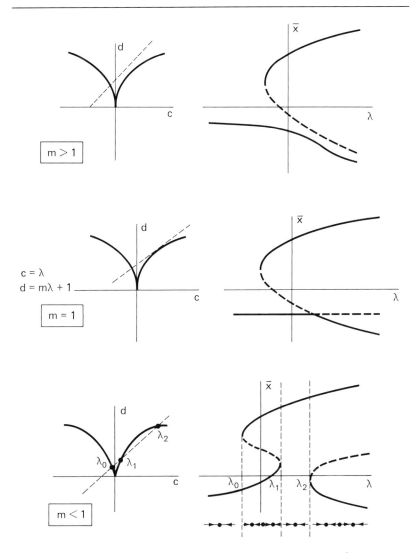

Figure 2.16.   *Bifurcation diagrams of $\dot{x} = \lambda + (m\lambda + 1)x - x^3$ for $m > 1$, $m = 1$, and $m < 1$.*

With these examples at our disposal, we now turn to generalities.

Exercises _____ ♣ ♡ ♠ ◇

2.1. Identify the groups of examples in Exercise 1.5 with the same orbit structure.

2.2. Provide the details of the computation of the bifurcation diagrams in Example 2.7.

2.3. *Constant harvesting:* Suppose that a population grows according to the "*logistic model*" and is harvested at a constant rate. The dynamics of the density $x(t)$ of such a population is governed by the differential equation

$$\dot{x} = kx - cx^2 - h,$$

where all the coefficients are positive; k and c reflect the intrinsic growth rate of the population and h is the rate of harvesting. Notice that the population grows at a rate proportional to its size when the population density is small. However, when the population gets large, growth is impaired because of, for example, overcrowding; the x^2 term reflects this behavior.

Now, the problem is, for fixed k and c, to determine the effect of harvesting on the population. Since the population density cannot be negative, we are interested in the solutions of this equation for $x \geq 0$. For a positive initial population density, the population is *exterminated* if there is a finite value of t such that $\varphi(t, x_0) = 0$. Without finding explicit solutions of the differential equation, show the following:

(a) If the harvesting rate h satisfies $0 < h \leq k^2/(4c)$, then there is threshold value of the initial size of the population such that if the initial size is below the threshold value, then the population is exterminated. On the other hand, if the initial size is above the threshold value, then the population approaches an equilibrium point.

(b) If the harvesting rate h satisfies $h > k^2/(4c)$, then the population is exterminated regardless of its initial size.

2.4. *Proportional harvesting:* Suppose that a population grows according to the "*logistic model*" as in the previous exercise, but is harvested at a rate proportional to the size of the population:

$$\dot{x} = kx - cx^2 - hx,$$

where k, c, and h are positive constants. Show that if $k < h$, then, regardless of the initial density $x_0 > 0$, such a population tends toward extermination as $t \to +\infty$, but is not exterminated in finite time. Also, analyze the fate of the population in the cases $k = h$ and $k > h$.

2.5. *Hydroplane:* The rectilinear motion of a hydroplane, ignoring pitching and rolling, is determined by a scalar differential equation of the form

$$m\dot{v} = T(v) - W(v),$$

where v is the velocity of the hydroplane, m is its mass, T is the thrust of the driving mechanism, and W is the resistance. It is reasonable to assume, for simplicity, that the thrust is approximately constant. The resistance, on the other hand, should increase with small and large v, but can be negative for intermediate values of the velocity due to rising of the hydroplane and the decrease of the wetted area. Discuss the possible motions of the hydroplane for various values of the constant thrust.

2.2. The Implicit Function Theorem

In this section, we state a fundamental result from mathematical analysis known as the Implicit Function Theorem, which turns out to be an indispensable tool in bifurcation theory. The simplified version presented below is tailored for the study of bifurcations of equilibria of scalar differential equations. A more general form of this important theorem is given in the Appendix.

Let $\lambda \equiv (\lambda_1, \ldots, \lambda_k)$ be a vector in $\mathbb{R}^k$. For now, we take the "norm" $\|\lambda\|$ of λ to be

$$\|\lambda\| = (\lambda_1^2 + \cdots + \lambda_k^2)^{1/2},$$

which we may interpret as the length of λ. We will say more about norms of vectors in Chapter 7.

Theorem 2.8. *Suppose that* $F : \mathbb{R}^k \times \mathbb{R} \to \mathbb{R}; \ (\lambda, x) \mapsto F(\lambda, x)$, *is a* C^1 *function satisfying*

$$F(\mathbf{0}, 0) = 0 \quad \text{and} \quad \frac{\partial F}{\partial x}(\mathbf{0}, 0) \neq 0.$$

Then there are constants $\delta > 0$ *and* $\eta > 0$, *and a* C^1 *function*

$$\psi : \{\lambda : \|\lambda\| < \delta\} \to \mathbb{R}$$

such that

$$\psi(\mathbf{0}) = 0 \quad \text{and} \quad F(\lambda, \psi(\lambda)) = 0 \quad \text{for } \|\lambda\| < \delta.$$

Moreover, if there is a $(\lambda_0, x_0) \in \mathbb{R}^k \times \mathbb{R}$ *such that* $\|\lambda_0\| < \delta$ *and* $|x_0| < \eta$, *and satisfies the equation* $F(\lambda_0, x_0) = 0$, *then* $x_0 = \psi(\lambda_0)$. $\diamondsuit$

The Implicit Function Theorem can be used to study equilibria in the following context. Let $\dot{x} = F(\lambda, x)$ be a differential equation depending on k parameters $\lambda \equiv (\lambda_1, \ldots, \lambda_k)$. If $\bar{x} = 0$ is a hyperbolic equilibrium point of the differential equation $\dot{x} = F(\lambda, x)$ at $\lambda = 0$, then the conditions of the Implicit Function Theorem are satisfied. This guarantees that the equation $F(\lambda, x) = 0$ may be solved locally for $x = \psi(\lambda)$ as a function of the parameters $(\lambda_1, \ldots, \lambda_k)$; see Figure 2.17. Furthermore, $\partial F(\lambda, \psi(\lambda))/\partial x \neq 0$ for λ sufficiently small. Thus, the qualitative structure of the flow does not change near $x = 0$. Consequently, there are no bifurcations in the neighborhood of $x = 0$ for sufficiently small values of the parameters. In the next section, we will discuss this situation in detail as well as the important role that the Implicit Function Theorem plays in bifurcation theory.

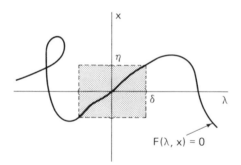

Figure 2.17. *The Implicit Function Theorem.*

To apply the Implicit Function Theorem in a specific situation, you may need to transform variables in such a way that the origin in the (λ, x)-space becomes a zero of the given function. For example, if $G : \mathbb{R}^k \times \mathbb{R} \to \mathbb{R}$; $(\lambda, x) \mapsto G(\lambda, x)$, has $G(\lambda_0, x_0) = 0$, then we can set $F(\lambda, x) = G(\lambda_0 + \lambda, x_0 + x)$ and $F(0, 0) = 0$.

Exercises _____ ♣♡♠♢

2.6. For the function $F(\lambda, x) = \lambda + (1 + \lambda)x + x^2$, determine the function $\psi(\lambda)$ whose existence is guaranteed by the Implicit Function Theorem.
Answer: $\psi(\lambda) = -\lambda$.

2.7. Apply the Implicit Function Theorem to show that there is a unique solution of the equation $\mu + (1 - \mu)y + y^3 = 0$ near $(\mu, y) = (1, -1)$.

2.8. Show that the conditions of the Implicit Function Theorem are satisfied for the following functions:
(a) $\lambda + \sin x$;   (b) $1/2 + \lambda - \cos(\pi/6 + x)$;   (c) $\sin \lambda + \tan x$.

2.3. Local Perturbations Near Equilibria

Earlier in Section 2.1 we considered three specific differential equations, linear [Eq. (2.1)], quadratic [Eq. (2.2)], cubic [Eq. (2.6)], and investigated the stability and bifurcation of equilibrium points as a function of the parameters. In this section we consider a given differential equation $\dot{x} = f(x)$ with an equilibrium point $\bar{x}$. We show that, if the first term of the Taylor expansion of the vector field f at $\bar{x}$ is linear, quadratic, or cubic, then under rather general perturbations the bifurcation of equilibria near $\bar{x}$ is essentially the same as in the examples mentioned above. It should be emphasized that in the following analysis we will consider the effects of quite arbitrary, but small, perturbations. Thus, our results are valid only in a sufficiently small neighborhood around the equilibrium point. For simplicity, we shall assume that $\dot{x} = f(x)$ has an equilibrium point at 0; if not, we always can translate to a new coordinate system, as in the proof of Theorem 1.14.

Case I: Hyperbolic equilibria. Suppose that f is a C^1 function with $f(0) = 0$ and $f'(0) \neq 0$. We saw in Theorem 1.14 that the stability properties of the equilibrium point 0 of the differential equation $\dot{x} = f(x)$ is determined by the linear approximation of the vector field near 0, that is, the higher order perturbations in the Taylor expansion of the vector field do not effect the qualitative structure of the flow near zero. However, the question remains: what happens if we make perturbations that influence the constant and the linear terms? We will show below that the situation in Example 2.1 also prevails in the general case.

To be precise, consider the perturbed differential equation

$$\dot{x} = F(\lambda, x), \tag{2.12}$$

where $F : \mathbb{R}^k \times \mathbb{R} \to \mathbb{R};\ (\lambda, x) \mapsto F(\lambda, x)$, is a C^1 function satisfying

$$F(0, x) = f(x) \quad \text{and} \quad \frac{\partial F}{\partial x}(0, 0) = f'(0) \neq 0. \tag{2.13}$$

Let us first investigate the existence of equilibria of the perturbed equation (2.12). If $F(\lambda, 0) \neq 0$, then the origin will no longer be an equilibrium point. However, from Eq. (2.13) and the fact that $f(0) = 0$, we have

$$F(0, 0) = 0 \quad \text{and} \quad \frac{\partial F}{\partial x}(0, 0) = f'(0) \neq 0.$$

Hence, the Implicit Function Theorem implies that there are constants $\delta > 0$ and $\eta > 0$, and a C^1 function $\psi(\lambda)$ defined for $\|\lambda\| < \delta$ with $\psi(0) = 0$ such that

$$F(\lambda, \psi(\lambda)) = 0.$$

Moreover, every (λ, x) with $\|\lambda\| < \delta$ and $|x| < \eta$ satisfying $F(\lambda, x) = 0$ is given by $(\lambda, \psi(\lambda))$. Therefore, for each $\|\lambda\| < \delta$, and $|x| < \eta$, there is a unique equilibrium $x = \psi(\lambda)$ of Eq. (2.12) satisfying $|x| < \eta$.

The stability behavior of the equilibrium $\psi(\lambda)$ can easily be determined from Theorem 1.14. To do so, we need to compute the sign of the derivative

$$\frac{\partial F}{\partial x}(\lambda, \psi(\lambda)). \tag{2.14}$$

From Eq. (2.13) and the fact that $\psi(0) = 0$, we have $\frac{\partial F}{\partial x}(0, \psi(0)) = f'(0) \neq 0$. Thus, there is a $\delta > 0$ such that, for $\|\lambda\| < \delta$, the sign of Eq. (2.14) is the same as that of $f'(0)$. Therefore, the stability type of the equilibrium $\psi(\lambda)$ of the perturbed equation (2.12) is the same as the stability type of the equilibrium 0 of the unperturbed equation $\dot{x} = f(x)$.

We can summarize the discussion above by saying that the flow near a hyperbolic equilibrium point is insensitive to small perturbations of the vector field.

Case II: Equilibria with quadratic degeneracy. Suppose that f is a C^2 function with $f(0) = 0$, $f'(0) = 0$, but $f''(0) \neq 0$. This is the next order of complication that occurs when we cannot make a decision about the stability of an equilibrium point based on the linearization.

Let us consider the perturbed differential equation

$$\dot{x} = F(\lambda, x), \tag{2.15}$$

where $F : \mathbb{R}^k \times \mathbb{R} \to \mathbb{R}$; $(\lambda, x) \mapsto F(\lambda, x)$, is a C^2 function satisfying

$$F(0, x) = f(x), \quad \frac{\partial F}{\partial x}(0, 0) = 0, \quad \frac{\partial^2 F}{\partial x^2}(0, 0) = f''(0) \neq 0. \tag{2.16}$$

These conditions together with $f(0) = 0$ imply that the Taylor expansion of F about the origin has the following form:

$$F(\lambda, x) = a(\lambda) + b(\lambda)x + c(\lambda)\frac{x^2}{2} + G(\lambda, x),$$

with $a(0) = 0$, $b(0) = 0$, $c(0) = f''(0) \neq 0$, and, for any $\epsilon > 0$, there are $\delta > 0$ and $\eta > 0$ such that the function G satisfies $|G(\lambda, x)| < \epsilon|x|^2$ for $\|\lambda\| < \delta$ and $|x| < \eta$.

As an instance of Case II, let us recall the differential equation $\dot{x} = f(x) = x^2$ and its perturbation (2.2) depending on one parameter ($k = 1$):

$$\dot{x} = F(\lambda, x) = \lambda + f''(0)\frac{x^2}{2}.$$

We saw in Example 2.2 that the flow of this equation changes at $\lambda = 0$ from two equilibria when $\lambda f''(0) < 0$ to no equilibrium when $\lambda f''(0) > 0$; see Figure 2.2. We will show below that the bifurcation behavior of this simple example occurs for the general case [Eqs. (2.15) and (2.16)] as well.

To verify the assertion above, it is only necessary to demonstrate that the function $F(\lambda, x)$ near $x = 0$ for any small λ has a graph which is like a parabola. This can be accomplished by showing the existence of a unique extreme point of the function $F(\lambda, x)$ near $x = 0$ for small λ. The extreme points of F correspond to the solutions x of the equation

$$\frac{\partial F}{\partial x}(\lambda, x) = 0. \tag{2.17}$$

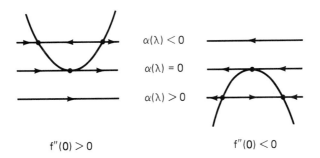

Figure 2.18. *Local bifurcations of a nonhyperbolic equilibrium point with quadratic degeneracy.*

Let $H(\lambda, x) \equiv \partial F(\lambda, x)/\partial x$. Then Eq. (2.16) implies that

$$H(0, 0) = 0 \quad \text{and} \quad \frac{\partial H}{\partial x}(0, 0) = f''(0) \neq 0.$$

Thus, the Implicit Function Theorem implies that there are constants $\delta > 0$, $\eta > 0$, and a C^1 function $\psi(\lambda)$ defined for $\|\lambda\| < \delta$ such that $\psi(0) = 0$ and $H(\lambda, \psi(\lambda)) = 0$, that is,

$$\frac{\partial F}{\partial x}(\lambda, \psi(\lambda)) = 0$$

and, moreover, every solution (λ, x) of Eq. (2.17) with $\|\lambda\| < \delta$ and $|x| < \eta$ is given by $x = \psi(\lambda)$.

For each fixed λ, the function $F(\lambda, x)$ has a minimum at $x = \psi(\lambda)$ if $f''(0) > 0$, or a maximum if $f''(0) < 0$. The number of equilibrium points of Eq. (2.15) depends upon the extreme value $\alpha(\lambda) \equiv F(\lambda, \psi(\lambda))$ of the function F. In Figure 2.18 we have drawn the flows for these two cases with several values of $\alpha(\lambda)$ (compare with Figure 2.2). These results can be summarized analytically by saying that when $\alpha(\lambda)f''(0) < 0$ there are two hyperbolic equilibria near the origin, $\alpha(\lambda) = 0$ implies that there is a nonhyperbolic equilibrium at the origin, and when $\alpha(\lambda)f''(0) > 0$ there are no equilibrium points about the origin.

It is important to observe in the discussion above that the qualitative structure of the flow of the perturbed equation (2.15) is determined from a single function of the parameter λ, namely, the function $\alpha(\lambda)$ corresponding to the extreme value of $F(\lambda, x)$. Thus, even though there may be k components of the (vector) parameter $\lambda = (\lambda_1, \lambda_2, \ldots, \lambda_k)$, the bifurcation behavior of the perturbed equation (2.15) depends on a single number, $\alpha(\lambda)$. When this situation occurs in a bifurcation problem, we say that the original vector field f is a *codimension-one bifurcation*.

Example 2.9. *Two parameters but codimension one:* As an example of a codimension-one bifurcation of a vector field depending on two parameters, consider the differential equation

$$\dot{x} = \lambda_1 + \lambda_2 x + x^2,$$

where $\lambda = (\lambda_1, \lambda_2)$ are two small parameters. The function $\alpha(\lambda)$ for this example corresponds to the minimum value of the function $F(\lambda, x) = \lambda_1 + \lambda_2 x + x^2$, and it is given by

$$\alpha(\lambda_1, \lambda_2) = \lambda_1 - \tfrac{1}{4}\lambda_2^2.$$

Thus the bifurcations occur as we cross the curve $\lambda_1 = \lambda_2^2/4$ in the (λ_1, λ_2)-plane: there are two equilibrium points if $\lambda_1 < \lambda_2^2/4$, and none if $\lambda_1 > \lambda_2^2/4$. $\Diamond$

Example 2.10. In applications, a single parameter may affect several terms in the Taylor expansion of a perturbation. For instance, consider the following perturbation of the vector field $f(x) = x^2$ given by

$$\dot{x} = \lambda^2 + 2a\lambda x + x^2,$$

where λ is a *scalar* parameter, and a is a given constant. As before, the bifurcation behavior of the perturbed equation depends on the minimum value of the function $F(\lambda, x) = \lambda^2 + 2a\lambda x + x^2$, which is equal to $\alpha(\lambda) = \lambda^2(1 - a^2)$. The sign of $\alpha(\lambda)$ is determined by the linear term in the perturbation if $|a| > 1$, and by the constant term if $|a| < 1$. The bifurcation takes place at $\lambda = 0$ in both cases, but the flow is different depending on whether $|a| < 1$ or $|a| > 1$. You are invited to draw the bifurcation diagram and representative pictures of the flow. $\Diamond$

Case III: Equilibria with cubic degeneracy. Suppose that f is a C^3-function with $f(0) = 0$, $f'(0) = 0$, $f''(0) = 0$, but $f'''(0) \neq 0$. This is the level of complication that occurs if the properties of the flow near the equilibrium point at zero cannot be determined from the linear or second-order terms of the Taylor expansion of $f(x)$.

We have seen previously (Section 2.1) in the specific example $f(x) = -x^3$ that at least two parameters were needed to capture all of the possible types of behavior of f under perturbations. In particular, we have studied the bifurcations in the perturbed equation $\dot{x} = c + dx - x^3$; see Example 2.6 and Figures 2.12 and 2.14. We will show below that this example is representative of what can happen in the general case near the equilibrium point $\bar{x} = 0$ of f satisfying the conditions above when f is subjected to "nice" perturbations that are small together with their derivatives up through order three.

More specifically, consider the perturbed differential equation

$$\dot{x} = F(\lambda, x), \tag{2.18}$$

where $F : \mathbb{R}^k \times \mathbb{R} \to \mathbb{R}$; $(\lambda, x) \mapsto F(\lambda, x)$, is a C^3 function satisfying

$$F(0, x) = f(x), \qquad \frac{\partial F}{\partial x}(0, 0) = 0,$$
$$\frac{\partial^2 F}{\partial x^2}(0, 0) = 0, \qquad \frac{\partial^3 F}{\partial x^3}(0, 0) = f'''(0) \neq 0. \tag{2.19}$$

A complete bifurcation analysis of Eqs. (2.18) and (2.19) is rather complicated and would take us too far afield. Therefore, we confine ourselves to the two-parameter case. In the presentation below, we first study an important particular two-parameter perturbation, and then show how to reduce a general two-parameter perturbation to this special perturbation. We conclude this somewhat lengthy and technical section by giving a simple example of a three-parameter perturbation.

We first present the bifurcation analysis for a two-parameter, $\lambda = (\lambda_1, \lambda_2)$, perturbation $F(\lambda, x)$ with the following Taylor expansion about the origin:

$$F(\lambda, x) = \lambda_1 + \lambda_2 x + c(\lambda)\frac{x^2}{2} + d(\lambda)\frac{x^3}{6} + G(\lambda, x), \tag{2.20}$$

with $c(0) = 0$, $d(0) = f'''(0) \neq 0$, and, for any $\epsilon > 0$, there are $\delta > 0$ and $\eta > 0$ such that the function G satisfies $|G(\lambda, x)| < \epsilon|x|^3$ for $\|\lambda\| < \delta$ and $|x| < \eta$.

As usual, we begin by finding the bifurcation values of the parameters, that is, the values of $\lambda = (\lambda_1, \lambda_2)$ for which the function $F(\lambda, x)$ given by Eq. (2.20) has multiple zeros. To accomplish this we must solve the following two equations:

$$0 = F(\lambda, x) = \lambda_1 + \lambda_2 x + c(\lambda)\frac{x^2}{2} + d(\lambda)\frac{x^3}{6} + G(\lambda, x), \tag{2.21}$$

$$0 = \frac{\partial F}{\partial x}(\lambda, x) = \lambda_2 + c(\lambda)x + d(\lambda)\frac{x^2}{2} + \frac{\partial G}{\partial x}(\lambda, x). \tag{2.22}$$

We view Eqs. (2.21) and (2.22) as a system of two equations defining λ_1 and λ_2 parametrically in terms of x. To obtain the solutions locally, we can use essentially the method of Gaussian elimination in conjunction with repeated applications of the Implicit Function Theorem: using Eq. (2.22) solve locally for λ_2 as a function of λ_1 and x, then substitute the result into Eq. (2.21) to determine λ_1 in terms of x.

In preparation for the Implicit Function Theorem, let

$$\frac{\partial F}{\partial x}(\lambda,\, x) \equiv H(\lambda_1,\, x,\, \lambda_2) = \lambda_2 + c(\lambda)x + d(\lambda)\frac{x^2}{2} + \frac{\partial G}{\partial x}(\lambda,\, x).$$

The reordering of the variables is meant to be suggestive because we will apply the Implicit Function Theorem to $H(\lambda_1,\, x,\, \lambda_2)$ treating $(\lambda_1,\, x)$ as parameters and λ_2 as the dependent variable. Observe that $H(0,\, 0,\, 0) = 0$ and $\partial H(0,\, 0,\, 0)/\partial \lambda_2 = 1 \neq 0$. Thus, by the Implicit Function Theorem, there is a C^2 function $\psi_2(\lambda_1,\, x)$ defined for λ_1 and x small and

$$H(\lambda_1,\, x,\, \psi_2(\lambda_1,\, x)) = 0.$$

Moreover, every small $(\lambda_1,\, x,\, \lambda_2)$ satisfying $H(\lambda_1,\, x,\, \lambda_2) = 0$ is given by $\lambda_2 = \psi_2(\lambda_1,\, x)$. From the special form of H, it is not difficult to see, with a little bit of computing, of course, that

$$\psi_2(\lambda_1,\, 0) = \frac{\partial \psi_2}{\partial x}(0,\, 0) = 0, \quad \frac{\partial^2 \psi_2}{\partial x^2}(0,\, 0) = -d(0) = -f'''(0) \neq 0.$$

Thus, the Taylor expansion of the function $\psi_2(\lambda_1,\, x)$ with respect to x has the form

$$\psi_2(\lambda_1,\, x) = C(\lambda_1)x + D(\lambda_1)\frac{x^2}{2} + \cdots, \tag{2.23}$$

$$C(0) = 0, \quad D(0) = -f'''(0) \neq 0.$$

We now substitute $\lambda_2 = \psi_2(\lambda_1,\, x)$ into Eq. (2.21) and solve for λ_1 in terms of x. Let

$$0 = J(x,\, \lambda_1) \equiv \lambda_1 + \psi_2(\lambda_1,\, x)x + c(\lambda_1,\, \psi_2(\lambda_1,\, x))\frac{x^2}{2}$$

$$+ d(\lambda_1,\, \psi_2(\lambda_1,\, x))\frac{x^3}{6} + G(\lambda_1,\, \psi_2(\lambda_1,\, x),\, x)$$

and observe that $J(0,\, 0) = 0$ with $\partial J(0,\, 0)/\partial \lambda_1 = 1 \neq 0$. Thus, by the Implicit Function Theorem, there is a C^3 function $\psi_1(x)$ defined for x small so that

$$J(x,\, \psi_1(x)) = 0.$$

Moreover, every small $(x,\, \lambda_1)$ satisfying $J(x,\, \lambda_1) = 0$ is given by $\lambda_1 = \psi_1(x)$. From the special form of J, it is not difficult to see, again with some computing, that

$$\psi_1(0) = \psi_1'(0) = \psi_1''(0) = 0, \quad \psi_1'''(0) = 2f'''(0).$$

Thus, the Taylor expansion of the function $\psi_1(x)$ has the form

$$\psi_1(x) = \tfrac{1}{3}f'''(0)x^3 + \cdots. \tag{2.24}$$

We can now summarize the results of the computations above by saying that the local solutions $\lambda_1(x)$ and $\lambda_2(x)$ of Eqs. (2.21) and (2.22) near the origin are given by

$$\lambda_1(x) = \psi_1(x) = \tfrac{1}{3}f'''(0)x^3 + \cdots,$$
$$\lambda_2(x) = \psi_2(\psi_1(x), x) = -\tfrac{1}{2}f'''(0)x^2 + \cdots. \tag{2.25}$$

These equations are the parametric representation of a cusp in the (λ_1, λ_2)-plane which near the origin approximately coincides with the curve

$$\lambda_2^3 = -\tfrac{9}{8}f'''(0)\lambda_1^2. \tag{2.26}$$

If $f'''(0) < 0$, then the equation above is essentially that of the cusp in Eq. (2.8). Therefore, the qualitative structure of the flows of the perturbed equation (2.20), for small values of λ, are the same as the ones given in Figure 2.12. For practice, you should draw the flows for the case when $f'''(0) > 0$.

Equation (2.20) is an example of a "good" two-parameter perturbation of a vector field with cubic degeneracy in the sense that the bifurcations are determined only by the constant and the linear terms of the Taylor expansion of the vector field. The term $c(\lambda)x^2/2$ did not enter into the first approximation to the cusp in the (λ_1, λ_2)-plane; see Eqs. (2.25) and (2.26). This is because the function $c(\lambda)$ is differentiable in λ and $c(0) = 0$. Thus, the Taylor expansion for $c(\lambda)$ must be given by

$$c(\lambda) = c_1\lambda_1 + c_2\lambda_2 + \cdots,$$

where c_1 and c_2 are constants; hence, the term $c(\lambda)x^2$ has the form

$$c(\lambda)x^2 = c_1\lambda_1 x^2 + c_2\lambda_2 x^2 + \cdots.$$

Observe that, when x is small, the term $c_1\lambda_1 x^2$ is smaller than λ_1, and the term $c_2\lambda_2 x^2$ is smaller than λ_2. Therefore, it is to be expected that the term $c(\lambda)x^2$ has little influence on the bifurcations of Eq. (2.20).

We now show that an arbitrary two-parameter perturbation of the cubic degeneracy, under certain reasonable conditions, can be reduced to the special two-parameter perturbation (2.20). Let $\mu \equiv (\mu_1, \mu_2)$ be two parameters and consider the perturbation $F(\mu, x)$ with the following Taylor expansion about the origin:

$$F(\mu, x) = a(\mu) + b(\mu)x + \widehat{c}(\mu)\frac{x^2}{2} + \widehat{d}(\mu)\frac{x^3}{6} + \widehat{G}(\mu, x), \tag{2.27}$$

with $a(0) = b(0) = \widehat{c}(0) = 0$, $\widehat{d}(0) = f'''(0) \neq 0$, and, for any $\epsilon > 0$, there are $\delta > 0$ and $\eta > 0$ such that the function $\widehat{G}$ satisfies $|\widehat{G}(\mu, x)| < \epsilon|x|^3$ for $\|\mu\| < \delta$ and $|x| < \eta$.

To make certain that (μ_1, μ_2) cover all possible small values of the constant and the linear terms, that is, the range of the vector-valued function $(a(\mu), b(\mu))$ cover a neighborhood of zero for small μ, we suppose that the *Jacobian* of $(a(\mu), b(\mu))$ with respect to μ at $\mu = \mathbf{0}$ is not zero:

$$\det \left. \frac{\partial(a(\mu), b(\mu))}{\partial(\mu_1, \mu_2)} \right|_{(\mu_1, \mu_2)=(0, 0)} \neq 0. \tag{2.28}$$

Next, we introduce the transformation of parameters

$$\lambda_1 \equiv a(\mu_1, \mu_2), \qquad \lambda_2 \equiv b(\mu_1, \mu_2) \tag{2.29}$$

in a neighborhood of $\lambda = (\lambda_1, \lambda_2)$ and $\mu = (\mu_1, \mu_2)$ equal to zero. For any μ, the constant λ is uniquely defined by Eq. (2.29). To know that relation (2.29) is a good transformation of parameters we need to make certain that, given λ, a constant μ is uniquely defined by Eq. (2.29). Condition (2.29) and the Implicit Function Theorem imply that this is the case in a neighborhood of zero. In fact, if we define the functions

$$P_1(\lambda, \mu) \equiv a(\mu) - \lambda_1, \qquad P_2(\lambda, \mu) \equiv b(\mu) - \lambda_2,$$

then $P_1(0, 0) = P_2(0, 0) = 0$ and

$$\det \left. \frac{\partial(P_1, P_2)}{\partial(\mu_1, \mu_2)} \right|_{\lambda=\mu=0} = \det \left. \frac{\partial(a(\mu), b(\mu))}{\partial(\mu_1, \mu_2)} \right|_{\mu=0} \neq 0.$$

Thus, μ is uniquely defined by λ and is C^1 in λ.

Once the transformation (2.29) of parameters is made, by putting $c(\lambda) \equiv \hat{c}(\mu(\lambda))$, $d(\lambda) \equiv \hat{d}(\mu(\lambda))$, and $G(\lambda, x) \equiv \hat{G}(\mu(\lambda), x))$, the general perturbation (2.27) becomes the special perturbation (2.20).

To illustrate the discussion above, we now give examples of a "good" and a "bad" two-parameter perturbation of the cubic degeneracy.

Example 2.11. Consider the two-parameter perturbation of the vector field $f(x) = -x^3/6$ given by

$$\dot{x} = \mu_1 + \mu_2 + (\mu_2 + \mu_1^3) x + \mu_1 \frac{x^2}{2} - \frac{x^3}{6}, \tag{2.30}$$

where μ_1 and μ_2 are two real parameters. For the perturbed vector field (2.30), the functions in Eq. (2.27) are given by $a(\mu) = \mu_1 + \mu_2$, $b(\mu) = \mu_2 + \mu_1^3$, $\hat{c}(\mu) = \mu_1$, $\hat{d}(\mu) = -1$, and $\hat{G}(\mu, x) = 0$. The Jacobian (2.28) of these functions $(a(\mu), b(\mu))$ at $\mu = \mathbf{0}$ is one—"good." Thus, we introduce the new parameters $\lambda = (\lambda_1, \lambda_2)$ defined by

$$\lambda_1 \equiv \mu_1 + \mu_2, \qquad \lambda_2 \equiv \mu_2 + \mu_1^3$$

so that Eq. (2.30) attains the special form of Eq. (2.20) as follows:

$$\dot{x} = \lambda_1 + \lambda_2 x + c(\lambda)\frac{x^2}{2} - \frac{x^3}{6}, \tag{2.31}$$

where the function $c(\lambda)$ is the unique solution $\mu_1 = c(\lambda)$ of the cubic equation

$$-\mu_1^3 + \mu_1 - \lambda_1 + \lambda_2 = 0$$

which vanishes for $(\lambda_1, \lambda_2) = (0, 0)$. The Implicit Function Theorem guarantees that such a solution exists and is unique.

The bifurcation curve of Eq. (2.31) in the (λ_1, λ_2)-plane near the origin is approximately the cusp $8\lambda_2^3 = 9\lambda_1^2$, which follows from Eq. (2.26). Thus, in the original (μ_1, μ_2)-plane this curve becomes

$$8(\mu_2 + \mu_1^3)^3 = 9(\mu_1 + \mu_2)^2$$

which, near the origin, is again a cusp with approximately the same shape as the one in the (λ_1, λ_2)-plane. $\Diamond$

Example 2.12. Consider the two-parameter perturbation of the vector field $f(x) = -x^3/6$ given by

$$\dot{x} = \mu_1 + \mu_2^2 x + \mu_2 \frac{x^2}{2} - \frac{x^3}{6}, \tag{2.32}$$

where μ_1 and μ_2 are two real parameters. As a special case of Eq. (2.27), we have $a(\mu) = \mu_1$, $b(\mu) = \mu_2^2$, $\widehat{c}(\mu) = \mu_2$, and $\widehat{d}(\mu) = -1$. Notice that the condition (2.28) is not satisfied. However, we can analyze Eq. (2.32) in the following way.

From our previous discussion of the cusp, it is natural to eliminate the x^2 term. If we introduce the new variable $y \equiv \mu_2 - x$, then

$$\dot{y} = \mu_1 + \tfrac{4}{3}\mu_2^3 + \tfrac{3}{2}\mu_2^2 y - \tfrac{1}{6}y^3.$$

Now, if we make the transformation of the parameters

$$\lambda_1 \equiv \mu_1 + \tfrac{4}{3}\mu_2^3, \qquad \lambda_2 \equiv \tfrac{3}{2}\mu_2^2, \tag{2.33}$$

then the differential equation above is put into the form of the special perturbation (2.20):

$$\dot{y} = \lambda_1 + \lambda_2 y - \tfrac{1}{6}y^3.$$

The bifurcation curve of this equation in the (λ_1, λ_2)-plane is given by Eq. (2.26), which is the cusp $8\lambda_2^3 = 9\lambda_1^2$.

To find the bifurcation curve of Eq. (2.32) in the original (μ_1, μ_2)-plane, we substitute the transformation (2.33) into the cusp and obtain

$$8 \left(\tfrac{3}{2}\mu_2^2\right)^3 = 9 \left(\mu_1 + \tfrac{4}{3}\mu_2^3\right)^2.$$

A simple computation yields the formula

$$\mu_1 = \left(-\tfrac{4}{3} \mp \sqrt{3}\right)\mu_2^3.$$

Thus, the bifurcation values of Eq. (2.32) in the (μ_1, μ_2)-plane form cubic curves, not a cusp. Of course, the reason for this deviant behavior is that the parameters (μ_1, μ_2) do not enter into the equation in a nice way, that is, the determinant in Eq. (2.28) is zero—"bad." ◊

As we mentioned earlier, a complete analysis of the cubic degeneracy is beyond the intended scope of our book. Therefore, we end this rather long section with an example of a three-parameter perturbation.

Example 2.13. *Three parameters:* Consider the three-parameter perturbation of $f(x) = -x^3/6$ given by

$$\dot{x} = \mu_1 + \mu_2 x + \mu_3 \frac{x^2}{2} - \frac{x^3}{6},$$

where μ_1, μ_2, and μ_3 are three parameters. As in the previous example, the x^2 term can be eliminated by introducing the new variable $y \equiv \mu_3 - x$. Then, the differential equation above in the new variable becomes

$$\dot{y} = \mu_1 + \mu_2\mu_3 + \tfrac{1}{3}\mu_3^3 + (\mu_2 + \tfrac{1}{2}\mu_3^2)y - \frac{y^3}{6}.$$

Therefore, if we define two new parameters

$$\lambda_1 \equiv \mu_1 + \mu_2\mu_3 + \tfrac{1}{3}\mu_3^3 \quad \text{and} \quad \lambda_2 \equiv \mu_2 + \tfrac{1}{2}\mu_3^2,$$

we obtain the following two-parameter differential equation:

$$\dot{y} = \lambda_1 + \lambda_2 y - \frac{y^3}{6}.$$

The values of λ_1 and λ_2 which correspond to bifurcation points are given by the cusp $8\lambda_2^3 = 9\lambda_1^2$, as we have computed before. Therefore, the values of the original parameters (μ_1, μ_2, μ_3) corresponding to bifurcation points must lie on the surface

$$8 \left(\mu_2 + \tfrac{1}{2}\mu_3^2\right)^3 = 9 \left(\mu_1 + \mu_2\mu_3 + \tfrac{1}{3}\mu_3^3\right)^2$$

in the three-dimensional (μ_1, μ_2, μ_3)-space. As one crosses this surface, the number of equilibrium points changes from one to three. $\Diamond$

Exercises —————————————————————

2.9. *On quadratic degeneracy:* For each of the vector fields below, draw the bifurcation diagrams with the corresponding phase portraits:

(a) $F(c, e, x) = c + ex + x^2$. This is Example 2.9, but this time use the transformation $x = y - e/2$.

(b) $F(\lambda, x) = a + (\lambda - a)x + x^2$. Draw x versus λ for various values of a. Notice the difference between $a < 0$ and $a > 0$.

(c) $F(\lambda, x) = \lambda^2 + 2a\lambda x + x^2$. Draw x versus λ for various values of a. Notice the difference between $|a| < 1$ and $|a| > 1$.

(d) $F(\lambda, x) = \lambda + 2a\lambda x + x^2$, for any fixed value of a.

(e) $F(\lambda, x) = \lambda^4 + 2a\lambda x + x^2$, for any fixed value of a.

2.10. *On cubic degeneracy:* Obtain the bifurcation curves of the following one- and two-parameter perturbations of the cubic vector field $f(x) = -x^3$ and sketch some representative phase portraits:

(a) $\dot{x} = 1 + \mu_1 + 2\mu_1 x - x^3$;   (b) $\dot{x} = 1 + \mu_1^2 + \mu_1 x - x^3$;

(c) $\dot{x} = \mu_1 + \mu_2 + (\mu_1 - \mu_2)x - x^3$;   (d) $\dot{x} = \mu_1 + \mu_2 + (\mu_1 - \mu_2^2)x - x^3$;

(e) $\dot{x} = \mu_1 + \mu_2^2 + \mu_2 x - x^3$;   (f) $\dot{x} = \mu_1 + \mu_2 + \mu_2^2 x + \mu_2 x^2 - x^3$.

2.11. *On quartic degeneracy:* This exercise is on various one- and two-parameter perturbations of the quartic vector field $f(x) = x^4$. A complete discussion of the most general perturbation of the quartic is very difficult. For each of the following perturbations, draw the bifurcation diagrams along with the corresponding vector fields:

(a) $F(c, x) = c + x^4$;   (b) $F(d, x) = dx + x^4$;

(c) $F(e, x) = ex^2 + x^4$;   (d) $F(c, d, x) = c + dx + x^4$;

(e) $F(c, e, x) = c + ex^2 + x^4$;   (f) $F(d, e, x) = dx + ex^2 + x^4$.

2.12. Obtain the bifurcation curves of the following one- and two-parameter differential equations and sketch some representative phase portraits:

(a) $\dot{x} = \mu_1 - x^2/(1 + x^2)$;   (b) $\dot{x} = \mu_1 - x^2/(1 + x^2)^2$;

(c) $\dot{x} = \mu_1 - x^3/(1 + x^3)$ for $x > -1$;

(d) $\dot{x} = \mu_1 + \mu_2 x - x^3/(1 + x^3)$ for $x > -1$.

2.13. *Unfolding a pitchfork:* As we have seen in Example 2.5, the pitchfork bifurcation occurs for the vector field $F(\mu, x) = -x^3 + \mu x$ at $\mu = 0$ where the number of equilibria changes from one to three or vice versa. An important practical question is what happens to the bifurcation diagram if this one-parameter vector field is subjected to small perturbations. Surprising as it may seem, one can in fact write down a "most general" perturbation of this vector field near the origin using only two additional parameters:

$$F(\mu, \lambda_1, \lambda_2, x) = -x^3 + \mu x + \lambda_1 + \lambda_2 x^2.$$

Any other perturbation, however many parameters it may contain, can be transformed to the particular perturbation above. For this reason, the three-parameter vector field above is referred to as the *unfolding* of the pitchfork bifurcation. The purpose of this exercise is to determine the bifurcation diagrams x versus μ near zero for fixed values of the unfolding parameters λ_1 and λ_2 near zero.

(a) To eliminate the quadratic term, let $x = +\lambda_2/3$ and transform the vector field to $-y^3 + \nu y + \delta_1 + \lambda_2\nu/3$, where $\nu = \mu + \lambda_2^2/3$ and $\delta_1 = \lambda_1 - \lambda_2^3/27$. Notice that this is a special case of Example 2.6.

(b) For $\lambda_1 = 0$ and fixed $\lambda_2 < 0$, obtain the bifurcation diagram x versus μ and compare your result with Figure 2.16.

(c) For $\delta_1 = 0$, that is, $\lambda_1 < \lambda_2^3/27$ and $\lambda_2 < 0$, obtain the bifurcation diagram and, again, refer to Figure 2.16. Do the same analysis for $\lambda_1 < \lambda_2^3/27$ and $\lambda_2 > 0$.

(d) Discuss some of the remaining cases.

Reference: For more details on this example, and the general theory of unfoldings of bifurcations, see Golubitsky and Schaeffer [1985].

2.4. An Example on a Circle

In this brief section, we study the bifurcations in a two-parameter perturbation of the periodic differential equation $\dot{x} = \sin x$ of Section 1.4. By considering the perturbed periodic equation on a circle, we will obtain a new type of a limit set different from an equilibrium point, namely, the circle itself.

Example 2.14. Consider the two-parameter perturbation of the periodic vector field $f(x) = \sin x$ given by

$$\dot{x} = c + d\sin x, \tag{2.34}$$

where c and d are real parameters. Of course, the perturbed vector field (2.34) is still periodic with period 2π for all values of the parameters. When $d = 0$, the vector field is actually periodic with any period. Furthermore, if $c = d = 0$, then the flow consists entirely of equilibrium points. These two cases are not very interesting. In Figure 2.19 we have drawn the orbits of Eq. (2.34) both on the real line $\mathbb{R}$ and the circle S^1 for various parameter values.

Observe that the α- and the ω-limit sets of Eq. (2.34) on S^1 are either equilibrium points (case $|d| \geq c$) or S^1 itself (case $|d| < |c|$). Although it is not apparent from Figure 2.19, in the latter case, each solution is also *periodic in t* when considered as a motion on S^1. Let us first make this notion precise.

Definition 2.15. *A solution $\varphi(t, x_0)$, with $\varphi(0, x_0) = x_0$, of a scalar differential equation $\dot{x} = f(x)$ satisfying $f(x+P) = f(x)$ is called a periodic solution of period T, if there is a $T > 0$ such that*

$$\varphi(t + T, x_0) = \varphi(t, x_0) + P$$

for all $t \geq 0$. If, moreover, $\varphi(t + \tau, x_0) \neq \varphi(t, x_0) + P$ for any $0 \leq \tau < T$, then T is called the minimal period.

Returning to our example, for the differential equation (2.34), we have $P = 2\pi$. We now show that if $|d| < |c|$, then every solution of Eq. (2.34) is

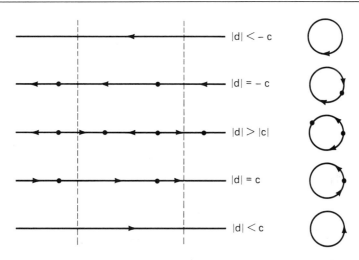

Figure 2.19. *Phase portraits of $\dot{x} = c + d\sin x$ on $\mathbb{R}$ and S^1.*

periodic on S^1. Since $\varphi(t, x_0)$ is implicitly defined by

$$t = \int_{x_0}^{\varphi(t, x_0)} \frac{dx}{c + d\sin x}$$

and $c + d\sin x \neq 0$ for all x, there is a unique T such that $\varphi(T, x_0) = x_0 + 2\pi$, that is,

$$T = \int_{x_0}^{x_0 + 2\pi} \frac{dx}{c + d\sin x} = \int_0^{2\pi} \frac{dx}{c + d\sin x}.$$

If $\varphi(t, x_0)$ is a solution of Eq. (2.34), then so are $\varphi(t, x_0) + 2\pi$ and $\varphi(t + T, x_0)$. At $t = 0$ these two solutions are equal; thus, by uniqueness (Theorem 1.4), they are identical for all $t \geq 0$. This proves the assertion above and, moreover, shows that the period T is independent of the initial condition.

The time periodicity of the solutions of Eq. (2.34), when $|d| < |c|$, can easily be observed if we plot trajectories in the (t, x)-plane; see Figure 2.20. $\diamond$

Exercises ━━━━━━━━━━━━━━━━━━━━━━━━━━━━━━━ ♣♡♠◇

2.14. Draw on the circle the phase portraits of the periodic differential equation

$$\dot{x} = -\lambda + 2 + \cos(2x) - 3\cos x$$

and verify the qualitative features in the indicated ranges of the scalar parameter λ:

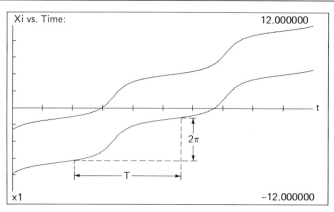

Figure 2.20. *Time periodicity of solutions of* $\dot{x} = c + d \sin x$.

(a) All solutions are periodic if $\lambda < -1/8$ and $\lambda > 6$.
(b) There are four equilibria if $-1/8 < \lambda < 0$.
(c) There are two equilibria if $0 < \lambda < 6$.

2.15. *No equilibria:* Suppose that f is a 2π-periodic C^1 function with the property that $f(x) \neq 0$ for all $x \in \mathbb{R}$. Show that every solution of the differential equation $\dot{x} = f(x)$ is periodic on the circle obtained by identifying the end points of an interval of length 2π. Derive a formula for the period(s) of the solutions.

2.5. Computing Bifurcation Diagrams

In exploring the dynamics of a differential equation $\dot{x} = f(x)$, a first key step is to locate its equilibrium points, equivalently, the zeros of the function f. For most functions, this is usually a formidable task for the pencil; however, there are effective approximation procedures that can be performed on the computer rather easily. We will see one such procedure—Newton's Method—in Chapter 3. For the moment, we can also use the observation in Section 1.2 that the equilibrium points of a scalar differential equation $\dot{x} = f(x)$ are the α- and ω-limit sets of orbits. Thus, we can obtain approximate values of the zeros of f by following a few selected orbits in forward or reverse time. Numerical integration of such orbits can be carried out on the computer using one of the standard numerical methods for initial-value problems such as Euler's Method (see Section 3.1).

In the case where the vector field depends on a parameter λ, one could apply the same procedure to the equation $\dot{x} = F(\lambda, x)$ for each fixed value of λ. By performing many numerical experiments for sufficient number of values of λ, one could then arrive at an approximate representation of the bifurcation diagram of the differential equation. For the simplest of examples, however, this process is very cumbersome and inefficient. It is the purpose of this section to describe a practical procedure for determining the zeros of $F(\lambda, x)$ using differential equations.

Let us suppose that we are given a C^1 function, depending on a scalar parameter λ,

$$F : \mathbb{R} \times \mathbb{R} \to \mathbb{R}; \quad (\lambda, x) \mapsto F(\lambda, x)$$

and wish to determine the set of points (λ, x) satisfying the equation

$$F(\lambda, x) = 0.$$

Using the observation described earlier, if necessary, let us first determine a pair of values (λ_0, x_0) with $F(\lambda_0, x_0) = 0$. Let us assume that in a neighborhood of (λ_0, x_0) on the (λ, x)-plane the zeros of $F(\lambda, x)$ lie on a smooth curve. We can parametrize this curve by t so that there will be two C^1 functions $\lambda(t)$ and $x(t)$ satisfying $\lambda(0) = \lambda_0$, $x(0) = x_0$, and $F(\lambda(t), x(t)) = 0$ for t in a neighborhood of zero.

To represent this parametrized curve as an orbit of a differential equation, we differentiate the equation $F(\lambda(t), x(t)) = 0$ with respect to t:

$$\frac{\partial F}{\partial \lambda}(\lambda, x)\,\dot{\lambda} + \frac{\partial F}{\partial x}(\lambda, x)\,\dot{x} = 0.$$

To satisfy this identity, $(\dot{\lambda}, \dot{x})$ must be a constant multiple of the vector $\left(-\frac{\partial F}{\partial x}(\lambda, x), \frac{\partial F}{\partial \lambda}(\lambda, x)\right)$. If we choose the parametrization of the curve $(\lambda(t), x(t))$ in such a way that this constant is one, then we arrive at the pair of differential equations

$$\begin{aligned}
\dot{\lambda} &= -\frac{\partial F}{\partial x}(\lambda, x) \\
\dot{x} &= \frac{\partial F}{\partial \lambda}(\lambda, x)
\end{aligned} \tag{2.35}$$

satisfying the initial conditions

$$\lambda(0) = \lambda_0, \quad x(0) = x_0.$$

These differential equations, unfortunately, are not just a pair of independent scalar equations because their right-hand sides depend on two variables λ and x simultaneously. Consequently, the orbits of such a pair must be investigated on the (λ, x)-plane. We will devote Part III of our book to a comprehensive study of differential equations of this type. At this point, we will confine our discussions to a step-by-step description of a constructive numerical procedure for obtaining the zero set of $F(\lambda, x)$:

- **Step 1.** Fix $\lambda = \lambda_0$ and use the scalar differential equation $\dot{x} = F(\lambda_0, x)$ to find the set E_{λ_0} of zeros of $F(\lambda_0, x)$.

- **Step 2.** For each $x_0 \in E_{\lambda_0}$, compute numerically the solution of the differential equations (2.35) with initial value $\lambda(0) = \lambda_0$ and $x(0) = x_0$, in both forward and reverse time (using negative step size) directions.

- **Step 3.** Repeat Step 1 as necessary for other values of the parameter λ because there may be some components of the curves of zeros of $F(\lambda, x)$ that never intersect the vertical line $\lambda = \lambda_0$.

Let us now implement these steps on several examples and attempt to clarify the ambiguity ("as necessary") that is present in Step 3.

Example 2.16. *Computing hysteresis:* Consider the equation

$$F(\lambda, x) = \lambda + x - x^3. \tag{2.36}$$

The zero set of this function is, of course, the cubic curve $\lambda = -x + x^3$. However, for the sake of practice, let us try to implement the three steps above.

In Step 1, let us fix, for example, $\lambda_0 = -5$. Then, using the scalar differential equation $\dot{x} = -5 + x - x^3$, we find that the set of its equilibria consists of a single point: $E_{-5} = \{-1.904\ldots\}$.

In Step 2, differential equations (2.35) for the function (2.36) become

$$
\begin{aligned}
\dot{\lambda} &= -(1 - 3x^2) \\
\dot{x} &= 1
\end{aligned}
\tag{2.37}
$$

with the initial data $\lambda(0) = -5$ and $x(0) = -1.904\ldots$. Numerical integration of this initial-value problem gives the cubic curve shown in Figure 2.21. Forward integration (using positive time and positive step size) of the positive orbit gives the piece of the cubic above the line $\lambda_0 = -5$ and the backward integration (using negative time and negative step size) for the negative orbit yields the remaining lower part.

Although there is no need for Step 3 in this simple example, let us see what happens for another value of λ_0. If we fix $\lambda_0 = 0$, for example, then we have $E_0 = \{-1, 0, 1\}$. Numerical computation of the orbits of Eq. (2.37) through any one of the points $(0, -1)$, $(0, 0)$, or $(0, 1)$ gives the same cubic curve. $\Diamond$

Example 2.17. Consider the equation

$$F(\lambda, x) = 1 + \lambda x - x^3. \tag{2.38}$$

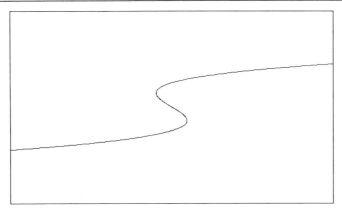

Figure 2.21. *Numerically computed bifurcation diagram for the differential equation* $\dot{x} = \lambda + x - x^3$.

The pair of differential equations (2.35) for the function (2.38) are

$$\dot{\lambda} = -\lambda + 3x^2$$
$$\dot{x} = x. \tag{2.39}$$

If we choose $\lambda_0 = 0$, then $E_0 = \{1\}$. The orbit of Eq. (2.39) through the point $(0, 1)$ gives the upper curve shown in Figure 2.22. Unlike the example above, the zero set of Eq. (2.38) is not obvious. So, let us experiment with another choice of λ_0. For $\lambda_0 = 2$, computations give $E_2 = \{-1, -0.618\ldots, 1.618\ldots\}$. The orbits of Eq. (2.39) through the points $(2, -1)$ and $(2, -0.618\ldots)$ coincide and they both give the lower curve. The orbit through $(2, 1.618\ldots)$ yields the same upper curve we obtained with the choice $\lambda_0 = 0$. Further experimentation with other choices of λ_0 does not alter Figure 2.22. $\Diamond$

Example 2.18. *Computing pitchfork:* As the final example of this section, let us try to recover numerically the bifurcation diagram of the pitchfork bifurcation, that is, compute the set of zeros of the equation

$$F(\lambda, x) = \lambda x - x^3. \tag{2.40}$$

The pair of differential equations (2.35) for this example are

$$\dot{\lambda} = -\lambda + 3x^2$$
$$\dot{x} = x. \tag{2.41}$$

Notice that Eq. (2.41) is the same as Eq. (2.39), but the bifurcation curves are determined by special choices of initial values. To compute the bifurcation diagram of Eq. (2.40), if we choose $\lambda_0 = -1$, then $E_{-1} = \{0\}$. The orbit of Eq. (2.41) through $(-1, 0)$ is the negative part of the λ-axis; see Figure 2.23.

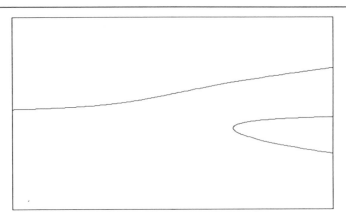

Figure 2.22. *Numerically computed bifurcation diagram of the differential equation* $\dot{x} = 1 + \lambda x - x^3$.

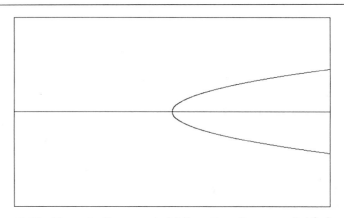

Figure 2.23. *Numerically computed bifurcation diagram, pitchfork, of the differential equation* $\dot{x} = \lambda x - x^3$.

For $\lambda_0 = 1$, we get $E_1 = \{-1, 0, 1\}$. The orbits of Eq. (2.41) through the points $(1, -1)$ and $(1, 1)$ give the lower and the upper parts, respectively, of the parabola $\lambda = 3x^2$. The orbit through $(1, 0)$ is the positive half of the λ-axis. The point $(0, 0)$ is not obtained as a solution of Eq. (2.41) unless we choose $\lambda_0 = 0$. In this case, $E_0 = \{0\}$ and the point $(0, 0)$ is a (equilibrium) solution of Eq. (2.41). $\diamond$

Exercises _____ ♣ ♡ ♠ ◇

2.16. Use the numerical procedure described in this section, and PHASER, to draw the bifurcation diagrams in the (λ, x)-plane of the following differential

equations:
(a) $\dot{x} = x^2 - \lambda$;   (b) $\dot{x} = \lambda^2 + x^2 - 1$;
(c) $\dot{x} = (\lambda + 2)^2 + x^2 - 1$;   (d) $\dot{x} = [\lambda x - x^3][(\lambda + 2)^2 + x^2 - 1]$;
(e) $\dot{x} = 2\lambda^3 + 3\lambda^2 x - x^3$;   (f) $\dot{x} = 2\lambda^3 + 2.9\lambda^2 x - x^3$;
(g) $\dot{x} = 0.1 + 2\lambda^2 + 3\lambda^2 x - x^3$;   (h) $\dot{x} = (\lambda + x - x^3)(\lambda^2 + x^2 - 1)$.

2.6. Equivalence of Flows

We agreed in Section 2.1 to consider two flows to be qualitatively equivalent
if they have the same orbit structure, that is, if they have equal number of
orbits and the directions of the flows on the corresponding orbits are the
same. For the case of scalar differential equations this notion of equivalence
of flows has been sufficient for our purposes. In the case of more than one
dimension, which we will consider in later chapters, this somewhat simple-
minded definition of equivalence proves to be inadequate. In this section
we give another definition of flow equivalence which generalizes naturally
to higher dimensions. The old and the new notions of equivalence will, of
course, turn out be the same for scalar differential equations, as we will
show below.

Let $\varphi(t, x_0)$ be the flow of $\dot{x} = f(x)$, and $\psi(t, x_0)$ be the flow of
$\dot{x} = g(x)$. To compare the orbits of these flows, it is natural to determine
a change of variables

$$h : \mathbb{R} \to \mathbb{R}; \quad x \mapsto h(x)$$

which, for each t, at least takes one flow to the other, that is,

$$h\left(\varphi(t, x_0)\right) = \psi\left(t, h(x_0)\right) \tag{2.42}$$

for all t as long as the flows are defined. Furthermore, since we are looking
for an equivalence relation, it is clear that h should be an invertible map
so that h^{-1} takes ψ to φ. What additional properties should we require of
h so as to capture the qualitative features of the flows? This is a rather
delicate question because if we do not restrict h sufficiently, we may not be
able to distinguish two qualitatively different flows; at the other extreme,
if we restrict h severely, then two flows with the same orbit structure may
not be equivalent. One "natural" choice would be to require both h and
h^{-1} to be C^1 functions.

Definition 2.19. *A C^1 function $h : \mathbb{R} \to \mathbb{R}$ with a C^1 inverse is called a
C^1 diffeomorphism of $\mathbb{R}$.*

To appreciate some of the possible consequences of requiring h to be a
C^1 diffeomorphism, let us study a specific example. Consider the following
linear vector fields

$$\dot{x} = -x, \qquad \dot{x} = -2x. \tag{2.43}$$

These two vector fields have the same orbit structure because they each have one asymptotically stable equilibrium point. For these two vector fields, Eq. (2.42) becomes

$$h\left(e^{-t}x_0\right) = e^{-2t}h(x_0).$$

If we differentiate this equation with respect to x_0 and evaluate the result at $x_0 = 0$, then we obtain

$$e^{-t}h'(0) = e^{-2t}h'(0).$$

Since we require h to be invertible, $h'(0) \neq 0$ and thus we arrive at the disturbing implication that $-1 = -2$. Consequently, we cannot have h to be a C^1 function with a C^1 inverse if we are to consider the flows of $\dot{x} = -x$ and $\dot{x} = -2x$ to be qualitatively equivalent. We settle for the next best thing.

Definition 2.20. *A continuous map $h : \mathbb{R} \to \mathbb{R}$ with a continuous inverse is called a homeomorphism of $\mathbb{R}$.*

Definition 2.21. *Two scalar differential equations $\dot{x} = f(x)$ and $\dot{x} = g(x)$ are said to be topologically equivalent if there is a homeomorphism h of $\mathbb{R}$ such that h takes the orbits of one differential equation to the orbits of the other and preserves the sense of direction in time.*

For the purposes of comparing the qualitative features of flows of scalar differential equations, it is not a loss to require h to be merely a homeomorphism.

Theorem 2.22. *Two scalar differential equations $\dot{x} = f(x)$ and $\dot{x} = g(x)$ each with a finite number of equilibrium points are topologically equivalent if and only if they have the same orbit structure.*

Proof. Let us first point out that if two vector fields are topologically equivalent, then the corresponding homeomorphism takes an equilibrium point of one vector field to an equilibrium point of the other. With this observation, it is clear that topological equivalence implies that the two vector fields have the same orbit structure. We will indicate how to prove the converse implication. Let $\bar{x}_1, \ldots, \bar{x}_n$ be the equilibrium points of the vector field f with their ordering on the line, and similarly, let $\hat{x}_1, \ldots, \hat{x}_n$ be the equilibrium points of g. Let us choose points $\alpha_1, \ldots, \alpha_{n+1}$ so that $\alpha_1 < \bar{x}_1$, $\bar{x}_n < \alpha_{n+1}$, and α_{i+1} lies in between the consecutive equilibria $(\bar{x}_i, \bar{x}_{i+1})$. Similarly, choose the points $\beta_1, \ldots, \beta_{n+1}$ so that $\beta_1 < \hat{x}_1$, $\hat{x}_n < \beta_{n+1}$, and $\beta_{i+1} \in (\hat{x}_i, \hat{x}_{i+1})$; see Figure 2.24.

We will first construct a homeomorphism $h : (-\infty, \bar{x}_1) \to (-\infty, \hat{x}_1)$ of the two open intervals. For any point x_0 in $(-\infty, \bar{x}_1)$, there is a unique value t_{x_0} of time depending on x_0 such that $\varphi(t_{x_0}, x_0) = \alpha_1$. If we let $h(x_0) = \psi(-t_{x_0}, \beta_1)$, then h is a homeomorphism. To extend h, let $h(\bar{x}_1) = \hat{x}_1$. Since $h(x_0) \mapsto \hat{x}_1$ as $x_0 \mapsto \bar{x}_1$, now the map $h : (-\infty, \bar{x}_1] \to (-\infty, \hat{x}_1]$ so defined is a homeomorphism.

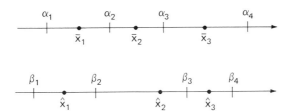

Figure 2.24. *Constructing homeomorphisms of scalar vector fields.*

Similarly, using the flows of f and g, we can construct a homeomorphism $h : (\bar{x}_i, \bar{x}_{i+1}) \to (\hat{x}_i, \hat{x}_{i+1})$ of open intervals and then extend h to the closed intervals by letting $h(\bar{x}_i) = \hat{x}_i$ and $h(\bar{x}_{i+1}) = \hat{x}_{i+1}$. This construction yields a homeomorphism of $\mathbb{R}$ which establishes the topological equivalence of the vector fields f and g. $\Diamond$

Example 2.23. Let us now return to the vector fields (2.43) and, using the proof of the theorem above, construct a homeomorphism to establish the topological equivalence of their flows. Since these vector fields have a single equilibrium point at the origin, let us choose, for example, $\alpha_1 = \beta_1 = -1$ and $\alpha_2 = \beta_2 = 1$. For $x_0 < 0$, we first find t_{x_0} such that $e^{-t_{x_0}}x_0 = -1$, that is, $t_{x_0} = \ln(-x_0)$. We then define $h(x_0) = e^{2\ln(-x_0)}(-1) = -x_0^2$. Similarly, for $x_0 > 0$, we have $t_{x_0} = \ln x_0$, and $h(x_0) = e^{2\ln(x_0)}1 = x_0^2$. Consequently, the homeomorphism

$$h(x) = \begin{cases} -x^2 & \text{if } x < 0 \\ 0 & \text{if } x = 0 \\ x^2 & \text{if } x > 0 \end{cases}$$

establishes the topological equivalence of the two vector fields $\dot{x} = -x$ and $\dot{x} = -2x$; see Figure 2.25. Notice that h fails to have a differentiable inverse only at the equilibrium point zero. $\Diamond$

For vector fields depending on parameters, we can also capture the notion of stable orbit structure in terms of the concept of topological equivalence.

Definition 2.24. *Let $\dot{x} = F(\lambda, x)$ be a vector field that depends on k parameters $\lambda = (\lambda_1, \ldots, \lambda_k)$. For a fixed value $\lambda = \bar{\lambda}$, the vector field $\dot{x} = F(\bar{\lambda}, x)$ is called structurally stable if there is an $\varepsilon > 0$ such that $\dot{x} = F(\lambda, x)$ is topologically equivalent to $\dot{x} = F(\bar{\lambda}, x)$ for all values of λ satisfying $\|\lambda - \bar{\lambda}\| < \varepsilon$.*

In light of Theorem 2.22, the notions of structural stability and stable orbit structure are evidently the same, as stated below, and thus a parameter value at which the vector field is not structurally stable is a bifurcation value.

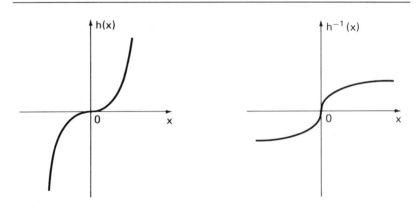

Figure 2.25. *Graphs of homeomorphism h and its inverse for the topological equivalence of the flows of $\dot{x} = -x$ and $\dot{x} = -2x$.*

Theorem 2.25. *A scalar differential equation $\dot{x} = F(\bar{\lambda}, x)$ with a finite number of equilibrium points is structurally stable if and only if it has stable orbit structure.* ◇

Here we conclude this admittedly lengthy chapter on scalar autonomous differential equations. Next, we undertake the study of another sort of dynamical system on the line—maps.

Exercises ━━━━━━━━━━━━━━━━━━━━━━━━━━━━━━━━━━━━━━━

2.17. Identify the functions below as homeomorphisms, diffeomorphisms, or neither in their domains of definition:
  (a) $h(x) = 2x + 1$;   (b) $h(x) = 2x^2$;   (c) $h(x) = x^3$;
  (d) $h(x) = \frac{5}{3}x^{5/3}$;   (e) $h(x) = e^x$;   (f) $h(x) = \arctan x$.

2.18. Construct a homeomorphism of the real line to establish the topological equivalence of the differential equations $\dot{x} = 2x$ and $\dot{x} = x + 2$.

2.19. Consider the equation $\dot{x} = x^2 - 1 + \lambda$ depending on a real parameter λ. Show that the flow of this differential equation is topologically equivalent to that of $\dot{x} = x^2 - 1$ if $-\infty < \lambda < 1$, to $\dot{x} = x^2$ if $\lambda = 1$, and to $\dot{x} = x^2 + 1$ if $\lambda > 1$.

2.20. Show that if a vector field $\dot{x} = F(\lambda, x)$ has a finite number of equilibria and they are hyperbolic, then the vector field is structurally stable.

Bibliographical Notes ━━━━━━━━━━━━━━━━━━━━━━━━━━━━━━━━━

Bifurcations of equilibria of vector fields have close connections with singularity theory of functions, popularly known as catastrophe theory. A light survey of some of these connections is Arnold [1984]; comprehensive sources are Golubitsky and Schaeffer [1985] and Golubitsky, Stewart, and Schaeffer [1988]. Sources on catastrophe theory are many; some of the originals are Whitney [1955], Thom [1975], and Zeeman [1974]. Golubitsky and Guillemin [1973], and Poston and Stewart [1978] contain detailed expositions. For applications to mechanics, see Thompson and Hunt [1973].

Numerical computation of bifurcation diagrams of one-parameter vector fields is an important exploratory tool. In dimensions greater than one, approximation of the bifurcation curves of one-parameter families of vector fields requires new ideas because they can no longer be obtained as solutions of appropriate differential equations. Some sample sources are Allgower and Georg [1979], Doedel [1986], Li and Yorke [1979], Kubicek and Marek [1986], Peitgen and Prüfer [1979], and Smale [1976] and [1987].

Poincaré [1881] was the first one to view an ordinary differential equation as defining a family of orbits, thereby establishing the geometric or qualitative theory. In spite of the outstanding contributions of Liapunov [1892] on stability theory and Birkhoff [1927] on dynamical systems, a formal definition of topological equivalence did not appear until Andronov and Pontrjagin [1937]. As we have seen, this concept led to a natural definition of bifurcation. We will say more about topological equivalence and bifurcations of planar vector fields in Chapters 8 and 13.

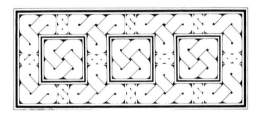

3

Scalar Maps

 With the current proliferation of computers, numerical simulations have become common practice, suggesting new mathematical discoveries and new areas of applications. Despite the power of numerical approximation schemes as "experimental" tools and their ease of implementation on the computer, there is always the difficulty of deciding on the accuracy of computations. Even in the case of a scalar differential equation, one can be confronted with rather strange mathematical phenomena. This is largely due to the fact that numerical approximation of a differential equation leads to a difference equation, and that difference equations, despite their innocuous appearance, can have amazingly complicated dynamics. In this chapter, we first illustrate how difference equations, also called maps, arise in numerical approximations. Because of their importance in other contexts, we then undertake the study of dynamics and bifurcations of maps. In particular, we investigate local bifurcations of a class of maps, monotone maps, which will later play a prominent role in our study of differential equations. We end the chapter with a brief exposition of a landmark quadratic map, the logistic map.

3.1. Euler's Algorithm and Maps

Most differential equations do not have "closed-form" solutions. To obtain approximate solutions, especially in applications, one must often resort to numerical methods. An initial value-problem for a differential equation is "solved" on the computer as follows. We take a discrete set of points $t_0, t_1, \ldots, t_n, \ldots$. For simplicity, we will require that the distance $h \equiv t_{n+1} - t_n$ between two consecutive points, called *step size*, be constant. Next, we calculate approximate values of the solution $x(t)$ at these equally spaced points, much like the tabulated values of trigonometric or logarithmic functions found in older calculus books. Now, given $\dot{x} = f(x)$ and $x(t_0)$, our task is to find an algorithm to approximate $x(t_1)$. Then, knowing $x(t_1)$, we determine $x(t_2)$, and so on with the same algorithm. We will denote the approximate value of $x(t_n)$ by x_n.

The simplest such algorithm, attributed to Euler, is to take $t_n = nh$ and to replace $\dot{x}(t)$ with the difference quotient $(x_{n+1} - x_n)/h$. Then the differential equation $\dot{x} = f(x)$ becomes the following *difference equation*:

$$x_{n+1} = x_n + hf(x_n). \tag{3.1}$$

Given x_0, all other approximate values $x_1, x_2, \ldots, x_n, \ldots$ can be computed in succession using this formula. This sequence of numbers is considered to be the solution of the difference equation with initial value x_0.

Example 3.1. *Logistic and Euler:* Let us now illustrate the procedure above on the *logistic* equation

$$\dot{x} = ax(1 - x), \tag{3.2}$$

where a is a positive parameter. This differential equation represents a simple model for the density of a population that grows almost exponentially for small initial density and saturates at $x = 1$. The parameter a reflects the intrinsic growth rate. In Figure 3.1 we have drawn the flow of Eq. (3.2) in the region $0 \leq x \leq 1$, which is the meaningful region in the model.

Let us approximate the logistic equation (3.2) with the algorithm of Euler (3.1) and set

$$b \equiv ha$$

to obtain the following difference equation:

$$x_{n+1} = bx_n \left(\frac{1+b}{b} - x_n \right). \tag{3.3}$$

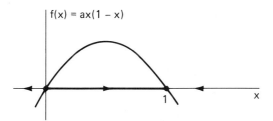

Figure 3.1. *Scalar logistic differential equation* $\dot{x} = ax(1-x)$.

Given an initial value x_0, it is rather easy to calculate the solution x_0, x_1, x_2, ... of Eq. (3.3) on the computer. So, let us perform several simple numerical experiments. If b is very small, that is, h is very small relative to a, then, for any initial value in the interval $(0, 1)$, the solution of Eq. (3.3) converges monotonically to 1 as $n \to +\infty$. For example, the solution of Eq. (3.3) with $b = 0.3$ and $x_0 = 0.567$ is tabulated in Figure 3.2a. This is exactly the same qualitative behavior that is exhibited by the orbit of the differential equation (3.2).

The difficulties begin to occur when $b > 1$. Since $b = ha$, this happens, for example, for $a = 1000.0$ and $h = 0.002$. Numerically, $h = 0.002$ appears on the surface to be a small step size, but it is overcompensated by the value of a in the differential equation. For instance, the solution of the difference equation (3.3) with $b = 1.3$ and $x_0 = 0.567$ tabulated in Figure 3.2b has little resemblance to the corresponding orbit of the differential equation (3.2). Although the solution of the difference equation converges to 1 as $n \to +\infty$, it is not monotone and leaves the interval $[0, 1]$. As the final numerical experiment, let us consider the solution of Eq. (3.3) with $b = 2.8$ and $x_0 = 0.567$; see Figures 3.2c and 3.2d. In this case, it is not even clear that the approximate solution converges to 1 as $n \to +\infty$. We will reveal its asymptotic fate at the end of this chapter. $\Diamond$

The dynamics of the difference equation (3.3) is surprisingly complicated. In fact, a variant of this difference equation, which comes up naturally in biology when modeling a seasonally breeding population, has largely been responsible for the recent surge of activities in this area. We will try to convey some of this exciting development in Section 3.5. Let us proceed with our discussion on the role of difference equations in numerical mathematics with a more familiar example—Newton's Method for computing zeros of functions.

Example 3.2. *Roots with Newton:* In Section 2.5, we indicated that the zeros of a function f could be found as the α- or ω-limit sets of the differential equation $\dot{x} = f(x)$, and that this process could be implemented on

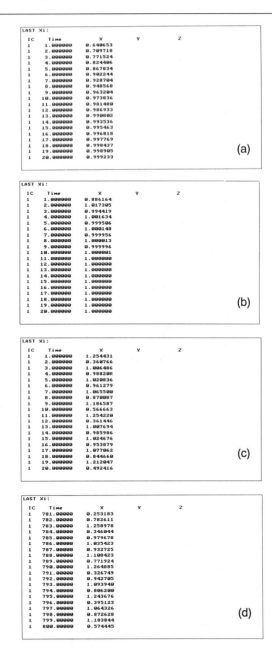

Figure 3.2. *Table of solutions of difference equation* (3.3) *with initial value* $x_0 = 0.567$: (a) $x_1 - x_{20}$ *for* $b = 0.3$, (b) $x_1 - x_{20}$ *for* $b = 1.3$, (c) $x_1 - x_{20}$ *for* $b = 2.8$, *and* (d) $x_{781} - x_{800}$ *for* $b = 2.8$.

the computer using an approximation procedure such as Euler's algorithm. Unfortunately, this method is not always "fast," and, more importantly, does not generalize to vector-valued functions which will be the case when we study differential equations on the plane.

There are other methods, for example, *Newton's Method*, designed specifically for solving $f(x) = 0$ which overcome the difficulties mentioned above. You are undoubtedly familiar with this method, but here is another view of Newton's Method in the spirit of this section. Given a function f, consider the scalar differential equation

$$\dot{x} = -\frac{f(x)}{f'(x)}.$$

If we approximate this differential equation using Euler's algorithm with step size $h = 1$, then the resulting difference equation

$$x_{n+1} = x_n - \frac{f(x_n)}{f'(x_n)} \tag{3.4}$$

is the method of Newton. As an example, consider the function $f(x) = x^2 - 2$ which has two zeros at $\pm\sqrt{2}$. For this function, the difference equation (3.4) becomes

$$x_{n+1} = \frac{x_n}{2} + \frac{1}{x_n}. \tag{3.5}$$

If we start with an initial value, for example, $x_0 = 3$, then the solution of this difference equation converges to $\sqrt{2}$:

$$3.00000, \quad 1.83333, \quad 1.46212, \quad 1.41499, \quad 1.41421, \quad \ldots.$$

In fact, any solution with initial value $x_0 > 0$ converges to $\sqrt{2}$, and with $x_0 < 0$ converges to $-\sqrt{2}$. For this particular function, an Euler approximation with a rather large step size gives the expected result. You should experiment on the computer with various functions and step sizes. $\Diamond$

In order to explore the exciting dynamics of difference equations a bit further, we now turn to the task of developing some basic concepts and theorems.

Exercises  ♣♡♠♢

3.1. *Euler on linear equations:* Using Euler's algorithm derive a difference equation for approximating the solutions of the linear differential equation $\dot{x} = \lambda x$. For a given value of the parameter λ, find the largest value of the step size for which the difference equation has the same qualitative behavior as that of the differential equation.

3.2. *Cube root:* Determine a difference equation for computing the cube root of a real number. Put your equation for $\sqrt[3]{2}$ into PHASER and iterate until the first five digits settle down. How many iterations do you need?

3.2. Geometry of Scalar Maps

Given a function $f : \mathbb{R} \to \mathbb{R}$ and an initial value x_0, consider the sequence of *iterates* of x_0 under the function f:

$$x_0, \quad f(x_0), \quad f(f(x_0)), \quad f(f(f(x_0))), \quad \ldots .$$

In this section, we will explore the geometry of such sequences.

In the sequel, we will use the notation f^n to denote the n-fold composition of a function f with itself, e.g., $f^2(x_0) = f(f(x_0))$. Note that $f^n(x_0)$ does not mean $f(x_0)$ raised to the nth power. Now, let $x_n = f^n(x_0)$. Then the iterates of x_0 under f can be conveniently written as the solution of the first-order difference equation

$$x_{n+1} = f(x_n). \tag{3.6}$$

For the sake of brevity, we will sometimes refer to such a difference equation as a *map* f.

Definition 3.3. *A positive orbit of x_0 is the set of points x_0, $f(x_0)$, $f^2(x_0)$, $\ldots$, and is denoted by $\gamma^+(x_0)$.*

It is important to realize that a positive orbit of Eq. (3.6) is a set of discrete points, not an interval. In fact, this is the main reason for the rich dynamics of difference equations even in one dimension.

Analogous to equilibrium points of differential equations, difference equations have simple distinguished orbits as well.

Definition 3.4. *The point $\bar{x}$ is called a fixed point for f if $f(\bar{x}) = \bar{x}$.*

Notice that the fixed points of f remain fixed under iterations of the map, as in the case of the equilibrium points of a differential equation being solutions that are independent of t. However, in computing fixed points, one must determine the zeros of the function $f(x) - x$, not $f(x)$. This remark will be important when we study bifurcations of maps in the next section.

We now turn to a geometric method, called *stair-step diagrams*, for following solutions of one-dimensional difference equations. We first plot the graph of the function f as well as the diagonal, the $45°$ line. Since $x_{n+1} = f(x_n)$, we think of the horizontal axis as x_n and the vertical axis as x_{n+1}. The vertical line from x_0 meets the graph of f at $(x_0, f(x_0)) = (x_0, x_1)$. The horizontal line from this point intersects the diagonal at

(x_1, x_1). The vertical line from this point intersects the horizontal axis at x_1. By repeating the same steps we can obtain x_2, x_3, etc. Notice that this procedure is equivalent to visualizing the phase portrait of Eq. (3.6) on the diagonal. Also, it is important to observe that the fixed points of Eq. (3.6) correspond to the points of intersection of the graph of f with the diagonal. Let us now practice drawing stair-step diagrams on linear maps.

Example 3.5. *Linear maps:* Consider the linear difference equation

$$x_{n+1} = ax_n, \qquad (3.7)$$

where a is a real parameter. It is easy to see that the positive orbit of an initial value x_0 is the set of points $x_n = a^n x_0$, for $n = 0, 1, 2, \ldots$. Typical stair-step diagrams for the parameter values $a = 2.0$, $a = 0.5$, $a = -0.5$, and $a = -2.0$ are shown in Figure 3.3 (you should consider the cases $a = \pm 1$). Notice that when $a > 0$, the positive orbit is monotonically increasing or decreasing on one side of the fixed point, much like an orbit of a differential equation. When $a < 0$, however, a positive orbit jumps alternately to either side of the origin and is no longer monotonic, a behavior with no counterpart in scalar differential equations. $\Diamond$

After this graphical diversion, let us return to fixed points and investigate their stability properties. Analogous to the notions of stability and asymptotic stability of equilibria of differential equations, we make the following definitions:

Definition 3.6. *A fixed point $\bar{x}$ of f is said to be stable if, for any $\epsilon > 0$, there is a $\delta > 0$ such that, for every x_0 for which $|x_0 - \bar{x}| < \delta$, the iterates of x_0 satisfy the inequality $|f^n(x_0) - \bar{x}| < \epsilon$ for all $n \geq 0$. The fixed point $\bar{x}$ is said to be unstable if it is not stable.*

Definition 3.7. *A fixed point $\bar{x}$ of f is said to be asymptotically stable if it is stable and, in addition, there is an $r > 0$ such that $f^n(x_0) \to \bar{x}$ as $n \to +\infty$ for all x_0 satisfying $|x_0 - \bar{x}| < r$.*

In analogy with the linearization about an equilibrium point of a differential equation given in Theorem 1.14, we expect, under certain conditions, that the stability type of the fixed point $\bar{x}$ of a map $f(x)$ to be the same as the stability type of the fixed point at the origin of the linear map $f'(\bar{x})x$. It is evident from the stair-step diagrams of the linear map in Figure 3.3 that the fixed point at the origin is asymptotically stable if $|a| < 1$, and unstable if $|a| > 1$. This suggests the following linearization theorem about a fixed point.

Theorem 3.8. *Let f be a C^1 map. A fixed point $\bar{x}$ of f is asymptotically stable if $|f'(\bar{x})| < 1$, and it is unstable if $|f'(\bar{x})| > 1$.*

Proof. For convenience, we first translate the point $(\bar{x}, \bar{x}) = (\bar{x}, f(\bar{x}))$ to the origin $(0, 0)$. Let u be the new variable defined by $u \equiv x - \bar{x}$. Then

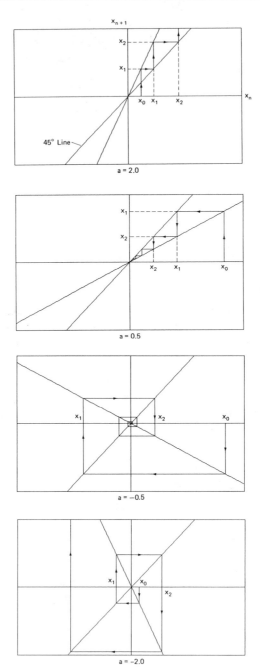

Figure 3.3. Typical stair-step diagrams of linear map $x_{n+1} = ax_n$.

the map f in the new coordinates becomes

$$g(u) \equiv f(\bar{x} + u) - f(\bar{x}).$$

Clearly, $g(0) = 0$ and studying the stability of the fixed point $\bar{x}$ of f is equivalent to studying the stability of the fixed point zero of g. Also, notice that $g'(u) = f'(\bar{x} + u)$.

Now, fix $\epsilon > 0$ and define

$$m_\epsilon \equiv \min_{|s| \leq \epsilon} |f'(\bar{x} + s)|, \qquad M_\epsilon \equiv \max_{|s| \leq \epsilon} |f'(\bar{x} + s)|.$$

Since $g(u) = \int_0^u f'(\bar{x} + s)ds$, if $|u| \leq \epsilon$, then

$$m_\epsilon |u| \leq |g(u)| \leq M_\epsilon |u|.$$

Consequently, by repeated applications of the chain rule, we have

$$m_\epsilon^n |u| \leq |g^n(u)| \leq M_\epsilon^n |u| \qquad \text{for } n \geq 0.$$

If $|f'(\bar{x})| < 1$, then there is an $\epsilon > 0$ such that $M_\epsilon < 1$. Furthermore, if $|u| < \epsilon$, then

$$|g^n(u)| \leq M_\epsilon^n |u| \leq M_\epsilon^n \epsilon < \epsilon \qquad \text{for } n \geq 0.$$

This shows, by taking $\delta = \epsilon$ in Definition 3.6, that the fixed point zero of g is stable. Also, since $M_\epsilon < 1$, $M_\epsilon^n \to 0$ as $n \to +\infty$. Thus, $g^n(u) \to 0$ as $n \to +\infty$ and so the fixed point is asymptotically stable.

To prove the second part of the theorem observe that, if $|f'(\bar{x})| > 1$, then there are $\epsilon_0 > 0$ and $\delta_0 > 0$ such that $m_{\epsilon_0} > 1 + \delta_0$. Suppose that $u \neq 0$ and $|u| \leq \epsilon_0$. Then, we have

$$|g^n(u)| \geq m_{\epsilon_0}^n |u| \geq (1 + \delta_0)^n |u|,$$

as long as $|g^n(u)| \leq \epsilon_0$. This inequality shows that there must be a value of n, say $\hat{n}$, such that $|g^{\hat{n}}(u)| \geq \epsilon_0$. Since u can be taken arbitrarily close to zero, this implies that the fixed point zero of g is unstable. $\diamond$

Example 3.9. *Computing $\sqrt{2}$:* Let us return to computing $\sqrt{2}$ with Newton in Example 3.2. The solutions of the difference equation (3.5) is equivalent to iterations of the map

$$f(x) = \frac{x}{2} + \frac{1}{x}.$$

We have drawn in Figure 3.4 the graph of this map and the stair-step diagram of two of its positive orbits, one with $x_0 < 0$ and the other with $x_0 > 0$. Observe that there are two fixed points at $\bar{x} = \pm\sqrt{2}$. Since $|f'(\bar{x})| = 0 < 1$, it follows from Theorem 3.8 that both fixed points are asymptotically stable. It is, of course, this fact which facilitates the computation of $\sqrt{2}$ with only an approximate initial guess. $\diamond$

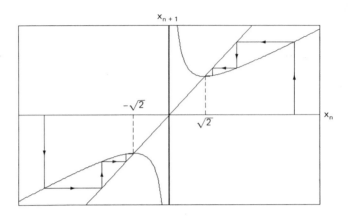

Figure 3.4. *Asymptotically stable fixed points of $f(x) = x/2 + 1/x$.*

We introduce the following notion for fixed points of differentiable maps to which Theorem 3.8 is applicable:

Definition 3.10. *A fixed point $\bar{x}$ of f is said to be hyperbolic if $|f'(\bar{x})| \neq 1$.*

From Theorem 3.8, if a fixed point $\bar{x}$ of f is hyperbolic, then it must be either asymptotically stable or unstable and the stability type is determined from $f'(\bar{x})$. However, we should emphasize that the behavior of orbits near a fixed point $\bar{x}$ is completely different depending on whether $f'(\bar{x}) > 0$ or $f'(\bar{x}) < 0$. In fact, as evident in the linear map (3.6), if $f'(\bar{x}) > 0$, then in a neighborhood of $\bar{x}$, positive orbits remain on one side of $\bar{x}$. If $f'(\bar{x}) < 0$, then positive orbits oscillate about $\bar{x}$.

The stability type of a nonhyperbolic fixed point cannot be determined from the first derivative of the map. To see instances of this, let us iterate several specific maps on the computer.

Example 3.11. *A nonhyperbolic fixed point:* Consider the quadratic map

$$f(x) = x + x^2. \tag{3.8}$$

The point $\bar{x} = 0$ is a nonhyperbolic fixed point of this map with $f'(0) = 1$. As seen in Figure 3.5, the fixed point is unstable because it is attracting from the left, but repelling from the right. If we consider the full map, not just its linear term, this behavior is evident. The nonlinear term x^2 is always positive. Consequently, near the origin, the subsequent iterates of x_0 are increasing. $\Diamond$

Example 3.12. *Another nonhyperbolic fixed point:* Consider the quadratic map

$$f(x) = -x - 3x^2. \tag{3.9}$$

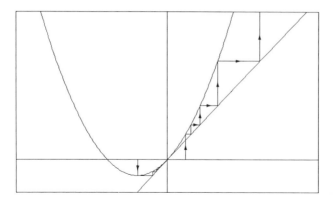

Figure 3.5. *Nonhyperbolic fixed point $\bar{x} = 0$ of $f(x) = x + x^2$ is unstable.*

The fixed point at $\bar{x} = 0$ is nonhyperbolic with $f'(0) = -1$. As seen in Figure 3.6a, the origin is a stable fixed point. Observe that the iterates move alternately to either side of the fixed point. Thus, to account for the effect of the nonlinear term in this example, it is more revealing to follow every other iterate of points by considering the second iterate of the map $f^2(x) = x - 18x^3 - 27x^4$; see Figure 3.6b. Now, it is clear that, for small x_0, the subsequent iterates converge to the origin. $\lozenge$

Example 3.13. *More nonhyperbolic fixed points:* Consider the cubic maps

$$f(x) = x - x^3, \tag{3.10}$$

$$f(x) = x + x^3. \tag{3.11}$$

The point $\bar{x} = 0$ is a nonhyperbolic fixed point of these maps with $f'(0) = 1$. It is stable in Eq. (3.10), but unstable in Eq. (3.11); see Figure 3.7. Now, consider two other cubic maps

$$f(x) = -x - x^3, \tag{3.12}$$

$$f(x) = -x + x^3. \tag{3.13}$$

Again, the point $\bar{x} = 0$ is a nonhyperbolic fixed point of these maps, but with $f'(0) = -1$. The origin is unstable in Eq. (3.12) and stable in Eq. (3.13); see Figure 3.7. To uncover the effects of the nonlinear terms in these two maps, you should consider their second iterates f^2. $\lozenge$

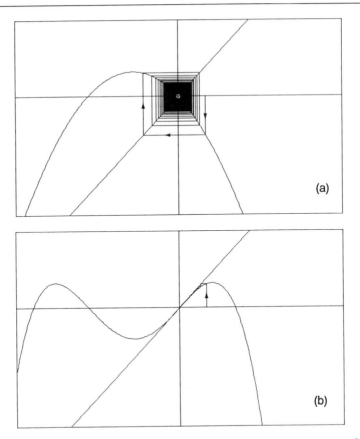

Figure 3.6. *Nonhyperbolic fixed point $\bar{x} = 0$ of map $f(x) = -x - 3x^2$ is stable: (a) graph of f, and (b) its second iterate $f^2(x) = x - 18x^3 - 27x^4$.*

You might suspect from these examples that bifurcation behavior of nonhyperbolic fixed points could be quite different depending on whether $f'(\bar{x}) = 1$ or $f'(\bar{x}) = -1$. This is indeed the situation. Therefore, we will investigate these two cases separately in the following two sections.

Exercises _____ ♣ ♡ ♠ ◇

3.3. *Existence and uniqueness:* Consider the the difference equation

$$x_{n+1} = -\sqrt{x_n}.$$

Are the solutions defined for all n? What is an appropriate existence and uniqueness theorem for initial value problems of difference equations?

3.4. Do the following for each map below: locate fixed points, find the values of the parameter at which fixed points are not hyperbolic, determine the

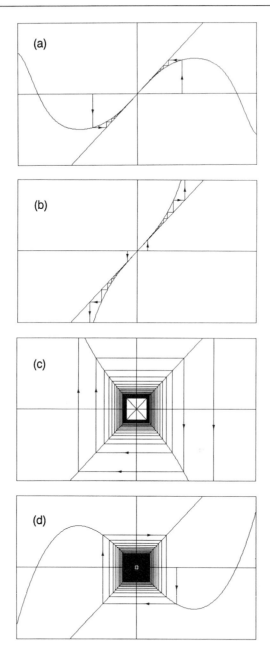

Figure 3.7. *Nonhyperbolic fixed point at the origin:* (a) *stable in* $f(x) = x - x^3$, (b) *unstable in* $f(x) = x + x^3$, (c) *unstable in* $f(x) = -x - x^3$, *and* (d) *stable in* $f(x) = -x + x^3$.

stability types of fixed points, draw typical stair-step diagrams near each fixed point. To create good pictures, you should use PHASER.

(a) $f(\lambda, x) = \lambda x(1 - x)$ for $\lambda > 1$;

(b) $f(\lambda, x) = \lambda x(1 - x) + 0.1$ for $0 \leq x \leq 1$;

(c) $f(\lambda, x) = \lambda - x^2$; (d) $f(\lambda, x) = \lambda^2 - x^2$;

(e) $f(\lambda, x) = e^x - \lambda$; (f) $f(\lambda, x) = -(\lambda/2) \arctan x$;

(g) $f(\lambda, x) = x - \lambda x(x - \frac{1}{3})(x - \frac{2}{3})(x - 1)$. In this example, do not forget to consider the parameter values 1, 4.6, 9.1, 13.6, and 27.1.

3.5. Show that the fixed point of the map obtained from Newton's method for computing $\sqrt[3]{2}$ is asymptotically stable. Accomplish this by estimating an interval in which $\sqrt[3]{2}$ is assured to be.

3.6. Compute the fixed points of the map (3.3),

$$x_{n+1} = bx_n \left(\frac{1+b}{b} - x_n \right).$$

Determine the values of the parameter b for which a given fixed point is unstable or asymptotically stable.

3.7. *A unique asymptotically stable fixed point:* Here is a useful setting for establishing the existence of such a fixed point. Let $f : [a, b] \to [a, b]$ be a map of an interval into itself. Show the following:

(a) If f is continuous, then it has at least one fixed point in the interval $[a, b]$.

(b) If, in addition, f is differentiable with $|f'(x)| < 1$ for all x in $[a, b]$, then f has a unique fixed point in $[a, b]$. This fixed point is, of course, asymptotically stable.

Hint: To show existence, apply the Intermediate Value Theorem to the function $f(x) - x$. For uniqueness, suppose that there are two fixed points and use the Mean Value Theorem to arrive at a contradiction.

3.8. Show that if $f : \mathbb{R} \to \mathbb{R}$ is continuous and there is an x_0 such that $f^n(x_0) \to \bar{x}$ as $n \to +\infty$, then $\bar{x}$ is a fixed point of f.

3.9. *Comparing rates:* Consider the difference equation

$$x_{n+1} = x_n - \frac{1}{4}(x_n^2 - 2).$$

Determine the appropriate initial values x_0 whose iterates converge to $\sqrt{2}$. Compare the rate of convergence of this method of computing $\sqrt{2}$ with that of Newton's Method.

3.10. *A redundancy:* If f is continuous, show that the requirement in Definition 3.7 that $\bar{x}$ be stable is redundant.

3.11. *Implicit Function Theorem is constructive:* Since we did not present a proof of the Implicit Function Theorem in Section 2.2, it is not clear how to obtain the function $x = \psi(\lambda)$. In this problem, we indicate a constructive method for obtaining this function as a "fixed point" of a map by converting the equation $F(\lambda, x) = 0$ to an equivalent equation $x = T(\lambda, x)$ and then defining the successive approximations

$$x_0 = 0, \qquad x_{n+1} = T(\lambda, x_n).$$

For a given value of λ, the sequence x_n converges to $\psi(\lambda)$ if λ is sufficiently small.

To define $T(\lambda, x)$, we use the Taylor expansion of $F(\lambda, x)$:

$$F(\lambda, x) = a(\lambda) + b(\lambda)x + G(\lambda, x).$$

The hypotheses of the theorem imply that $b(\lambda) \neq 0$ for λ sufficiently small. Therefore, $F(\lambda, x) = 0$ is equivalent to the equation $x = T(\lambda, x)$ with

$$T(\lambda, x) = -b(\lambda)^{-1}a(\lambda) - b(\lambda)^{-1}G(\lambda, x).$$

Here is a specific example illustrating the convergence of the iterates. Suppose that

$$F(\lambda, x) = \lambda + (1 + \lambda)x + x^2,$$

where λ is a scalar parameter. Then the function $\psi(\lambda)$ of the Implicit Function Theorem is $\psi(\lambda) = -\lambda$. Recover this function using the method of successive approximations described above. For this purpose, compute that $T(\lambda, x) = -\lambda(1 + \lambda)^{-1} - (1 + \lambda)^{-1}x^2$. Then, take $\lambda = 0.1$, $\lambda = 0.3$, $\lambda = 0.5$, etc., and iterate with initial value $x_0 = 0$. Do you observe any difference in the rate of convergence for different values of λ?

3.3. Bifurcations of Monotone Maps

A restricted class of maps, called monotone maps, play a central role in certain aspects of differential equations. In this section we undertake the study of this class. We will apply the results to differential equations in Chapter 5.

We begin by introducing some terminology. A positive orbit $\gamma^+ = \{x_0, x_1, x_2, \ldots, x_n, \ldots\}$ of f, where $x_{n+1} = f(x_n)$ for $n \geq 0$, is said to be monotone nondecreasing if the sequence $\{x_n\}$ is nondecreasing, that is, $x_{n+1} \geq x_n$ for every positive integer n. Similarly, $\gamma^+(x_0)$ is said to be monotone nonincreasing if $x_{n+1} \leq x_n$ for every positive integer n. Combining these two notions, we simply say that $\gamma^+(x_0)$ is monotone if it is either monotone nondecreasing or monotone nonincreasing. Finally, we call f a *monotone map* if every positive orbit $\gamma^+(x_0)$ of f is a monotone sequence.

Lemma 3.14. *If f is a C^1 function with $f'(x) > 0$ for all x in the domain of definition of f, then f is a monotone map, that is, the positive orbit $\gamma^+(x_0)$ of any initial condition x_0 is a monotone sequence.*

Proof. From the Mean Value Theorem, we have

$$x_{n+1} - x_n = f(x_n) - f(x_{n-1}) = f'(\hat{x}_n)(x_n - x_{n-1})$$

for some $\hat{x}_n$. Therefore, $x_{n+1} - x_n$, for every positive integer n, has the same sign as that of $x_1 - x_0$. ◇

For the purposes of dynamics, we will require f to be at least C^1 with positive derivative, and refer to such an f simply as a *monotone map*.

If f is monotone, then f^{-1}, the inverse of f, exists. We will use the notation f^{-n} to denote the n-fold composition of f^{-1} with itself.

Definition 3.15. *If f is monotone, then the negative orbit of x_0 is the set of points $x_0, f^{-1}(x_0), f^{-2}(x_0), \ldots$, and is denoted by $\gamma^-(x_0)$. The orbit γ of x_0 is defined to be $\gamma(x_0) \equiv \gamma^+(x_0) \cup \gamma^-(x_0)$.*

The geometry of orbits of a monotone map is very similar to that of a scalar differential equation: the fixed points act like equilibria, and we can use arrows to indicate the direction of other orbits under forward iteration. Consequently, to study bifurcations of fixed points of monotone maps we need only to reinterpret the results in Section 2.3, as we shall do now.

For simplicity of notation, let us assume that the map f has a fixed point at $\bar{x} = 0$; if not, we can change coordinates to make it so. Furthermore, suppose that $f'(0) > 0$ so that f is monotone in a sufficiently small neighborhood of the origin. Consider the perturbed map $F(\lambda, x)$ depending on k parameters $\lambda \equiv (\lambda_1, \lambda_2, \ldots, \lambda_k)$:

$$F : \mathbb{R}^k \times \mathbb{R} \to \mathbb{R}; \quad (\lambda, x) \mapsto F(\lambda, x) \quad \text{with} \quad F(0, x) = f(x).$$

If $F(\lambda, x)$ is a C^1 function, then it follows that $F(\lambda, x)$ is also monotone in x for each small value of λ. Now, for each fixed λ, the key observation is that the analysis of fixed points of $F(\lambda, x)$ is equivalent to the analysis of the zeros of the function

$$F(\lambda, x) - x.$$

In Section 2.3 we have analyzed the bifurcations of zeros of a function, or, equivalently, the bifurcations of equilibria, under various types of hypotheses on the linear, quadratic, and cubic terms. We now translate those results for bifurcations of fixed points of monotone maps.

Case I: Hyperbolic Fixed Points. Suppose that f is a monotone, C^1 map with $f(0) = 0$ and $f'(0) \neq 1$. Consider a C^1 map $F(\lambda, x)$ satisfying

$$F(\mathbf{0}, x) = f(x) \quad \text{and} \quad \frac{\partial F}{\partial x}(\mathbf{0}, 0) = f'(0) \neq 1.$$

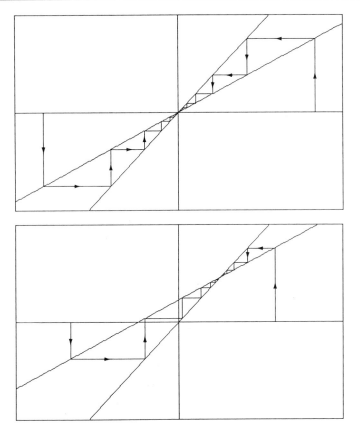

Figure 3.8. *Asymptotically stable hyperbolic fixed point of $F(\lambda, x) = \lambda + 0.5x$ persists as λ is varied.*

Then, for $\|\lambda\|$ sufficiently small, the perturbed map F has a unique fixed point near zero whose stability type is the same as the stability of the fixed point zero of the unperturbed map f.

Example 3.16. *One-parameter linear map:* Let us consider the linear map $f(x) = 0.5x$ and its one-parameter perturbation given by

$$F(\lambda, x) = \lambda + 0.5x.$$

For each value of the parameter λ, there is a unique hyperbolic fixed point whose stability type is the same as the fixed point for $\lambda = 0$. See Figure 3.8 for the stair-step diagrams of the perturbed map F for several values of the parameter λ. $\diamondsuit$

Case II: Fixed Points with Quadratic Degeneracy. Suppose that f is a monotone, C^2 map with $f(0) = 0$, $f'(0) = 1$, but $f''(0) \neq 0$. Consider a C^2 map $F(\lambda, x)$ satisfying

$$F(0, x) = f(x), \quad \frac{\partial F}{\partial x}(0, 0) = 1, \quad \frac{\partial^2 F}{\partial x^2}(0, 0) = f''(0) \neq 0.$$

Then there is a function $\alpha(\lambda)$, satisfying $\alpha(0) = 0$, which corresponds to the extreme value of the function $F(\lambda, x) - x$ such that

$$\alpha(\lambda) f''(0) < 0 \quad \Longrightarrow \quad \text{two fixed points of F,}$$
$$\alpha(\lambda) f''(0) = 0 \quad \Longrightarrow \quad \text{one fixed point of F,}$$
$$\alpha(\lambda) f''(0) > 0 \quad \Longrightarrow \quad \text{no fixed point of F}$$

for sufficiently small values of λ.

Example 3.17. *One-parameter quadratic map:* Consider the map $f(x) = x + x^2$ in a sufficiently small neighborhood of zero so that it is monotone. The local bifurcations of the one-parameter perturbation of f given by

$$F(\lambda, x) = \lambda + x + x^2$$

are illustrated in Figure 3.9. In this case, we have $\alpha(\lambda) = \lambda$ and $f''(0) = 2$. Therefore, when $\lambda < 0$, there are two fixed points; at $\lambda = 0$, there is one fixed point; for $\lambda > 0$, there is no fixed point of $F(\lambda, x)$. $\Diamond$

Case III: Fixed Points with Cubic Degeneracy. Suppose that f is a monotone, C^3 map with $f(0) = 0$, $f'(0) = 1$, $f''(0) = 0$, but $f'''(0) \neq 0$. A complete bifurcation analysis of an arbitrary C^3 perturbation $F(\lambda, x)$ of f is difficult. Therefore, we confine our discussion to the analysis of a "typical" example of a two-parameter perturbation. As in Case III of Section 2.3, under mild hypotheses, study of local bifurcations of fixed points of any two-parameter map F can be reduced to this example. The details are identical to the ones given previously.

Example 3.18. *Two-parameter cubic map:* Consider the map $f(x) = x - x^3$ in a small neighborhood of the origin so that f is monotone. For the two-parameter, $\lambda = (\lambda_1, \lambda_2)$, perturbation of f given by

$$F(\lambda, x) = \lambda_1 + (1 + \lambda_2)x - x^3$$

there is a cusp in the (λ_1, λ_2)-plane such that

- there are three fixed points of $F(\lambda, x)$ for values of (λ_1, λ_2) inside the cusp,
- there is one fixed point of $F(\lambda, x)$ for values of (λ_1, λ_2) outside the cusp,
- there are two fixed points of $F(\lambda, x)$ for values of (λ_1, λ_2) on the cusp if $(\lambda_1, \lambda_2) \neq (0, 0)$, and one fixed point if $(\lambda_1, \lambda_2) = (0, 0)$.

The possible local bifurcations of fixed points of F listed above are illustrated in Figure 3.10. $\Diamond$

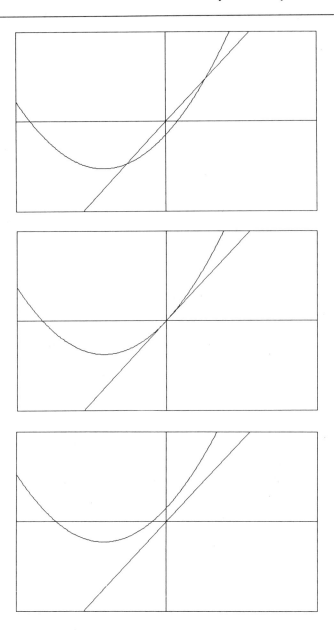

Figure 3.9. *Bifurcations of fixed points of* $F(\lambda, x) = \lambda + x + x^2$ *near the origin: values of* λ *are* -0.1, 0, *and* 0.1.

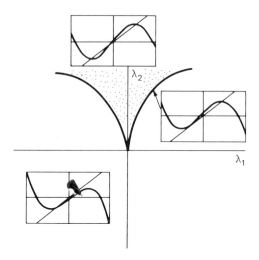

Figure 3.10. *Bifurcations of $F(\lambda, x) = \lambda_1 + (1 + \lambda_2)x - x^3$.*

Even if f is not monotone, its hyperbolic fixed points persist under small perturbation. Nonhyperbolic fixed points of f, however, can undergo bifurcations with no counterparts in our catalog of bifurcations of equilibria of scalar differential equations. We now turn to one such bifurcation of great importance.

Exercises _____ ♣ ♡ ♠ ◇

3.12. *A transcritical bifurcation:* Show that the map $F(\lambda, x) = (1 + \lambda)x + x^2$ undergoes a transcritical bifurcation at the parameter value $\lambda = 0$. Compare this map with the differential equation in Example 2.3.

3.13. *A saddle-node bifurcation:* Show that the map $F(\lambda, x) = e^x - \lambda$ undergoes a saddle-node bifurcation at the parameter value $\lambda = 1$.

3.14. Find a value of the parameter λ at which the map $F(\lambda, x) = \lambda - x^2$ undergoes a local bifurcation. Identify the bifurcation and draw three representative stair-step diagrams to illustrate your bifurcation.

3.4. Period-doubling Bifurcation

In this section, we investigate an important bifurcation that a nonhyperbolic fixed point $\bar{x}$ with $f'(\bar{x}) = -1$ is likely to undergo when f is subjected to perturbations.

As we saw in previous examples, when $f'(\bar{x}) < 0$, the map f is not monotone and it flips a point close to $\bar{x}$ to the other side of $\bar{x}$. If the fixed point $\bar{x}$ becomes unstable, an orbit cannot approach to $\bar{x}$. However, if the iterates remain bounded, it is plausible that the odd iterates converge to a limit point, say, $x^\star$ and the even iterates converge to $f(x^\star)$. If this is the case, then $f^2(x^\star) = x^\star$ with $f(x^\star) \neq x^\star$, that is $x^\star$ is a periodic point of period 2. This bifurcation is called *period-doubling* or *flip* bifurcation. A typical bifurcation diagram of this important bifurcation is shown in Figure 3.11. Despite its resemblance, this diagram should not be confused with that of pitchfork bifurcation of equilibrium points.

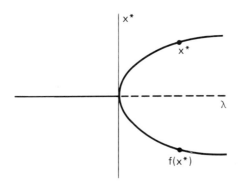

Figure 3.11. *Bifurcation diagram of supercritical period-doubling.*

Let us begin our exploration of period-doubling bifurcation by defining the concept of a general periodic point and some related remarks.

Definition 3.19. *A point $x^\star$ is called a periodic point of minimal period n if $f^n(x^\star) = x^\star$ and n is the least such positive integer. The set of all iterates of a periodic point is called a periodic orbit.*

A periodic point $x^\star$ of minimal period n is a fixed point of the map $f^n(x)$. Consequently, the notion of stability of $x^\star$ follows that of a fixed point and the linearization result in Theorem 3.8 can be applied to f^n to determine the stability type of $x^\star$.

Definition 3.20. *A periodic point $x^\star$ of minimal period n is said to sta-ble, asymptotically stable, or unstable if $x^\star$ is, respectively, a stable, an asymptotically stable, or an unstable fixed point of f^n.*

Let us now establish that our intuitive observation on the possibility of period-doubling bifurcation in fact occurs for typical perturbations of a nonhyperbolic fixed point with $f'(\bar{x}) = -1$. The statement of the theorem below is more general than it may appear, as we shall point out shortly.

Theorem 3.21. *Let $f(x)$ be a C^3 function with a fixed point at the origin satisfying the following conditions:*

    (i) $f(0) = 0, \quad f'(0) = -1;$

    (ii) $[f^2(0)]''' \neq 0.$

Consider a one-parameter perturbation $F(\lambda, x)$ of $f(x)$ with

    (iii) $F(0, x) = f(x), \quad F(\lambda, 0) = 0,$

    (iv) $\frac{\partial F(\lambda, 0)}{\partial x} = -(1 + \lambda).$

Then there is a neighborhood of $(\lambda, x) = (0, 0)$ in which, for each value of λ with $\lambda\left[f^2(0)\right]''' < 0$, there exists a unique periodic orbit $\{\, x^\star_\lambda, F(\lambda, x^\star_\lambda)\,\}$ of minimal period 2 of the perturbed function $F(\lambda, x)$; for values of λ with $\lambda\left[f^2(0)\right]''' > 0$, there is no periodic orbit of minimal period 2. Furthermore, the period-2 orbit is asymptotically stable [respectively, unstable] if the origin is an unstable [respectively, asymptotically stable] fixed point at this value of λ.

Proof. Periodic points of period 2 of the map $F(\lambda, x)$ correspond to the fixed points of $F^2(\lambda, x) = F(\lambda, F(\lambda, x))$, equivalently, to the zeros of $F^2(\lambda, x) - x$. However, because of condition *(iii)*, the fixed point $x = 0$ is a zero of this equation but its minimal period is 1. Therefore, to avoid this point and locate only the periodic points of *minimal* period 2, we need to analyze the zeros of the function

$$\frac{1}{x}\left[F^2(\lambda, x) - x\right].$$

To accomplish this, we begin, as usual, by determining the first several terms of its Taylor expansion. Let us use the notation "prime" to denote the derivative with respect to x and compute some derivatives:

$$\left[F^2(\lambda, x)\right]' = F'(\lambda, F(\lambda, x))\, F'(\lambda, x),$$

$$\left[F^2(\lambda, x)\right]'' = F''(\lambda, F(\lambda, x))\left[F'(\lambda, x)\right]^2 + F'(\lambda, F(\lambda, x))\, F''(\lambda, x).$$

In particular, at the origin, we have

$$\left[F^2(0, 0)\right]' = \left[f^2(0)\right]' = 1, \qquad \left[F^2(0, 0)\right]'' = \left[f^2(0)\right]'' = 0.$$

It follows from these formulae that the Taylor expansion of $F^2(\lambda, x)$ about the origin is given by

$$F^2(\lambda, x) = (1 + \lambda)^2 x + \frac{a(\lambda)}{2} x^2 + \frac{b(\lambda)}{6} x^3 + \cdots,$$

where the functions $a(\lambda)$ and $b(\lambda)$ satisfy

$$a(0) = 0, \qquad b(0) = \left[f^2(0) \right]'''.$$

Therefore, the Taylor expansion we have been seeking is

$$\frac{1}{x} \left[F^2(\lambda, x) - x \right] = \lambda(2 + \lambda) + \frac{a(\lambda)}{2} x + \frac{b(\lambda)}{6} x^2 + \cdots. \qquad (3.14)$$

Since $b(0) \neq 0$, the analysis of the zeros of this function is identical to the one we have already given in Case II of Section 2.3 for bifurcations of nonhyperbolic equilibrium points with quadratic degeneracy. In fact, if $\lambda \left[f^2(0) \right]''' > 0$, then there is no zero of Eq. (3.14). If $\lambda \left[f^2(0) \right]''' < 0$, then there are two zeros of Eq. (3.14) which correspond to a single period-2 orbit, $\{ x_\lambda^\star, F(\lambda, x_\lambda^\star) \}$.

In the case $\lambda \left[f^2(0) \right]''' < 0$, to determine the stability type of the period-2 orbit, we consider the cubic function $F^2(\lambda, x) - x$ and its three simple zeros, in a neighborhood of the origin, given by 0, $x_\lambda^\star$, and $F(\lambda, x_\lambda^\star)$. If the slope $(1 + \lambda)^2$ of $F^2(\lambda, x)$ at $x = 0$ is less than one ($\lambda < 0$ and 0 is stable), then the slope of $F^2(\lambda, x) - x$ at $x_\lambda^\star$ and $F(\lambda, x_\lambda^\star)$ must be greater than one and thus the period-2 orbit is unstable. Similarly, if $(1 + \lambda)^2 > 1$, that is, $\lambda > 0$ and 0 is unstable, then the period-2 orbit is stable. $\diamond$

Example 3.22. *Continuation of Example 3.12:* Let us now return to the map in Example 3.12 and consider its one-parameter perturbation given by

$$f(x) = -x - 3x^2, \qquad F(\lambda, x) = -(1 + \lambda)x - (3 + \lambda)x^2.$$

Since $f^2(x) = x - 18x^3 - 27x^4$, we have $\left[f^2(0) \right]''' = -108 \neq 0$. It is easy to verify that all of the remaining conditions of Theorem 3.21 are satisfied. Thus, for each small positive value of λ, there is a unique periodic orbit of minimal period 2 which is asymptotically stable; see Figures 3.11 and 3.12. $\diamond$

We conclude this section with several remarks regarding Theorem 3.21.

1. For notational simplicity, we assumed $\bar{x} = 0$; this can always be achieved by translation.

2. The nonvanishing assumption on the third derivative cannot be replaced by a condition on the second derivative because $\left[f^2(0) \right]'' = 0$ is always satisfied.

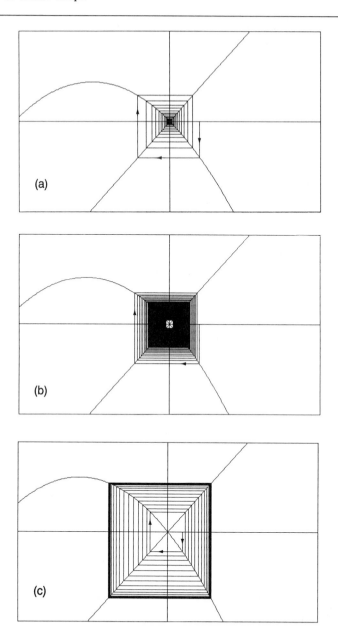

Figure 3.12. *Graphs of the map $F(\lambda, x) = -(1 + \lambda)x - (3 + \lambda)x^2$ near the origin: (a) for $\lambda < 0$, asymptotically stable fixed point at the origin, (b) for $\lambda = 0$, the origin is not hyperbolic but still attracting, (c) for $\lambda > 0$, there is a unique asymptotically stable periodic orbit of period 2 (continued).*

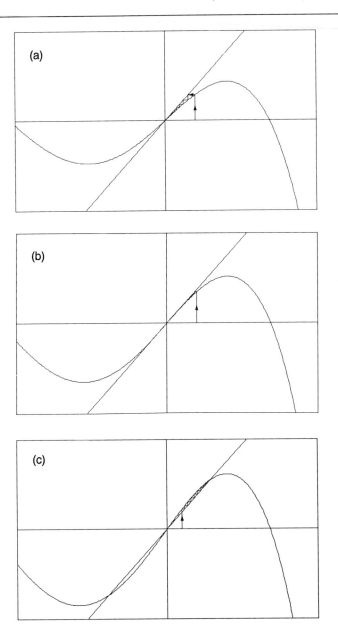

Figure 3.12 Continued.   *The same animation sequence as in the previous page but for* F^2, *the second iterate of the map.*

3. There is no loss of generality in assuming $F(\lambda, 0) = 0$. In fact, if we let $G(\lambda, x) = F(\lambda, x) - x$, then $G(0, 0) = 0$ and $\partial G(0, 0)/\partial x = -2$. Thus, the Implicit Function Theorem implies that there is a unique function $\psi(\lambda)$, with $\psi(0) = 0$, in a neighborhood of $(\lambda, x) = (0, 0)$ such that $G(\lambda, \psi(\lambda)) = 0$, that is, there is a unique fixed point of $F(\lambda, x)$. If we let $x \mapsto \psi(\lambda) + x$, then we obtain $F(\lambda, 0) = 0$.

4. Hypothesis (iv) says that 0 is asymptotically stable for $\lambda < 0$ but unstable for $\lambda > 0$ — the stability type changes.

5. The theorem can be used to establish period-doubling bifurcations of periodic points of any period, not just fixed points, by considering an appropriate iterate of the map. If the period is large, however, computing derivatives could get rather cumbersome.

Exercises _____

3.15. *Two maps:* Find a value of the parameter λ at which the map $F(\lambda, x) = \lambda x(1 - x)$ is likely to undergo a period-doubling bifurcation. Substantiate your guess numerically using PHASER. Can you prove your observation? Perform a similar numerical and theoretical investigation on the map $F(\lambda, x) = \lambda - x^2$. Do you observe any essential difference between the dynamics of the two maps?

3.16. Show that the map $F(\lambda, x) = -(\lambda/2)\arctan x$ undergoes a period-doubling bifurcation at $\lambda = 2$. Can you determine the stability type of the resulting period-two orbit?

3.17. *Doubling period-three:* At approximately $\lambda = 3.83$, the map $F(\lambda, x) = \lambda x(1 - x)$ has an asymptotically stable periodic orbit of minimal period three. Verify this on the computer using PHASER. Increase the parameter gradually and observe that the period three orbit gives up its stability to a periodic orbit of minimal period six. Can you prove the result of your experiment?

3.5. An Example: The Logistic Map

Let us reconsider the difference equation $x_{n+1} = bx_n[(1 + b)/b - x_n]$, Eq. (3.3) of Section 3.1. If we scale x_n and x_{n+1} by $(1 + b)/b$, that is, use the change of variables $x_n \to [(1 + b)/b]x_n$ and $x_{n+1} \to [(1 + b)/b]x_{n+1}$, then the equation becomes $x_{n+1} = (1+b)x_n(1-x_n)$. Furthermore, if we let $(1 + b) \equiv \lambda$, then the difference equation (3.3) is equivalent to the iteration of the following one-parameter map:

$$f(\lambda, x) = \lambda x(1 - x), \qquad \lambda > 1. \tag{3.15}$$

This map is known as the *logistic map* and has been the subject of recent intensive studies. Despite its innocuous looks, the logistic map exhibits many important phenomena encountered in dynamical systems. In this section, we will touch upon several of its basic properties. Since the results below will not be used elsewhere in the book, our exposition is descriptive without full details.

We start our study of the logistic map (3.15) by finding its fixed points: there are two fixed points, 0 and $\bar{x}_\lambda \equiv 1 - 1/\lambda$. Since we assume $\lambda > 1$, the latter fixed point always lies in the interval $(0, 1)$. For future reference, we note that

$$f'(\lambda, 0) = \lambda, \qquad f'(\lambda, \bar{x}_\lambda) = 2 - \lambda. \tag{3.16}$$

When $1 < \lambda < 4$, the following lemma shows that the interesting dynamics of the logistic map occurs for initial conditions in the interval $(0, 1)$. Of course, $f(\lambda, 1) = f(\lambda, 0) = 0$.

Lemma 3.23. *Consider the logistic map* (3.15):

(i) *Suppose that* $\lambda > 1$. *If* $x_0 < 0$ *or* $x_0 > 1$, *then* $f^n(\lambda, x_0) \to -\infty$ *as* $n \to +\infty$.

(ii) *Suppose that* $1 < \lambda < 4$. *If* $x_0 \in (0, 1)$, *then* $f^n(\lambda, x_0) \in (0, 1)$ *for any positive integer* n.

Proof. (i) If $x_0 < 0$, then $f(\lambda, x_0) < x_0$. Thus $f^n(\lambda, x_0)$ is a decreasing sequence. This sequence cannot converge because f has no negative fixed points. If $x_0 > 1$, then $f(\lambda, x_0) < 0$; hence the same argument applies. (ii) The maximum value of $f(\lambda, x)$ is less than 1. $\diamond$

Let us now investigate the dynamics of the logistic map (3.15) for various ranges of the parameter λ.

$\Rightarrow$ $\overline{1 < \lambda < 3}$: It follows from Theorem 3.8 and Eq. (3.16) that the fixed point zero is unstable and the other fixed point $\bar{x}_\lambda \equiv 1 - 1/\lambda$ is asymptotically stable.

We can say more about the stability of $\bar{x}_\lambda$. In fact, if $0 < x_0 < 1$, then $\lim_{n \to +\infty} f^n(\lambda, x_0) = \bar{x}_\lambda$, that is, $\bar{x}_\lambda$ is attracting globally. However, the manner in which solutions approach $\bar{x}_\lambda$ depends on the value of λ. To see this, first suppose that $0 < x_0 \leq 0.5$. Then $|f(\lambda, x_0) - \bar{x}_\lambda| < |x_0 - \bar{x}_\lambda|$, if $x_0 \neq \bar{x}_\lambda$ and $f'(\lambda, x) > 0$. Thus, $f^n(\lambda, x_0) \to \bar{x}_\lambda$ monotonically as $n \to +\infty$. The case $0.5 < x_0 < 1$ follows from the same argument by noting that the first iterate $f(\lambda, x_0)$ lies in the interval $(0, 0.5)$. Consequently, it is evident from Eq. (3.16) when $1 < \lambda < 2$, that after at most one iteration, solutions approach $\bar{x}_\lambda$ monotonically; see Figure 3.13.

If $2 < \lambda < 3$, then $\bar{x}_\lambda$ is still attracting globally but the approach to it is no longer monotonic (this is expected from linearization); see Figure 3.14. The proof in this case is somewhat more difficult. You should determine what happens in the intermediate case $\lambda = 2$.

$\Rightarrow$ $\overline{\lambda = 3}$: The fixed point zero is unstable, but the stability type of $\bar{x}_\lambda$ cannot be determined from Theorem 3.8 because $f'(\bar{x}_\lambda) = -1$ (it can be shown

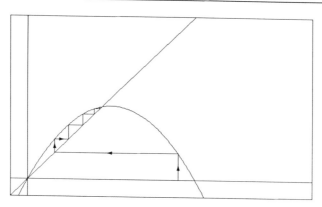

Figure 3.13. *Asymptotically stable fixed point of the logistic map for the parameter value* $\lambda = 1.8$.

that $\bar{x}_\lambda$ is still globally attracting). This suggests that $\lambda = 3$ is a bifurcation value.

$\Rightarrow$ $\overline{3 < \lambda < 1 + \sqrt{6}}$: In this case, the fixed points zero and $\bar{x}_\lambda$ are both unstable, therefore, the iterates $f^n(x_0)$ cannot approach to either of the fixed points. However, they remain in the interval $(0, 1)$. It is plausible that as λ is increased through 3, the fixed point $\bar{x}_\lambda$ undergoes a period-doubling bifurcation giving rise to an asymptotically stable periodic orbit of minimal period 2; see Figure 3.14.

This is in fact the case. The existence and the asymptotic stability of such a periodic orbit can be established by analyzing the zeros of the function $f^2(\lambda, x) - x$. Indeed, using the fact that $x = 0$ and $x = \bar{x}_\lambda$ are roots of $f^2(\lambda, x) - x$, one shows that the periodic orbit of period 2 is located at the roots of the equation $\lambda^2 x^2 - \lambda(\lambda+1)x + \lambda + 1 = 0$. For $3 < \lambda < 1 + \sqrt{6}$, this period-2 orbit is asymptotically stable. It can further be shown that it is globally attracting, except for countably many initial conditions which land on the fixed point $\bar{x}_\lambda$ after a finite number of iterations.

We should, of course, be able to conclude the presence of the period-doubling bifurcation from Theorem 3.21 as well. However, to apply this theorem to the logistic map, we need to translate the fixed point $\bar{x}_\lambda$ to the origin, and also translate the bifurcation value $\lambda = 3$ to zero. To this end, introduce the new variable y and the parameter μ, defined by

$$x = 1 - \frac{1}{\lambda} + y, \qquad \lambda = 3 + \mu.$$

Then the logistic map becomes

$$f(\mu, y) = -(1 + \mu)y - (3 + \mu)y^2.$$

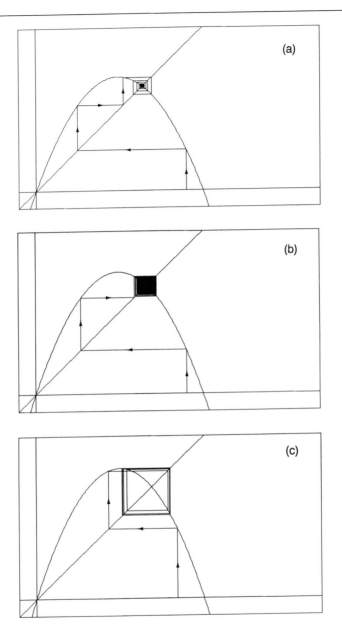

Figure 3.14. *The logistic map near the first period-doubling bifurcation: (a) $\lambda = 2.8$, (b) $\lambda = 3$, and (c) $\lambda = 3.2$ (continued).*

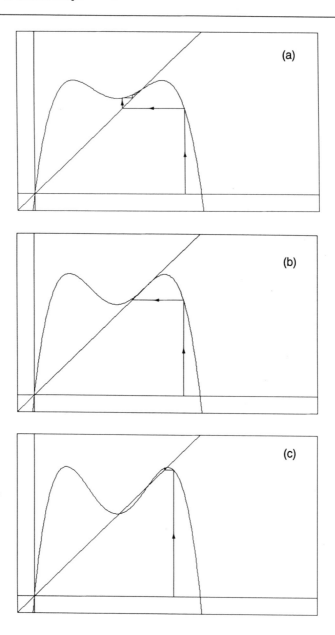

Figure 3.14 Continued. *The second iterates of the logistic map for the parameter values as in the previous page.*

Now, this map is precisely the one we already studied in Example 3.22.

⇒ $\overline{3.449 < \lambda < 3.570}$: The appearance of the period-two orbit as λ is increased through 3 is only the beginning of a fascinating sequence of bifurcations that lead to very complicated dynamics. For example, at $\lambda = 3.449$ (approximately) the period-two orbit loses its stability and gives rise to an asymptotically stable period-four orbit, as seen in Figure 3.15. In fact, there is an increasing sequence of parameter values $\lambda_1 < \lambda_2 < \lambda_3 < \ldots$ at which the logistic map repeatedly undergoes a *period-doubling bifurcation*: as λ is increased through λ_k the asymptotically stable periodic orbit of period 2^k becomes unstable and a stable periodic orbit of twice the period bifurcates from the orbit of the lower period; see Figure 3.15. The first several of the bifurcation values approximately are

$$\lambda_1 = 3, \quad \lambda_2 = 3.449, \quad \lambda_3 = 3.544, \quad \lambda_4 = 3.564, \quad \ldots .$$

This sequence converges to a number λ_∞ as $k \to +\infty$ and λ_∞ is 3.5699456 (approximately). Furthermore, the ratio of the distances of the parameter values between successive period-doubling bifurcations approaches a constant

$$\lim_{k \to +\infty} \frac{\lambda_k - \lambda_{k-1}}{\lambda_{k+1} - \lambda_k} = 4.6692\ldots . \tag{3.17}$$

Because of the geometric nature of this sequence, it is not easy by experimentation on the computer to locate the bifurcation values after the first several. However, using Eq. (3.17), this is a trivial matter. In fact, it is a remarkable discovery of Feigenbaum that the number above turns out to be the same for a large class of maps, not just the logistic map.

⇒ $\overline{\lambda > 3.570}$: In this range of the parameter the dynamics of the logistic map becomes quite complicated. For some values of λ, iterates move in very erratic ways, a behavior often called *chaos*; see Figure 3.16. In fact, long before the dynamics of the logistic map was unraveled, numerical analysts were using it as a way to generate random numbers. For other values of λ, there are periodic points which undergo sequences of period-doubling bifurcations as the parameter is varied. All of this remarkable dynamics is encapsulated in the numerically computed bifurcation diagram in Figure 3.17. We refrain from telling the full story, but cannot resist to point out what happens in the large white band.

⇒ $\overline{\lambda = 3.839}$: There is a unique asymptotically stable periodic orbit of minimal period 3. This orbit is easy to locate on the computer with accuracy by simply iterating almost any initial value long enough:

$$x^\star = 0.149888, \quad f(x^\star) = 0.489172,$$
$$f^2(x^\star) = 0.959299, \quad f^3(x^\star) = 0.149888.$$

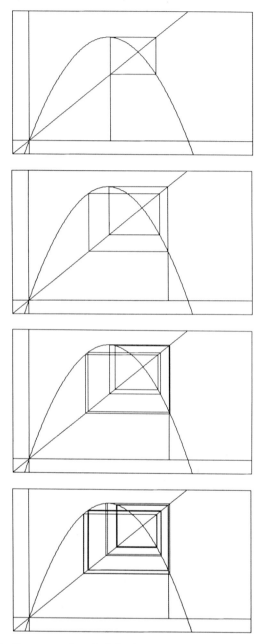

Figure 3.15. *Sequence of period doublings in the logistic map: asymptotically stable periodic orbits of period 2, 4, 8, and 16 at values of $\lambda = 3.2$, 3.52, 3.55, and 3.567, respectively. First several hundred iterates are not plotted.*

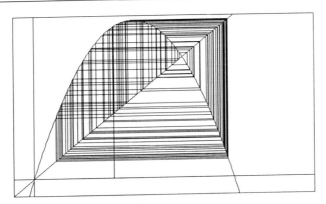

Figure 3.16. *Apparent chaos, or a periodic orbit with a very long period, in the logistic map for $\lambda = 3.891$. Iterates between 1000–3000 are plotted.*

However, there is more to the dynamics of the logistic map at $\lambda = 3.839$ than what is readily visible on the computer, as a startling theorem of Sharkovskii points out.

To state this theorem, let us order the positive integers in the following way, called the *Sharkovskii ordering*:

$$3 \triangleright 5 \triangleright 7 \triangleright \cdots \triangleright 2 \cdot 3 \triangleright 2 \cdot 5 \triangleright 2 \cdot 7 \triangleright \cdots \triangleright 2^2 \cdot 3 \triangleright 2^2 \cdot 5 \triangleright 2^2 \cdot 7 \triangleright \cdots$$

$$\triangleright 2^3 \cdot 3 \triangleright 2^3 \cdot 5 \triangleright 2^3 \cdot 7 \triangleright \cdots \quad \cdots \triangleright 2^3 \triangleright 2^2 \triangleright 2 \triangleright 1.$$

To describe in words, write all the odd numbers except 1, then 2 times the odd numbers, 2^2 times the odd numbers, 2^3 times, etc. Finally, write the powers of 2 in decreasing order, with 1 at the end. Despite its strangeness, this list includes all of the positive integers.

Theorem 3.24. (Sharkovskii) *Let $f : \mathbb{R} \to \mathbb{R}$ be a continuous map. Suppose that f has a periodic point of minimal period m. If $m \triangleright n$ in the Sharkovskii ordering, then f also has a periodic point of minimal period n.* $\diamond$

A noteworthy consequence of this theorem is that if f has a periodic point of minimal period 3, then it has periodic points of every minimal period. In particular, the logistic map at $\lambda = 3.839$, in addition to the period-3 orbit, has periodic orbits of all minimal periods. However, because of their instability, they are not readily detectable on the computer.

$\Rightarrow$ $\overline{\lambda > 3.839 :}$ As the parameter λ is increased, the period-3 orbit undergoes a period-doubling bifurcation and gives up its stability to an asymptotically stable period-6 orbit; see Figure 3.18. If λ is increased further, there

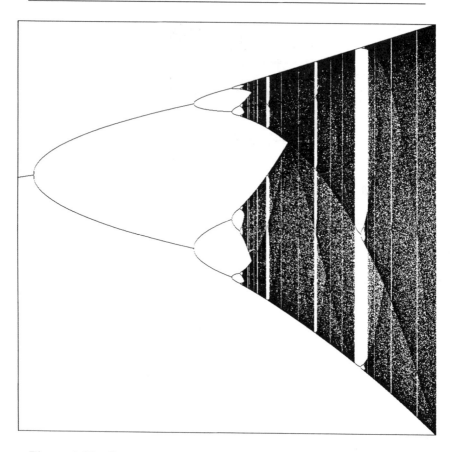

Figure 3.17. *Computer-generated complete bifurcation diagram of the logistic map.*

is a sequence of period-doubling bifurcations. Remarkably, the ratio of the distances between these successive period-doubling bifurcations again approaches the same constant 4.6692.... .

Here we end the interlude on maps and turn to the task of putting some of these results to good use in our study of nonautonomous scalar differential equations in the next chapter.

Exercises ———————————————————— ♣ ♡ ♠ ◇

3.18. *Another form for logistic:* In the mathematical literature, the logistic map $x_{n+1} = \lambda x_n(1 - x_n)$ is often transformed to the map $x_{n+1} = \mu - x_n^2$. Find the transformation. Identify the values of the parameter μ at which the latter map undergoes a saddle-node or a period-doubling bifurcation.

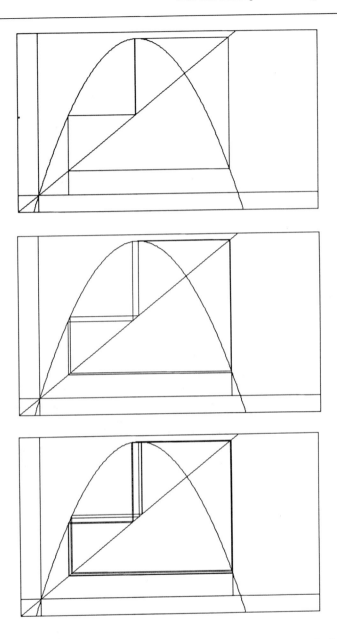

Figure 3.18. *Sequence of period doublings of a period-3 orbit in the logistic map: asymptotically stable periodic orbits of period 3, 6, and 12 at the values of $\lambda = 3.839$, 3.845, and 3.849, respectively.*

3.19. *A cubic map:* The one-parameter cubic map

$$x_{n+1} = (1 - \lambda)x_n + \lambda x_n^3$$

is used in genetics involving one locus with two alleles. This map undergoes a sequence of period-doubling bifurcations which accumulate at the value of the parameter $\lambda = 3.5980\ldots$. Experiment on the machine to determine several successive values of the parameter at which the map undergoes a period-doubling bifurcation and attempt to find a universal constant of Feigenbaum type. Be warned that this could be tedious experiment because of the geometric nature of the sequence of bifurcation values.

Help: This map is stored in the library of PHASER under the name *discubic*; just set the parameters. Also, see Rogers and Whitley [1983].

3.20. *A map with a period 5 but no period 3 orbit:* However unusual it may appear, the ordering of periods in Sharkovskii's theorem is sharp. Consider, for example, the map $f : [1, 5] \to [1, 5]$ defined by

$$f(1) = 3, \quad f(3) = 4, \quad f(4) = 2, \quad f(2) = 5, \quad f(5) = 1$$

with f linear between these integers. Draw the graph of f. Notice that 1 is a periodic point of minimal period 5. Following the suggestions below, show that f has no periodic point of minimal period 3. Verify that $f^3[1, 2] = [2, 5]$, $f^3[2, 3] = [3, 5]$, $f^3[4, 5] = [1, 4]$ so that f^3 has no fixed points in these intervals. However, $f^3[3, 4] = [1, 5]$ and thus f^3 has a fixed point in $[3, 4]$. Now, show that this fixed point is unique and thus it is the fixed point of f, not a periodic point of minimal period 3. To verify the uniqueness of this fixed point you should observe that f^3 is monotonically decreasing on $[3, 4]$. To see all this, you may want to draw the graph of f^3.

This example is due to Li and Yorke [1975]. The construction above can be generalized to verify the sharpness of Sharkovskii's ordering for odd periods. Can you find a map with a period 7 but not a period 5 point? The even periods require a different idea.

3.21. *Topological equivalence of maps:* Two maps f and g are said to be *topologically equivalent*, or *topologically conjugate*, if there is a homeomorphism h such that $h \circ f \circ h^{-1} = g$.

(a) Show that h takes orbits of f to orbits of g, that is, $h \circ f^n \circ h^{-1} = g^n$.

(b) Here is a famous example of a topological conjugacy. Consider the *tent map* defined on the unit interval by the formula

$$g(x) = \begin{cases} ax & \text{if } x < 0.5 \\ a(1.0 - x) & \text{if } x \geq 0.5, \end{cases}$$

where a is a parameter. Using the homeomorphism

$$h(x) = \frac{2}{\pi} \arcsin \sqrt{x},$$

verify that the tent map for the parameter value $a = 2.0$ is topologically conjugate to the logistic map $f(x) = \lambda x(1.0 - x)$ for the parameter value

$\lambda = 4$. Notice that the tent map is piecewise linear while the logistic map is quadratic. It is usually the case that the mathematical analysis of a piecewise linear map turns out to be somewhat easier than that of a nonlinear map even if they are topologically conjugate. Indeed, Ulam and von Neumann [1947] showed the existence of an "invariant measure" for these maps, at the indicated parameter values, using the topological conjugacy above. Even if you do not know what an invariant measure is, you may still want to look up this famous paper.

(c) The tent map is stored in the library of PHASER under the name *tent*. Compute some orbits numerically for the two maps at the indicated parameter values where the two maps are topologically conjugate. Do you notice a rather strange thing happening in the numerical computations with the tent map? Try the parameter value $a = 1.9999$; what is going on this time? The answer may elude you, but ponder about it anyway.

3.22. *A one-hump map with two attractors:* Consider the quartic map

$$f(a, x) = a(7.86x - 23.31x^2 + 28.75x^3 - 13.30x^4),$$

where a is a scalar parameter. This map has a unique maximum on the unit interval; visually it looks very much like the logistic map or the tent map, the so-called "one-hump" maps. Initially, it was hoped that a one-hump map depending on a parameter could be shown to have at most one stable periodic orbit for a given value of the parameter. However, the Singer map has both an asymptotically stable fixed point and an asymptotically stable periodic orbit of minimal period 2 for the parameter value $a = 1.0$. Find them on the computer by trying various initial conditions. Also, explore the dynamics of this map by changing the parameter a.

Help: This map is stored in the library of PHASER under the name *singer*. Also, see Singer [1978] where this map first appeared.

Bibliographical Notes ⊙⌢⊙

The dynamics community has been dealing with monotone maps since the days of Poincaré, as we shall see in the next chapter. The study of noninvertible real scalar maps, however, became popular in the early seventies, and eventually turned into a large industry. The review article by Whitley [1983], and the books by Collet and Eckmann [1980] and Devaney [1986] cover some of the basics and more.

It is interesting to note that one of the most remarkable theorems on scalar maps was already proved in Sharkovskii [1964] before the subject became vogue; see also Stefan [1977]. A special case was rediscovered in an article with a provocative title by Li and Yorke [1975].

The surprising geometric nature of the successive period-doubling bifurcations in logistic-like maps was observed numerically by Feigenbaum [1978 and 1980]. This *"universal"* property was later proved by Lanford

[1982 and 1987], using the machine in a novel way. Universality can be used to make quantitative predictions about bifurcations in a physical system where an explicit formula for a map governing the system may not be available. Such a prediction has been demonstrated by Libchaber and Mauer [1982] in experiments on Rayleigh-Benard convection.

An elementary exposition of the algorithms of Euler and Newton are contained in, for example, Conte and deBoor [1972]. Recently, ideas from dynamical systems have begun to open new avenues in numerical analysis, as exemplified in Smale [1981 and 1987]; see also Saari and Urenko [1984] and Ushiki [1986].

Iterating a number on a computer 64 times, let alone $1,000,000$ times, can lead to spurious results because of finite floating point arithmetic. It has long been known that certain well-behaved maps exhibit *shadowing* property: near a numerically computed orbit, there exists a true orbit which, however, may be the orbit of a different initial value than the one intended; see Anosov [1967] and Bowen [1975]. This important, and comforting, property has been established at some parameter values of the logistic map in Hammel et al. [1987].

$1\tfrac{1}{2}$ D

4

Scalar Nonautonomous Equations

 In this chapter, we begin our study of nonautonomous scalar differential equations and develop a geometric theory analogous to the one given in Chapter 1 for autonomous equations. After a brief general introduction, we focus our attention on equations with coefficients that are periodic in time. For such equations, we show that α- and ω-limit sets, if they exist, are periodic solutions. With the help of this fundamental result, we then illustrate in several specific examples how to establish the existence of periodic solutions. Finally, we investigate the stability of periodic solutions using the theory of scalar maps developed in Chapter 3.

4.1. General Properties of Solutions

Let $f : \mathbb{R} \times \mathbb{R} \to \mathbb{R}$; $(t, x) \mapsto f(t, x)$ be a given continuous function and C^1 in the variable x. Differential equations of the form

$$\dot{x} = f(t, x) \tag{4.1}$$

will be the subject of this and the following chapter. Such an equation is called *nonautonomous* because the function f depends explicitly on t as well as x. In this section, we point out some of the basic similarities and the important differences between solutions of Eq. (4.1) and the autonomous case discussed in Sections 1.1, and 1.2.

The notion of a solution of the nonautonomous differential equation (4.1) and the initial-value problem for Eq. (4.1) are defined in the same way as in the autonomous case. However, for given initial data $(t_0, x_0) \in \mathbb{R} \times \mathbb{R}$, it will be necessary to denote a solution of Eq. (4.1) through x_0 at t_0 by $\varphi(t, t_0, x_0)$ with $\varphi(t_0, t_0, x_0) = x_0$. The necessary generalization of the existence and uniqueness theorem for an initial-value problem holds, as stated in the Appendix, that is, $\varphi(t, t_0, x_0)$ is uniquely defined and is continuous together with its first derivatives with respect to t, t_0, and x_0 in all variables (t, t_0, x_0).

The direction field of Eq. (4.1) is defined as before, and the trajectory through (t_0, x_0) is defined to be the set

$$\{(t, \varphi(t, t_0, x_0)) : t \in I_{t_0, x_0}\} \subset \mathbb{R} \times \mathbb{R},$$

where I_{t_0, x_0} is the maximal interval of definition of the solution $\varphi(t, t_0, x)$.

It is, of course, also possible to define the orbit of Eq. (4.1) through x_0 at t_0 as the set $\{(\varphi(t, t_0, x_0)) : t \in I_{t_0, x_0}\} \subset \mathbb{R}$. However, orbits of nonautonomous equations do not enjoy the same properties as those of autonomous ones. More precisely, in the autonomous case, there is a unique orbit through x_0 due to the fact that the solution $\varphi(t, t_0, x_0)$ through x_0 at t_0 satisfies $\varphi(t, t_0, x_0) = \varphi(t - t_0, 0, x_0)$. In the nonautonomous case, as illustrated by Example 4.2 below, orbits are not uniquely defined by x_0.

The main implication of the discussion above is that, in the qualitative study of nonautonomous differential equations, we cannot effectively use the orbits in $\mathbb{R}$, but must consider the trajectories in $\mathbb{R} \times \mathbb{R}$. The trajectories are, of course, uniquely defined by (t_0, x_0). Although, they are not monotone in t, solutions of nonautonomous differential equations still have the useful property of monotonicity with respect to the initial data x_0. This fact is the content of the lemma below which is a direct consequence of the uniqueness of solutions (compare with Lemma 1.7).

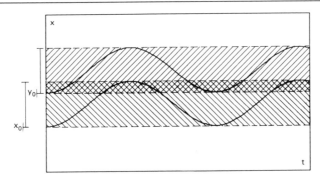

Figure 4.1. *Two trajectories of* $\dot{x} = \sin t$.

Lemma 4.1. *Let* $\varphi(t, t_0, x_0)$ *and* $\varphi(t, t_0, y_0)$ *be two solutions of Eq. (4.1) through* x_0 *at* t_0 *and* y_0 *at* t_0, *respectively. Then*

$$\varphi(t, t_0, x_0) < \varphi(t, t_0, y_0) \quad \text{for } t \geq t_0 \qquad \text{if } x_0 < y_0. \; \Diamond$$

Let us now give several examples of nonautonomous differential equations to illustrate some of their differences from autonomous ones.

Example 4.2. *Orbits do not suffice:* Consider the simple differential equation

$$\dot{x} = \sin t.$$

Two trajectories of this equation are shown in Figure 4.1 to illustrate the monotonicity with respect to the initial data x_0. Notice that the orbit of the solution $\varphi(t, 0, x_0) = x_0 + 1 - \cos t$ is the interval $[x_0, x_0 + 2]$ for any x_0. Thus, if x_0 is near y_0, then the orbits of the solutions $\varphi(t, 0, x_0)$ and $\varphi(t, 0, y_0)$ overlap but are not equal. Also, observe that every solution is periodic with period 2π, the same period as the vector field. $\Diamond$

Example 4.3. *Finite time:* Consider the differential equation

$$\dot{x} = (\cos t)x^2.$$

If $x(t_0) = 0$, then the solution is $x(t) = 0$, which is defined on all of $\mathbb{R}$. When $x(t_0) \neq 0$, using the method of separation of variables, it is easy to see that the solution is given by

$$\varphi(t, t_0, x_0) = \frac{1}{(\sin t_0 + \frac{1}{x_0}) - \sin t}.$$

If $|\sin t_0 + 1/x_0| > 1$, then the solution is again defined for all t. Otherwise, the solution is defined only on a finite interval; see Figure 4.2. For $t_0 = 0$, notice that every solution with $|x_0| < 1$ is periodic with a period of 2π. $\Diamond$

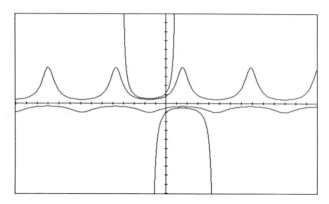

Figure 4.2. Trajectories of $\dot{x} = (\cos t)x^2$.

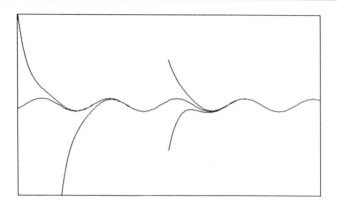

Figure 4.3. Trajectories of $\dot{x} = -x + \cos t$.

Example 4.4. *Periodic limit:* Consider the differential equation

$$\dot{x} = -x + \cos t.$$

The flow of this equation is given by

$$\varphi(t, t_0, x_0) = e^{-(t-t_0)}[x_0 - \tfrac{1}{2}(\sin t_0 + \cos t_0)] + \tfrac{1}{2}(\sin t + \cos t).$$

Notice that all solutions have the same asymptotic fate for $t \to +\infty$: they approach the periodic solution $\tfrac{1}{2}(\sin t + \cos t)$; see Figure 4.3. $\Diamond$

Example 4.5. *Limit is no solution:* Consider the differential equation

$$\dot{x} = -x + \frac{1}{t} - \frac{1}{t^2}$$

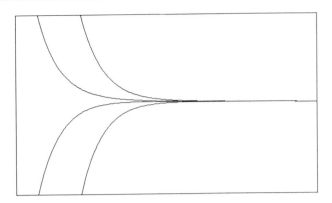

Figure 4.4. *Trajectories of $\dot{x} = -x + 1/t - 1/t^2$.*

for $x \in \mathbb{R}$ and $t \geq 1$. The solution $\varphi(t, t_0, x_0)$ is given by

$$\varphi(t, t_0, x_0) = e^{-(t-t_0)} \left(x_0 - \frac{1}{t_0} \right) + \frac{1}{t}.$$

Notice that every solution approaches zero as $t \to +\infty$, but $x(t) = 0$ is not a solution of the differential equation; see Figure 4.4. $\diamondsuit$

There is very little that can be said about the general qualitative properties of solutions of arbitrary nonautonomous scalar differential equations. However, when the function $f(t, x)$ is periodic in t, we shall see shortly that the qualitative behavior of solutions is very similar to the autonomous case if we replace equilibrium points with periodic solutions, as is evident in Example 4.4. Before delving into the details of this subject, we will derive a useful explicit formula for solutions of a general linear nonautonomous differential equation.

Example 4.6. *Variation of the constants formula.* Consider the linear (in x) nonautonomous equation

$$\dot{x} = a(t)x + b(t), \tag{4.2}$$

where $a(t)$ and $b(t)$ are scalar continuous functions. The solution $\varphi(t, t_0, x_0)$ of Eq. (4.2) is given by

$$\varphi(t, t_0, x_0) = e^{\int_{t_0}^{t} a(u)\, du} \left[x_0 + \int_{t_0}^{t} e^{-\int_{t_0}^{s} a(u)\, du} b(s)\, ds \right]. \tag{4.3}$$

This imposing formula is called the *variation of the constants* formula because when $b(t) = 0$, the homogeneous case, the solution is an exponential function multiplied by a constant. In the nonhomogeneous case, the constant is varied with a function of t.

Despite its looks, formula (4.3) is in fact rather easy to derive. To eliminate the term $a(t)x$ in Eq. (4.2), we introduce the new variable y defined by

$$x = e^{\int_{t_0}^t a(u)\,du}\, y.$$

If $x(t_0) = x_0$, then $y(t_0) = y_0$ and the differential equation (4.2) in the new variable becomes

$$\dot{y} = e^{-\int_{t_0}^t a(u)\,du}\, b(t).$$

To obtain the solution of this differential equation, we simply integrate with respect to t :

$$y(t) = y_0 + \int_{t_0}^t e^{-\int_{t_0}^s a(u)\,du}\, b(s)\,ds.$$

Now, returning back to the variable x, we recover the solution (4.3). If you enjoy performing integrations, you may wish to use this formula to obtain some of the explicit solutions we have used in the examples above. $\Diamond$

Exercises

4.1. In Example 4.3, $\dot{x} = (\cos t)x^2$, take $t_0 = 0$ and discuss the maximal interval of existence of solutions as a function of x_0.

4.2. Write the variation of constants formula for the equation $\dot{x} = b(t)$, where $b(t)$ is a continuous function.

4.3. *Formula for linear equations:* Show that the solution of $\dot{x} = a(t)x + f(x)$ with $x(0) = x_0$ satisfies

$$x(t) = e^{\int_{t_0}^t a(u)\,du}\, x_0 + \int_0^t e^{\int_s^t a(u)\,du}\, f(x(s))\,ds.$$

4.4. Discuss the behavior of the solutions of $\dot{x} = -(\sin t)x + 1$ as $t \to +\infty$.

4.5. *A boundary-value problem:* Show that there is a nontrivial solution of the boundary-value problem $\dot{x} = a(t)x$, with $0 < t < 1$, satisfying $x(0) = x(1)$ if and only if $\int_0^1 a(u)\,du = 0$.

4.6. *Another boundary-value problem:* Consider the boundary-value problem $\dot{x} = a(t)x + b(t)$, with $0 < t < 1$, satisfying $x(0) = x(1)$. Prove the following:
(a) If $\int_0^1 a(u)\,du \neq 0$, then there is a unique solution.
(b) If $\int_0^1 a(u)\,du = 0$, then there is a solution if and only if

$$\int_0^1 e^{\int_s^1 a(u)\,du}\, b(s)\,ds = 0.$$

How many solutions exist in this case?

4.7. *More boundary-value problems:* Discuss the solutions of the boundary-value
problem in the previous exercise for the following specific coefficients:
(a) $a(t) = 1, \quad b(t) = e^{t-1}$;
(b) $a(t) = \sin t, \quad b(t) = \sin t$;
(c) $a(t) = \sin 2\pi t, \quad b(t) = \sin 2\pi t$;
(d) $a(t) = \sin 2\pi t, \quad b(t) = \sin t$.

4.2. Geometry of Periodic Equations

In this section, we begin our study of an important special class of nonautonomous differential equations where the function $f(t, x)$ has the additional property that it is periodic in t with a period of 1. More specifically,
we will consider the differential equation

$$\dot{x} = f(t, x) \quad \text{with} \quad f(t+1, x) = f(t, x). \tag{4.4}$$

Briefly, we say $f(t, x)$ is *1-periodic* in t. For simplicity of notation, we
assume the period to be 1; if not, we can always rescale t to make it so.
Let us proceed with certain general observations; specific examples will be
forthcoming in the subsequent sections.

Because of the periodicity of f, the solutions of Eq. (4.4) possess certain properties which are useful in determining the asymptotic behavior of
the solutions. In order to take advantage of these properties listed below,
we will assume throughout this chapter that the solutions of Eq. (4.4) are
defined for all $t \in \mathbb{R}$. It is easy to see by direct substitution that if $x(t)$ is
a solution of Eq. (4.4), then for any integer k, $x(t + k)$ is also a solution.
Let $\varphi(t, t_0, x_0)$ be the solution of Eq. (4.4) through x_0 at t_0. The observation above, in conjunction with the uniqueness of solutions of initial-value
problems, implies that

$$\varphi(t + 1, t_0 + 1, x_0) = \varphi(t, t_0, x_0), \tag{4.5}$$

$$\varphi(t + 1, t_0, x_0) = \varphi(t, t_0, \varphi(t_0 + 1, t_0, x_0)). \tag{4.6}$$

The geometric interpretation of Eq. (4.6) is illustrated in Figure 4.5: when
translated horizontally, the piece of the solution $\varphi(t, t_0, x_0)$ on the inter-
val $[t_0 + 1, t_0 + 2]$ coincides with the piece of the solution $\varphi(t, t_0, \varphi(t_0 + 1, t_0, x_0))$ on the interval $[t_0, t_0 + 1]$.

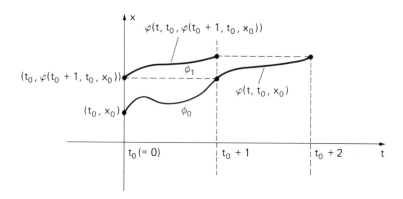

Figure 4.5. *Translation invariance of solutions of 1-periodic equations.*

In the qualitative study of Eq. (4.4) certain special solutions, namely, the 1-periodic ones, are of great significance. As we shall see shortly, 1-periodic solutions of the nonautonomous equation (4.4) play a role similar to that of equilibria of the autonomous equation $\dot{x} = f(x)$. Analogous to our presentation in Section 2.4, we begin with the following definition:

Definition 4.7. *A solution $\varphi(t, t_0, x_0)$ of a 1-periodic differential equation $\dot{x} = f(t, x)$ is called a T-periodic solution if*

$$\varphi(t + T, t_0, x_0) = \varphi(t, t_0, x_0) \quad \text{for all } t.$$

If, moreover, $\varphi(t + \tau, t_0, x_0) \neq \varphi(t, t_0, x_0)$ for any $0 < \tau < T$, then T is called the minimal period.

In our subsequent discussion of 1-periodic differential equations, we will be interested chiefly in 1-periodic solutions. Let us proceed with a simple lemma.

Lemma 4.8. *A solution $\varphi(t, t_0, x_0)$ of a 1-periodic differential equation $\dot{x} = f(t, x)$ is 1 periodic if and only if $\varphi(t_0 + 1, t_0, x_0) = x_0$.*

Proof. If $\varphi(t + 1, t_0, x_0) = \varphi(t, t_0, x_0)$, then evaluating this equation at $t = t_0$ gives $\varphi(t_0 + 1, t_0, x_0) = x_0$. Conversely, if $\varphi(t_0 + 1, t_0, x_0) = x_0$, then Eq. (4.6) implies that $\varphi(t, t_0, x_0)$ is 1 periodic. ◇

The notion of stability of a periodic solution of Eq. (4.4) can be defined in an analogous way to that of the stability of an equilibrium point of an autonomous equation given in Section 1.3.

Definition 4.9. *A periodic solution $\varphi(t, t_0, x_0)$ of Eq. (4.4) is said to be stable if, for any given $\epsilon > 0$, there is a $\delta > 0$ (depending only on*

ϵ and not on t_0) such that, for every y_0 for which $|y_0 - x_0| < \delta$, the solution $\varphi(t, t_0, y_0)$ of Eq. (4.4) through y_0 at t_0 satisfies the inequality $|\varphi(t, t_0, y_0) - \varphi(t, t_0, x_0)| < \epsilon$ for all $t \geq t_0$. The periodic solution $\varphi(t, t_0, x_0)$ is said to be *unstable* if it is not stable.

Definition 4.10. *A periodic solution $\varphi(t, t_0, x_0)$ of Eq. (4.4) is said to be* asymptotically stable *if it is stable and, in addition, there is an $r > 0$, independent of t_0, such that $|\varphi(t, t_0, y_0) - \varphi(t, t_0, x_0)| \to 0$ as $t \to +\infty$ for all y_0 satisfying $|y_0 - x_0| < r$.*

As an illustration of an asymptotically stable periodic solution, you may wish to refer back to the examples in the previous section, in particular, Example 4.4.

We now begin our investigation of the asymptotic behavior of solutions of Eq. (4.4). For this purpose, we will suppose that $t_0 = 0$. It is evident that the asymptotic behavior of the solution $\varphi(t, 0, x_0)$ can be studied by looking at the following sequence of functions on the interval $[0, 1]$:

$$\phi_k \colon [0, 1] \to \mathbb{R}, \quad k = 0, 1, 2, \ldots, \quad \text{defined by}$$
$$\phi_0(t) = \varphi(t, 0, x_0),$$
$$\phi_1(t) = \varphi(t + 1, 0, x_0),$$
$$\vdots \tag{4.7}$$
$$\phi_k(t) = \varphi(t + k, 0, x_0), \quad 0 \leq t \leq 1,$$
$$\vdots$$

The uniqueness of solutions and the periodicity in the differential equation imply that this sequence of functions is monotone, that is, if $\phi_1(0) \geq \phi_0(0) = x_0$, then $\phi_{k+1}(t) \geq \phi_k(t)$ [or, if $\phi_1(0) \leq \phi_0(0) = x_0$, then $\phi_{k+1}(t) \leq \phi_k(t)$]. Consequently, if the sequence is bounded, then there must be a limit function $\Phi(t)$ such that $\phi_k(t) \to \Phi(t)$ monotonically as $k \to +\infty$. In particular, $\phi_k(0) \to \Phi(0)$ monotonically as $k \to +\infty$. Since $\phi_k(t) = \varphi(t, 0, \phi_k(0))$ from Eq. (4.6), and $\varphi(t, 0, x_0)$ is continuous in t_0, x_0, it follows that $\varphi(t, 0, \Phi(0))$ exists for $0 \leq t \leq 1$ and $\Phi(t) = \varphi(t, 0, \Phi(0))$, that is, $\Phi(t)$ is a solution of the differential equation (4.4) on the interval $0 \leq t \leq 1$. Furthermore, using Eq. (4.6) we obtain

$$\Phi(1) = \lim_{k \to +\infty} \varphi(1, 0, \phi_k(0)) = \lim_{k \to +\infty} \phi_{k+1}(0) = \Phi(0).$$

Hence, it follows from Lemma 4.8 that the solution of Eq. (4.4) through $\Phi(0)$ at $t_0 = 0$ is 1 periodic.

The asymptotic behavior of a solution $\varphi(t, 0, x_0)$ of Eq. (4.4) as $t \to -\infty$ can be studied by considering the sequence of functions defined by $\psi_k \equiv \varphi(t - k, 0, x_0)$. If this sequence is bounded, then there must be a limit function $\Psi(t)$ such that $\psi_k(t) \to \Psi(t)$ as $k \to +\infty$, and $\Psi(1) = \Psi(0)$.

We now record the results of these observations in the theorem below.

Theorem 4.11. *If a solution $\varphi(t, 0, x_0)$ of a 1-periodic differential equation (4.4) is bounded for $t \geq 0$, then there is a 1-periodic solution $\Phi(t)$ of Eq. (4.4) such that*

$$\varphi(t + k, 0, x_0) \to \Phi(t) \qquad \text{as the integer} \quad k \to +\infty$$

monotonically and uniformly for $0 \leq t \leq 1$. Similarly, if $\varphi(t, 0, x_0)$ is bounded for $t \leq 0$, then there is a 1-periodic solution $\Psi(t)$ of Eq. (4.4) such that

$$\varphi(t - k, 0, x_0) \to \Psi(t) \qquad \text{as the integer} \quad k \to +\infty$$

monotonically and uniformly for $0 \leq t \leq 1$. $\Diamond$

In words, this theorem guarantees that the limit of a solution of a 1-periodic differential equation is always a solution and, in addition, is 1 periodic. However, as seen in Example 4.5, if the differential equation is not 1 periodic, then it is possible that the limit need not be a solution.

There is an important idea used in the arguments leading to the proof of Theorem 4.11 and we would like to explore it further. In studying properties of the functions $\phi_k(t)$, it was only necessary to discuss the behavior of the sequence of points $\phi_k(0)$. These points can be thought of as an orbit of a scalar map on the x-axis of the (t, x)-plane.

Definition 4.12. *The Poincaré map (also called time-one or period map) of a 1-periodic differential equation (4.4) is the scalar mapping*

$$\Pi : \mathbb{R} \to \mathbb{R} \, ; \ x_0 \mapsto \varphi(1, 0, x_0).$$

In words, the Poincaré map takes the initial value x_0 at $t_0 = 0$ to the value of the solution $\varphi(t, 0, x_0)$ at $t = 1$. Thus, it follows from Eq. (4.6) that the kth iterate of x_0 under Π is given by

$$\Pi^k(x_0) = \varphi(k, 0, x_0).$$

Of course, the point $\Pi^k(x_0)$ is the same as the point $\phi_k(0)$; hence, the Poincaré map is monotone. Also, it is important to notice that, from differentiable dependence on initial values as stated in the Appendix, the Poincaré map is differentiable with nonnegative derivative.

The chief importance of the Poincaré map stems from the following restatement of Lemma 4.8: a point x_0 is the initial value of a 1-periodic solution of Eq. (4.4) if and only if x_0 is a fixed point of the Poincaré map, that is, $\Pi(x_0) = x_0$. Naturally, the stability properties of a 1-periodic solution of Eq. (4.4) are the same as the stability properties of the corresponding fixed point of the Poincaré map. Thus, we can apply the previous results on monotone maps (Section 3.3) to the Poincaré map of Eq. (4.4) in our investigation of periodic orbits, as we shall do subsequently.

In concluding this section, we should remark that during the foregoing discussion of asymptotic behavior of solutions of Eq. (4.4), particularly in the definition of Poincaré map, we have assumed t_0 to be 0. If we take a value of t_0 other than 0, the resulting Poincaré map will not be identical to the one at $t_0 = 0$. However, the qualitative dynamics of the two Poincaré maps, that is, the number and the stability properties of their fixed points, will be the same.

Exercises

4.8. *Scaling time:* Rescale the time variable so that the 2π-periodic differential equation $\dot{x} = -x + \cos t$ becomes 1 periodic.

4.9. Suppose that $b(t)$ is a 1-periodic continuous function and let $b_0 = \int_0^1 b(s)\, ds$. Show that all solutions of $\dot{x} = b(t)$ are 1 periodic if $b_0 = 0$; otherwise, they are all unbounded.
    Hint: The Poincaré map is $\Pi(x_0) = x_0 + b_0$.

4.10. Consider a 1-periodic scalar differential equation $\dot{x} = f(t, x)$ satisfying $f(t, 0) = 0$. If there is an $r > 0$ such that for $|x_0| < r$ the solutions $\varphi(t, t_0, x_0) \to 0$ as $t \to +\infty$, then show that the zero solution $x = 0$ is stable.

4.11. *Fredholm's Alternative:* Suppose that $a(t)$ and $b(t)$ are 1-periodic continuous functions and let $a_0 = \int_0^1 a(s)\, ds$. Show the following properties of the differential equation $\dot{x} = a(t)x + b(t)$:
    1. If $a_0 \neq 0$, then there is a unique 1-periodic orbit which is asymptotically stable when $a_0 < 0$ and unstable when $a_0 > 0$.
    2. Let $c_0 = \int_0^1 \exp\{-\int_s^1 a(u)\, du\}b(s)\, ds$. If $a_0 = 0$, then every solution is 1-periodic if and only if $c_0 = 0$.
    3. If $a_0 = 0$, then every solution is unbounded if $c_0 \neq 0$.
    4. Why do you think this problem is called the Fredholm's Alternative? You may want to look it up in a mathematical encyclopedia or in a good book on linear algebra.
    Hint: From the variations of the constants formula, the Poincaré map is given by
    $$\Pi(x_0) = e^{a_0}x_0 + \int_0^1 \exp\{\int_s^1 a(u)\, du\}b(s)\, ds$$
    and $\Pi(x_0) = x_0$ if and only if $(e^{-a_0} - 1)x_0 = c_0$.

4.12. Let $b(t)$ be a 1-periodic continuous function and consider the differential equation
    $$\dot{x} = -\left[a_0 + \frac{1}{2\pi}\sin(2\pi t)\right]x + e^{-\cos(2\pi t)}b(t).$$

    Show the following properties of the solutions:
    1. If $a_0 \neq 0$, then there is unique 1-periodic solution.
    2. If $a_0 = 0$ and $b_0 = \int_0^1 b(s)\, ds = 0$, then every solution is 1 periodic.
    3. If $a_0 = 0$ and $b_0 \neq 0$, then every solution is unbounded.

4.13. Can you relate the Fredholm Alternative for 1-periodic solutions to the results on the boundary-value problem in the exercises at the end of Section 4.1?

Hint: Consider $a(t)$ and $b(t)$ for the boundary-value problem as defined only on $[0, 1]$, and extend them as 1-periodic functions to $\mathbb{R}$. Even though the new coefficients are discontinuous in t, the theory remains valid if we allow discontinuities in the derivatives at the integers.

4.3. Periodic Equations on a Cylinder

In this section, we describe an interesting way of visualizing the trajectories, functions ϕ_k, and the Poincaré map of a 1-periodic nonautonomous differential equation.

The solutions of Eq. (4.4) can be represented as orbits of a two-dimensional autonomous differential equation on a cylinder. To accomplish this, let us first convert Eq. (4.4) to the following equivalent pair of autonomous differential equations:

$$\dot{\theta} = 1$$
$$\dot{x} = f(\theta, x). \tag{4.8}$$

It is clear from the form of the first equation that the orbits of Eq. (4.8) on the (x, θ)-plane correspond to the trajectories of $\dot{x} = f(t, x)$. We will undertake a detailed study of a pair autonomous differential equations on the plane in Part III of our book. However, Eq. (4.8) is rather special, so let us examine it more closely.

Observe that the first equation is periodic with any period. Let us take this period to be 1. Thus, from the considerations in Section 1.4, it can be viewed as a differential equation on the circle S^1. Since f is 1-periodic, $f(\theta + 1, x) = f(\theta, x)$, if we identify the points θ with points $\theta + k$, for any integer k, the second equation remains unchanged. Therefore, the orbits of Eq. (4.8), hence the trajectories of Eq. (4.4), can be conveniently viewed on $S^1 \times \mathbb{R}$, a cylinder.

Since $\theta(t) = t_0 + t$ is an increasing function of t, an orbit of Eq. (4.8) continues to wind around the cylinder as time goes on. The function $\phi_k(t)$, $0 \leq t \leq 1$, defined in Eq. (4.7), represents one revolution around the cylinder. Because $\phi_k(1) = \phi_{k+1}(0)$, the sequence of functions $\{\phi_k\}$ fit together to make a "spiral" on the cylinder. Iterations of an initial point under the Poincaré map is the successive intersection of this spiral with a vertical line on the cylinder.

Also, Theorem 4.11 can be reinterpreted as follows: if a positive orbit $(t \geq 0)$ of Eq. (4.8) is bounded, then it must approach a periodic orbit as $t \to +\infty$. Of course, a periodic orbit wraps around the cylinder only once.

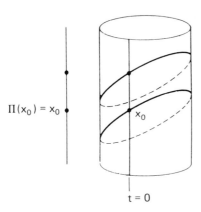

Figure 4.6. *Poincaré map and two solutions of the 1-periodic differential equation* $\dot{x} = \sin 2\pi t$ *on the cylinder.*

Similarly, if a negative orbit $(t \leq 0)$ is bounded, then it must do the same as $t \to -\infty$. In other words, the α- and ω-limit sets of Eq. (4.8), if they exist, are periodic orbits. We will explore the geometry of limit sets of systems of autonomous differential equations in Part III. For the moment, let us just illustrate these ideas on a couple of simple examples that we have examined previously.

Example 4.13. In order to make its period 1, let us rescale t in Example 4.2 and consider the equation

$$\dot{x} = \sin 2\pi t.$$

This 1-periodic differential equation is equivalent to the following pair of autonomous equations on the plane:

$$\dot{\theta} = 1$$
$$\dot{x} = \sin 2\pi\theta.$$

We have drawn in Figure 4.6 the flow of this system on the cylinder. Notice that all orbits are periodic with period 1 which wrap around the cylinder. Consequently, the Poincarè map is the identity map, $\Pi(x_0) = x_0$, and thus all points are fixed points. $\lozenge$

Example 4.14. Consider the 1-periodic differential equation

$$\dot{x} = -x + \cos 2\pi t - 2\pi \sin 2\pi t,$$

which is a variation of Example 4.4. Any solution of this linear equation satisfying $x(0) = x_0$ is given by

$$\varphi(t, 0, x_0) = e^{-t}(x_0 - 1) + \cos 2\pi t.$$

It is evident that there is a unique periodic solution ($x_0 = 1$) and it is asymptotically stable; see Figure 4.7. This observation is, of course, reflected in the Poincaré map

$$\Pi(x_0) = e^{-1}(x_0 - 1) + 1.$$

There is a unique fixed point of this map at $x_0 = 1$,

$$\Pi(1) = 1,$$

which corresponds to the periodic solution $\varphi(t, 0, 1) = \cos 2\pi t$. Furthermore, since $\Pi'(1) = e^{-1} < 1$, the fixed point, hence the corresponding periodic solution, is asymptotically stable.

The qualitative properties of the solutions of the 1-periodic nonautonomous differential equation above can also be seen in the following equivalent pair of autonomous equations on the cylinder:

$$\dot{\theta} = 1$$
$$\dot{x} = -x + \cos 2\pi\theta - 2\pi \sin 2\pi\theta.$$

We have drawn in Figure 4.7 a typical orbit of this system on the cylinder as the orbit spirals towards its ω-limit set, the asymptotically stable periodic orbit. ◇

Exercises

4.14. To obtain numerical solutions of a 1-periodic scalar differential equation, especially on PHASER, it is often necessary to convert such an equation to a pair of autonomous equations using Eq. (4.8). Convert the 1-periodic equation in Example 4.3 and reproduce Figure 4.2 using PHASER. To recover the entire figure, you will have to run orbits of the planar differential equations backward in time.

4.15. Experiment numerically, on PHASER, of course, with the following periodic differential equations and try to locate as many periodic solutions as you can; also to sharpen your imagination, when you get an exciting picture, visualize it as a flow on the cylinder:
   (i) $\dot{x} = x^2 + a + b \sin t$;   (ii) $\dot{x} = a \sin t + (b + c \cos t)x - x^3$;
   (iii) $\dot{x} = a \sin t - \sin x$;   (iv) $\dot{x} = a \sin t - x^5$;
   (v) $\dot{x} = a \sin t + b \sin(\sqrt{2}\, t) - \sin x$.
   Suggestions: You should first convert these equations to pairs of autonomous differential equations. When entering your equations into PHASER, do so with parameters. Then systematically fix the values of the parameters and experiment with different initial values. You will easily locate the asymptotically stable periodic solutions with forward integration. To locate the unstable periodic solutions, however, integrate backwards in time.

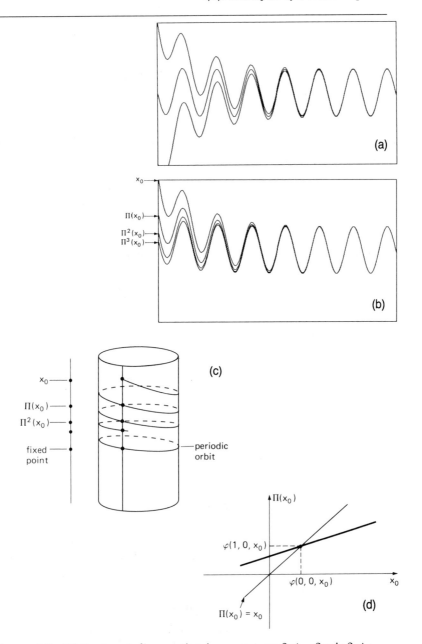

Figure 4.7. *Of the 1-periodic equation $\dot{x} = -x + \cos 2\pi t - 2\pi \sin 2\pi t$:*
(a) *several solutions,* (b) *one solution and its translations to the left by integer amounts,* (c) *one solution on the cylinder, and* (d) *the graph of the Poincaré map.*

4.4. Examples of Periodic Equations

In this section, we consider several specific examples of 1-periodic nonautonomous differential equations illustrating the concepts and the results
introduced so far. We also show certain practical ramifications of Theorem 4.11 in demonstrating the existence of periodic solutions and their
stability.

The minimal period of a 1-periodic solution of a 1-periodic differential
equation may not always be 1. The minimal period is either $1/n$, where n
is some positive integer, or the 1-periodic solution is in fact an equilibrium
solution. The next two examples illustrate these possibilities.

Example 4.15. *Period 1/2:* Consider the 1-periodic differential equation

$$\dot{x} = 4\pi \cos 4\pi t + (x - \sin 4\pi t) \sin 2\pi t.$$

It is easy to verify by direct substitution that $x(t) = \sin 4\pi t$ is a solution
whose prime period is $1/2$. $\Diamond$

Example 4.16. *Any period:* Consider the 1-periodic differential equation

$$\dot{x} = (\cos^2 2\pi t)x.$$

It is easy to see that any solution of this equation satisfying $x(0) = x_0$ is
given by

$$\varphi(t, 0, x_0) = x_0 \exp\left[\frac{t}{2} + \frac{\sin 4\pi t}{8\pi}\right],$$

where exp is the exponential function. If $x_0 \neq 0$, then the solution is
unbounded as $t \to +\infty$, but tends to zero as $t \to -\infty$. So the 1-periodic
solution whose existence is guaranteed by Theorem 4.11 is periodic with
any period; in fact, it is an (unstable) equilibrium point; see Figure 4.8.
This dynamics is reflected in the Poincaré map

$$\Pi(x_0) = \varphi(1, 0, x_0) = x_0\sqrt{e};$$

there is a unique fixed point $x_0 = 0$ which is unstable because $\Pi'(x_0) = \sqrt{e} > 1$. $\Diamond$

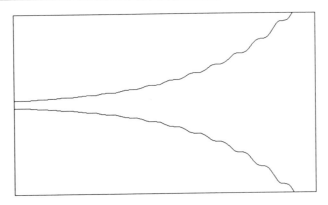

Figure 4.8. *Solutions of* $\dot{x} = (\cos^2 2\pi t)x$.

Example 4.17. *Qualitative analysis of a linear equation:* Consider the following generalization of the linear (in x) 1-periodic equations in Examples 4.4 and 4.14:

$$\dot{x} = -x + c(t), \qquad c(t+1) = c(t), \qquad\qquad (4.9)$$

where $c(t)$ is an arbitrary continuous function of period 1. Using the Fourier series, it is not difficult to show that the linear equation (4.9) has a unique periodic solution $\Phi(t)$. In fact, if the Fourier series of $c(t)$ is

$$\sum_{n=-\infty}^{+\infty} c_n e^{i2\pi n t},$$

then

$$\Phi(t) = \sum_{n=-\infty}^{+\infty} \frac{c_n}{1 + i2\pi n} e^{i2\pi n t}$$

is a 1-periodic solution (of course, one can obtain a real solution from the complex one above, if so desired). Since the solution of Eq. (4.9) through x_0 at $t = 0$ is

$$\varphi(t, 0, x_0) = e^{-t}\big[x_0 - \Phi(0)\big] + \Phi(t),$$

it follows that $\Phi(t)$ is the unique 1-periodic solution. In addition, this formula for the solution through x_0 shows that the Poincaré map is given by

$$\Pi(x_0) = e^{-1}\big[x_0 - \Phi(0)\big] - \Phi(0),$$

and its graph is similar to the one given in Figure 4.7d. Thus, the 1-periodic solution $\Phi(t)$ is asymptotically stable.

Let us give another proof of the existence and uniqueness of the 1-periodic solution of this example without using the Fourier series. Since $c(t)$ is bounded, there is a positive constant M such that $|c(t)| \leq M$ for all t. Thus, it is clear from the form of Eq. (4.9) that any solution $\varphi(t, 0, x_0)$ of Eq. (4.9) satisfies the inequality

$$-\varphi - M \leq \dot{\varphi} \leq -\varphi + M \quad \text{for all } t. \tag{4.10}$$

This implies, as we will show below, that

$$e^{-t}(x_0 + M) - M \leq \varphi \leq e^{-t}(x_0 - M) + M. \tag{4.11}$$

Therefore, $\varphi(t, 0, x_0)$ is bounded for $t \geq 0$ and, by Theorem 4.11, it approaches a 1-periodic solution $\Phi(t)$ as $t \to +\infty$. To indicate how to establish, for instance, the left part of this inequality, first multiply the left part of Eq. (4.10) by e^t and obtain

$$e^t(\dot{\varphi} + \varphi + M) \geq 0.$$

Then observe that

$$\frac{d}{dt}\left(e^t \varphi\right) + e^t M \geq 0.$$

Now, integration from 0 to t yields

$$e^t \varphi - x_0 + e^t M - M \geq 0.$$

Finally, multiply through with e^{-t} and rearrange the terms to arrive at the inequality (4.11).

The uniqueness of $\Phi(t)$, and thus its asymptotic stability, is easy to establish. If $x(t)$ and $y(t)$ are two solutions, then $z(t) = x(t) - y(t)$ satisfies the linear autonomous differential equation $\dot{z} = -z$, whose flow is given by $z(t) = e^{-t}z(0)$. If $x(t)$ and $y(t)$ are 1-periodic, then so is $z(t)$. Consequently, $z(0) = 0$ and thus $z(t) = 0$ for all t. This establishes the uniqueness of $\Phi(t)$. Its asymptotic stability also follows from the form of the flow of $\dot{z} = -z$. $\diamondsuit$

Since Eq. (4.9) is linear, it may appear that the latter argument above is unnecessarily convoluted. After all, one could use the explicit formula given in Example 4.6. Although no such formula exists for nonlinear equations, qualitative reasoning similar to the one above, however, can be used successfully to exhibit 1-periodic solutions of certain nonlinear 1-periodic differential equations. We now give an instance of such a situation.

Example 4.18. *Qualitative analysis of a nonlinear equation:* Consider the nonlinear differential equation

$$\dot{x} = -x^3 + c(t), \qquad c(t + 1) = c(t), \tag{4.12}$$

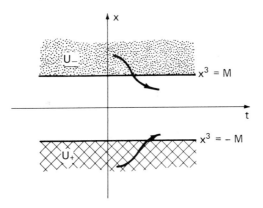

Figure 4.9. *Regions U_- and U^+ where solutions of $\dot{x} = -x^3 + c(t)$ are decreasing and increasing.*

where $c(t)$ is an arbitrary continuous function of period 1. Since $c(t)$ is bounded, there is a positive constant M such that $|c(t)| \le M$ for all $t \in \mathbb{R}$. Thus, any solution $\varphi(t, 0, x_0)$ of Eq. (4.12) satisfies the inequality

$$-\varphi^3 - M \le \dot{\varphi} \le -\varphi^3 + M \quad \text{for all } t.$$

From the form of this inequality, it is natural to determine the sets where $\dot{\varphi} > 0$ and $\dot{\varphi} < 0$, that is, the subsets of the (t, x)-plane defined by

$$U_+ = \{\, (t, x) : -x^3 - M > 0 \,\}, \qquad U_- = \{\, (t, x) : -x^3 + M < 0 \,\}.$$

A solution is increasing in U_+ and decreasing in U_- as t increases; hence, it remains bounded for all $t \ge 0$; see Figure 4.9. Therefore, by Theorem 4.11, any solution φ approaches a periodic solution. Consequently, there exists at least one 1-periodic solution $\Phi(t)$ of Eq. (4.12). The same conclusion can be stated in terms of the Poincaré map $\Pi(x_0)$. Notice that when $|x_0|$ is large, we have $\Pi(x_0) > x_0$ if $x_0 < 0$, and $\Pi(x_0) < x_0$ if $x_0 > 0$. Therefore, the graph of Π must cross the diagonal, that is, Π has a fixed point corresponding to a 1-periodic solution $\Phi(t)$.

We now show the uniqueness of Φ. If we let $y(t) \equiv x(t) - \Phi(t)$, then Eq. (4.12) becomes

$$\dot{y} = -(\Phi + y)^3 + \Phi^3 = -y\{(\Phi + y)^2 + (\Phi + y)\Phi + \Phi^2\}.$$

The expression inside the braces is a positive definite quadratic form in Φ and $\Phi + y$, that is, it is nonnegative and vanishes only when both variables are zero. If you have no previous knowledge of quadratic forms, consider the expression inside the braces as a function of two independent variables $u = \Phi + y$ and $v = \Phi$, and find its zeros. This shows that the expression inside the braces is positive if $y \neq 0$. This implies that $y(t) \to 0$ as $t \to +\infty$. Therefore, the 1-periodic solution $\Phi(t)$ is asymptotically stable and every solution of Eq. (4.12) tends to $\Phi(t)$ as $t \to +\infty$. Thus, $\Phi(t)$ is the unique 1-periodic solution of Eq. (4.12). This result implies that the Poincaré map of Eq. (4.12) looks qualitatively like that of Example 4.11; see Figure 4.7. $\diamondsuit$

Example 4.19. *In a fluctuating environment:* In this example we investigate a generalization of the logistic equation $\dot{x} = ax(1 - x)$. Let $r(t)$, $k(t)$ be 1-periodic, continuous, positive functions and consider the equation

$$\dot{x} = r(t)\, x \left[1 - \frac{x}{k(t)} \right].$$

One can regard this differential equation as a model for the growth of a population where the intrinsic growth rate $r(t)$ and the carrying capacity $k(t)$ exhibit periodic (for example, seasonal) fluctuations.

Let us first show that this equation has at least two 1-periodic solutions. The trivial solution $x(t) \equiv 0$ is obviously a 1-periodic solution. To look for a second solution, observe that if $x(t)$ is a solution with $x(0) < 0$, then $\dot{x}(t) < 0$ for all $t \geq 0$ and thus there is no 1-periodic solution with negative initial data. Consequently, we must show that there is a 1-periodic solution with positive initial data. To this end, let $K_m \leq k(t) \leq K_M$ for some constants K_m, K_M. Then for any solution $x(t)$ satisfying $x(t) > K_M$, we have $\dot{x}(t) < 0$. Also, if $x(t)$ is a solution satisfying $0 < x(t) < K_m$, then $\dot{x}(t) > 0$. Therefore, any solution $x(t)$ with initial data $0 < x(0) < K_m$ must be bounded for all $t \in \mathbb{R}$, approach zero as $t \to -\infty$, and approach a 1-periodic solution $\Phi(t)$ as $t \to +\infty$. Furthermore, this 1-periodic solution has the property $K_m \leq \Phi(t) \leq K_M$.

It is possible to show that $\Phi(t)$ is the only 1-periodic solution with positive initial data; hence, it is asymptotically stable. It is not entirely trivial to prove this fact; see the exercises. $\diamondsuit$

We end this section with some remarks and problems related to possible generalizations of the examples above. For instance, the *Riccati* equation with periodic coefficients,

$$\frac{dx}{d\tau} = \hat{b}(\tau) + \hat{a}(\tau)x - \hat{c}(\tau)x^2,$$

where $\hat{a}$, $\hat{b}$, and $\hat{c}$ are continuous, 1-periodic functions with $\hat{c}(\tau) > 0$ for all τ, generalizes Example 4.19. The form of this equation can be simplified somewhat if we change the independent variable τ to t through the formula $\tau = \int_0^t \hat{c}^{-1}(s)ds$. Then the Riccati equation becomes

$$\dot{x} = b(t) + a(t)x - x^2, \tag{4.13}$$

where $a = \hat{a}/\hat{c}$ and $b = \hat{b}/\hat{c}$. If the initial data for Eq. (4.13) is very large in absolute value, then the solution is decreasing. In the case where the 1-periodic solutions are hyperbolic [that is, $\Pi'(x_0) \neq 1$], there must be a finite number of them because they are isolated; see the exercises. Furthermore, the number is even. It is possible to prove that there are no more than two such solutions; see the exercises.

Consider the following generalization of Example 4.18:

$$\dot{x} = -x^3 + d(t)x + c(t), \qquad c(t+1) = c(t), \quad d(t+1) = d(t), \qquad (4.14)$$

where $c(t)$ and $d(t)$ are arbitrary continuous functions of period 1. Using arguments similar to the ones given in Example 4.17, one can show that every solution of Eq. (4.14) is bounded for $t \geq 0$. However, the argument for uniqueness is no longer valid for this generalization. In fact, we saw in Section 2.1 that for c and d constants the uniqueness did not hold and there were sometimes three solutions. If the 1-periodic solutions of Eq. (4.14) are hyperbolic, then there must be an odd number of solutions, but are there no more than three 1-periodic solutions? The answer is yes, but a complete discussion of Eq. (4.14) is difficult; special cases are contained in the exercises.

Exercises _____

4.16. Show that the equation $\dot{x} = -x^5 + c(t)$, where $c(t)$ is a continuous 1-periodic function, has a unique 1-periodic solution and is asymptotically stable.

4.17. If $c(t)$, $d(t)$, and $e(t)$ are continuous 1-periodic functions, show that the equation

$$\dot{x} = -x^3 + c(t)x^2 + d(t)x + e(t)$$

has at least one 1-periodic solution. Also, if this 1-periodic solution is unstable, show that there must be another 1-periodic solution.
Hint: Show that $\dot{x} < 0$ if x is large enough, and $\dot{x} > 0$ if $-x$ is large enough.

4.18. Suppose that f is a C^1 function with $f'(x) > 0$ for all x and satisfying $f(x) \to +\infty$ as $x \to +\infty$, and $f(x) \to -\infty$ as $x \to -\infty$. Show that the differential equation $\dot{x} = -f(x) + c(t)$, where $c(t)$ is a continuous 1-periodic function, has a unique 1-periodic solution which is asymptotically stable.
Hint: Use the Mean Value Theorem.

4.19. *Hyperbolic is isolated:* A 1-periodic solution $\varphi(t, 0, x_0)$ is called *hyperbolic* if $\Pi'(x_0) \neq 1$. Show that hyperbolic 1-periodic solutions are isolated from other 1-periodic solutions, that is, the corresponding hyperbolic fixed points of the Poincaré map are isolated.

4.20. *Riccati Equation:* If $a(t)$ and $b(t)$ are 1-periodic continuous functions, prove that the Riccati equation

$$\dot{x} = b(t) + a(t)x - x^2$$

has at most two 1-periodic solutions.

Hint: Suppose that $\varphi(t)$ is a 1-periodic solution and introduce the transformation of the variable $x = \varphi + y$. Then in the new variable y, the Riccati equation becomes

$$\dot{y} = c(t)y - y^2, \qquad c(t) = a(t) - 2\varphi(t).$$

If we further let $w = y^{-1}$, then $\dot{w} = -c(t)w + 1$. Now, discuss separately the two cases $\int_0^1 c(t)\,dt \neq 0$ and $\int_0^1 c(t)\,dt = 0$ using the Fredholm Alternative.

4.21. *Periodic Logistic Equation:* In Example 4.19, we observed that the logistic equation $\dot{x} = r(t)x[1 - x/k(t)]$ with 1-periodic coefficients $r(t) > 0$ and $k(t) > 0$ has as least one positive 1-periodic solution. Use the suggestion below to establish that there is exactly one such solution.

Hint: Suppose that $\bar{x}(t)$ and $\hat{x}(t)$ are two positive 1-periodic solutions with $\bar{x}(t) - \hat{x}(t) > 0$ for all $t \in [0, 1]$. Let $v(t) = \bar{x}(t) - \hat{x}(t)$. Then show that

$$\dot{v} = r(t)\,\bar{x}(t)\left[\hat{x}(t) - \bar{x}(t)\right]\bar{x}(t)/k(t) + r(t)\left[1 - \hat{x}(t)/k(t)\right]v$$
$$< r(t)[1 - \hat{x}(t)/k(t)]v$$

and

$$v(1) < v(0)\exp\left\{\int_0^1 r(s)(1 - \hat{x}(s)/k(s))\,ds\right\}.$$

Since $\hat{x}(t)$ is a 1-periodic solution of the equation, the exponential term is equal to 1 and thus $v(1) < v(0)$, which is a contradiction.

4.22. *Generalized Logistic:* Consider the differential equation $\dot{x} = xf(t, x)$, where f satisfies the following ecologically reasonable conditions: C^1, 1-periodic in t, decreasing in x for $x > 0$ and all t, and $M > 0$ such that $f(t, x) < 0$ for $x \geq M$ and all t. Prove that
   (a) If $x(0) \geq 0$, then $x(t) \geq 0$ for all t.
   (b) If $\int_0^1 f(t, 0)\,dt > 0$, then there is a unique positive 1-periodic solution to which any other solution with $x(0) > 0$ approaches.
Hint: Use the suggestions in the previous exercise. For further information, see de Mottoni and Schiaffino [1981].

4.23. *Periodic Harvesting:* Consider the logistic equation with harvesting:

$$\dot{x} = r(t)x\left[1 - \frac{x}{k(t)}\right] - h(t)x,$$

where $h(t)$ is a continuous, nonnegative, 1-periodic function. Furthermore, assume that $r(t) - h(t) > 0$ so that the harvested population will not eventually die out. As usual, $r(t)$ and $k(t)$ are continuous, positive, and 1-periodic. Show that there is a 1-periodic solution $\bar{x}(t)$, which is also unique, with the property

$$\frac{r_m - h_M}{r_M}k_m \leq \bar{x}(t) \leq \frac{r_M - h_m}{r_m}k_M,$$

where the subscripts m and M denote the minima and maxima of the appropriate functions, respectively.

Hint: Note that the equation can be written as

$$\dot{x} = [r(t) - h(t)]\, x \left[1 - \frac{x}{\frac{r(t)-h(t)}{r(t)}k(t)}\right].$$

For further information on periodic harvesting, see Sanchez [1982].

4.5. Stability of Periodic Solutions

We have remarked in Section 4.2 that a 1-periodic solution $\varphi(t, 0, x_0)$ of Eq. (4.4) corresponds to a fixed point x_0, $\Pi(x_0) = x_0$, of the Poincaré map Π. Furthermore, the stability properties of φ are the same as those of the fixed point x_0. The Poincaré map is differentiable (see the Appendix) and also monotone nondecreasing; thus, $\Pi'(x_0) \geq 0$ for all x_0. Consequently, it follows from Theorem 3.8 that φ is asymptotically stable if $\Pi'(x_0) < 1$, and unstable if $\Pi'(x_0) > 1$. It may appear that to use this result one would need an explicit formula for the Poincaré map which is a difficult object to compute unless the general solution of Eq. (4.4) is available. To circumvent this difficulty, in this section we derive a formula for the derivative of the Poincaré map in terms of only the 1-periodic solution φ and the vector field $f(t, x)$. In doing so, we will also discover some other properties of differential equations which are of independent interest.

Lemma 4.20. *If $\varphi(t, 0, x_0)$ is the solution of a 1-periodic equation $\dot{x} = f(t, x)$ with $\varphi(0, 0, x_0) = x_0$, then $\partial\varphi(t, 0, x_0)/\partial x_0$ is the solution of the following initial-value problem for a linear differential equation:*

$$\dot{z} = \frac{\partial f}{\partial x}(t, \varphi(t, 0, x_0))\, z, \qquad z(0) = 1, \tag{4.15}$$

that is,

$$\frac{\partial\varphi}{\partial x_0}(t, 0, x_0) = \exp\left[\int_0^t \frac{\partial f}{\partial x}(s, \varphi(s, 0, x_0))\, ds\right]. \tag{4.16}$$

Proof. The solution $\varphi(t, 0, x_0)$ is given by

$$\varphi(t, 0, x_0) = x_0 + \int_0^t f(s, \varphi(s, 0, x_0))\, ds.$$

Differentiating both sides of this equation with respect to x_0, and using the chain rule, yields

$$\frac{\partial\varphi}{\partial x_0}(t, 0, x_0) = 1 + \int_0^t \frac{\partial f}{\partial x}(s, \varphi(s, 0, x_0))\, \frac{\partial\varphi}{\partial x_0}(s, 0, x_0)\, ds.$$

Now, if we let $z(t) \equiv \partial \varphi(t, 0, x_0)/\partial x_0$, then the equation above becomes

$$z(t) = 1 + \int_0^t \frac{\partial f}{\partial x}(s, \varphi(s, 0, x_0)) z(s) \, ds.$$

By differentiating this equation with respect to t, we obtain the linear differential equation (4.15). Also, since at $t = 0$ the integral is zero, we have the initial value $z(0) = 1$. $\diamond$

The differential equation (4.15) is called the *linear variational equation* about the solution $\varphi(t, 0, x_0)$. The reason for this terminology is the following: if we consider the "variation" $z(t)$ about the solution $\varphi(t, 0, x_0)$, that is, $x(t) = \varphi(t, 0, x_0) + z(t)$, then we obtain

$$\dot{z}(t) = f(t, \varphi(t, 0, x_0) + z(t)) - f(t, \varphi(t, 0, x_0)). \qquad (4.17)$$

Retaining only the first term of the Taylor expansion of this vector field yields the differential equation (4.15).

Lemma 4.21. *Let $\varphi(t, 0, x_0)$ be the solution of a 1-periodic equation $\dot{x} = f(t, x)$ with $\varphi(0, 0, x_0) = x_0$ and Π be the Poincaré map. Then the derivative of the Poincaré map is given by*

$$\Pi'(x_0) = \exp\left[\int_0^1 \frac{\partial f}{\partial x}(t, \varphi(t, 0, x_0)) \, dt\right]. \qquad (4.18)$$

Proof. From the definition of the Poincaré map, $\Pi(x_0) = \varphi(1, 0, x_0)$. Differentiating both sides with respect to x_0 yields

$$\Pi'(x_0) \equiv \frac{d\Pi}{dx_0}(x_0) = \frac{\partial \varphi}{\partial x_0}(1, 0, x_0).$$

Now, the conclusion of the lemma follows from Eq. (4.16). $\diamond$

Theorem 4.22. *Let $\varphi(t, 0, x_0)$ be a 1-periodic solution of the 1-periodic equation (4.4) and define*

$$a_0 \equiv \int_0^1 \frac{\partial f}{\partial x}(t, \varphi(t, 0, x_0)) \, dt.$$

Then,
(i) $\varphi(t, 0, x_0)$ is asymptotically stable if $a_0 < 0$, and
(ii) $\varphi(t, 0, x_0)$ is unstable if $a_0 > 0$.

Proof. Since $\varphi(t, 0, x_0)$ is a 1-periodic solution, the point x_0 is a fixed point of the Poincaré map Π and, from Lemma 4.21, $\Pi'(x_0) = e^{a_0}$. Thus, the conclusion follows from Theorem 3.6 because $e^{a_0} < 1$ if $a_0 < 0$, and $e^{a_0} > 1$ if $a_0 > 0$. $\diamond$

Analogous to Definition 3.10 for the case of autonomous equations, we introduce the following terminology for 1-periodic solutions satisfying the hypotheses of the theorem above.

Definition 4.23. *A periodic solution* $\varphi(t, 0, x_0)$ *of* (4.4) *is called hyperbolic if* $\Pi'(x_0) \neq 1$.

We conclude this chapter with examples of hyperbolic and nonhyperbolic 1-periodic solutions. You also may wish to determine if some of the 1-periodic solutions of earlier examples are hyperbolic.

Example 4.24. *A hyperbolic periodic solution:* Consider the differential equation

$$\dot{x} = -x^3 + \sin^3 2\pi t + 2\pi \cos 2\pi t$$

and its 1-periodic solution $\varphi(t, 0, 0) = \sin 2\pi t$. Notice that this equation is a particular instance of Example 4.13. A simple computation yields

$$a_0 = \int_0^1 -3\sin^2 2\pi t \, dt = -\frac{3}{2} < 0.$$

Therefore, it follows from Theorem 4.21 that the 1-periodic solution is asymptotically stable. ◊

Example 4.25. *A nonhyperbolic periodic solution:* Consider the differential equation

$$\dot{x} = x^2 - \cos^2 2\pi t - 2\pi \sin 2\pi t \qquad (4.19)$$

and its 1-periodic solution $\varphi(t, 0, 1) = \cos 2\pi t$. This periodic solution corresponds to the fixed point of the Poincaré map at $x_0 = 1$. A simple computation shows that

$$\Pi'(1) = e^{a_0} = \exp\left[\int_0^1 2\cos 2\pi t \, dt\right] = 1.$$

Therefore, the periodic solution is not hyperbolic and thus its stability type cannot be determined from the linearization of the Poincaré map. We will return to this example in the next chapter. ◊

Exercises ―――――――――――――――――――――――――― ♣ ♡ ♠ ◊

4.24. Verify that the differential equation

$$\dot{x} = x^2 - \frac{1}{4\pi^2}\cos^4 2\pi t - \sin 4\pi t$$

has the 1-periodic solution $\varphi(t, 0, 1/(2\pi)) = [1/(2\pi)]\cos^2 2\pi t$. Show that this solution is hyperbolic and $\Pi'(1/(2\pi)) = \exp(1/(2\pi))$. There is another 1-periodic solution which is asymptotically stable. Can you find it explicitly? Are there any more 1-periodic solutions?
Hint: See the hint for the Riccati equation.

4.25. Verify that the differential equation

$$\dot{x} = -x^3 + 2x + \sin^3 2\pi t - 2\sin 2\pi t + 2\pi \cos 2\pi t$$

has the 1-periodic solution $\varphi(t, 0, 0) = \sin 2\pi t$. Show that this solution is unstable with $\Pi'(0) = \exp(1/2)$. How many more 1-periodic solutions can you guarantee?

4.26. Suppose that $a(t)$ is 1-periodic with $0 < a(t) < 1$. Verify that the equation $\dot{x} = x(1-x)[x - a(t)]$ has at least three 1-periodic solutions.
Hint: Show that $x(t) = 0$ and $x(t) = 1$ are asymptotically stable and therefore there is a 1-periodic solution in the interval $(0, 1)$.

Bibliographical Notes

As we have seen in this chapter, the qualitative behavior of the solutions of scalar 1-periodic differential equations is nearly as simple as the behavior of scalar autonomous equations once the dynamics is reduced to that of the Poincaré map. If equilibria are replaced by the fixed points of the Poincaré map, the orbit structure is easy to classify because the Poincaré map is monotone. On the other hand, determining the precise number of 1-periodic solutions for a given 1-periodic equation is not an easy task because the differential equation must be integrated to obtain the Poincaré map. For a given polynomial vector field with 1-periodic coefficients, the maximal number of its 1-periodic solutions is not known. Pliss [1966] gives an example of a fourth degree polynomial with at least five 1-periodic solutions. See also Lloyd [1972] for an extensive discussion of these matters.

An important extension of the theory of 1-periodic scalar equations is to take $f(t, x)$ to be quasiperiodic in t; for example, $f(t, x) = \sin t + \sin \sqrt{2} t$. Although there is an extensive theory for such equations (see, for example, Fink [1974], Sell [1971], and Yoshizawa [1974]), very little is known about the number of quasiperiodic solutions that can exist. If $f(t, x) = g(x) + c(t)$ and $g'(x) \neq 0$ for all x, there is a unique quasiperiodic solution, nicely generalizing the result in Section 4.3. On the Riccati equation with quasiperiodic coefficients, see Johnson and Sell [1981].

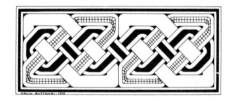

5

Bifurcations of Periodic Equations

In this chapter, we investigate the bifurcation behavior of periodic solutions of 1-periodic differential equations. We first point out that, in terms of Poincaré maps, the study of local bifurcations of periodic solutions is equivalent to the study of bifurcations of fixed points of monotone maps given in Section 3.3. Next, we develop selected ideas from the "method of averaging" and show how to compute higher-order derivatives of the Poincaré map about a fixed point. For practical purposes, we accomplish this solely in terms of the vector field and the 1-periodic solution. We then employ these results to determine local bifurcations of nonhyperbolic 1-periodic solutions of 1-periodic differential equations depending on a scalar parameter. We should warn you that this chapter, by necessity, is more technical than the previous ones.

5.1. Bifurcations of Poincaré Maps

Let us consider a sufficiently differentiable perturbation of a 1-periodic differential equation $\dot{x} = f(t, x)$ given by

$$\dot{x} = F(\lambda, t, x), \tag{5.1}$$

where

$$F : \mathbb{R}^k \times \mathbb{R} \times \mathbb{R} \to \mathbb{R}; \quad (\lambda, t, x) \mapsto F(\lambda, t, x)$$

satisfying

$$F(0, t, x) = f(t, x) \quad \text{and} \quad F(\lambda, t+1, x) = F(\lambda, t, x).$$

Let $\Pi(\lambda, x)$ denote the Poincaré map of the perturbed equation (5.1). It is clear that, regardless of the specific form of Eq. (5.1), its Poincaré map is a monotone map. Thus, to study the local bifurcations of a 1-periodic solution $\varphi(t, 0, x_0)$ of $\dot{x} = f(t, x)$, we need only to recall the results in Section 3.3 for $\Pi(\lambda, x)$ in the neighborhood of the fixed point x_0 for sufficiently small values of the parameter λ.

It may appear that this remark is of no practical use unless an explicit formula for $\Pi(\lambda, x)$ is available, which is rarely the case. Luckily, it is possible to circumvent the need for such a formula. For instance, we have already seen in Section 4.5 how to compute the first derivative of the Poincaré map at a fixed point in terms of the 1-periodic solution and the vector field. Consequently, if the 1-periodic solution is hyperbolic, it is easy to conclude that there will be no local bifurcations. But, what if the 1-periodic solution is not hyperbolic? We now turn to this difficult question which will occupy our attention for the remainder of this chapter.

Exercises ———————————————————————— ♣ ♡ ♠ ◇

5.1. Consider the autonomous differential equation $\dot{x} = f(x)$ with an equilibrium point $\bar{x}$, that is, $f(\bar{x}) = 0$. When such an autonomous differential equation is viewed, trivially, as a 1-periodic differential equation in t, the equilibrium point of the autonomous equation becomes a 1-periodic solution of the 1-periodic equation. Show that, as such, the Poincaré map $\Pi(x_0) = \varphi(1, 0, x_0)$ of the 1-periodic differential equation satisfies $\Pi(\bar{x}) = \bar{x}$ and $\Pi'(\bar{x}) = e^{f'(\bar{x})}$.

5.2. Show that the differential equation $\dot{x} = x^2 - 1 + \lambda \cos 2\pi t$ has at least two 1-periodic solutions for each λ when $|\lambda|$ is sufficiently small. Using the linear variational equation, discuss the stability properties of these 1-periodic solutions. Finally, show that there are exactly two 1-periodic solutions for each λ when $|\lambda|$ is sufficiently small.

5.3. Consider the differential equation $\dot{x} = a_0 + c_0 x^2 + G(\lambda, t, x)$, where λ is a scalar parameter, a_0 and c_0 are constants satisfying $a_0 c_0 < 0$, and

G is a continuous 1-periodic function in t with $|G(\lambda, t, x)| \leq |\lambda|$ for all $(t, x) \in \mathbb{R} \times \mathbb{R}$. Show that there is a $\lambda_0 > 0$ such that, for $|\lambda| < \lambda_0$, there are exactly two 1-periodic solutions $\psi_1(\lambda, t)$ and $\psi_2(\lambda, x)$ which have the properties $\psi_1(0, t) = \sqrt{-a_0/c_0}$ and $\psi_2(0, t) = -\sqrt{-a_0/c_0}$. Show also that these 1-periodic solutions are hyperbolic and that ψ_1 is unstable and ψ_2 is asymptotically stable if $c_0 > 0$.

Hint: Use the first exercise above and the properties of the flow when $\lambda = 0$.

5.4. Consider the differential equation

$$\dot{x} = -x^3 + (d + \lambda \cos 2\pi t)x + (c + \lambda \sin 2\pi t),$$

where c and d are constants, and λ is a parameter.

(i) Suppose that for some values of c, d, and $\lambda = 0$ there is a unique equilibrium point which is hyperbolic. Show that with the same values of c and d there is a unique 1-periodic solution for each $|\lambda|$ small.

(ii) For given values of c, d, and $\lambda = 0$, suppose that there are three equilibrium points. Show that for these values of c and d there are three 1-periodic solutions for each $|\lambda|$ small.

(iii) Discuss the stability properties of these 1-periodic solutions.

5.5. Consider the differential equation $\dot{x} = f(x) + \lambda g(t, x)$, where the function g is continuous and 1-periodic with $|g(t, x)| \leq 1$, and the function f satisfies $f(\bar{x}) = 0$ with $f'(\bar{x}) \neq 0$ for some $\bar{x}$. Show that there are constants $\varepsilon > 0$ and λ_0 such that, for $|\lambda| < \lambda_0$, the differential equation has a 1-periodic solution $\psi(\lambda, t)$ with $\psi(0, t) = \bar{x}$. Moreover, show that this is the only 1-periodic solution within ε of $\bar{x}$.

Hint: Consider the Poincaré map. Use the first exercise above, and some results from Chapter 3.

5.2. Stability of Nonhyperbolic Periodic Solutions

As we have seen in Example 4.25, if a 1-periodic solution of a 1-periodic differential equation (4.4) is not hyperbolic, then the stability type of the solution cannot be determined from the linearization of the Poincaré map using Theorem 4.22. In this case, we need to compute the higher-order derivatives of the Poincaré map at the fixed point corresponding to the periodic solution. The purpose of this section is to show how to do this in terms of the vector field and the 1-periodic solution. We first give the statement of the main result, and then use it to determine the stability of the 1-periodic solution in Example 4.25. Finally, we present the details of the proof. The relief sign $\diamond$ does not appear for some time; however, the proof contains important ideas such as transformation of variables.

Let $\varphi(t, 0, x_0)$ be a 1-periodic solution of Eq. (4.4) and consider the variation $z(t)$ about $\varphi(t, 0, x_0)$, that is,

$$x(t) = \varphi(t, 0, x_0) + z(t). \tag{5.2}$$

Then the first several terms of the variational differential equation (4.17) for $\dot{z}$ will have the form

$$\dot{z} = b(t)z + \hat{c}(t)z^2 + \hat{d}(t)z^3 + O(z^4),\tag{5.3}$$

where the coefficients are 1-periodic functions, and the notation $O(z^4)$ denotes the terms of order z^4 and higher; see the Appendix. Notice that the study of the neighborhood of the zero solution of Eq. (5.3) is equivalent to the study of the neighborhood of the 1-periodic solution φ. Now, let us assume that φ is nonhyperbolic, that is, $\Pi'(0) = 1$, or equivalently $\int_0^1 b(s)\,ds = 0$. Then the function $y(t) \equiv \int_0^t b(s)\,ds$ is 1-periodic in t. To wit, consider $y(t+1) - y(t) = \int_t^{t+1} b(s)\,ds$. Since the integral over a period of a periodic function is invariant under translation, $\int_t^{t+1} b(s)\,ds = \int_0^1 b(s)\,ds = 0$. Thus, $y(t)$ is 1-periodic.

In order to eliminate the linear term in the variational equation (5.3), introduce the 1-periodic change of variable $u(t)$ defined by

$$z(t) = e^{\int_0^t b(s)\,ds}\, u(t).\tag{5.4}$$

The differential equation (5.3) for the new variable u will have the form

$$\dot{u} = c(t)u^2 + d(t)u^3 + O(u^4),\tag{5.5}$$

where $c(t)$ and $d(t)$ are 1-periodic functions given by

$$c(t) = e^{\int_0^t b(s)ds}\,\hat{c}(t) \quad \text{and} \quad d(t) = e^{2\int_0^t b(s)ds}\,\hat{d}(t).\tag{5.6}$$

With this notation, we now state our main theorem.

Theorem 5.1. *Suppose that x_0 is a fixed point of the Poincaré map Π with $\Pi'(x_0) = 1$ of a 1-periodic differential equation (4.4). Then*

(i) *The second derivative of Π at the fixed point is given by*

$$\Pi''(x_0) = 2\int_0^1 c(t)\,dt \equiv 2c_0,$$

where the 1-periodic function $c(t)$ is as in Eq. (5.6). Thus, the fixed point at x_0; hence, the corresponding periodic solution $\varphi(t, 0, x_0)$ of Eq. (4.4), is unstable if $c_0 \neq 0$.

(ii) *If $c_0 = 0$, then the third derivative of Π at the fixed point x_0 is given by*

$$\Pi'''(x_0) = 6\int_0^1 d(t)\,dt \equiv 6d_0,$$

where $d(t)$ is as in Eq. (5.6). Thus, the fixed point at x_0, hence, the corresponding periodic solution $\varphi(t, 0, x_0)$ of Eq. (4.4), is unstable if $d_0 > 0$ and asymptotically stable if $d_0 < 0$.

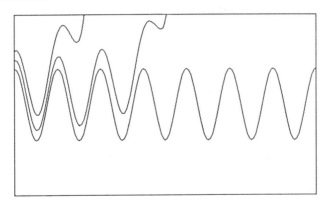

Figure 5.1. *The nonhyperbolic 1-periodic solution* $\varphi(t, 0, 1) = \cos 2\pi t$ *of the differential equation* $\dot{x} = x^2 - \cos^2 2\pi t - 2\pi \sin 2\pi t$ *is unstable.*

Before embarking on the difficult proof of this theorem, let us return to Example 4.25 and determine whether its nonhyperbolic 1-periodic solution is asymptotically stable or unstable.

Example 5.2. *Continuation of Example 4.25.* Consider the 1-periodic differential equation $\dot{x} = x^2 - \cos^2 2\pi t - 2\pi \sin 2\pi t$ and its nonhyperbolic 1-periodic solution $\varphi(t, 0, 1) = \cos 2\pi t$. If we let $x(t) = (\cos 2\pi t) + z(t)$, then the variational equation is given by

$$\dot{z} = 2(\cos 2\pi t)z + z^2.$$

Since $\int_0^t 2(\cos 2\pi s)ds = (\sin 2\pi t)/\pi$, we introduce the new variable $u(t)$ defined by

$$z(t) = e^{(\sin 2\pi t)/\pi} u(t)$$

and obtain the new differential equation

$$\dot{u} = e^{(\sin 2\pi t)/\pi} u^2.$$

Now, observe that

$$c_0 = \int_0^1 e^{(\sin 2\pi t)/\pi} dt > 0.$$

Thus, it follows from the theorem above that the nonhyperbolic 1-periodic solution $\varphi(t, 0, 1) = \cos 2\pi t$ of Example 4.25 is unstable; see Figure 5.1. $\Diamond$

Proof of Theorem 5.1. The key idea of the proof is to reduce the variational equation (5.3) to a differential equation with constant coefficients by using successive transformation of variables. In the choice of the transformations,

however, care must be taken. The transformations must be 1-periodic in
t so as to have the differential equation in the transformed variables be
1-periodic.

As a consequence of the assumption $\int_0^1 b(s)ds = 0$, the nonhyperbolicity of φ, we have shown that $\int_0^t b(s)\,ds$ is 1-periodic. Therefore, the function
$\exp\{\int_0^t b(s)\,ds\}$ in the transformation (5.4) is 1-periodic. Consequently, the
functions $c(t)$ and $d(t)$ in Eq. (5.4) are also 1-periodic.

Now, our objective is to determine one final transformation of the
variables to convert Eq. (5.5) to a differential equation with constant coefficients in the lower-order terms. The form of Eq. (5.5) suggests that we
consider the transformation

$$u(t) = w(t) + \beta(t)w^2(t) + \gamma(t)w^3(t), \tag{5.7}$$

where $w(t)$ is the new variable and $\beta(t)$ and $\gamma(t)$ are 1-periodic functions to
be determined. For simplicity of notation, in the formulae below, we will
omit the variable t if there is no danger of confusion. A few calculations
show that the differential equation for w is given by

$$\dot{w} = (1 + 2\beta w + 3\gamma w^2)^{-1}\left[(-\dot\beta + c)w^2 + (-\dot\gamma + d + 2\beta c)w^3 + O(w^4)\right].$$

Using the series expansion $(1 - x)^{-1} = 1 + x + x^2 + \cdots$ for $|x| < 1$, the
differential equation above for $\dot{w}$ can be simplified to the following form:

$$\dot{w} = (-\dot\beta + c)w^2 + (-\dot\gamma + d + 2\beta\dot\beta)w^3 + O(w^4). \tag{5.8}$$

Let us now choose β and γ in a such a way that the differential equation (5.8) becomes as simple as possible. It is tempting to put $\dot\beta = c$
so that the coefficient of the w^2 term will vanish. Unfortunately, this
choice makes $\beta(t)$ a 1-periodic function if and only if $c_0 = \int_0^1 c(s)ds = 0$.
In fact, the solutions of $\dot\beta = c$ are given by $\beta(t) = \beta_0 + \int_0^t c(s)\,ds$ and
$\beta(t + 1) = \beta(t)$ if and only if $0 = \int_t^{t+1} c(s)\,ds = \int_0^1 c(s)\,ds$. However, if
we chose $\beta(t) = (constant) + \int_0^t \left[c(s) - c_0\right]ds$, that is, $\beta(t)$ satisfies the
differential equation

$$\dot\beta = c(t) - c_0, \tag{5.9}$$

then $\beta(t)$ is 1-periodic. So, we choose β satisfying the differential equation (5.9). We may also require that $\int_0^1 \beta(s)\,ds = 0$.

Suppose now that β has been chosen as above. To simplify the coefficient of the w^3 term in Eq. (5.8), we can determine the function $\gamma(t)$,
satisfying $\int_0^1 \gamma(s)\,ds = 0$, from the differential equation

$$\dot\gamma = m(t) - \mu, \tag{5.10}$$

where

$$m = d + 2\beta\dot{\beta}, \qquad \mu = \int_0^1 m(s)\,ds.$$

We note that

$$\mu = d_0 \equiv \int_0^1 d(s)\,ds \tag{5.11}$$

since $\mu - d_0 = \int_0^1 2\beta\dot{\beta}\,dt = 2\int_{\beta(0)}^{\beta(1)} \beta\,d\beta$ and $\beta(0) = \beta(1)$.

For the choices of $\beta(t)$ and $\gamma(t)$ given by Eqs. (5.9) and (5.10), respectively, the transformation of variables

$$u = w + \beta(t)w^2 + \gamma(t)w^3 \tag{5.12}$$

when applied to Eq. (5.5) yields the following differential equation we have been seeking:

$$\dot{w} = c_0 w^2 + \mu w^3 + O(w^4). \tag{5.13}$$

For initial data w_0 small, the Poincaré map of Eq. (5.13) can be approximated by computing the Taylor expansion of $w(t)$ in w_0 near $w_0 = 0$. We now indicate how to do this for the first two terms. The solution of Eq. (5.13) satisfying $w(0) = w_0$ is given by

$$w(t) = w_0 + \int_0^t \left[c_0 w^2(s) + O(w^3)\right] ds.$$

Thus, using the chain rule and changing the order of integration and differentiation we have

$$\frac{\partial w(t)}{\partial w_0} = 1 + \int_0^t \left[2c_0 w(s)\frac{\partial w(s)}{\partial w_0} + O(w^2)\right] ds. \tag{5.14}$$

Evaluating Eq. (5.14) at $w_0 = 0$ yields

$$\frac{\partial w(t)}{\partial w_0} = 1. \tag{5.15}$$

Differentiating Eq. (5.14) once more gives the second derivative

$$\frac{\partial^2 w(t)}{\partial w_0^2} = \int_0^t \left[2c_0\left(\frac{\partial w(s)}{\partial w_0}\right)^2 + 2c_0 w(s)\frac{\partial^2 w(s)}{\partial w_0^2} + O(w^2)\right] ds.$$

Evaluating this at $w_0 = 0$ and using the fact (5.15) results in

$$\frac{\partial^2 w(t)}{\partial w_0^2} = 2c_0 t.$$

With similar computations, the third-order derivative of w with respect to w_0 evaluated at $w_0 = 0$ can be seen to be

$$\frac{\partial^3 w(t)}{\partial w_0^3} = 6\mu t.$$

Thus the Taylor series of $w(t)$ in w_0 near $w_0 = 0$ starts as

$$w(t) = w_0 + \frac{1}{2!} 2 c_0 t w_0^2 + \frac{1}{3!} 6 \mu t w_0^3 + O(w_0^4).$$

Finally, we evaluate this series at $t = 1$ to obtain the Poincaré map of Eq. (5.13):

$$\Pi(w_0) = w_0 + c_0 w_0^2 + \mu w_0^3 + O(w_0^4). \tag{5.16}$$

The transformation (5.12) implies that

$$u_0 = w_0 + \beta(0) w_0^2 + \gamma(0) w_0^3.$$

If we let $\widetilde{\Pi}$ be the Poincaré map of Eq. (5.5), then the transformation (5.12) and the fact that β and γ are 1-periodic imply that, for initial data u_0 small,

$$\begin{aligned}
\widetilde{\Pi}(u_0) &= \Pi(w_0) + \beta(0) \left[\Pi(w_0) \right]^2 + \gamma(0) \left[\Pi(w_0) \right]^3 \\
&= u_0 + c_0 u_0^2 + \mu u_0^3 + O(u_0^4).
\end{aligned} \tag{5.17}$$

Now, the conclusions of the theorem follow immediately if we differentiate Eq. (5.17) and use the fact in Eq. (5.11). $\diamondsuit$

Exercises ——————————————————————— ♣♡♠♢

5.6. Consider the 1-periodic equation

$$\dot{x} = -x^3 + \tfrac{3}{2} x + \sin^3 2\pi t - \tfrac{3}{2} \sin 2\pi t + 2\pi \cos 2\pi t.$$

Verify that $\varphi(t, 0, 0) = \sin 2\pi t$ is a 1-periodic solution. Show that $\Pi'(0) = 1$, $\Pi''(0) = 0$, and $\Pi'''(0) < 0$. Establish that this 1-periodic solution is asymptotically stable. You may wish to experiment numerically to see if there are any additional 1-periodic solutions.

5.7. Show that the zero solution of $\dot{x} = \left[(\cos^2 t)(\sin^2 t) \right] x^3$ is asymptotically stable.

5.8. Show that the zero solution of $\dot{x} = [\sin t] x^2$ is unstable. Notice that Theorem 5.1 is not applicable.

5.3. Perturbations of Vector Fields

In this section, we extend our foregoing computations of higher-order approximations of Poincaré maps to 1-periodic equations depending on a scalar parameter. This extension, in conjunction with the remarks in Section 5.1, will enable us to determine local bifurcations of nonhyperbolic 1-periodic solutions. A specific example will be forthcoming, immediately following our presentation of the general setting.

Let us reconsider the perturbed equation (5.1),

$$\dot{x} = F(\lambda, t, x), \qquad F(0, t, x) = f(t, x), \tag{5.18}$$

with the additional assumptions that λ is a scalar parameter and that the unperturbed 1-periodic equation $\dot{x} = f(t, x)$ has a nonhyperbolic periodic solution $\varphi(t, 0, x_0)$. In terms of the Poincaré map of Eq. (5.18), let us assume that $\Pi(0, x_0) = x_0$, $\Pi'(0, x_0) = 1$, and $\Pi''(0, x_0) = 2c_0 \neq 0$. We now want to study the behavior of solutions of the perturbed equation (5.18) near the fixed point x_0 and $\lambda = 0$ via its Poincaré map $\Pi(\lambda, x_0)$.

Using the transformation theory given in the proof of Theorem 5.2, we can assume that the Taylor expansion of Eq. (5.18) is put in the form

$$\dot{w} = A(\lambda, t) + B(\lambda, t)w + C(\lambda, t)w^2 + O(w^3) \tag{5.19}$$

with

$$A(0, t) = B(0, t) = 0, \quad C(0, t) = c_0 \neq 0.$$

Notice that Eq. (5.19) is a perturbation of Eq. (5.13). The Poincaré map of Eq. (5.19) is of the form

$$\Pi(\lambda, w_0) = \hat{A}(\lambda) + [\hat{B}(\lambda) + 1]w_0 + \hat{C}(\lambda)w_0^2 + O(w_0^3) \tag{5.20}$$

with

$$\hat{A}(0) = \hat{B}(0) = 0, \quad \hat{C}(0) = c_0 \neq 0.$$

The number of fixed points of the Poincaré map (5.20) depends on the local behavior of the functions $\hat{A}(\lambda)$, $\hat{B}(\lambda)$, and $\hat{C}(\lambda)$ near $\lambda = 0$. Thus it becomes important to have a procedure for computing approximate values of these functions. One possible way of accomplishing this is to determine the first several terms of their Taylor expansions.

Suppose for the moment that we can compute the partial derivatives

$$\frac{\partial \Pi}{\partial \lambda}(0, 0) \equiv A_0, \qquad \frac{\partial^2 \Pi}{\partial \lambda \partial w_0}(0, 0) \equiv B_0. \tag{5.21}$$

Then, the first several terms in the Taylor series of the Poincaré map is given by

$$\begin{aligned} \Pi(\lambda, w_0) = A_0\lambda + O(\lambda^2) + \left[1 + \tfrac{1}{2}B_0\lambda + O(\lambda^2)\right]w_0 \\ + \left[c_0 + O(\lambda)\right]w_0^2 + O(w_0^3). \end{aligned} \tag{5.22}$$

The following theorem—saddle-node bifurcation for 1-periodic differential equations—is an easy consequence of the remarks in Section 5.1.

Theorem 5.3. *Suppose that $A_0 c_0 \neq 0$. Then, for λ near zero, the differential equation (5.18) has*
  (i) *no 1-periodic solution if $\lambda A_0 c_0 > 0$;*
  (ii) *one 1-periodic solution if $\lambda A_0 c_0 = 0$;*
  (iii) *two 1-periodic solutions if $\lambda A_0 c_0 < 0$.* $\Diamond$

We now turn to the somewhat difficult task of computing the partial derivatives of Eq. (5.21). In order to give the idea of the technique for calculating such things let us start with a general lemma.

Lemma 5.4. *If $\varphi(\lambda, t, 0, x_0)$ is the solution of Eq. (5.18) with initial data $\varphi(\lambda, 0, 0, x_0) = x_0$, then $\partial \varphi(\lambda, t, 0, x_0)/\partial \lambda$ is the solution of the following initial-value problem for a linear differential equation:*

$$
\dot{z} = \frac{\partial F}{\partial x}(\lambda, t, \varphi(\lambda, t, 0, x_0))z + \frac{\partial F}{\partial \lambda}(\lambda, t, \varphi(\lambda, t, 0, x_0)),
$$
$$
z(0) = 0.
\tag{5.23}
$$

Proof. The proof is similar to that of Lemma 4.20. Observe that

$$
\varphi(\lambda, t, 0, x_0) = x_0 + \int_0^t F(\lambda, s, \varphi(\lambda, s, 0, x_0))\, ds.
$$

Now, differentiate both sides of this equation with respect to λ and then imitate the proof of Lemma 4.20. $\Diamond$

In the case of the perturbed equation (5.19) under study, the lemma above yields the following initial-value problem:

$$
\dot{z} = \left[B(\lambda, t) + 2C(\lambda, t)\varphi(\lambda, t, 0, w_0) + O(\varphi^2) \right] z
$$
$$
+ \frac{\partial A}{\partial \lambda}(\lambda, t) + \frac{\partial B}{\partial \lambda}(\lambda, t)\varphi(\lambda, t, 0, w_0),
\tag{5.24}
$$
$$
z(0) = 0.
$$

This is a nonhomogeneous linear equation and thus can be solved exactly; see Example 4.6. Since $\varphi(0, t, 0, 0) = 0$ and $B(0, t) = 0$, we obtain from the solution of Eq. (5.24) that

$$
\frac{\partial \varphi}{\partial \lambda}(0, t, 0, 0) = \int_0^t \frac{\partial A}{\partial \lambda}(0, s)\, ds.
\tag{5.25}
$$

Therefore,

$$
\frac{\partial \Pi}{\partial \lambda}(0, 0) = \int_0^1 \frac{\partial A}{\partial \lambda}(0, s)\, ds \equiv A_0.
\tag{5.26}
$$

The computation of the mixed partial derivative of the solution φ can also be obtained from the solution of an appropriate initial-value problem. Rather than proving a general lemma, we just give the result for

Eq. (5.19). Differentiating Eq. (5.24) with respect to w_0 and knowing that $z = \partial\varphi(\lambda, t, 0, w_0)/\partial\lambda$, we see that $\partial^2\varphi(\lambda, t, 0, w_0)/\partial\lambda\partial w_0$ is the solution of the initial-value problem (for simplicity of notation, variables are omitted)

$$\dot{u} = \left[B + 2C\varphi + O(\varphi^2)\right]u + 2C\frac{\partial\varphi}{\partial w_0}\frac{\partial\varphi}{\partial\lambda} + \frac{\partial B}{\partial\lambda}\frac{\partial\varphi}{\partial w_0}, \qquad (5.27)$$

$$u(0) = 0.$$

Since, from Lemma 4.20, $\partial\varphi/\partial w_0$ satisfies the differential equation

$$\dot{v} = \left[B + 2C\varphi + O(\varphi^2)\right]v, \qquad (5.28)$$

$$v(0) = 1,$$

and $B(0, t) = \varphi(0, t, 0, 0) = 0$, we have

$$\frac{\partial\varphi}{\partial w_0}(0, t, 0, 0) = 1. \qquad (5.29)$$

Therefore, from the solution of the differential equation (5.27) in u, using the fact $C(0, t) = c_0$ and Eqs. (5.25) and (5.29), we obtain

$$\frac{\partial^2\varphi}{\partial\lambda\partial w_0}(0, t, 0, 0) = \int_0^t \left[2c_0\int_0^s \frac{\partial A}{\partial\lambda}(0, \tau)d\tau + \frac{\partial B}{\partial\lambda}(0, s)\right]ds. \qquad (5.30)$$

Consequently,

$$\frac{\partial^2\Pi}{\partial\lambda\partial w_0}(0, 0) = \int_0^1 \left[2c_0\int_0^s \frac{\partial A}{\partial\lambda}(0, \tau)d\tau + \frac{\partial B}{\partial\lambda}(0, s)\right]ds \equiv B_0. \qquad (5.31)$$

Admittedly, the sequence of ideas and the computational details of the foregoing bifurcation analysis are rather complicated. So, let us now illustrate the necessary steps on a specific equation.

Example 5.5. *A saddle-node bifurcation:* Consider the 1-parameter perturbation of Example 4.25 given by

$$\dot{x} = \lambda(1 + \cos 2\pi t) + x^2 - \cos^2 2\pi t - 2\pi \sin 2\pi t, \qquad (5.32)$$

where λ is a scalar parameter.

If we let $x(t) = (\cos 2\pi t) + z(t)$, then the variational equation is given by

$$\dot{z} = \lambda(1 + \cos 2\pi t) + 2(\cos 2\pi t)z + z^2.$$

Since $\int_0^t 2(\cos 2\pi s)ds = (\sin 2\pi t)/\pi$, we introduce the new variable $u(t)$ defined by

$$z(t) = e^{(\sin 2\pi t)/\pi}u(t) \tag{5.33}$$

and obtain the new differential equation

$$\dot{u} = \lambda(1 + \cos 2\pi t)e^{-(\sin 2\pi t)/\pi} + e^{(\sin 2\pi t)/\pi}u^2. \tag{5.34}$$

To put this equation into the form of Eq. (5.19), that is, to make the coefficient of the u^2 term constant, we introduce the transformation

$$z(t) = w(t) + \beta(t)w^2(t),$$

where

$$\dot{\beta} = e^{(\sin 2\pi t)/\pi} - c_0,$$

$$c_0 = \int_0^1 e^{(\sin 2\pi t)/\pi}\,dt.$$

Imitating the calculations in the proof of Theorem 5.1, we obtain the following differential equation for w:

$$\begin{aligned} \dot{w} = \ &\lambda(1 + \cos 2\pi t)e^{-(\sin 2\pi t)/\pi} \\ &- 2\lambda\beta(t)(1 + \cos 2\pi t)e^{-(\sin 2\pi t)/\pi}w + [c_0 + O(\lambda)]w^2 + O(w^3). \end{aligned} \tag{5.35}$$

Now, observe that

$$A_0 \equiv \int_0^1 (1 + \cos 2\pi t)e^{-(\sin 2\pi t)/\pi}dt > 0,$$

and, from Example 5.2, $c_0 > 0$. Thus, it follows from Theorem 5.3 above that near the periodic solution $\cos 2\pi t$ the perturbed equation (5.32) undergoes a saddle-node bifurcation: it has two 1-periodic solutions if $\lambda < 0$, and no 1-periodic solution if $\lambda > 0$; see Figure 5.2. You may wish to compute B_0 and determine the Poincaré map (5.22) of the perturbed equation (5.35). $\Diamond$

We end this section with a remark on the general theory. In Theorem 5.3 we assumed that both A_0 and c_0 were not zero. If $A_0 = 0$ but $c_0 \neq 0$, for example, then we need further computations to determine the nature of the bifurcation point $\lambda = 0$. In particular, it will be necessary to compute the partial derivative $\partial^2\Pi(0, 0)/\partial\lambda^2$ in order to determine the quadratic terms in λ of the function $\Pi(\lambda, 0)$. This is a relatively routine but tedious task, thus we refrain from giving the details.

Exercises _____ ♣ ♡ ♠ ◇

5.9. Augment the statement of Theorem 5.3 by identifying the stability types of the periodic solutions therein.

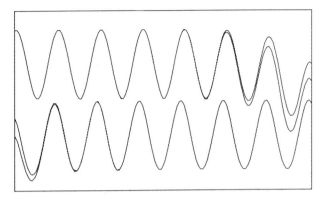

Figure 5.2. *Saddle-node bifurcation of periodic solutions of the differential equation* $\dot{x} = \lambda(1 + \cos 2\pi t) + x^2 - \cos^2 2\pi t - 2\pi \sin 2\pi t$.

5.10. Consider the 1-periodic equation depending on a parameter λ:

$$\dot{x} = \lambda a(t) + c_0 x^2 + O(x^3),$$

where the function $a(t)$ is 1-periodic continuous. Moreover, suppose that $a_0 c_0 \neq 0$ where $\int_0^1 a(s)\,ds = a_0$. Identify the bifurcation of 1-periodic solutions at $\lambda = 0$.

5.11. Consider the 1-periodic equation

$$\dot{x} = -x^3 + \left(\tfrac{3}{2} + \lambda\right)x + \sin^3 2\pi t - \tfrac{3}{2}\sin 2\pi t + 2\pi \cos 2\pi t$$

depending on a scalar parameter λ. Knowing that $\varphi(t, 0, 0) = \sin 2\pi t$ is a 1-periodic solution at $\lambda = 0$, obtain the bifurcation diagram of 1-periodic solutions of this differential equation near $\varphi(t, 0, 0)$.

5.12. Consider the 1-periodic equation depending on two parameters λ_1 and λ_2:

$$\dot{x} = \lambda_1 a(t) + \lambda_2 b(t) + c_0 x^3 + O(x^4),$$

where $c_0 \neq 0$, and $a(t)$ and $b(t)$ are 1-periodic continuous functions satisfying $\int_0^1 a(s)\,ds = \int_0^1 b(s)\,ds = 1$. Show that there is a cusp in the (λ_1, λ_2)-plane near $(\lambda_1, \lambda_2) = (0, 0)$ for which there are three 1-periodic solutions inside the cusp and one 1-periodic solution outside the cusp. Find approximate formulas for the cusp and for the solutions.

Suggestions: Integrate the differential equation formally to obtain the following expression for the Poincaré map:

$$\Pi(\lambda_1, \lambda_2, x_0) = x_0 + \lambda_1 + \lambda_2 \int_0^1 b(s)x(s)\,ds - \int_0^1 x^3(s)\,ds + O(x^4).$$

Observe that

$$\Pi(0, 0, x_0) = x_0 - x_0^3, \quad \Pi(\lambda_1, 0, 0) = \lambda_1, \quad \frac{\partial \Pi}{\partial \lambda_2 \partial x_0}(0, 0, 0) = \lambda_2.$$

Then, obtain the expansion for the Poincaré map near zero:

$$\Pi(\lambda_1, \lambda_2, x_0) = x_0 + \lambda_1 + o(\|\lambda\|) + [\lambda_2 + o(\|\lambda\|)]x + O(\|\lambda\|)x^2 \\ - [1 + O(\|\lambda\|)]x^3 + O(x^4)$$

as $\lambda \to 0$, where $\lambda = (\lambda_1, \lambda_2)$. Now, get the cusp.

Bibliographical Notes

The method of using an appropriate change of variables to transform a non-linear differential equation into a form which exhibits more clearly the interaction between the linear and nonlinear terms, especially the resonances that can occur, goes under the name of *normal form* theory. The origins of this powerful technique can be traced back to Liapunov and Poincaré; Birkhoff formalized the procedure and made extensive use of it in Hamiltonian mechanics. Normal form theory, with its many variations, commands a large literature; see, for example, Arnold [1983], Bibikov [1979], Birkhoff [1927], Chow and Hale [1982], Guckenheimer and Holmes [1983], Kirchgraber and Stiefel [1978], Sanders and Verhulst [1985], Rand and Armbruster [1987], and references therein. In particular, performing time-varying change of variables with the effect of reducing a nonautonomous differential equation to an autonomous one is called the *method of averaging;* a standard exposition is in Hale [1980].

6

On
Tori and Circles

 In this chapter, as a generalization of the ideas from Section 4.3, we show that if a 1-periodic nonautonomous differential equation is also periodic in x, then it gives rise to a differential equation on a torus (the surface of a doughnut). The dynamics of such equations are explored most conveniently in terms of their Poincaré maps, which happen to be maps on a circle. Accordingly, in the spirit of Chapter 3, we include a brief discussion of such maps and study a landmark example, the standard circle map. Poincaré, in conjunction with his work on classical mechanics, was the first to study vigorously the subject of differential equations on a torus, in particular circle maps. Since his days, a deep analytical theory of circle maps has emerged. The purpose of this chapter is merely to point out a few rudimentary facts and some highlights. We will return to this subject in Part IV and explore several seminal examples from the theory of oscillations and Hamiltonian mechanics, where tori are naturally omnipresent.

6.1. Differential Equations on a Torus

In this section, we consider nonautonomous scalar differential equations that are 1-periodic in both t and x,

$$\dot{x} = f(t, x) \quad \text{with} \quad f(t+1, x) = f(t, x) = f(t, x+1), \qquad (6.1)$$

and describe how the solutions of Eq. (6.1) can be viewed as orbits of a pair of autonomous differential equations on a torus, $S^1 \times S^1$. It is not a restriction to assume the periods to be 1; if not, we can always rescale the variables t and x to make it so.

We begin our discussion by pointing out two key properties of solutions of Eq. (6.1) which facilitate the viewing of solutions on a torus. Let $\varphi(t, t_0, x_0)$ be a solution of Eq. (6.1) with $\varphi(t_0, t_0, x_0) = x_0$. Then, from the uniqueness of the solution of the initial-value problem and the periodicity properties of f, we have

$$\varphi(t+1, t_0, x_0) = \varphi(t, t_0, \varphi(t_0+1, t_0, x_0)), \qquad (6.2)$$

$$\varphi(t, t_0, x_0) + 1 = \varphi(t, t_0, x_0 + 1). \qquad (6.3)$$

To proceed, we convert the scalar nonautonomous equation (6.1) to the following equivalent pair of autonomous differential equations:

$$\begin{aligned} \dot{\theta} &= 1 \\ \dot{x} &= f(\theta, x). \end{aligned} \qquad (6.4)$$

It is clear from the form of the first equation that the orbits of Eq. (6.4) correspond to the trajectories of Eq. (6.1). The first equation is periodic with any period; let us take its period to be 1. As we saw in Section 4.3, the relation (6.2) implies that if we identify θ with $\theta + k$, for any integer k, then the orbits of Eq. (6.4), hence the trajectories of Eq. (6.1), can be viewed as smooth curves on the cylinder $S^1 \times \mathbb{R}$. Using the relation (6.3), if we further identify x with $x + k$, for any integer k, then the orbits of Eq. (6.4) become smooth curves on the two-torus $S^1 \times S^1$ which we denote by T^2. For simplicity, let us take $\theta_0 = 0$ and summarize this construction: Translate all unit squares whose corners lie at the integers to the unit square $[0, 1] \times [0, 1]$, then glue the top of this square to its bottom, and its left side to its right, while preserving their orientations; see Figure 6.1.

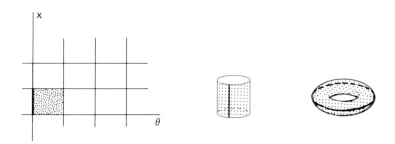

Figure 6.1. *Gluing the opposite sides of a square, while preserving their orientations, gives rise to a torus.*

Next, we define the notion of the Poincaré map of the flow of Eq. (6.4) on the torus. If we modify Definition 4.12 to take into account the periodicity in x, then we are led to define the *Poincaré map* Π of the flow of Eq. (6.1) to be

$$\Pi(x_0) = \varphi(1, 0, x_0) \quad (\text{mod } 1),$$

where (mod 1) means that only the fractional part of the value of the solution is retained. In terms of the orbits of Eq. (6.4) on the (θ, x)-plane this has the following geometric meaning: After all the unit squares with integer corners are translated to the first unit square, the image of a point $(0, x_0)$ on the left side of the first unit square under the Poincaré map is the x-coordinate of the first point at which the orbit through $(0, x_0)$ leaves the right side of the first unit square. As such, the Poincaré map is a map of the unit interval into itself, $\Pi : [0, 1] \to [0, 1]$; it usually has a jump discontinuity, but is convenient to iterate on the computer. If we now perform the identification described earlier by gluing the opposite sides of the first unit square, then the unit interval becomes a circle of circumference 1 (the dark circle on the torus in Figure 6.1). Hence, the Poincaré map becomes a diffeomorphism of the circle into itself, $\Pi : S^1 \to S^1$. The properties of the orbits of Eq. (6.4) on the torus are, of course, reflected in the properties of the iterates of the Poincaré map on the circle. In fact, the periodic points, not just the fixed points, of the Poincaré map correspond to the orbits of Eq. (6.4) that are closed curves.

For the purposes of visualization, it may sometimes be more convenient to denote a point on the circle by its usual angle ϕ measured in radians. In this case, a point is determined up to (mod 2π), that is, by any angle of the form $\phi + 2\pi k$ for any integer k.

Let us now illustrate these ideas on the simplest possible example.

Example 6.1. *Parallel flow:* Consider the differential equation $\dot{x} = c$, where c is a given real number, which is trivially 1-periodic both in x in t.

This equation is equivalent to the system

$$\dot\theta = 1$$
$$\dot x = c. \tag{6.5}$$

For reasons which will become self-evident momentarily, the flow of this system on the torus is called a *parallel flow*.

The orbit of Eq. (6.5) through the point $(0, x_0)$ is the line $x = c\theta + x_0$. By taking $\theta = 1$, we see that its Poincaré map is given by $\Pi(x_0) = c + x_0$ (mod 1), which is a rotation by arclength of c around the circle. We will subsequently analyze the geometry of the orbits of Eq. (6.5) on the torus and its Poincaré map for all possible values of c. For the moment, let us consider a particular case.

For $c = 1/2$, the orbit of Eq. (6.5) through the point $(0, x_0)$ is the line

$$x = \tfrac{1}{2}\theta + x_0.$$

As seen in Figure 6.2a, this orbit repeats itself in every other unit square it passes through. So, when these squares are translated to the first unit square the entire orbit consists of two line segments. If we now identify the opposite sides of this square, we obtain a closed curve on the torus which goes through the hole twice before returning to its initial position; see Figure 6.2b. All orbits are closed curves of this type.

For $c = 1/2$, the Poincaré map of Eq. (6.5) is given by

$$\Pi(x) = \tfrac{1}{2} + x \qquad (\text{mod } 1).$$

Every point is a periodic orbit of period 2 because $\Pi^2(x) = x$; see Figures 6.2c and 6.2d. It is easy to determine that these period-2 points of the Poincaré map, hence the closed orbits of the corresponding differential equation, are nonhyperbolic. $\Diamond$

Our next example concerns the description of a flow on the torus coming from a nonlinear 1-periodic differential equation for which an explicit formula for the Poincaré map is not readily available.

Example 6.2. Let us consider the scalar differential equation

$$\dot x = \sin(2\pi x) + \lambda \sin(2\pi t), \tag{6.6}$$

where λ is a real parameter, and try to determine the flow of the equivalent system

$$\dot\theta = 1$$
$$\dot x = \sin(2\pi x) + \lambda \sin(2\pi\theta) \tag{6.7}$$

on the torus.

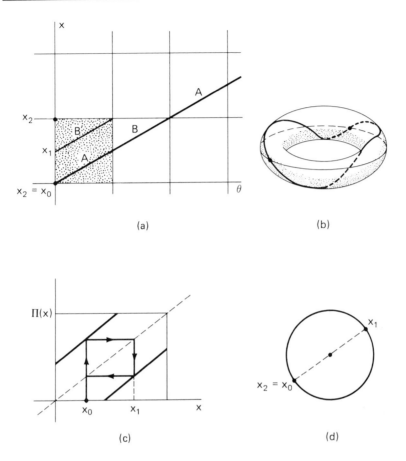

Figure 6.2. For $\dot{\theta} = 1$, $\dot{x} = 1/2$: (a) a solution on the (θ, x)-plane and its translation to the first unit square, (b) a solution on the torus, (c) its Poincaré map on the unit interval, and (d) its Poincaré map on the circle.

Let us first set $\lambda = 0$ and examine the flow of

$$\dot{\theta} = 1$$
$$\dot{x} = \sin(2\pi x) \tag{6.8}$$

on the torus. Notice that these two equations are independent of each other. Therefore, we can determine the flow of each equation on S^1 and then put the two together on the torus $S^1 \times S^1$. Recall from Section 1.4 that $\dot{x} = \sin(2\pi x)$ has two equilibria on S^1, an unstable one at $x = 0$, and an asymptotically stable one at $x = 1/2$. Moreover, all orbits approach one of these equilibria in either forward or reverse time. On the torus, these equilibria correspond to the fixed points of the Poincaré map of Eq. (6.8). From the linear variational equation in Section 4.4, it is not difficult to determine that these fixed points are hyperbolic with the same stability type as the corresponding equilibria. Consequently, Eq. (6.8) has two hyperbolic periodic orbits on the torus, one asymptotically stable and the other unstable, and all the remaining orbits approach one of these periodic orbits either in forward or reverse time.

For λ sufficiently small, the fixed points of the Poincaré map of the unperturbed equation retain their stability types because they are hyperbolic. It is possible to show that under small perturbations the entire flow of the perturbed system (6.7) remains qualitatively the same as the flow of the unperturbed equation (6.8) on the torus; see Figure 6.3. $\Diamond$

Let us now return to Example 6.1 and determine its flow and the Poincaré map for all values of the parameter c.

Theorem 6.3. *Let Π be the circle map defined by*

$$\Pi(x) = x + c \qquad (\text{mod } 1), \tag{6.9}$$

where c is a real parameter. Then
  (i) *there is a periodic point of minimal period q (and then all points are q-periodic) if and only if $c = p/q$ where p and q are integers with no common factors;*
 (ii) *each orbit is dense if and only if c is an irrational number.*

Proof. (i) If $c = p/q$, then $\Pi^q(x) = x + p = x$ and thus all points are periodic of period q. It is clear that q is the minimal period. Conversely, if $\Pi^q(x_0) = x_0$ for some x_0 and integer q, then $\Pi^q(x_0) = qc + x_0$ which implies that qc is an integer, hence c is rational. If q is the minimal period, then $c = p/q$ with p and q having no common factors.

(ii) Let us begin by recalling the definition of "dense." A subset U of S^1 is said to be *dense* in S^1 if, for any $\varepsilon > 0$ and any point x in S^1, there is a point $\bar{x}$ in U such that $|\bar{x} - x| < \varepsilon$. To proceed, fix $x_0 \in S^1$ and consider the set U consisting of the orbit of x_0, that is, $U = \{\Pi^n(x_0)\}$ where n runs over the integers. We will show that if c is irrational, then the set U is dense in S^1.

The points $\Pi^n(x_0)$, where n runs over the integers, are certainly distinct for otherwise there would be a periodic orbit of Eq. (6.9) and we know this is not the case since c is irrational. This infinite sequence of points has a limit point because S^1 is closed and bounded. Thus, for any $\varepsilon > 0$, there are integers $k > 0$ and m such that $|\Pi^{k+m}(x_0) - \Pi^m(x_0)| < \varepsilon$. Since Π preserves distances on S^1, it follows that $|\Pi^k(x_0) - x_0| < \varepsilon$. Also, the map Π^k takes the arc connecting the points x_0 and $\Pi^k(x_0)$ to the arc connecting $\Pi^k(x_0)$ to $\Pi^{2k}(x_0)$ and this arc is of length less than ε. Thus, the points

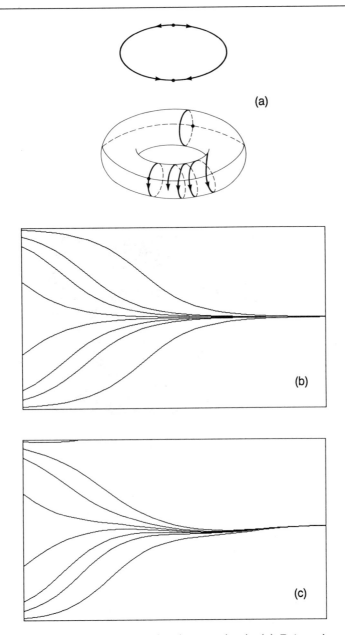

Figure 6.3. For equation $\dot{x} = \sin(2\pi x) + \lambda \sin(2\pi t)$: (a) Poincaré map and the flow on the torus for $\lambda = 0$, (b) flow on the unit square for $\lambda = 0$, and (c) flow on the unit square for $\lambda = 0.15$.

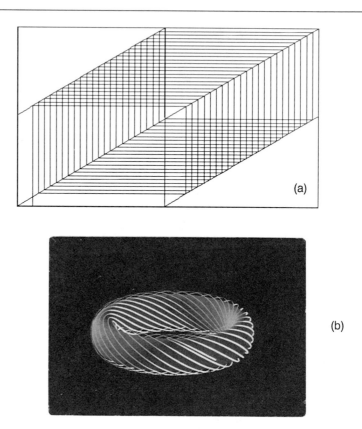

Figure 6.4. *For* $\dot{x} = 0.51234\ldots$ *(irrational!):* (a) *its Poincaré map on the unit interval, and* (b) *a dense orbit on the torus.*

x_0, $\Pi^k(x_0)$, $\Pi^{2k}(x_0)$, ... divide S^1 into arcs of length less than ε. Since ε was arbitrary, it follows that the orbit $\{\Pi^n(x_0)\}$ is dense in S^1. $\Diamond$

The implications of this theorem for the geometry of the parallel flow of Eq. (6.5) are easy to deduce. When c is rational of the form $c = p/q$, the orbits are closed curves on the torus which go around the torus p times while going through the hole of the torus q times. When c is irrational, the orbits never close up and any single orbit comes arbitrarily close to any given point on the torus; see Figure 6.4.

Exercises ⎯⎯⎯⎯⎯⎯⎯⎯⎯⎯⎯⎯⎯⎯⎯ ♣ ♡ ♠ ♢

6.1. *Rational or not?:* Draw the solutions of $\dot{x} = 2/3$ on the unit square and on the torus. Now, draw the solutions of $\dot{x} = 0.666666\ldots$ on the unit square and on the torus.

6.2. *No Sharkovskii on the circle:* Show by an example that Theorem 3.24 of Sharkovskii does not hold on the circle.

6.3. Investigate the flows of the following differential equations on the torus:
  (a) $\dot{x} = \sin 2\pi t;$ (b) $\dot{x} = 1 + \sin 2\pi t;$
  (c) $\dot{x} = \lambda + \sin 2\pi t$, where λ is a real parameter.

6.4. *A numerical puzzle:* Consider the circle map

$$x_{n+1} = 2x_n \qquad (\text{mod } 1).$$

Iterate the initial point $x_0 = 0.33$ under this map several times by hand; yes, by hand. Then iterate the same point in the computer. Do you see a difference in the asymptotic behavior of the orbit?

Aid: Use the one-dimensional map named *mod* stored in the library of PHASER.

6.2. Rotation Number

In Lemma 6.3, we saw that, for c irrational, each orbit under the Poincaré map is dense in S^1. Also, in Example 6.2, we observed that the Poincaré map has (two) isolated hyperbolic periodic points to which the remaining orbits approach either in forward or backward iteration. In this section, we state several theorems establishing the remarkable fact that, for C^2 *vector fields* on the torus, these are the "typical" orbit structures, and that the two possibilities can be distinguished in terms of a single number. We begin with a precise definition of this magic number.

Definition 6.4. *Let* $\dot{x} = f(t, x)$ *be 1-periodic in* t *and* x*, and* Π *its Poincaré map. The rotation number of* Π*, denoted by* $\rho(\Pi)$*, is defined as*

$$\rho(\Pi) = \lim_{|n| \to +\infty} \frac{1}{n} \varphi(n, 0, x_0),$$

where n is an integer.

Geometrically, the number $\rho(\Pi)$ can be thought of as the "average" rotation of a point x_0 on the circle under the iterates of the Poincaré map Π. The rotation number characterizes much of the qualitative features of a sufficiently differentiable flow on the torus, as the following fundamental theorem of Denjoy asserts:

Theorem 6.5. *The rotation number* $\rho(\Pi)$ *is well-defined, that is, the limit exists, and is independent of the initial point* x_0*. Furthermore, if* Π *is a* C^2 *map, then*
  (i) $\rho(\Pi)$ *is rational if and only if* Π *has a periodic orbit of some period;*

(ii) $\rho(\Pi)$ *is irrational if and only if every orbit of* Π *is dense on* S^1. $\Diamond$

In applications, one must often consider Poincaré maps of flows on the torus which depend on parameters. Let Π_λ be the Poincaré map of $\dot{x} = F(\lambda, t, x)$, where $\lambda \in \mathbb{R}^k$, and F is a C^2 function that is 1-periodic in t and x. An important fact about Π_λ is the following:

Theorem 6.6. *Rotation number* $\rho(\Pi_\lambda)$ *is a continuous function of the parameter* λ. $\Diamond$

We will refrain from presenting proofs of these theorems. Instead, let us return to our previous examples and determine their rotation numbers. It is evident that the rotation number of the differential equation $\dot{x} = c$ in Example 6.1 is $\rho(\Pi) = c$, and the contents of Theorem 6.3 is just a precursor of the results in Theorem 6.5. Next, we compute the rotation number of Example 6.2. To begin, let us set $\lambda = 0$. Since the rotation number is independent of the initial point, it is convenient to choose the equilibrium solution $\varphi(t, 0, 0) = 0$. With this choice, it is obvious that $\rho(\Pi_0) = 0$. Thus, on the average, there is no rotation around the circle. This is consistent with the flow depicted in Figure 6.3 because all points under the iterates of the Poincaré map eventually approach one of the two fixed points. Since $x_0 = 0$ is a hyperbolic fixed point of Π_0, there is a unique fixed point of the Poincaré map Π_λ near zero for $|\lambda|$ small. Thus $\rho(\Pi_\lambda)$ is a rational number for $|\lambda|$ small. Since the rotation number is continuous in λ, it follows that $\rho(\Pi_\lambda) = 0$ for $|\lambda|$ small.

In light of these examples and the theorem of Denjoy, let us now turn our attention to the task of characterizing structurally stable flows on the torus. The notions of topological equivalence and structural stability for scalar autonomous differential equations given in Definitions 2.22 and 2.24 can be extended naturally to the case of flows on the torus by using a homeomorphism $h : T^2 \to T^2$. If the rotation number is irrational, then every orbit of the Poincaré map is dense in S^1. Therefore, one should be able to make a small change in the vector field to create some closed orbits, possibly with very long periods. But, in the presence of periodic orbits the rotation number becomes rational. Moreover, if these periodic orbits are hyperbolic, then they persist under small perturbations and the rotation number remains constant. Consequently, the following theorem, which we state without proof, is plausible:

Theorem 6.7. *Consider the flow on the torus of a* C^2 *differential equation* $\dot{x} = F(\lambda, t, x)$ *that is 1-periodic in* t *and* x, *where* λ *is a (vector) parameter. For a fixed value* $\lambda = \bar{\lambda}$, *the differential equation* $\dot{x} = F(\bar{\lambda}, t, x)$ *is structurally stable if and only if it has rational rotation number and all of its periodic orbits are hyperbolic.* $\Diamond$

This result seems to suggest that in applications one is most likely to encounter rational rotation numbers despite the ubiquity of the irrationals.

Exercises ──────────────────────────────── ♣ ♡ ♠ ◇

6.5. Find the rotation numbers of the following vector fields on the torus:
  (a) $f(t, x) = \cos 2\pi t$;
  (b) $f(t, x) = 1 + \cos 2\pi x$;
  (c) $f(t, x) = \lambda x$ (mod 1), where λ is a real parameter. Is the rotation number rational or irrational for $\lambda = 1$? You may have to do this problem numerically. Set $\lambda = 1$ and use an explicit solution to compute the iterates of the Poincaré map to estimate the rotation number. Also, try several different initial values; do you get the same rotation number?

6.6. Suppose that Π_λ satisfies the conditions of Theorem 6.6 and that all fixed points of Π_0 are hyperbolic. Show that $\rho(\Pi_\lambda)$ is constant for $|\lambda|$ small.

6.7. Suppose that ω is an irrational number and $g(t, x)$ is a C^2 function which is 1-periodic in t and x. Let Π_g be the Poincaré map of the differential equation $\dot{x} = \omega + g(t, x)$ on the torus. For any $\varepsilon > 0$, show that there is a function g such that $\max \{ |g(t, \theta)| : 0 \leq t \leq 1, \ 0 \leq \theta \leq 1 \} < \varepsilon$ and $\rho(\Pi_g)$ is rational, and that not all orbits are closed on the torus. For a given positive integer n, can you choose g so that there are exactly n closed orbits?
Hint: Try $g(t, x) = a \sin(b2\pi t + c2\pi x)$ with appropriate constants a, b, and c. Also, use the fact from number theory that an irrational number ω can be approximated by rationals p/q in such a way that $|\omega - p/q| < \gamma/q^2$, where γ is a fixed constant.

6.3. An Example: The Standard Circle Map

Our encounters with circle maps have thus far been as the Poincaré map of the flow on the torus of 1-periodic differential equations (6.1). Since Poincaré maps arise from differential equations, they are necessarily diffeomorphisms of the circle. However, the theory of general circle maps that are not necessarily diffeomorphisms has its own prominence in the same way that nonmonotone interval maps do, as we have seen in Chapter 3. In this section, we describe some highlights of the dynamics of a landmark circle map—the standard map—which has played an exemplary role in theoretical and numerical investigations, much like the role of the logistic map in interval maps.

  To introduce our map in a natural way let us reconsider the scalar differential equation (2.34) from Chapter 2,

$$\dot{x} = a + b \sin(2\pi x), \tag{6.10}$$

where a and b are real parameters, and attempt to investigate its dynamics numerically. For this purpose, if we use Euler's algorithm with step size h

to approximate the solutions of Eq. (6.10), then we arrive at the difference equation

$$x_{n+1} = \omega + x_n + \frac{\varepsilon}{2\pi} \sin(2\pi x_n), \tag{6.11}$$

where $\omega = ah$ and $\varepsilon = 2\pi bh$. We have intentionally introduced the factor 2π so that an important change occurs at $\varepsilon = 1$. Of course, Eq. (6.11) is equivalent to the iteration of the two-parameter map

$$F(\omega, \varepsilon, x) = \omega + x + \frac{\varepsilon}{2\pi} \sin(2\pi x). \tag{6.12}$$

To investigate the dynamics of the differential equation (6.10) on the circle, we need to study Eq. (6.12) as a map on the circle, that is,

$$F(\omega, \varepsilon, x) = \omega + x + \frac{\varepsilon}{2\pi} \sin(2\pi x) \qquad (\text{mod } 1) \tag{6.13}$$

with the end points 0 and 1 of the unit interval identified. As such, this map is referred to in the mathematical literature as the *standard* or the *canonical* circle map. To see the effect of the choice of the step size, you may wish to compare the dynamics of this circle map described below with that of the differential equation (6.10) on the circle.

For $0 \le \varepsilon < 1$, the map (6.13) is a diffeomorphism of the circle because $F(\omega, \varepsilon, x + 1) = F(\omega, \varepsilon, x) + 1$ and $(d/dx)F(\omega, \varepsilon, x) > 0$. At $\varepsilon = 1$, it is a homeomorphism, and for $\varepsilon > 1$, it is no longer one-to-one.

The notion of rotation number can be generalized for homeomorphisms of the circle. Indeed, when $0 \le \varepsilon \le 1$, the rotation number $\rho(\omega, \varepsilon)$ of the standard map is defined to be the following limit:

$$\rho(\omega, \varepsilon) \equiv \lim_{|n| \to +\infty} \frac{F^n(\omega, \varepsilon, x_0)}{n}, \tag{6.14}$$

where n is integer and the values of the iterates of x_0 are used without (mod 1). It can be proved that all the conclusions in Theorem 6.5, which we stated for Poincaré maps only, remain valid with this definition of the rotation number. When $\varepsilon > 1$, rotation number is not defined as the limit above no longer exists; we will say more about this later.

Let us first focus our attention on the parameter range $0 \le \varepsilon \le 1$ where the notion of rotation number is well defined and try to describe the behavior of the rotation number for various values of the parameters ω and ε. We begin with two very special cases where only one of the parameters is present. When $\varepsilon = 0$, this map is simply the rotation around the circle by arclength ω and we have already studied its dynamics in Theorem 6.3. At the other extreme, when $\omega = 0$, there are two fixed points at $x = 0$ and $x = 1/2$. They are both hyperbolic and the first one is unstable while

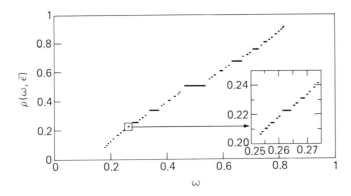

Figure 6.5. *The graph of the rotation number $\rho(\omega, \bar{\varepsilon})$ as a function of ω for fixed $0 < \bar{\varepsilon} \leq 1$.*

the second is asymptotically stable as long as $0 < \varepsilon < 1$; compare with the Poincaré map of Example 6.2.

For each fixed $\varepsilon = \bar{\varepsilon}$, with $0 < \bar{\varepsilon} \leq 1$, one can prove the following properties of the rotation number $\rho(\omega, \bar{\varepsilon})$:

- $\rho(\omega, \bar{\varepsilon})$ is a nondecreasing and continuous function of ω;
- for each rational number p/q, there is an interval $I_{p/q}$ with nonempty interior such that for all $\omega \in I_{p/q}$ we have $\rho(\omega, \bar{\varepsilon}) = p/q$;
- for each irrational number α, there is a unique ω such that $\rho(\omega, \bar{\varepsilon}) = \alpha$.

A typical graph of $\rho(\omega, \bar{\varepsilon})$ for fixed $0 < \bar{\varepsilon} \leq 1$ is shown in Figure 6.5, where the properties listed above are readily visible. This striking graph, which is an example of a *Cantor function*, has also been dubbed as the "devil's staircase."

In Figure 6.6, we have plotted some of the important features of the bifurcation diagram of the standard map on the (ω, ε)-plane. From each rational number on the ω-axis, there originates a sharp widening wedge with nonempty interior in which the rotation number is this constant rational. Furthermore, none of these wedges with rational rotation numbers overlap when $0 \leq \varepsilon \leq 1$. From each irrational number on the ω-axis, however, there originates a continuous curve with no interior and extends to $\varepsilon = 1$.

The dynamics of the standard map within each wedge is rather simple. Let us examine, for instance, the wedge emanating from the origin. Since the rotation number is zero in this wedge, the map must have a fixed point. Indeed, for $\varepsilon \neq 0$, the fixed points of the map are given by

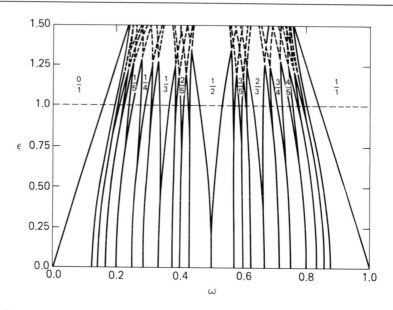

Figure 6.6. *Bifurcation diagram of the standard map. Inside each wedge, the so-called "Arnold tongues," the rotation number is rational.*

$$\sin(2\pi x) = \frac{-2\pi\omega}{\varepsilon}.$$

If we fix ε and increase ω from 0, we see that two fixed points, the intersections of the graphs of $\sin(2\pi x)$ and a horizontal line, one stable and the other unstable, eventually coalesce and disappear. At this moment, we move out of the wedge with rotation number 0 and the orbits on S^1 become dense. We have illustrated this saddle-node bifurcation of fixed points in Figure 6.7, which should be compared with Figure 2.19. A similar dynamical phenomenon occurs in the other wedges, but the role of fixed points is replaced by that of periodic points of appropriate periods. Figure 6.8 shows the saddle-node bifurcation of period-2 points in the large wedge with rotation number $1/2$.

For $\varepsilon > 1$, the dynamics of the standard map becomes rather complicated and some of the interesting observations still remain to be numerical. As we observed before, in this parameter range the standard map is no longer a homeomorphism and the rotation number is not defined. The qualitative dynamics of the map cannot be captured by the rationality or the irrationality of a single number. The limit in Eq. (6.14) is not unique but takes values on an interval of real numbers. Consequently, if we try to extend the bifurcation diagram above the line $\varepsilon = 1$, we notice that the

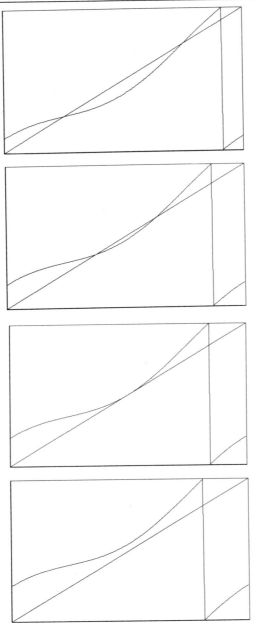

Figure 6.7. *Graphs of the standard map depicting saddle-node bifurcation of fixed points in and out of the wedge with rotation number 0. The parameter ε is fixed and ω is varied past the edge of the wedge.*

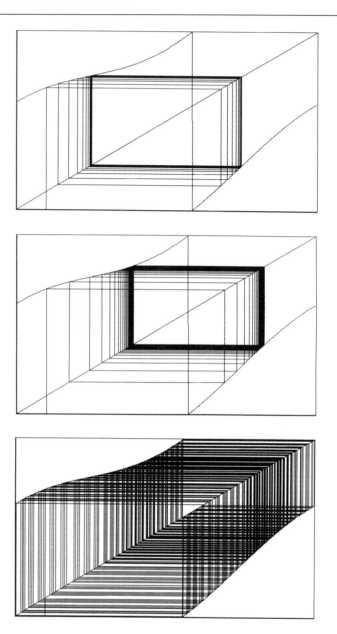

Figure 6.8. *Stair-step diagrams of F and the graphs of F^2 (see the following page) depicting the saddle-node bifurcation of period-2 orbits of the standard map in and out of the wedge with rotation number 1/2.*

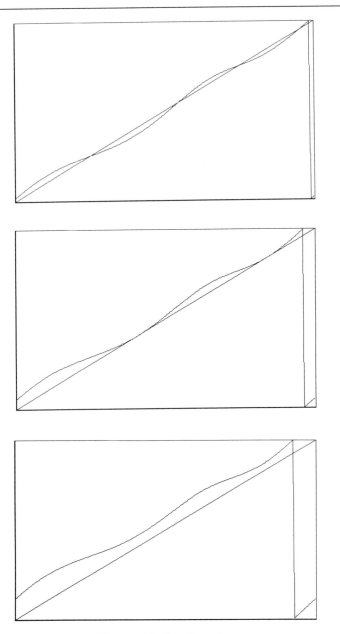

Figure 6.8 Continued.

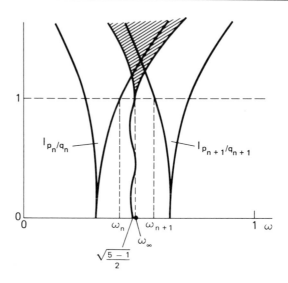

Figure 6.9. *Constructing a universal constant for the standard circle map.*

rational wedges overlap and new wedges grow from irrational rotation numbers; see Figures 6.6 and 6.9. This signals the presence of very complicated orbits, a situation often called chaos.

In the mist of this bewildering complexity, however, there are several "universal constants" which mark the transition to chaos for a large class of maps, including the standard map. Since these constants are independent of the precise form of a map, they could be of practical use in applications, much like the universal constant of Feigenbaum for interval maps. We now describe one such universal constant. Consider the irrational number $(\sqrt{5} - 1)/2$, the *golden mean*, and its continued fraction expansion

$$\frac{\sqrt{5}-1}{2} = \cfrac{1}{1 + \cfrac{1}{1 + \cfrac{1}{\ddots}}}.$$

By truncating this expansion at each stage we obtain a sequence of rational numbers p_n/q_n converging to the golden mean. The integers p_n and q_n turn out to be the Fibonacci numbers and satisfy the recursion relations $q_{n+1} = q_n + q_{n-1}$ and $p_n = q_{n-1}$ with the initial values $p_0 = 0$, $q_0 = 1$. Now, let ω_∞ denote the value of the parameter ω such that $\rho(\omega_\infty, 1) = (\sqrt{5}-1)/2$ and let ω_n denote the value of ω closest to ω_∞ such that $\rho(\omega_n, 1) = p_n/q_n$;

see Figure 6.9. Then, the numerical experiments point to the fact

$$\lim_{n \to +\infty} \frac{\omega_n - \omega_{n-1}}{\omega_{n+1} - \omega_n} = -2.834\ldots.$$

It has been proved that this number turns out to be the same for a large class maps of the circle at their points of transition to chaos. However, the standard map is not yet one of them.

Exercises ♣♡♠♢

6.8. *On the shape of a tongue:* For the standard circle map, in a region of the (ω, ε)-plane with a rational rotation number there is an asymptotically stable periodic orbit. To determine an approximate shape of such a region, one can follow the fixed point while changing the parameters. Using this observation, determine numerically the shapes of the two large regions with rotation numbers 0 and 1/2.
   Help: The standard circle map is stored in the library of PHASER under the name *arnold*.

Bibliographical Notes ◎◯◎

Since its inception by Poincaré, the study of toral flows, circle diffeomorphism and rotation number in particular, has evolved into a deep theory due to the efforts of Denjoy, Arnold, Herman, and others. Arnold [1983], a good source on this topic, contains proofs of the basic properties rotation numbers, as well the structural stability result in Theorem 6.7; see also Coddington and Levinson [1955], Devaney [1986], Hale [1980], Hartman [1964], and Nitecki [1971].

   For C^2 vector fields, there are essentially only two types of qualitative behavior as described in Theorem 6.5. It is a remarkable fact that there may be other types of behavior if the vector field is only a C^1 function; see Denjoy [1932].

   It is difficult to determine rotation numbers in specific equations and one must often resort to numerical computations. Such computations require special algorithms and a great deal of care; see, for example, Van Veldhuizen [1988]. After all, what is irrational in floating point arithmetic?

   Since the seminal paper of Arnold [1965], the two-parameter standard map has captured the attention of many mathematicians and scientists. Prompted by the discoveries of Feigenbaum on interval maps, numerical experiments (Shenker [1982]) eventually led to the universal scaling properties of the standard map (Rand [1988]). Circle maps similar to the standard map are encountered in applications; indeed, the phenomenon of "Arnold tongues" is used as a paradigm to explain frequency locking phenomena in certain oscillatory systems (Bak [1986]).

If a circle map is not a diffeomorphism, one can still consider the limit to define a rotation number, but the limit depends on the initial point. For piecewise monotone circle maps, one can generalize the concept of rotation number to one of rotation interval; see Newhouse et al. [1976] and Levi [1981].

Our investigation of scalar equations that are 1-periodic in t and x led us naturally to flows on the torus which do not have any equilibrium points. Toral flows with equilibria, however, do occur in other contexts and they may exhibit quite different dynamics; see Cherry [1938] and Palis and de Melo [1982].

2D

HANS BJUTTANRI-1991

7

Planar Autonomous Systems

 With this chapter we commence our investigation of the geometry of planar autonomous differential equations. After pointing out how such equations arise in applications, we develop some necessary generalizations of certain geometric ideas which are reminiscent of the ones explored earlier for scalar equations. Because the simplest examples of planar systems are constructed by bundling a pair of scalar equations—product systems—we present a discussion of such systems, including the Flow Box Theorem. We also analyze the geometry of conservative systems as another class of vector fields with special properties. Finally, to give a hint of things to come, we present multiple examples of autonomous differential equations illustrating various bifurcations on the plane.

7.1. "Natural" Examples of Planar Systems

Before we embark on the mathematical details of planar equations, let us first point out several instances in which planar systems arise naturally.

Example 7.1. *Planar pendulum:* The behavior of many mechanical systems is governed by Newton's Law, "$F = ma$," which is a second-order differential equation (acceleration a is the second derivative of displacement). For example, consider the motion of a simple pendulum that moves on a vertical plane; see Figure 7.1. In the absence of friction, the displacement angle θ from the vertical rest position of the pendulum satisfies the following second-order differential equation:

$$\frac{d^2\theta}{dt^2} + \frac{g}{l}\sin\theta = 0, \tag{7.1}$$

where g is the gravitational constant and l is the length of the pendulum.

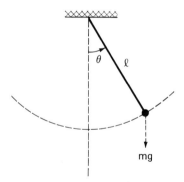

Figure 7.1. *Planar pendulum.*

For the purposes of geometrical analysis, as well as numerical simulations, it is desirable to convert a second-order differential equation to an equivalent planar system of first-order equations. This can be accomplished, for example, by introducing the variables

$$x_1(t) = \theta(t), \quad x_2(t) = \frac{d\theta}{dt}.$$

Then, in the new variables, Eq. (7.1) becomes

$$\dot{x}_1 = x_2$$
$$\dot{x}_2 = -\frac{g}{l}\sin x_1. \tag{7.2}$$

We will analyze Eq. (7.2) of the pendulum further in Section 7.4.

Example 7.2. *Linear harmonic oscillator:* For a moment, let us consider only "small" oscillations of the pendulum and approximate $\sin\theta$ by θ. As a further simplification, let us also take $g = 1.0$ and $l = 1.0$. Then, the pair of equations (7.2) above becomes

$$\begin{aligned} \dot{x}_1 &= x_2 \\ \dot{x}_2 &= -x_1. \end{aligned} \tag{7.3}$$

This approximate system is known as the *linear harmonic oscillator.* Because of its simplicity we will use Eq. (7.3) to illustrate certain basic geometric concepts in the next section. $\Diamond$

The equations of the planar pendulum (7.2) and the linear harmonic oscillator (7.3) are examples of "conservative" systems. We will explain the meaning of this label later in this chapter and also, because of their importance, devote Chapter 14 to this important class of systems.

Example 7.3. *Competing species*: In certain applications the use of planar differential equations may in fact be necessary to describe models. For instance, let us consider two interacting populations: a prey species x_1 and its predator x_2.

For the sake of simplicity, it is plausible to assume that in the absence of any interactions between the two species, the prey could grow without bounds and the predator would become extinct. In the presence of interactions, however, the growth rate of x_1 should be impaired while the growth rate of x_2 improves. The simplest such mathematical model arising from these assumptions is called the *predator-prey* system of Volterra and Lotka:

$$\begin{aligned} \dot{x}_1 &= a_1 x_1 - a_2 x_1 x_2 \\ \dot{x}_2 &= -a_3 x_2 + a_4 x_1 x_2, \end{aligned} \tag{7.4}$$

where a_1, a_2, a_3, and a_4 are positive constants.

To obtain a somewhat more realistic model of two interacting species we can modify Eq. (7.4) to include the effects of competition of the prey among themselves for their limited amount of resources, and the competition among the predators for the limited amount of prey. The resulting system

$$\begin{aligned} \dot{x}_1 &= a_1 x_1 - a_2 x_1 x_2 - b_1 x_1^2 \\ \dot{x}_2 &= -a_3 x_2 + a_4 x_1 x_2 - b_2 x_2^2, \end{aligned} \tag{7.5}$$

where b_1 and b_2 are again positive constants, is called the *competing species* model.

Naturally, both of these systems for modeling populations are meaningful only in the positive quadrant of the (x_1, x_2)-plane. We will shortly

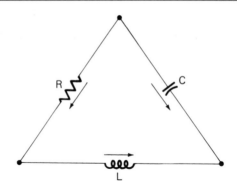

Figure 7.2. *Circuit diagram of Van der Pol's oscillator.*

give a pictorial description of typical solutions of Eqs. (7.4) and (7.5). We remark here that the predator-prey equation (7.4) is another example of a conservative system whereas the competing species equation (7.5) contains a subtle "dissipative" mechanism. ◊

Example 7.4. *Van der Pol's oscillator:* The study of electrical circuits provides another source of important differential equations. Let us consider, for example, the series RLC circuit depicted in Figure 7.2. The state of such an electrical circuit is determined by six time-varying quantities: i_R, i_C, i_L, which are the values of the currents through the resistor R, capacitor C, and the inductor L with the direction of the current flow indicated by the arrows in Figure 7.2; and v_R, v_C, v_L, which are the values of voltage differences across the three electrical components.

The behaviors of these three electrical components can be characterized mathematically. The generalized Ohm's law says that the voltage across a resistor is a function of the current running through it:

$$v_R = f(i_R).$$

The function f is called the *characteristic* of the resistor and its exact form depends on the type of the material the resistor is made from. The voltage and the current across an inductor satisfy Faraday's law

$$L\frac{di_L}{dt} = v_L,$$

and a capacitor is governed by the relation

$$C\frac{dv_C}{dt} = i_C,$$

where L and C are positive constants reflecting the physical characteristics of the inductor and capacitor, respectively.

The voltage and the current variables of an RLC circuit are interdependent and must obey certain "conservation" rules as asserted by Kirchkoff. For our RLC circuit, his current law implies that

$$i_R = i_L = -i_C,$$

and his voltage law yields

$$v_R + v_L - v_C = 0.$$

Using the relations above, we can eliminate all but two of the variables, i_L and v_L, which satisfy the system of differential equations

$$L\frac{di_L}{dt} = v_C - f(i_L)$$
$$C\frac{dv_C}{dt} = -i_L.$$

If we scale the time variable as $t \mapsto (CL)^{1/2}t$, and let $i_L = x_1$, $v_C = (L/C)^{1/2}x_2$, then the resulting equivalent system

$$\dot{x}_1 = x_2 - (C/L)^{1/2}f(x_1)$$
$$\dot{x}_2 = -x_1$$

is known as *Lienard's equation*.

A special type of resistor known as a tunnel diode exhibits a cubic characteristic function, say, $f(x_1) = x_1^3/3 - x_1$. In this case, the system above becomes

$$\dot{x}_1 = x_2 - (C/L)^{1/2}\left(\tfrac{1}{3}x_1^3 - x_1\right) \tag{7.6}$$
$$\dot{x}_2 = -x_1,$$

which is called the *Lienard form* of the famous Van der Pol's equation. This system of equations is equivalent, as indicated in the exercises, to the original second-order differential *equation of Van der Pol*:

$$\ddot{x} + (C/L)^{1/2}\left(1 - x^2\right)\dot{x} + x = 0, \tag{7.7}$$

which played a very important role in the development of the theory of nonlinear oscillations and the qualitative study of differential equations. We will draw in the next section a numerically computed phase portrait of Van der Pol's equation. However, a complete mathematical analysis of Van der Pol's oscillator is not trivial, as we shall see in Chapter 12. ◇

This concludes our collection of "natural" examples of planar differential equations. In order to facilitate geometric analyses of these equations, we now turn to some generalities.

Exercises _____ ♣♡♠♢

7.1. *Where is the mass?* Look up in a physics book to learn how the differential equation for the pendulum is derived. Does it bother you that the differential equation (7.1) is independent of the mass of the bob at the end of the pendulum? Consult Galileo.

7.2. *Lienard form:* Convert the second-order Lienard's equation $\ddot{y}+f(x)\dot{y}+y=0$ to the first-order system

$$\dot{x}_1 = x_2 - F(x_1)$$
$$\dot{x}_2 = -x_1$$

using the transformation $x_1 = y$ and $x_2 = \dot{y}-F(y)$, where $F(y) = \int_0^y f(s)\,ds$. As you see, there is more than one way to convert a second-order equation to an equivalent first-order system. Now, use this result to obtain the Lienard form of Van der Pol's equation in Example 7.4.

7.2. General Properties and Geometry

Let I be an open interval of the real line $\mathbb{R}$ and

$$x_i : I \to \mathbb{R}; \quad t \mapsto x_i(t) \qquad \text{for } i = 1, 2$$

be two C^1 functions of a real variable t. Also, let

$$f_i : \mathbb{R}^2 \to \mathbb{R}; \quad (x_1, x_2) \mapsto f_i(x_1, x_2) \qquad \text{for } i = 1, 2$$

be two given real-valued functions in two variables. In the following several chapters we will undertake a geometrical study of a pair of simultaneous differential equations of the form

$$\dot{x}_1 = f_1(x_1, x_2)$$
$$\dot{x}_2 = f_2(x_1, x_2). \tag{7.8}$$

Let us begin our study of the general planar system (7.8) by developing some basic notations and geometric concepts. In this discussion, it will be convenient to use **boldface** letters to denote vector quantities. For instance, if we let $\mathbf{x} = (x_1, x_2)$, $\dot{\mathbf{x}} = (\dot{x}_1, \dot{x}_2)$, and $\mathbf{f} = (f_1, f_2)$, then Eq. (7.8) can be written as

$$\dot{\mathbf{x}} = \mathbf{f}(\mathbf{x}). \tag{7.9}$$

This equation now looks the same as the scalar equation $\dot{x} = f(x)$ considered in Part I, but we must keep in mind that $\mathbf{x}$ is a two-vector and $\mathbf{f}$ is a vector-valued function. We will follow the convention of using subscripts to denote the components of a vector and superscripts to label different vectors, e.g., $\mathbf{x^1} = (x_1^1, x_2^1)$. In particular, an *initial-value problem* for Eq. (7.9) will be indicated by

$$\dot{\mathbf{x}} = \mathbf{f}(\mathbf{x}), \quad \mathbf{x}(t_0) = \mathbf{x^0}. \tag{7.10}$$

Since Eq. (7.9) is autonomous, there is no loss of generality, as for scalar equations, in assuming that the initial-value problem (7.10) is specified with $t_0 = 0$.

In our study of planar differential equations, it is necessary to measure distances between two vectors on the plane. To define distance, we first introduce the concept of a norm on $\mathbb{R}^2$, which is a generalization of the usual notion of length of vectors.

Definition 7.5. *A norm on $\mathbb{R}^2$ is a function*

$$\| \ \| : \mathbb{R}^2 \to \mathbb{R}; \quad \mathbf{x} \mapsto \|\mathbf{x}\|$$

that satisfies the following properties for any vectors $\mathbf{x}$, $\mathbf{x^1}$, and $\mathbf{x^2}$ in $\mathbb{R}^2$, and any scalar $a \in \mathbb{R}$:

$\|\mathbf{x}\| \geq 0$ *and* $\|\mathbf{x}\| = 0$ *if and only if* $\mathbf{x} = \mathbf{0}$;

$\|\mathbf{x^1} + \mathbf{x^2}\| \leq \|\mathbf{x^1}\| + \|\mathbf{x^2}\|$ *(triangle inequality)*;

$\|a\mathbf{x}\| = |a| \, \|\mathbf{x}\|$.

For a given norm $\| \ \|$, we define the *distance* between two vectors $\mathbf{x^1}$ and $\mathbf{x^2}$ to be $\|\mathbf{x^1} - \mathbf{x^2}\|$. The *Euclidean norm* (length) of a vector $\mathbf{x}$ defined by

$$\|\mathbf{x}\| = \sqrt{x_1^2 + x_2^2} \tag{7.11}$$

will usually be sufficient for most of our purposes. Sometimes, however, it may be more convenient to use other norms, hence other distances. For example, it is easy to verify that the *sup-norm* defined by

$$\|\mathbf{x}\|_{max} = \max\left(|x_1|, |x_2|\right)$$

satisfies all the requirements of a norm listed above. The sup-norm and the Euclidean norm, however, are considered to be *equivalent* to one another because for any $\mathbf{x} \in \mathbb{R}^2$ we have

$$(1/\sqrt{2}) \, \|\mathbf{x}\|_{max} \leq \|\mathbf{x}\| \leq \sqrt{2} \, \|\mathbf{x}\|_{max}. \tag{7.12}$$

One geometric implication of their equivalence is that a "circle" about the origin in one of the norms can be inscribed in between two appropriate "circles" in the other norm, and vice versa; see Figure 7.3.

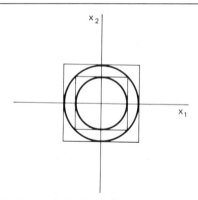

Figure 7.3. *Equivalence of the Euclidean norm* ∘ *and the sup norm* □.

Consequently, all of the qualitative results concerning differential equations will be independent of the choice of norm. Quantitative results, on the other hand, will of course depend on the norm. With the Euclidean norm, certain arguments become simpler. For example, if $\mathbf{x}(t)$ is a solution of Eq. (7.9), then the function $r(t) = \|\mathbf{x}(t)\|$ is a C^1 function. For the equivalent sup norm, the similar function $s(t) = \|\mathbf{x}(t)\|_{max}$ is not differentiable at any value of t for which $x_1(t)$ or $x_2(t)$ is zero.

The existence and uniqueness results from Theorem 1.4 can readily be generalized for the planar initial-value problem (7.10). Indeed, if $\mathbf{f}$ is a C^1 function, then, for any $\mathbf{x}^0 \in \mathbb{R}^2$, there is an interval (possibly infinite) $I_{\mathbf{x}^0} \equiv (\alpha_{\mathbf{x}^0}, \beta_{\mathbf{x}^0})$ containing $t_0 = 0$ and a unique solution $\varphi(t, \mathbf{x}^0)$ of the initial-value problem (7.10) defined for all $t \in I_{\mathbf{x}^0}$, satisfying the initial condition $\varphi(0, \mathbf{x}^0) = \mathbf{x}^0$. Moreover, $\varphi(t, \mathbf{x_0})$ is a C^1 function. For further information on these matters, you may consult the Appendix.

To begin our qualitative study, we now reconsider the system (7.9) and its flow $\varphi(t, \mathbf{x}^0)$ from a geometric point of view. At each point of the $(t, \mathbf{x})$-space where $\mathbf{f}(\mathbf{x})$ is defined, the right-hand side of Eq. (7.9) gives a value of the derivative $d\mathbf{x}/dt$ which can be considered as the slope of a line segment at that point. The collection of all such line segments is called the *direction field* of the differential equation (7.9).

The graph of the solution of Eq. (7.9) through $\mathbf{x}^0$, that is, the curve in the three-dimensional $(t, \mathbf{x})$-space defined by $\{ (t, \varphi(t, \mathbf{x}^0)) : t \in I_{\mathbf{x}^0} \}$ is called the *trajectory* through $\mathbf{x}^0$. Of course, at each point through which it passes, a trajectory is tangent to a line segment of the direction field. The spiral-like trajectory of the linear oscillator (7.3) through the point (5, 5) is shown in Figure 7.4.

Since the function $\mathbf{f}$ is independent of t, on any line parallel to the t-axis the segments of the direction field all have the same slope. Therefore, it is natural to consider the projections of the direction field and the trajectories

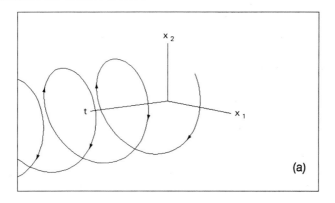

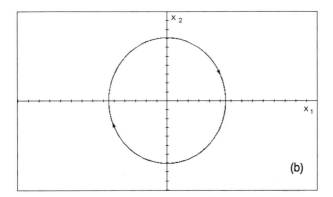

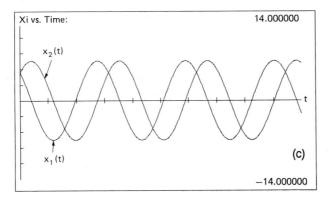

Figure 7.4. For the linear harmonic oscillator (7.3): (a) a trajectory in the three-dimensional (t, x_1, x_2)-space, (b) circular orbit resulting from projecting the helical trajectory onto the (x_1, x_2)-plane, and (c) graphs of $x_1(t)$ and $x_2(t)$ vs. t.

of Eq. (7.9) onto the (x_1, x_2)-plane. More precisely, to each point $\mathbf{x}$ on the (x_1, x_2)-plane, where $\mathbf{f}(\mathbf{x})$ is defined, we can associate the vector $\mathbf{f}(\mathbf{x}) = (f_1(\mathbf{x}), f_2(\mathbf{x}))$ which should be thought of as being based at $\mathbf{x}$. In other words, we can assign to the point $\mathbf{x}$ the directed line segment from $\mathbf{x}$ to $\mathbf{x} + \mathbf{f}(\mathbf{x})$. In the case of the linear harmonic oscillator (7.3), for example, at the point $(5, 5)$ we picture an arrow pointing from $(5, 5)$ to $(5, 5) + (5, -5) = (10, 0)$. The collection of all such vectors is called the *vector field* generated by Eq. (7.3), or simply the vector field $\mathbf{f}$. Projections of trajectories onto the (x_1, x_2)-plane are called *orbits*. More specifically, we make the following definition:

Definition 7.6. *The positive orbit* $\gamma^+(\mathbf{x}^0)$, *negative orbit* $\gamma^-(\mathbf{x}^0)$, *and orbit* $\gamma(\mathbf{x}^0)$ *of* $\mathbf{x}^0$ *are defined, respectively, as the following subsets of* $\mathbb{R}^2$ *[the (x_1, x_2)-plane]:*

$$\gamma^+(\mathbf{x}^0) = \bigcup_{t \in [0, \beta_{\mathbf{x}^0})} \varphi(t, \mathbf{x}^0),$$

$$\gamma^-(\mathbf{x}^0) = \bigcup_{t \in (\alpha_{\mathbf{x}^0}, 0]} \varphi(t, \mathbf{x}^0),$$

$$\gamma(\mathbf{x}^0) = \bigcup_{t \in (\alpha_{\mathbf{x}^0}, \beta_{\mathbf{x}^0})} \varphi(t, \mathbf{x}^0).$$

To compensate for the loss of time parametrization in orbits, on the orbit $\gamma(\mathbf{x}^0)$ we insert arrows to indicate the direction in which $\varphi(t, \mathbf{x}^0)$ is changing as t increases. The flow of a differential equation is then drawn as the collection of all its orbits together with the direction arrows; the resulting picture is called the *phase portrait* of the differential equation. Numerically computed phase portraits of the examples we have compiled in Section 7.1 are illustrated in various figures below.

Certain distinguished orbits play a prominent role in the qualitative theory of planar systems. The simplest of such orbits, an equilibrium point, is defined as in the case of scalar equations.

Definition 7.7. *A point* $\bar{\mathbf{x}} \in \mathbb{R}^2$ *is called an equilibrium point* (*also critical point, steady state solution, etc.*) *of* $\dot{\mathbf{x}} = \mathbf{f}(\mathbf{x})$, *if* $\mathbf{f}(\bar{\mathbf{x}}) = \mathbf{0}$, *that is, if* $\bar{\mathbf{x}} = (\bar{x}_1, \bar{x}_2)$, *then*

$$f_1(\bar{x}_1, \bar{x}_2) = 0, \quad f_2(\bar{x}_1, \bar{x}_2) = 0.$$

In planar systems there can be another orbit of special interest, called a periodic orbit, which has no counterpart among the scalar autonomous differential equations. Because of their importance we will devote a great deal of our attention to periodic orbits in later chapters. For the moment, however, here is its definition.

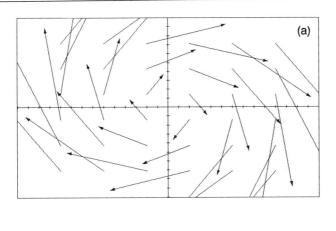

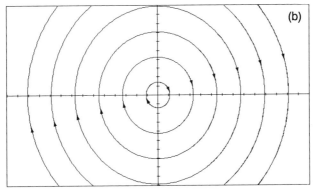

Figure 7.5. (a) *The vector field, and* (b) *the phase portrait of the linear harmonic oscillator* (7.3) *in the two-dimensional phase plane* (x_1, x_2). *A family of concentric periodic orbits encircling an equilibrium point is called a center.*

Definition 7.8. *A solution* $\varphi(t, \mathbf{x}^0)$ *of* $\dot{\mathbf{x}} = \mathbf{f}(\mathbf{x})$ *is called a periodic solution of period p, with p > 0, if* $\varphi(t + p, \mathbf{x}^0) = \varphi(t, \mathbf{x}^0)$ *for all* $t \in \mathbb{R}$. *The minimal period p is that period with the property that* $\varphi(t, \mathbf{x}^0) \neq \mathbf{x}^0$ *for* $0 < t < p$. *The orbit* $\gamma(\mathbf{x}^0) = \{\varphi(t, \mathbf{x}^0), \, t \in \mathbb{R}\}$ *of a periodic solution* $\varphi(t, \mathbf{x}^0)$ *with period p is said to be a periodic orbit (also closed orbit) of period p.*

It is evident from this definition that a periodic orbit is a closed curve on the (x_1, x_2)-plane. Also, any orbit of $\dot{\mathbf{x}} = \mathbf{f}(\mathbf{x})$ that is a closed curve must correspond to a periodic solution.

Example 7.9. *Linear harmonic oscillator continued:* We now return to

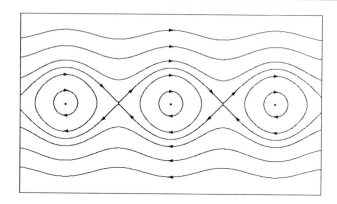

Figure 7.6. *Phase portrait of the planar pendulum* (7.3). *Notice the presence of centers.*

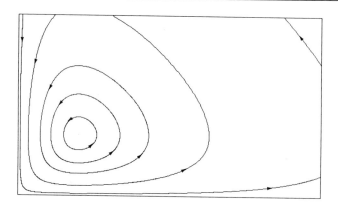

Figure 7.7. *Phase portrait of the predator-prey equations* (7.4) *in the positive quadrant for $a_1 = a_2 = a_3 = a_4 = 1$ is a center. Compare with Figure 7.5.*

the linear harmonic oscillator and rigorously justify the picture of its phase portrait as depicted in Figure 7.4b.

Notice first that the only equilibrium point is the origin. Let $\mathbf{x}(t) = (x_1(t), x_2(t))$ be the solution through $\mathbf{x}(0) = \mathbf{x}^0 \neq \mathbf{0}$ and consider the square of the distance of the solution from the origin, $\|\mathbf{x}(t)\|^2 = [x_1(t)]^2 + [x_2(t)]^2$, as a function of t. Then,

$$\frac{d}{dt}\|\mathbf{x}(t)\|^2 = 2x_1(t)\dot{x}_1(t) + 2x_2(t)\dot{x}_2(t) \equiv 0,$$

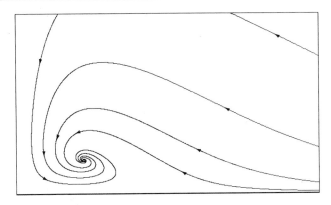

Figure 7.8. *Phase portrait of the competing species equations* (7.5) *in the positive quadrant for* $a_1 = a_2 = a_3 = a_4 = 1$ *and* $b_1 = 0.4$, $b_2 = 0.2$.

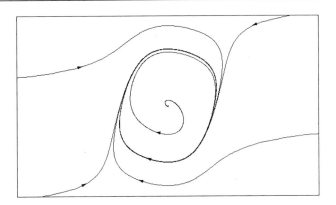

Figure 7.9. *Phase portrait of the oscillator of Van der Pol* [Eq. (7.6)]. *The periodic orbit attracting the nearby orbits is called a* limit cycle.

and so

$$\|\mathbf{x}(t)\|^2 = \|\mathbf{x}^0\|^2 \quad \text{for all } t.$$

Consequently, the solution $\mathbf{x}(t)$ must remain on the circle of radius $\|\mathbf{x}^0\|$ with its center at the origin. Since the circle contains no equilibrium points, the solution must be periodic of some period.

The direction of the arrows on the circular orbits can easily be determined from the differential equation. For example, on the positive x_1-axis the vector field points down, hence the solutions move in the clockwise direction. The resulting phase portrait consisting of a family of periodic orbits encircling an equilibrium point is called a *center*.

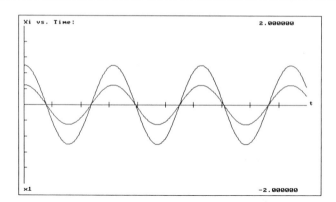

Figure 7.10. *Graphs of $x_1(t)$ vs. t of two different solutions of the linear harmonic oscillator. Notice that both solutions have the same period.*

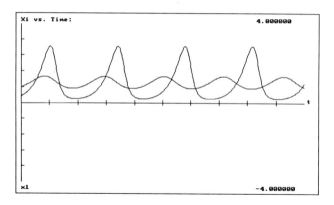

Figure 7.11. *Graphs of $x_1(t)$ vs. t of two solutions of the predator-prey equations. Notice that the solutions have different periods.*

We intentionally proceeded to analyze Eq. (7.3) without solving the differential equation in order to demonstrate that qualitative information can be obtained without the knowledge of explicit solutions. (Also, the idea above has far-reaching generalizations which we will discuss in Section 9.3.) In fact, the only thing that we do not know at this point is the period of a solution. This, of course, can be determined to be 2π directly from explicit solutions given in the next chapter, or indirectly as indicated in the exercises. ◇

One of our main objectives in the following chapters will be the study of the asymptotic behavior of solutions of Eq. (7.9). The concept, and the variety, of the set of limit points of an orbit of a planar system is considerably more complicated than in the case of scalar equations. Therefore, before giving any examples, we make several precise definitions.

Definition 7.10. *A point* **y** *is an* ω-*limit point of the orbit* $\gamma(\mathbf{x}^0)$ *if there is a sequence* t_j *with* $t_j \to \beta_{\mathbf{x}^0}$ *as* $j \to +\infty$ *such that* $\varphi(t_j, \mathbf{x}^0) \to \mathbf{y}$ *as* $j \to +\infty$. *That is,* **y** *is an* ω-*limit point of the orbit* $\gamma(\mathbf{x}^0)$ *if, for any* $\varepsilon > 0$, *there is a* $t(\varepsilon)$ *such that* $\|\mathbf{y} - \varphi(t(\varepsilon), \mathbf{x}^0)\| < \varepsilon$. *The set of all* ω-*limit points of the orbit* $\gamma(\mathbf{x}^0)$ *is called the* ω-*limit set of* $\gamma(\mathbf{x}^0)$ *and is denoted by* $\omega(\mathbf{x}^0)$.

An equivalent definition of $\omega(\mathbf{x}^0)$ which is geometrically easier to understand is

$$\omega(\mathbf{x}^0) = \bigcap_{\tau \geq 0} \overline{\gamma^+(\varphi(\tau, \mathbf{x}^0))}.$$

In narrative form, to find $\omega(\mathbf{x}^0)$ keep discarding the "tail end" of the closure of the positive orbit of $\mathbf{x}^0$. The equivalence of these two definitions of the ω-limit set is not immediately obvious; this you may want to ponder.

The concept of the α-*limit set* $\alpha(\mathbf{x}^0)$ of an orbit $\gamma(\mathbf{x}^0)$ can be defined similarly by reversing the direction of time. More precisely, a point $\mathbf{y} \in \alpha(\mathbf{x}^0)$ if there is a sequence t_j with $t_j \to \alpha_{\mathbf{x}^0}$ as $j \to +\infty$ such that $\varphi(t_j, \mathbf{x}^0) \to \mathbf{y}$ as $j \to +\infty$. The geometric definition is

$$\alpha(\mathbf{x}^0) = \bigcap_{\tau \leq 0} \overline{\gamma^-(\varphi(\tau, \mathbf{x}^0))}.$$

To illustrate the concepts of limit sets, let us consider several simple, but common, situations.

Let $\bar{\mathbf{x}}$ be a point in $\mathbb{R}^2$ such that $\varphi(t, \mathbf{x}^0) \to \bar{\mathbf{x}}$ as $t \to +\infty$. In this case, we can choose the sequence $\{t_j\}$ to be any increasing sequence with $t_j \to +\infty$ as $j \to +\infty$ to show that $\bar{\mathbf{x}} \in \omega(\mathbf{x}^0)$. Also, $\bar{\mathbf{x}}$ is an equilibrium point (why?) and $\omega(\mathbf{x}^0) = \bar{\mathbf{x}}$.

As the next example of a planar limit set, let us consider the case of a point on a periodic orbit. More specifically, suppose that $\varphi(t, \mathbf{x}^0)$ is a periodic solution of minimal period p. Then $\gamma(\mathbf{x}^0)$ is a closed curve and $\omega(\mathbf{x}^0) = \gamma(\mathbf{x}^0) = \alpha(\mathbf{x}^0)$. In fact, if $\mathbf{y} \in \gamma(\mathbf{x}^0)$, then there is a $t_{\mathbf{y}} \in [0, p)$ such that $\mathbf{y} = \varphi(t_{\mathbf{y}}, \mathbf{x}^0)$. If we now take the sequence $t_j = jp + t_{\mathbf{y}}$, where $j = 1, 2, \ldots$, then $\mathbf{y} = \varphi(t_j, \mathbf{x}^0)$ for all j and $\mathbf{y} \in \omega(\mathbf{x}^0)$. It is evident that no other points can be in $\omega(\mathbf{x}^0)$. A similar argument shows that $\alpha(\mathbf{x}^0) = \gamma(\mathbf{x}^0)$.

We will meet equilibria and periodic orbits as limit sets in many specific planar systems. Examples of more complicated limit sets and their complete classification will be given in Chapter 12.

Exercises  ♣ ♡ ♠ ◇

7.3. *Numerical dangers:* As we saw in Chapter 3, numerical solutions of scalar differential equations can sometimes be tricky. Similar difficulties may become more pronounced in planar systems. To appreciate some of the possible dangers, try to compute with PHASER one of the nontrivial periodic orbits of the linear harmonic oscillator using Euler's algorithm with step size $h = 0.1$. Do you believe what you see? What is the source of the difficulty? Recompute the same orbit with Euler, but use smaller step sizes, for example, 0.05, 0.01, and 0.001. Will you ever get the correct result in a *finite* time? Try these experiments with negative step sizes also. Finally, compute the same orbit using Runge–Kutta's algorithm with step size $h = 0.1$.

7.4. *Equivalent norms:* Even though it may not be an exciting exercise, try to establish the inequalities (7.12) regarding the equivalence of the Euclidean and sup norms.

7.5. *Recreating pictures:* The differential equations of the linear harmonic oscillator, predator-prey model, and planar pendulum are stored in the 2D library of PHASER under the names *linear2d, predprey,* and *pendulum,* respectively. Recreate the illustrations in this section using PHASER. Also, perform experiments to estimate the periods of the periodic orbits of these systems. Do the periods increase monotonically as you move away from an equilibrium point?

7.6. *Limit sets:* Suppose that the vector-valued function $\mathbf{x}(t)$ given in each case below is the solution of some differential equation on the plane. Find the α- and ω-limit sets if they exist:

$$\text{(a) } \mathbf{x}(t) = e^{-t} \begin{pmatrix} \cos t \\ -\sin t \end{pmatrix}; \qquad \text{(b) } \mathbf{x}(t) = e^{t} \begin{pmatrix} \cos t \\ -\sin t \end{pmatrix};$$

$$\text{(c) } \mathbf{x}(t) = e^{-t} \begin{pmatrix} \cos t \\ -\sin t \end{pmatrix} + \begin{pmatrix} 1 \\ 0 \end{pmatrix}; \qquad \text{(d) } \mathbf{x}(t) = \begin{pmatrix} \cos t \\ -\sin t \end{pmatrix}.$$

7.7. *Two interesting solutions:* Verify that the function

$$p(t) = 1 - \left(\frac{1 - e^{t}}{1 + e^{t}} \right)^{2}$$

is a solution of the second-order differential equation $\ddot{p} - p + \frac{3}{2}p^2 = 0$. Convert this equation to a first-order planar system and determine the α- and ω-limit sets of the solution. To see the orbit of this solution on the plane, enter this equation into PHASER and use an appropriate initial value.

Repeat these steps for the solution $q(t) = \sqrt{2}\,\mathrm{sech}\, t$ of the second-order differential equation $\ddot{q} - q + q^3 = 0$.

The two types of orbits represented by these two solutions are dubbed *homoclinic* and *heteroclinic,* respectively, and they are the source of much interesting and complicated dynamical behavior, as we shall see later.

7.3. Product Systems

The simplest examples of planar systems are constructed by taking a scalar equation on each axis and interpreting the flow of the two variables on the plane—*product systems*. In this section we present several such examples. Although this construction can be regarded as a way of manufacturing complicated examples from simple ones, we will see later that it also is an important technique for decomposing complicated equations into simple ones.

Example 7.11. *Simplest product:* Consider the "simplest" product system

$$\dot{x}_1 = 1$$
$$\dot{x}_2 = 0. \tag{7.13}$$

Notice first that there is no equilibrium point. Since $x_1(t)$ moves with uniform speed and $x_2(t)$ is constant, the orbits are the lines parallel to the x_1-axis. See Figure 7.12 for the phase portrait of Eq. (7.13). ◊

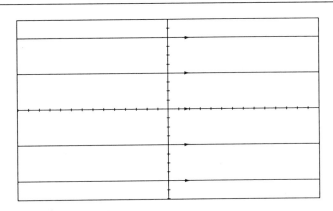

Figure 7.12. *Phase portrait of the product system* $\dot{x}_1 = 1$, $\dot{x}_2 = 0$.

Despite its simplicity, the product system (7.13) captures the local dynamics of any planar system away from an equilibrium point. A point $\hat{\mathbf{x}}$ in the plane is called an *ordinary point* of $\dot{\mathbf{x}} = \mathbf{f}(\mathbf{x})$ if $\mathbf{f}(\hat{\mathbf{x}}) \neq \mathbf{0}$. By the continuity of $\mathbf{f}$, there is a neighborhood of $\hat{\mathbf{x}}$ containing only ordinary points. In such a neighborhood, we have the following theorem:

Theorem 7.12. (Flow Box Theorem) *In a sufficiently small neighborhood of an ordinary point of the planar system* $\dot{\mathbf{x}} = \mathbf{f}(\mathbf{x})$ *there is a differentiable*

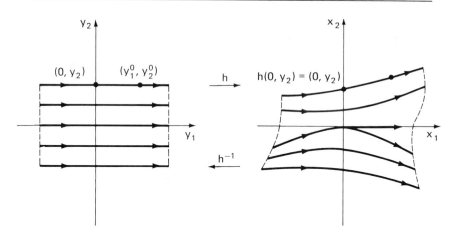

Figure 7.13. *Diffeomorphism h for constructing a flow box.*

change of coordinates $\mathbf{y} = \mathbf{y}(\mathbf{x})$ such that in the new coordinates the original system becomes the product system $\dot{y}_1 = 1$, $\dot{y}_2 = 0$.

Proof. We may first change coordinates (translation, rotation, scaling in time) so that the ordinary point of interest is moved to the origin, and the vector field at the origin is $\mathbf{f}(\mathbf{0}) = (1, 0)$, that is, the vector field at the origin is pointing along the x_1-axis. In a sufficiently small neighborhood of the origin, consider the mapping $h : \mathbb{R}^2 \to \mathbb{R}^2$ given by

$$h(y_1, y_2) \equiv \varphi(y_1, (0, y_2)),$$

where φ is the flow of $\dot{\mathbf{x}} = \mathbf{f}(\mathbf{x})$. From the fundamental theorem given in the Appendix, the map h is differentiable. Also, observe that h is the identity map on the y_2-axis, and on the y_1-axis its derivative at the origin is $\mathbf{f}(\mathbf{0}) = (1, 0)$. Hence, from the Inverse Function Theorem, h has a differentiable inverse, that is, h is a diffeomorphism. Consequently, h maps an open neighborhood of the origin in the (y_1, y_2)-plane, which we take to be the box shown in Figure 7.13, diffeomorphically to an open neighborhood of the origin on the (x_1, x_2)-plane.

Now, using the differentiable mapping h^{-1}, let us pull back the flow $\varphi(t, (x_1^0, x_2^0))$ to the (y_1, y_2)-coordinates. Let (y_1^0, y_2^0) be the unique point satisfying $h(y_1^0, y_2^0) = (x_1^0, x_2^0)$. Then,

$$\begin{aligned}
h^{-1}\varphi(t, (x_1^0, x_2^0)) &= h^{-1}\varphi(t, h(y_1^0, y_2^0)) \\
&= h^{-1}\varphi(t, \varphi(y_1^0, (0, y_2^0))) \\
&= h^{-1}\varphi(t + y_1^0, (0, y_2^0)) \\
&= h^{-1}h(t + y_1^0, y_2^0) \\
&= (t + y_1^0, y_2^0).
\end{aligned}$$

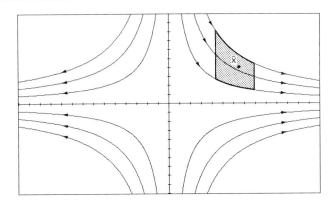

Figure 7.14. *Flow box for the product system* $\dot{x}_1 = x_1$, $\dot{x}_2 = -x_2$ *about an ordinary point* $\hat{\mathbf{x}}$.

Since the vector-valued function $(t + y_1^0, y_2^0)$ is the flow of the product system $\dot{y}_1 = 1$, $\dot{y}_2 = 0$, the desired coordinate system is (y_1, y_2). $\Diamond$

It is evident in the proof above that the Flow Box Theorem guarantees the existence of a box-like neighborhood of any ordinary point of any planar system such that the orbits of the system enter at one end of the box and flow out through the other—*flow box*. Moreover, no orbit leaves through the sides of the box. Let us illustrate these ideas on a simple example.

Example 7.13. *A flow box:* Consider the product system

$$\dot{x}_1 = x_1$$
$$\dot{x}_2 = -x_2. \tag{7.14}$$

Using the scalar differential equation $dx_2/dx_1 = -x_2/x_1$, it is easy to see that the orbits of Eq. (7.14) in the neighborhood of an ordinary point $\hat{\mathbf{x}}$ are the hyperbolas $x_1 x_2 = constant$. Therefore, if we introduce the new coordinates $y_1 = \ln x_1$ with $x_1 > 0$ and $y_2 = x_1 x_2$, then the sides of the flow box can be taken to be the lines $y_1 = constant$ and $y_2 = constant$; see Figure 7.14. It is easy to verify that in the new coordinates, Eq. (7.14) becomes the product system (7.13). $\Diamond$

To continue our discussion of product systems, let us consider a generalization of this simple example and determine its phase portrait.

Example 7.14. *Linear product:* Let a and b be two real constants and consider the pair of linear equations

$$\dot{x}_1 = ax_1$$
$$\dot{x}_2 = bx_2. \tag{7.15}$$

The solutions of this linear system are given by

$$x_1(t) = e^{at}x_1^0, \quad x_2(t) = e^{bt}x_2^0.$$

The phase portraits of (7.15) on the (x_1, x_2)-plane can be determined quite easily; see Figure 7.15. Let us fix $a < 0$ and consider all possible values of b. You might like to draw the phase portraits for the cases when $a > 0$.

If $b < a < 0$, then both $x_1(t), x_2(t) \to 0$ exponentially as $t \to +\infty$, that is, the ω-limit set of any orbit is the origin $(0, 0)$ which is the unique equilibrium point. Furthermore, all the orbits except the x_2-axis are asymptotically tangent, as $t \to +\infty$, to the x_1-axis because

$$\frac{dx_2}{dx_1} = \frac{bx_2}{ax_1} = \frac{x_2^0}{x_1^0}\frac{b}{a}e^{(b-a)t} \to 0 \quad \text{as } t \to +\infty.$$

If $b = a < 0$, then all orbits tend to the origin along straight lines as $t \to +\infty$. The phase portrait of the case $a < b < 0$ is similar to that of $b < a < 0$ with the roles of the axes interchanged. In all of these cases the equilibrium point $(0, 0)$ is called a stable *node* because the ω-limit sets of all the orbits is the origin. Typical phase portraits of these three cases are shown in Figures 7.15a–c. If $a < 0 = b$, then all the points on the x_2-axis are equilibrium points; see Figure 7.15d.

If $a < 0 < b$, then the equilibrium point at the origin is called a *saddle* and the flow is shown in Figure 7.15e (see also Example 7.13). It has a completely different structure than a node since there are two orbits whose ω-limit sets are $(0, 0)$ and two orbits whose α-limit sets are again the origin. All other orbits leave a neighborhood of the origin in the directions of both increasing and decreasing t.

The linear product system (7.15) has a nice geometric interpretation as a *gradient system*. Let us consider the function

$$F(x_1, x_2) = -\tfrac{1}{2}ax_1^2 - \tfrac{1}{2}bx_2^2$$

and its *gradient*

$$\nabla F(x_1, x_2) \equiv \left(\frac{\partial}{\partial x_1} F(x_1, x_2), \frac{\partial}{\partial x_2} F(x_1, x_2) \right).$$

It is easy to verify that Eq. (7.15) is equivalent to the system

$$\dot{x}_1 = -\frac{\partial}{\partial x_1} F(x_1, x_2)$$

$$\dot{x}_2 = -\frac{\partial}{\partial x_2} F(x_1, x_2),$$

(7.16)

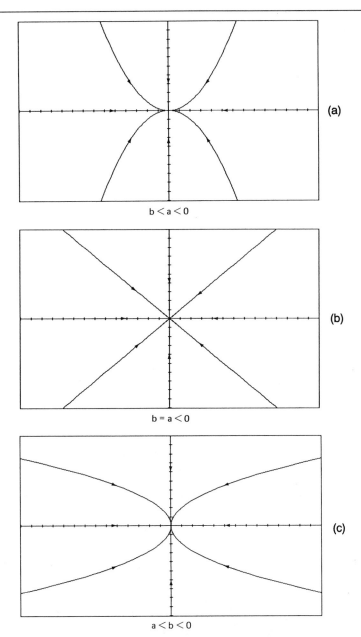

Figure 7.15. Typical phase portraits of the product linear system $\dot{x}_1 = ax_1$, $\dot{x}_2 = bx_2$ (continued).

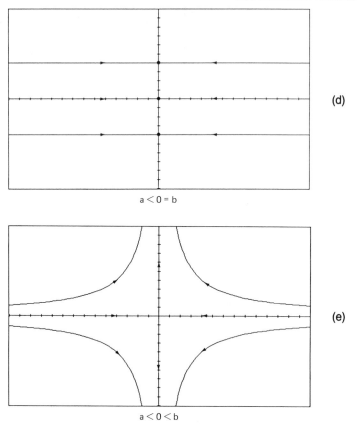

Figure 7.15 Continued.

where the vector field is now written as the negative of the gradient of F :

$$\dot{\mathbf{x}} = -\nabla F(\mathbf{x}).$$

If $(x_1(t), x_2(t))$ is a solution of Eq. (7.16), then using the chain rule we have

$$\frac{d}{dt}F(x_1(t),\, x_2(t)) = -\left[\frac{\partial}{\partial x_1}F\left(x_1(t),\, x_2(t)\right)\right]^2 - \left[\frac{\partial}{\partial x_2}F\left(x_1(t),\, x_2(t)\right)\right]^2$$
$$= -\|\nabla F(x_1(t),\, x_2(t))\|^2 \le 0.$$

Thus, F is always decreasing along the solutions of Eq. (7.16) and can be thought of as a "potential" function of Eq. (7.16). The graphs of the surface $z = F(x_1, x_2)$ for several choices of a and b are shown in Figure 7.16. A

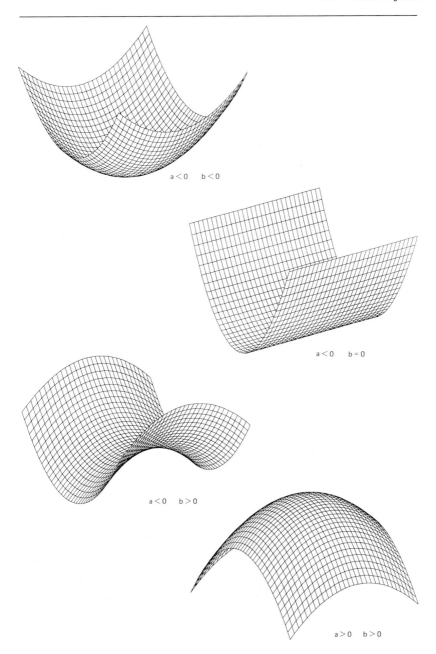

Figure 7.16. *Graphs of the potential function* $F(x_1, x_2) = -\frac{1}{2}(ax_1^2 + bx_2^2)$ *for several choices of* a *and* b.

particle starting on these surfaces moves "downhill" under the flow defined by Eq. (7.16). For example, when $a < 0$, $b < 0$, the particle goes to the minimum of F. However, when $a < 0 < b$, the surface has the shape of a saddle and the particle misses the critical point unless it is started at a special position. ◇

It is also easy to construct nonlinear examples on the plane using products of nonlinear scalar equations. We present one such example below. You might like to select some other nonlinear scalar equations from Part I and draw the phase portraits of their products.

Example 7.15. *Product logistic:* Let us consider the following system consisting of a copy of the logistic equation (3.2) on each axis:

$$\dot{x}_1 = x_1(1 - x_1)$$
$$\dot{x}_2 = x_2(1 - x_2). \tag{7.17}$$

We have plotted the phase portrait of Eq. (7.17) in Figure 7.17 which you are invited to decipher. ◇

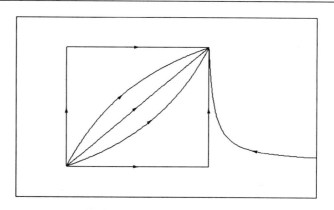

Figure 7.17. *Phase portrait of a pair of logistic equations as a planar product system.*

It is possible to construct product equations in coordinate systems other than cartesian coordinates. For instance, product systems in polar coordinates are particularly useful for constructing examples with periodic orbits. Of course, once constructed, such examples can be transformed into cartesian coordinates, if so desired. Here is a favorite example of this sort.

Example 7.16. *Polar product:* Consider the planar system

$$\dot{x}_1 = x_2 + x_1(1 - x_1^2 - x_2^2)$$
$$\dot{x}_2 = -x_1 + x_2(1 - x_1^2 - x_2^2). \tag{7.18}$$

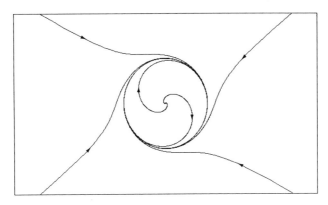

Figure 7.18. *Phase portrait of the system* (7.18).

The appearance of the term $x_1^2 + x_2^2$ in both equations suggests the presence of circular symmetry. Hence, if we introduce polar coordinates (r, θ) by letting $x_1 = r \cos \theta$ and $x_2 = r \sin \theta$, then Eq. (7.18) becomes equivalent to the decoupled system

$$\dot{r} = r(1 - r^2)$$
$$\dot{\theta} = -1. \tag{7.19}$$

Since this system consists of two independent scalar equations, its analysis is quite easy: the radial component is the logistic equation and the angular component moves with unit speed in the clockwise direction. Thus, the first equilibrium point of the radial equation at $r = 0$ is an equilibrium of Eq. (7.18), but the other equilibrium at $r = 1$ corresponds to a periodic orbit of the system. Furthermore, it is clear from the flow of the radial equation that all solutions of Eq. (7.18), except the origin, spiral onto the unit circle $x_1^2 + x_2^2 = 1$ as $t \to +\infty$. The phase portrait of Eq. (7.18) is drawn in Figure 7.18.

In fact, the general solution of Eq. (7.19) can be found explicitly:

$$r(t) = \frac{r_0}{\left[r_0^2 + (1 - r_0^2)e^{-2t}\right]^{1/2}}$$
$$\theta(t) = -t + \theta_0,$$

where $r_0 = r(0)$ and $\theta_0 = \theta(0)$. Notice that solutions are defined for all $t \in (-\infty, +\infty)$.

Let (r_0, θ_0) be an initial value with $r_0 \neq 0$. Then it is easy to see that any point $(1, \hat{\theta})$ on the unit circle is an ω-limit point of (r_0, θ_0) by simply taking the sequence t_j in Definition 7.10 to be $t_j = (\theta_0 - \hat{\theta}) + 2\pi j$. Consequently, the ω-limit set of any initial value with $r_0 \neq 0$ is the unit

circle. Hence, all the orbits of Eq. (7.19), except the equilibrium point at
the origin, spiral onto the unit circle with increasing time. The unit circle
itself is a periodic orbit with period 2π. Finally, the origin is the α-limit set
of orbits with initial value $r_0 < 1$ (what is an appropriate sequence t_j?). $\diamondsuit$

Exercises ——————————————————————————— ♣ ♡ ♠ ◇

7.8. Sketch the phase portraits and discuss the α- and ω-limit sets of orbits of
the following product systems:
   (a) $\dot{x}_1 = -x_1$, $\dot{x}_2 = x_2^2$;   (b) $\dot{x}_1 = x_1^2$, $\dot{x}_2 = -x_2$;
   (c) $\dot{x}_1 = -x_1$, $\dot{x}_2 = x_2 - x_2^3$;   (d) $\dot{x}_1 = x_1 - x_1^3$, $\dot{x}_2 = x_2 - x_2^3$.

7.9. Find a function F so that the vector field in Example 7.15 can be written
as a gradient system. Plot the graph of this function.

7.4. First Integrals and Conservative Systems

Certain natural systems possess special characteristics, such as symmetry
or conservation of energy, which facilitate an analytical investigation of the
differential equations modeling such systems. For example, the planar pendulum (7.2) without friction, albeit idealized, conserves its energy imparted
at the start of its motion. By using this physical information, the analysis
of the pendulum can be reduced to curve sketching and the dynamics of a
scalar differential equation. In this section, we briefly explain how to detect the presence of conserved quantities and, when such a quantity exists,
how to perform the reduction alluded to above. We begin with a precise
definition of a conserved quantity, or a *first integral*.

Definition 7.17. *A real-valued C^1 function*

$$H : \mathbb{R}^2 \to \mathbb{R}; \qquad \mathbf{x} \mapsto H(\mathbf{x})$$

*that is not constant on any open subset of $\mathbb{R}^2$ is called a first integral of
a planar differential equation $\dot{\mathbf{x}} = \mathbf{f}(\mathbf{x})$ if the function H is constant along
every solution, that is, for any solution $\mathbf{x}(t)$ with initial value $\mathbf{x}(0) = \mathbf{x}^0$,
the composite function satisfies $H(\mathbf{x}(t)) = H(\mathbf{x}^0)$ for all t for which the
solution is defined.*

We should remark that in the definition of a first integral above we
required the domain of H to be $\mathbb{R}^2$ (or, the domain of the definition of
the vector field); for this reason, H as defined above is often said to be a
global first integral. In our presentation in this chapter we will simply use
the term a first integral. The notion of a *local* first integral is discussed in
the exercises.

It is easy to check if a function is a first integral of a differential equation using just the vector field, without any knowledge about the solutions: H is a first integral if

$$\dot{H}(\mathbf{x}) \equiv \nabla H(\mathbf{x}) \cdot \mathbf{f}(\mathbf{x}) = \frac{\partial H}{\partial x_1}(\mathbf{x}) f_1(\mathbf{x}) + \frac{\partial H}{\partial x_2}(\mathbf{x}) f_2(\mathbf{x}) = 0. \qquad (7.20)$$

Here are some planar systems which possess first integrals:

Example 7.18. Let us reconsider three examples from Section 7.1.
Linear Harmonic Oscillator: The total energy

$$H(x_1, x_2) = \tfrac{1}{2}(x_1^2 + x_2^2)$$

is a first integral of the linear harmonic oscillator (7.3) because

$$\dot{H}(\mathbf{x}) = x_1 x_2 - x_2 x_1 = 0.$$

This is essentially the computation we have performed earlier in Example 7.9, except for the factor $1/2$.

Planar pendulum: The total energy function

$$H(x_1, x_2) = \tfrac{1}{2}ml^2 x_2^2 + mgl(1 - \cos x_1)$$

of the planar pendulum is a first integral, as a similar derivative computation shows.

Predator-prey: Although the vector field of the predator-prey equations (7.4) is defined on the entire plane, the biologically meaningful region is the positive first quadrant. In this restricted domain, the function

$$H(x_1, x_2) = a_1 \ln x_2 - a_2 x_2 + a_3 \ln x_1 - a_4 x_1$$

is a first integral. Three-dimensional plots of the three first integrals above are in Figure 17.19. $\Diamond$

We will shortly reveal how we found the first integrals in the example above. First, however, let us explain why it is beneficial to have a first integral. The main utility of a first integral stems from the simple observation that the orbit through any $\mathbf{x}^0$ of the differential equation lies on the *level set*

$$H^{-1}(H(\mathbf{x}^0)) \equiv \{ \mathbf{x} : H(\mathbf{x}) = H(\mathbf{x}^0) \}$$

of the function H. A function H that takes on a constant value on all of $\mathbb{R}^2$ trivially satisfies Eq. (7.20) for any differential equation. In this trivial case, the only level set is the entire plane; this, of course, yields no information about the orbits. However, if H is not constant on any

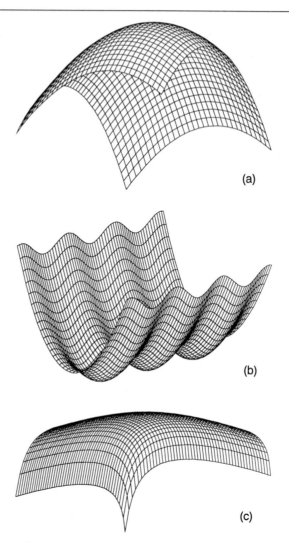

Figure 7.19. *Graphs of the first integrals of* (a) *the linear harmonic oscillator,* (b) *planar pendulum, and* (c) *predator-prey.*

open subset of the entire plane, then a level set of H is, in general, a one-dimensional subset of the plane and the orbit through $\mathbf{x}^0$ is a connected piece of this one-dimensional set. Now, by just sketching the level sets of a first integral, one can infer a considerable amount of information about the shapes of the orbits and hence the phase portrait. The exercise of sketching curves—algebraic geometry—can at times be difficult even for polynomial equations. For example, while it is clear that the level sets of the first integral of the linear harmonic oscillator are concentric circles about the origin, we will have to work a little to determine the level sets of the first integral of the pendulum.

Unlike the impression we may have given in the example above, "most" differential equations do not possess nontrivial first integrals. Here is one such example.

Example 7.19. *Nonexistence of a first integral:* Consider the planar system

$$
\begin{aligned}
\dot{x}_1 &= -x_1 \\
\dot{x}_2 &= -x_2.
\end{aligned}
\tag{7.21}
$$

As we saw in Example 7.14, the orbits of this system are rays terminating at the origin. Any first integral must take on a constant value on any one of these rays. However, the requirement that a first integral be continuous, in particular, at the origin, implies that the first integral must have the same constant value on all the orbits. Thus, any first integral of this system must be the constant function. ◇

We now turn to the important question of the existence of first integrals. Since a first integral gives information about the shapes of orbits and not their time parametrizations, it is convenient to eliminate the time variable from the system $\dot{\mathbf{x}} = \mathbf{f}(\mathbf{x})$. This is easy to do locally as we now explain. If $\mathbf{x}^0$ is not an equilibrium point, then at least one of $f_1(\mathbf{x}^0)$ or $f_2(\mathbf{x}^0)$ is not zero. Let us suppose that $f_1(\mathbf{x}^0) \neq 0$. Then, there is an open neighborhood of $\mathbf{x}^0$ such that $f_1(\mathbf{x}) \neq 0$ in this neighborhood. Therefore, the first component of the orbit $\varphi(t, \mathbf{x}^0)$ is strictly monotone in t and it makes sense to replace the t-parametrization of the orbit by a parametrization in the first coordinate, that is, in the neighborhood of $\mathbf{x}^0$, the orbit through $\mathbf{x}^0$ can be defined as a solution of the nonautonomous scalar equation

$$
\frac{dx_2}{dx_1} = \frac{f_2(x_1, x_2)}{f_1(x_1, x_2)}.
\tag{7.22}
$$

Similar remarks apply when $f_2(\mathbf{x}^0) \neq 0$. If $\mathbf{x}^0$ is an equilibrium point, then the orbit through $\mathbf{x}^0$ is the point $\mathbf{x}^0$ itself and no parametrization is necessary.

Now, suppose that

$$
H(x_1, x_2) = constant
\tag{7.23}
$$

is the (implicit) solution of the scalar equation (7.22). Then it is a simple computation to show that such a solution is indeed a first integral of $\dot{\mathbf{x}} = \mathbf{f}(\mathbf{x})$ in a suitable domain. To wit, differentiate Eq. (7.23),

$$
\frac{dH}{dx_1} \equiv 0 = \frac{\partial H}{\partial x_1} + \frac{\partial H}{\partial x_2}\frac{dx_2}{dx_1},
$$

and use Eq. (7.22) to obtain Eq. (7.20).

If we are lucky, the function H as constructed above will be defined on all of $\mathbb{R}^2$ to be a first integral of $\dot{\mathbf{x}} = \mathbf{f}(\mathbf{x})$. Here are two simple illustrations of this method.

Example 7.20. *Detecting first integrals:* Consider the linear harmonic oscillator (7.3) in a neighborhood of $\mathbf{x}^0 \neq \mathbf{0}$. Then the corresponding scalar equation (7.22) becomes

$$\frac{dx_2}{dx_1} = \frac{x_1}{-x_2}.$$

Since this differential equation is separable, it is easy to see by direct integration that its solutions satisfy the implicitly given equation $\frac{1}{2}\left(x_1^2 + x_2^2\right) = $ *constant*, thus recovering the first integral we have given earlier.

It is instructive to see how the method above for detecting first integrals fails for Eq. (7.21). The scalar equation (7.22) for the system (7.21) away from the origin is

$$\frac{dx_2}{dx_1} = \frac{x_2}{x_1}.$$

Integration of this separable equation yields the function

$$H(x_1,\, x_2) = \frac{x_2}{x_1},$$

which, unfortunately, is not C^1 on the plane and thus not a first integral of the differential equation (7.21). $\Diamond$

The most notable examples of differential equations possessing first integrals arise in mechanical systems without friction and the existence of a first integral is most apparent in their so-called Hamiltonian formulations. For a given C^1 function $H : \mathbb{R}^2 \to \mathbb{R}$, a planar system of differential equations of the form

$$\dot{x}_1 = \frac{\partial H}{\partial x_2}$$

$$\dot{x}_2 = -\frac{\partial H}{\partial x_1}$$

is called a *Hamiltonian system* with the Hamiltonian H. The total energy of a mechanical system, up to a multiplicative or additive constant, can often be taken as the Hamiltonian of the system. It is clear from the special form of the equations that the Hamiltonian function is a first integral— *conservation of energy.* A special class of Hamiltonian systems known as *conservative systems* comes from a second-order differential equation of the form $\ddot{y} + g(y) = 0$. Indeed, it is easy to verify that

$$H(x_1,\, x_2) = \tfrac{1}{2}x_2^2 + \int_0^{x_1} g(u)\, du$$

is the Hamiltonian of the equivalent first-order system

$$\dot{x}_1 = x_2$$
$$\dot{x}_2 = -g(x_1).$$

The theory of conservative, more generally, Hamiltonian, systems is one of the oldest yet most vigorous areas of dynamical systems. We will devote many pages to this important topic in later chapters. For the moment, however, we will be content to conclude this section with an analysis of the phase portrait of the planar pendulum, Figure 7.6, viewed as a Hamiltonian system.

Example 7.21. *Pendulum:* Let us begin our analysis of the pendulum by finding a first integral. The orbits of the pendulum away from the equilibrium points $(n\pi, 0)$, where n is any integer, are given by the solutions of the scalar equation

$$\frac{dx_2}{dx_1} = \frac{-(g/l)\sin x_1}{x_2}.$$

The direct integration of this separable equation yields the first integral

$$H(x_1, x_2) = \tfrac{1}{2}(x_2)^2 + \frac{g}{l}(1 - \cos x_1). \tag{7.24}$$

We now indicate how to obtain the level sets

$$H^{-1}\big((H(\mathbf{x}^0)\big) \equiv \{\,\mathbf{x} : H(\mathbf{x}) = H(\mathbf{x}^0)\,\}$$

of the first integral H. Since the function H is 2π-periodic in x_1, we will confine our analysis to the vertical strip of the plane with $-\pi \le x_1 \le \pi$. Moreover, again due to this periodicity, it suffices to take initial data on the x_2-axis. That is, we need to determine the shapes of the curves of the form

$$\tfrac{1}{2}(x_2)^2 + \frac{g}{l}(1 - \cos x_1) = \tfrac{1}{2}(x_2^0)^2. \tag{7.25}$$

For any x_2^0, the curve defined by Eq. (7.25) is symmetric with respect to the x_1-axis. Therefore, we need to plot only the curves

$$x_2 = \sqrt{(x_2^0)^2 - (2g/l)(1 - \cos x_1)} \tag{7.26}$$

and then reflect through the x_1-axis. Notice that the range of the values of x_1 needed to define the curve are given by the inequality

$$\left(x_2^0\right)^2 \ge (2g/l)(1 - \cos x_1).$$

With these observations, we can effectively construct the orbits of the pendulum by considering the values of x_2^0. The directions of the orbits can easily be inferred from the vector field.

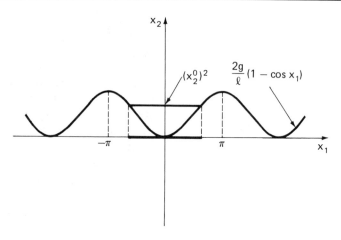

Figure 7.20. *Constructing the orbits of the planar pendulum from a first integral.*

For $x_2^0 = 0$, Eq. (7.26) gives the equilibrium points $(-\pi, 0)$, $(0, 0)$, and $(\pi, 0)$.

For $0 < \left(x_2^0\right)^2 < 4g/l$, the range of x_1 is an interval of length less than 2π and symmetric about the origin; see Figure 7.20. The curve defined by Eq. (7.26) when reflected about the x_1-axis yields a closed curve on the plane. Since there are no equilibrium points on it, this closed curve is a periodic orbit corresponding to the oscillation of the pendulum about the equilibrium position $(0, 0)$; see Figure 7.21.

For $(x_2^0)^2 = 4g/l$, the curve defined by Eq. (7.26) and its reflection about the x_1-axis is again a closed curve. However, on this closed curve there are several orbits. In particular, the equilibrium points $(-\pi, 0)$ and $(\pi, 0)$ are on this curve. These equilibria correspond to the vertical position of the pendulum while the pendulum is "sitting on its head." There are two other special orbits: one whose α-limit set is $(-\pi, 0)$ and ω-limit set is $(\pi, 0)$, the other, which is the reflection of this orbit, whose α-limit set is $(\pi, 0)$ and ω-limit set is $(-\pi, 0)$. These special orbits are called *heteroclinic orbits* and they correspond to the motions of the pendulum from one equilibrium point to the other, in infinite time. Because of its importance in dynamical systems, we record here the definition of a heteroclinic orbit for future reference.

Definition 7.22. *An orbit whose α-limit set is an equilibrium point and ω-limit set is another equilibrium point is called a heteroclinic orbit.*

If $(x_2^0)^2 > 4g/l$, then the range of x_1 is unrestricted and the curve defined by Eq. (7.26) is a 2π-periodic graph over the x_1-axis. The level set

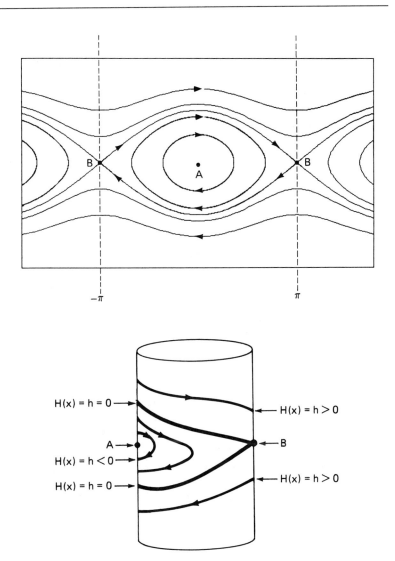

Figure 7.21. *Flow of the pendulum as viewed on the plane and on the cylinder.*

consists, of course, of this graph and its reflection about the x_1-axis. There are no equilibria on these curves and they correspond to the orbits of the motions of the pendulum with initial velocity so large that the pendulum revolves around and around without end.

Let us reexamine the qualitative features of the phase portrait of the pendulum and present a view of the flow on a cylinder. This is quite natural, and also convenient. Indeed, the physical state of the pendulum is determined by its angle of deviation from the vertical position and by its velocity. Two physical states of the pendulum that differ in θ by 2π should be considered the same. This, of course, is reflected in the differential equation describing the evolution of the states of the pendulum: Equation (7.2) remains the same under the change of variables $(x_1, x_2) \mapsto (x_1 + 2\pi, x_2)$. If we now take the first variable $x_1 \bmod 2\pi$, the flow of the pendulum takes place on the cylinder $S^1 \times \mathbb{R}$; see Figure 7.21. Notice that the orbits that go around and around now become periodic orbits encircling the cylinder. Besides its physical appeal, this observation has an important theoretical consequence, as we shall see in the next chapter: all orbits are bounded. Also, a heteroclinic orbit on the plane turns into another type of special orbit on the cylinder called a homoclinic orbit whose formal definition will appear in the following section.

The qualitative description of the flow of the equation of pendulum is now complete. However, there is one important quantitative question that still remains: what are the periods of the periodic orbits of the pendulum? Unlike in the case of the linear harmonic oscillator, the concentric periodic solutions, say, near the origin, of the pendulum have different periods; the period is a monotone function of the amplitude. Further details on the periods are contained in the exercises. ◇

Exercises _____

7.10. *On the line:* Define the notion of a first integral for a scalar autonomous differential equation. Does the scalar equation $\dot{x} = -x$ have a first integral that is not identically constant? Can you characterize all conservative scalar autonomous differential equations?

7.11. *A first integral for predator-prey equations:* Consider the predator-prey equations (7.4) in the positive quadrant. Its orbits away from the two equilibria $(0, 0)$ and $(a_3/a_4, a_1/a_2)$ are the solution curves of the scalar equation

$$\frac{dx_2}{dx_1} = \frac{-a_3 x_2 + a_4 x_1 x_2}{a_1 x_1 - a_2 x_1 x_2}.$$

Write this equation in the separable form

$$\frac{a_1 - a_2 x_2}{x_2} \frac{dx_2}{dx_1} = \frac{-a_3 + a_4 x_1}{x_1}$$

and integrate it to obtain the first integral

$$H(x_1, x_2) = a_1 \ln x_2 - a_2 x_2 + a_3 \ln x_1 - a_4 x_1.$$

Although it is not immediately obvious, show that the level sets of H are concentric closed curves encircling the equilibrium $(a_3/a_4, a_1/a_2)$, as depicted in the numerically computed phase portrait in Figure 7.7.

Hint: Write the level set $H(x_1, x_2) = k$ as the product $f(x_1) g(x_2) = K$, where the functions f and g are given by $f(x_1) = x_1^{a_3}/e^{a_4 x_1}$ and $g(x_2) = x_2^{a_1}/e^{a_2 x_2}$, and K is some constant. Now, determine the graphs of these functions by finding their critical points, etc.

7.12. *Local first integrals:* Show, using the Flow Box Theorem, that in a sufficiently small open neighborhood of a regular (nonequilibrium) point there always exists a local first integral. As seen in Example 7.19, however, a local first integral need not exist in an open neighborhood of an equilibrium point.

7.13. Draw the orbits and the direction of the flows of the following differential equations:
(a) $\dot{x}_1 = x_2(x_1^2 - x_2^2), \quad \dot{x}_2 = -x_1(x_1^2 - x_2^2);$
(b) $\dot{x}_1 = x_2(1 - x_1^2 - x_2^2), \quad \dot{x}_2 = -x_1(1 - x_1^2 - x_2^2).$
Warning: Watch your division when computing dx_2/dx_1.

7.14. Sketch the phase portraits of the following equations and discuss the α- and ω-limit sets of orbits:
(a) $\ddot{\theta} + \theta + \theta^3 = 0;$ (b) $\ddot{\theta} + \theta - \theta^3 = 0;$
(c) $\ddot{\theta} + \theta - \theta^2 = 0;$ (d) $\ddot{\theta} + \theta(1 - \theta)(\lambda - \theta) = 0, \quad 0 < \lambda < 1/2;$
(e) $\dot{x}_1 = \sin x_2, \quad \dot{x}_2 = -\sin x_1.$

7.15. *Period in conservative systems:* Consider the second-order equation $\ddot{y} + g(y) = 0$, or the equivalent first-order system

$$\dot{x}_1 = x_2$$
$$\dot{x}_2 = -g(x_1).$$

(a) Verify that this is a conservative system with the Hamiltonian function $H(x_1, x_2) = x_2^2/2 + G(x_1)$, where $G(x_1) = \int_0^{x_1} g(u)\, du$.
(b) Show that any periodic orbit of this system must intersect the x_1-axis at two points, say, $(a, 0)$ and $(b, 0)$ with $a < b$.
(c) Using the symmetry of the periodic orbits with respect to the x_1-axis, show that the minimal period T of a periodic orbit passing through two such points is given by

$$T = 2 \int_a^b \frac{du}{\sqrt{2[G(b) - G(u)]}}.$$

7.16. *Period of the pendulum:* For the pendulum:
(a) Show that every periodic orbit encircling the origin has $a = -b$ and the period is given by

$$T = \frac{4}{\sqrt{g/l}} \int_0^{\pi/2} \frac{d\xi}{\sqrt{1 - \sin^2(b/2) \sin^2 \xi}}.$$

Hint: Use the symmetry with respect to the x_2-axis, the half-angle formula $\cos u = 1 - 2\sin^2(u/2)$, and the change of variable $\sin(u/2) = \sin \xi \sin(b/2)$.

(b) The formidable integral above is known as an *elliptic integral*. Do not try to evaluate it. However, for b small, compute the following development of its Taylor expansion:

$$T = \frac{2\pi}{\sqrt{g/l}} \left[1 + \frac{1}{16}b^2 + \cdots \right].$$

Observe that the frequency $2\pi/T$ decreases with b. This behavior is referred to as a *soft spring*.

(c) What happens to the period as $b \to \pi$?

7.17. *A hard spring:* Consider the system

$$\dot{x}_1 = x_2$$
$$\dot{x}_2 = -x_1 - x_1^3.$$

(a) If you have not done so already, sketch the phase portrait.
(b) Obtain the integral formula for the period of the periodic orbits.
(c) Compute the first two terms of the Taylor expansion of the period when b is small and observe that the frequency increases with b. This behavior is referred to as a *hard spring*.
(d) Show that, for b large

$$T = \frac{4}{b} \int_0^1 \frac{dv}{\sqrt{1 - v^4}} \left[1 + O\left(\frac{1}{b}\right) \right].$$

(e) What happens to the frequency in the limit as $b \to +\infty$?

7.5. Examples of Elementary Bifurcations

In this section, we present several common examples of bifurcations in planar differential equations. The first pair of examples are product systems, and the bifurcations they exhibit are essentially the ones that we have studied in Section 1.3 in the context of autonomous scalar equations. The remaining examples, however, are specific to planar systems.

Example 7.23. *Saddle-node bifurcation:* Consider the product system depending on a scalar parameter λ:

$$\dot{x}_1 = \lambda + x_1^2$$
$$\dot{x}_2 = -x_2. \tag{7.27}$$

Observe that the second equation is linear with $x_2(t) \to 0$ as $t \to +\infty$. Thus, all the orbits of Eq. (7.27) eventually approach the x_1-axis where the dynamics of the system are governed by the first equation. We should point out, however, that the first equation is simply Example 2.2. The phase portraits of the flow of Eq. (7.27) for various parameter values are now easy to construct. For $\lambda < 0$, there are two equilibria. One of these equilibria is a *saddle* point because, as in the linear system in Example 7.13, there are two orbits near the origin such that the ω-limit sets of these orbits are the origin, and there are two orbits whose α-limit sets are again the origin. We will undertake a detailed study of the geometry of flows near a saddle point in Chapter 9. The other equilibrium point is a *node* because the ω-limit sets of all the orbits starting near this equilibrium point is the origin. At $\lambda = 0$ the two equilibria coalesce into one, and for $\lambda > 0$, the equilibrium point disappears; see Figure 7.22.

In the study of bifurcations of scalar autonomous equations, bifurcation diagrams were a very convenient way to portray much information about the qualitative dynamics of the flow. We want to continue to employ similar bifurcation diagrams in higher dimensions. For the present example, since the dynamics take place in the first equation we can follow the equilibrium points by drawing the x_1-coordinates of the equilibrium points as functions of the parameter λ. The resulting bifurcation diagram is then the same as Figure 2.3. $\diamondsuit$

Saddle-node bifurcations of equilibria occur quite commonly in non-product systems as well. The example above, despite its simplicity, captures the essential features of the general case. Unraveling the meaning of this remark will be the subject of Chapter 10.

Example 7.24. *Pitchfork bifurcation:* Consider the product system

$$\dot{x}_1 = -\lambda x_1 - x_1^3$$
$$\dot{x}_2 = -x_2. \tag{7.28}$$

As in the previous example, the dynamics of the system are contained in the first equation, which we have already analyzed as Example 2.5. The phase portraits of the system (7.28) for several values of the parameter λ are depicted in Figure 7.23. As in the previous example, if we represent the equilibrium solutions by their x_1-coordinates, then the resulting bifurcation diagram of Eq. (7.28) is the same as the one in Figure 2.10. $\diamondsuit$

Example 7.25. *Vertical bifurcation:* Consider the following one-parameter perturbation of the harmonic oscillator:

$$\dot{x}_1 = \lambda x_1 + x_2$$
$$\dot{x}_2 = -x_1 + \lambda x_2. \tag{7.29}$$

In polar coordinates (see Example 7.16), Eq. (7.29) is equivalent to the decoupled system

$$\dot{r} = \lambda r$$
$$\dot{\theta} = -1. \tag{7.30}$$

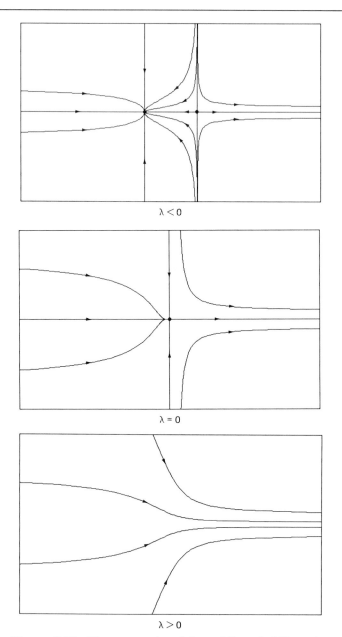

$\lambda < 0$

$\lambda = 0$

$\lambda > 0$

Figure 7.22. *Phase portraits of the saddle-node bifurcation.*

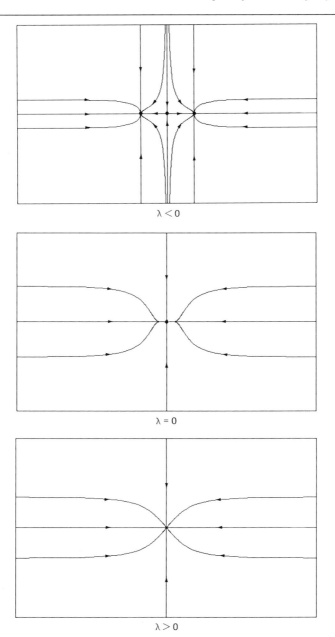

$\lambda < 0$

$\lambda = 0$

$\lambda > 0$

Figure 7.23. *Phase portraits of the pitchfork bifurcation.*

When $\lambda < 0$, all solutions spiral clockwise into the origin with increasing t. For $\lambda = 0$, this is the harmonic oscillator and as we have already seen, all solutions are periodic so that the origin is a center. Since at this value of the parameter the number of periodic orbits changes from none to many, we consider $\lambda = 0$ a bifurcation value. For $\lambda > 0$, all solutions spiral out clockwise without bounds; see Figure 7.24. Let us now plot a bifurcation diagram for the periodic orbits of Eq. (7.29). Since every periodic orbit encircles the origin, it is convenient to represent a periodic orbit by its "amplitude," that is, by the a at which point the periodic orbit intersects the x_1-axis. With this convention, the equilibrium point at the origin is viewed as a degenerate periodic orbit of zero amplitude. The resulting bifurcation diagram, which is sometimes aptly dubbed as *vertical bifurcation,* is shown in Figure 7.25. $\diamond$

Example 7.26. *Poincaré–Andronov–Hopf bifurcation:* Consider the following one-parameter system, which is a variant of Example 7.16:

$$\dot{x}_1 = x_2 + x_1(\lambda - x_1^2 - x_2^2)$$
$$\dot{x}_2 = -x_1 + x_2(\lambda - x_1^2 - x_2^2). \tag{7.31}$$

In polar coordinates, Eq. (7.31) is equivalent to the product system

$$\dot{r} = r(\lambda - r^2)$$
$$\dot{\theta} = -1. \tag{7.32}$$

From this polar representation, following the reasoning used in Example 7.15, it is easy to determine the phase portraits of Eq. (7.31) for various values of the parameter λ. For $\lambda \leq 0$, all solutions spiral clockwise to the origin with increasing time. When $\lambda > 0$, the origin becomes unstable and a periodic orbit of radius $\bar{r} = \sqrt{\lambda}$ appears. Furthermore, all the orbits, except the origin, spiral onto this periodic orbit, that is, the ω-limit set $\omega(\mathbf{x}^0)$ of any orbit is the periodic orbit if $\mathbf{x}^0 \neq \mathbf{0}$. Phase portraits of Eq. (7.31) for several values of the parameter are shown in Figure 7.26.

Using the same conventions as in the previous example, we obtain Figure 7.27 for the bifurcation diagram of Eq. (7.31). It is instructive to compare this diagram with that of the linear system in Figure 7.25. The line segment (the nonnegative a-axis) representing the periodic orbits of the linear system has now been deformed to the curve $a = \sqrt{\lambda}$ for the nonlinear problem.

The birth, or the death, of a periodic orbit through a change in the stability of an equilibrium point is known as the *Poincaré–Andronov–Hopf bifurcation.* This most celebrated bifurcation will be the subject of Chapter 11. $\diamond$

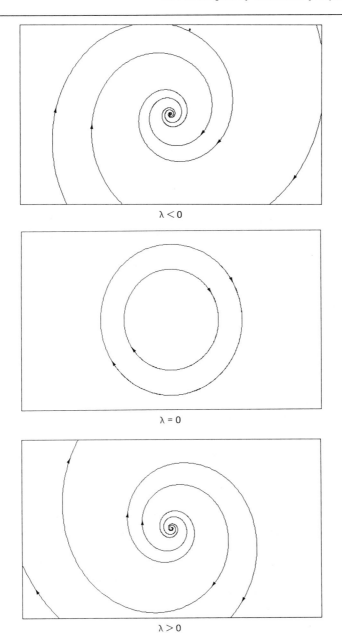

$\lambda < 0$

$\lambda = 0$

$\lambda > 0$

Figure 7.24. *Phase portraits of a bifurcation in a linear system.*

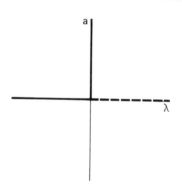

Figure 7.25. *Bifurcation diagram (amplitude of periodic orbits as a function of parameter) for periodic solutions of a linear system: vertical bifurcation.*

Example 7.27. *Homoclinic or saddle-loop bifurcation:* Consider the planar system depending on a real parameter λ:

$$\dot{x}_1 = x_2$$
$$\dot{x}_2 = x_1 + \lambda x_2 - x_1^2. \tag{7.33}$$

At this point in our book it is somewhat difficult to construct the phase portraits of this system except for $\lambda = 0$. Luckily, as you may suspect, $\lambda = 0$ will turn out to be a bifurcation value and thus is a good place to start.

For $\lambda = 0$, the system (7.33) is conservative with the first integral

$$H(x_1, x_2) = -\tfrac{1}{2}x_1^2 + \tfrac{1}{2}x_2^2 + \tfrac{1}{3}x_1^3. \tag{7.34}$$

A three-dimensional plot of this function is shown in Figure 7.28. The phase portrait of Eq. (7.33) at $\lambda = 0$ is sometimes referred to as "the fish" and it is not difficult to determine from the level sets of this first integral. The equilibrium point at $(1, 0)$ is a center locally surrounded by concentric periodic orbits. The other equilibrium point at the origin, when viewed locally, is a saddle; when viewed globally, however, one of the orbits emanating from the origin terminates again at the origin after going around the other equilibrium point. Indeed, the level set $H(x_1, x_2) = 0$ is rather special. It contains the equilibrium point at the origin and the orbit whose α- and ω-limit sets are again the origin. Such orbits, like the heteroclinic orbits of the pendulum, play a prominent role in dynamics.

Definition 7.28. *An orbit whose α- and ω-limit sets are both the same equilibrium point is called an homoclinic orbit.*

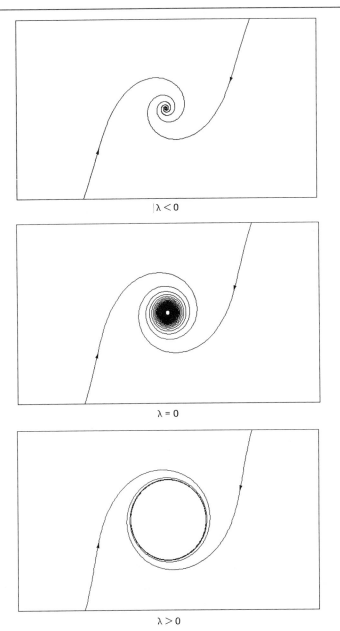

Figure 7.26. *Phase portraits of Poincaré–Andronov–Hopf bifurcation.*

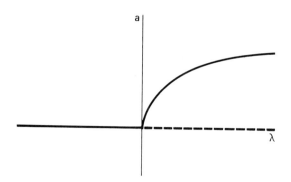

Figure 7.27. *Bifurcation diagram for Poincaré–Andronov–Hopf bifurcation (amplitude of periodic orbit as a function of parameter).*

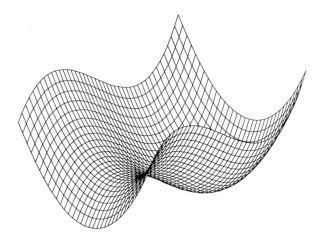

Figure 7.28. *The graph of $H(x_1, x_2) = -\frac{1}{2}x_1^2 + \frac{1}{2}x_2^2 + \frac{1}{3}x_1^3$.*

When $\lambda \neq 0$, the center is destroyed but the saddle remains. The loop consisting of the homoclinic orbit and the equilibrium point at the origin, however, is broken. The manner in which the loop breaks depends on the sign of the parameter λ; see Figure 7.29. $\Diamond$

Unlike in the case of scalar equations, the precise notion of qualitative change, or bifurcation, in planar systems is not easy to formulate. Sometimes bifurcations are clearly marked with a change either in the number of equilibrium points or the number of periodic orbits. At other times, there can be subtler bifurcations that are "global" in character and thus

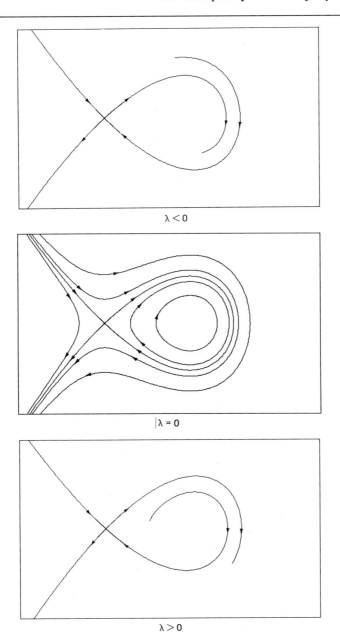

$\lambda < 0$

$|\lambda = 0$

$\lambda > 0$

Figure 7.29. *Breaking a homoclinic loop in Eq.* (7.33).

difficult to detect. We will explore these important and exciting issues in later chapters. Next, however, we explore the dynamics of a very special class of vector fields—linear.

Exercises ♣ ♡ ♠ ◇

7.18. *Recreating pictures:* The differential equations of saddle-node, pitchfork, vertical, and Poincaré-Andronov-Hopf bifurcations are stored in the 2D library of PHASER under the names *saddlenod, pitchfork, linear2d,* and *hopf.* Recreate the illustrations in this section using PHASER.

7.19. *A product system:* Draw some representative phase portraits of the product system

$$\dot{x}_1 = \lambda + x_1^2$$
$$\dot{x}_2 = \lambda + x_2^2$$

for negative, zero, and positive values of the scalar parameter λ.

7.20. *A product in polar:* Consider the following planar system depending on two real parameters λ and μ:

$$\dot{x}_1 = -x_2 + x_1 \left[\lambda + \mu(x_1^2 + x_2^2) - (x_1^2 + x_2^2)^2 \right]$$
$$\dot{x}_2 = x_1 + x_2 \left[\lambda + \mu(x_1^2 + x_2^2) - (x_1^2 + x_2^2)^2 \right].$$

Transform these equations into polar coordinates to obtain a product system. Then, discuss the bifurcations of periodic orbits as a function of the parameters. Draw the bifurcation curves in the (λ, μ)-plane. Also, draw some typical bifurcation diagrams for fixed λ, as well as some for fixed μ.

7.21. *Conservative perturbation of a conservative system:* Consider the following Hamiltonian function depending on a real parameter λ:

$$H(x_1, x_2) = \lambda x_1 + \tfrac{1}{2} x_2^2 + \tfrac{1}{3} x_2^3.$$

Write down the corresponding planar Hamiltonian system of differential equations and analyze its phase portraits for $\lambda < 0$, $\lambda = 0$, and $\lambda < 0$. Describe in words what happens to the homoclinic orbit as the parameter λ is increased through zero.

7.22. *Computing homoclinic orbits numerically:* Despite their importance, it is notoriously difficult to locate heteroclinic or homoclinic orbits in numerical simulations. Put Example 7.20 into PHASER and try to determine an initial value to compute the homoclinic loop. Can you determine the coordinates of the point at which the homoclinic orbit crosses the x_1-axis? Plan a similar numerical experiment for computing the heteroclinic orbits of the pendulum.

7.23. *Period near a homoclinic loop:* Consider Example 7.27 with $\lambda = 0$:

$$\dot{x}_1 = x_2$$
$$\dot{x}_2 = x_1 - x_1^2.$$

Obtain the period function and discuss its behavior in the limit near the homoclinic loop; and also near the origin.

Bibliographical Notes

We will reconsider many of the topics of this chapter in more detail later, where we will also provide a wide choice of references. For the moment, here are some sources pertaining mostly to applications which you might find interesting.

Natural examples of differential equations abound. Mechanics is discussed from a contemporary viewpoint in Arnold [1978]. The two species problem is analyzed in Hirsch and Smale [1974]. For various ecological models, consult D'Ancona [1954], May [1973], Maynard Smith [1968 and 1974], and Pielou [1969]. For Van der Pol's equation, the original source is Van der Pol [1927]; we will return to this famous equation in Chapter 12. A general mathematical formulation of electrical circuits is given in Hirsch and Smale [1974] and Smale [1972]; engineering details are available in, for example, Desoer and Kuh [1969]. For general applications, see Andronov, Vitt, and Xhaikin [1966].

Various norms on $\mathbb{R}^n$ and their equivalence can be found in, for instance, Smith [1983]. Appropriate formulations of the existence, uniqueness, and dependence on initial data of solutions of systems of ordinary differential equations are in the Appendix.

8

Linear Systems

 In this chapter, we undertake a detailed investigation of the rather special class of planar autonomous differential equations where the vector field is given by a linear map. By exploiting special properties of solutions of linear systems, with a small dose of linear algebra, we will be able to compute the flows of these systems explicitly and determine their phase portraits. After obtaining explicit solutions, we direct our attention to qualitative questions and classify linear systems up to flow equivalence. We also investigate certain bifurcation phenomena within the class of linear systems. We conclude the chapter with several useful facts about the solutions of nonautonomous linear differential equations. Admittedly, striving for explicit solutions may seem somewhat of a deviation from our earlier efforts. However, this is one of the few situations where such a complete answer, possessing its own mathematical appeal, exists. Furthermore, this information will be important in the local qualitative analysis of equilibrium points of nonlinear systems. The success of obtaining explicit solutions of linear systems is not, however, without an ironic disappointment. The important task of deciding the qualitative equivalence of two linear systems requires considerably more mathematics than mere formulas for their explicit solutions, as we shall see in this chapter.

8.1. Properties of Solutions of Linear Systems

In this section, we study the rather special class of planar autonomous equations $\dot{\mathbf{x}} = \mathbf{f}(\mathbf{x})$ in the case where the vector field $\mathbf{f} : \mathbb{R}^2 \rightarrow \mathbb{R}^2$ is given by a linear map. More specifically, we consider systems of the form

$$\dot{x}_1 = a_{11}x_1 + a_{12}x_2$$
$$\dot{x}_2 = a_{21}x_1 + a_{22}x_2, \tag{8.1}$$

where each a_{ij} is a real number. If we let

$$\mathbf{x} = \begin{pmatrix} x_1 \\ x_2 \end{pmatrix}, \qquad \mathbf{A} = \begin{pmatrix} a_{11} & a_{12} \\ a_{21} & a_{22} \end{pmatrix},$$

then the system (8.1) can be written in the equivalent but more convenient vector notation

$$\dot{\mathbf{x}} = \mathbf{A}\mathbf{x}. \tag{8.2}$$

A system of the form (8.2) is called a homogeneous autonomous linear system, or simply a *linear system*.

The existence and uniqueness of solution to initial-value problems for linear systems (8.2) is, of course, contained in the basic general theorem in the Appendix. The fact that Eq. (8.2) is linear, however, facilitates the following stronger conclusion whose proof is indicated in the exercises.

Lemma 8.1. *The solutions of a linear system $\dot{\mathbf{x}} = \mathbf{A}\mathbf{x}$ are defined for all $t \in \mathbb{R}$.* ◇

Another elementary but basic property enjoyed by the solutions of linear systems is that, if $\mathbf{x}^1(t)$ and $\mathbf{x}^2(t)$ are two solutions of Eq. (8.2), and c_1 and c_2 are any two real numbers, then the *linear combination* $c_1\mathbf{x}^1(t) + c_2\mathbf{x}^2(t)$ is also a solution of Eq. (8.2).

This property, called the *superposition principle*, enables us to determine the flow of Eq. (8.2) from any two particular solutions that are sufficiently different. To ensure that two solutions are different we need to generalize the usual notion of linear independence of (constant) vectors. Let us recall first that two vectors $\mathbf{v}^1$ and $\mathbf{v}^2$ are linearly independent if and only if $c_1\mathbf{v}^1 + c_2\mathbf{v}^2 = \mathbf{0}$ implies $c_1 = 0$ and $c_2 = 0$. This is equivalent to requiring that $\mathbf{v}^1$ is not a multiple of $\mathbf{v}^2$, or the determinant of the 2×2 matrix whose columns consist of these two vectors is nonzero: $\det\left(\mathbf{v}^1 \,|\, \mathbf{v}^2\right) \neq 0$. Otherwise, the two vectors are said to be *linearly dependent*.

Definition 8.2. *Two solutions* $\mathbf{x}^1(t)$ *and* $\mathbf{x}^2(t)$ *of Eq. (8.2) are said to be linearly independent if, for each* $t \in \mathbb{R}$, *the relation* $c_1\mathbf{x}^1(t) + c_2\mathbf{x}^2(t) = 0$ *implies that* $c_1 = 0$ *and* $c_2 = 0$.

Linear independence of $\mathbf{x}^1(t)$ and $\mathbf{x}^2(t)$ is equivalent to the fact that the determinant of the 2×2 matrix whose columns consist of these two vectors is nonzero:

$$\det\left(\mathbf{x}^1(t) \mid \mathbf{x}^2(t)\right) \neq 0 \qquad \text{for all } t \in \mathbb{R}. \tag{8.3}$$

In order to manipulate a pair of solutions effectively we introduce a bit of terminology.

Definition 8.3. *If* $\mathbf{x}^1(t)$ *and* $\mathbf{x}^2(t)$ *are two solutions of Eq. (8.2), then the* 2×2 *matrix* $\mathbf{X}(t) \equiv \left(\mathbf{x}^1(t) \mid \mathbf{x}^2(t)\right)$, *whose columns are the two solutions is called a matrix solution of Eq. (8.2). If, in addition,* $\det \mathbf{X}(t) \neq 0$ *for all* $t \in \mathbb{R}$, *then* $\mathbf{X}(t)$ *is said to be a fundamental matrix solution of Eq. (8.2). A special fundamental matrix solution satisfying the condition* $\mathbf{X}(0) = \mathbf{I}$, *where* $\mathbf{I}$ *is the* 2×2 *identity matrix, is called a principal matrix solution.*

We now state a lemma which provides the useful fact that it suffices to check $\det \mathbf{X}(t)$ at only one value of t, and gives an explicit formula for the flow of Eq. (8.2) in terms of any fundamental matrix solution.

Lemma 8.4. *Properties of fundamental solutions:*
 (i) *If* $\mathbf{X}(t)$ *is a matrix solution of Eq. (8.2) with* $\det \mathbf{X}(0) \neq 0$, *then* $\det \mathbf{X}(t) \neq 0$ *for all* $t \in \mathbb{R}$, *that is,* $\mathbf{X}(t)$ *is a fundamental solution of Eq. (8.2).*
 (ii) *If* $\mathbf{X}(t)$ *is a fundamental matrix solution, then the solution of Eq. (8.2) satisfying the initial condition* $\mathbf{x}(0) = \mathbf{x}^0$ *is given by*

$$\varphi(t, \mathbf{x}^0) = \mathbf{X}(t)\left[\mathbf{X}(0)\right]^{-1}\mathbf{x}^0. \tag{8.4}$$

Proof. (i) First observe that the solution of the initial-value problem $\mathbf{x}(0) = \mathbf{0}$ for Eq. (8.2) is identically zero: $\varphi(t, \mathbf{0}) = \mathbf{0}$. Now, suppose that there are constants c_1, c_2, and τ such that $c_1\mathbf{x}^1(\tau) + c_2\mathbf{x}^2(\tau) = \mathbf{0}$, where $\mathbf{x}^1(t)$ and $\mathbf{x}^2(t)$ are the columns of $\mathbf{X}(t)$. Then, $c_1\mathbf{x}^1(t + \tau) + c_2\mathbf{x}^2(t + \tau)$ is also a solution for Eq. (8.2), which for $t = 0$ is zero. Therefore, by uniqueness of solutions, we have

$$\varphi(t, \mathbf{0}) = c_1\mathbf{x}^1(t + \tau) + c_2\mathbf{x}^2(t + \tau).$$

Thus, if we take $t = -\tau$, then we obtain

$$\mathbf{0} = \varphi(-\tau, \mathbf{0}) = c_1\mathbf{x}^1(0) + c_2\mathbf{x}^2(0).$$

Now, the linear independence of the vectors $\mathbf{x}^1(0)$ and $\mathbf{x}^2(0)$ implies that $c_1 = c_2 = 0$.

(ii) The right-hand side of Eq. (8.4) is a solution of Eq. (8.2) because it is a linear combination of the solutions $\mathbf{x}^1(0)$ and $\mathbf{x}^2(0)$. Since $\mathbf{X}(0)\,\mathbf{X}(0)^{-1} = \mathbf{I}$, it also satisfies the initial condition. Thus, from the uniqueness theorem, it is the solution. $\diamondsuit$

The superposition principle implies that the set of all solutions of Eq. (8.2) is a vector space. The lemma above shows that the dimension of this vector space is two. After discovering some general facts about the flows of linear systems, we will determine explicit bases for the vector space of solutions of Eq. (8.2) for any given coefficient matrix $\mathbf{A}$.

As we saw in the first example of our book, the flow of the scalar linear differential equation $\dot{x} = ax$ is given by the exponential function $\varphi(t, x_0) = e^{at}x_0$. To obtain an analogous formula for the flow of linear planar systems we introduce the notation

$$e^{\mathbf{A}t} \equiv \mathbf{X}(t)\,\mathbf{X}(0)^{-1}, \tag{8.5}$$

where $\mathbf{X}(t)$ is any fundamental matrix solution of Eq. (8.2). Then, Eq. (8.4) for the flow of Eq. (8.2) can be written as the matrix exponential

$$\varphi(t, \mathbf{x}^0) = e^{\mathbf{A}t}\mathbf{x}^0, \tag{8.6}$$

hence establishing the desired analogy. Of course,

$$e^{\mathbf{A}0} = \mathbf{I},$$

and thus $e^{\mathbf{A}t}$ is a principal matrix solution of Eq. (8.2).

We now collect several important properties of the principal matrix solution $e^{\mathbf{A}t}$ and provide the reason for this choice of notation.

Lemma 8.5. *The principal matrix solution $e^{\mathbf{A}t}$ satisfies the following properties:*

(i) $e^{\mathbf{A}(t+s)} = e^{\mathbf{A}t}\,e^{\mathbf{A}s}$;

(ii) $\left(e^{\mathbf{A}t}\right)^{-1} = e^{-\mathbf{A}t}$;

(iii) $\frac{d}{dt}e^{\mathbf{A}t} = \mathbf{A}e^{\mathbf{A}t} = e^{\mathbf{A}t}\mathbf{A}$;

(iv) $e^{\mathbf{A}t} = \sum_{n=0}^{+\infty} \frac{1}{n!}\mathbf{A}^n t^n = \mathbf{I} + \mathbf{A}t + \frac{1}{2!}\mathbf{A}^2 t^2 + \cdots$.

Proof. (i) For any fixed s, the matrices $e^{\mathbf{A}(t+s)}$ and $e^{\mathbf{A}t}e^{\mathbf{A}s}$ are matrix solutions of Eq. (8.2) which coincide at $t = 0$. The uniqueness theorem implies the desired equality.

(ii) Take $s = -t$ in property (i).

(iii) The first equality follows from the definition of $e^{\mathbf{A}t}$ given in Eq. (8.5). The second one is again a consequence of the uniqueness theorem with the observation that both $\mathbf{A}e^{\mathbf{A}t}$ and $e^{\mathbf{A}t}\mathbf{A}$ are matrix solutions of Eq. (8.2).

(iv) This property is considerably more difficult to establish than the previous ones and a complete proof requires a certain amount of mathematical sophistication. Here, we will present the essential steps to make it convincing.

We should remark that it is possible to take property (iv) as the definition of $e^{\mathbf{A}t}$ provided that one establishes the convergence of this matrix power series. Then, one can demonstrate all the other properties of $e^{\mathbf{A}t}$ including the result that $e^{\mathbf{A}t}$ is the principal matrix solution of Eq. (8.2). Following our presentation in this chapter, however, we will establish the formula for its power series expansion from the fact that $e^{\mathbf{A}t}$ is the principal matrix solution of Eq. (8.2).

For the sake of brevity of notation, let us set $\mathbf{P}(t) \equiv e^{\mathbf{A}t}$. Now, observe that for each vector $\mathbf{x}^0 \in \mathbb{R}$, the matrix $\mathbf{P}(t)$ satisfies the integral equation

$$\mathbf{P}(t)\mathbf{x}^0 = \mathbf{I}\,\mathbf{x}^0 + \int_0^t \mathbf{A}\mathbf{P}(s)\mathbf{x}^0 \, ds. \tag{8.7}$$

Since $\mathbf{P}(t)$ appears on both sides of this equation, we attempt to find $\mathbf{P}(t)$ iteratively by using successive approximations. If we take as our initial guess

$$\mathbf{P}^{(0)}(t)\mathbf{x}^0 = \mathbf{I}\,\mathbf{x}^0,$$

and compute the successive iterates with

$$\mathbf{P}^{(k+1)}(t)\mathbf{x}^0 = \mathbf{I}\,\mathbf{x}^0 + \int_0^t \mathbf{A}\mathbf{P}^{(k)}(s)\mathbf{x}^0 \, ds, \tag{8.8}$$

then as the kth iterate, for $k = 0, 1, 2, \ldots$, we obtain the polynomial expression

$$\mathbf{P}^{(k)}(t)\mathbf{x}^0 = \mathbf{I}\,\mathbf{x}^0 + \mathbf{A}\mathbf{x}^0 t + \frac{1}{2!}\mathbf{A}^2\mathbf{x}^0 t^2 + \cdots + \frac{1}{k!}\mathbf{A}^k\mathbf{x}^0 t^k.$$

We now show that the sequence of vectors $\mathbf{P}^{(k)}(t)\mathbf{x}^0$ given by this formula converges as $k \to +\infty$. The first observation is that it suffices to show the convergence of $\mathbf{P}^{(k)}(t)\mathbf{x}^0$ for all $\mathbf{x}^0$ on the unit circle $S^1 \equiv \{\mathbf{x} : \|\mathbf{x}\| = 1\}$ only, because any vector in $\mathbb{R}^2$ can be written as a scalar multiple of some vector on S^1. The second observation is the fact that, since $\|\mathbf{A}\mathbf{x}^0\|$ as a function of $\mathbf{x}^0$ is continuous and S^1 is a closed bounded set, there exists a constant $\alpha > 0$ such that

$$\|\mathbf{A}\mathbf{x}^0\| < \alpha \qquad \text{for all } \mathbf{x}^0 \in S^1.$$

With these observations it is now easy to establish the following estimates on the norms of the iterates $\mathbf{P}^{(k)}(t)\mathbf{x}^0$:

$$\|\mathbf{P}^{(k)}(t)\mathbf{x}^0\| \leq 1 + \alpha t + \frac{1}{2!}\alpha^2 t^2 + \cdots + \frac{1}{k!}\alpha^k t^k \leq e^{\alpha t}$$

and

$$\|\mathbf{P}^{(k+1)}(t)\mathbf{x^0} - \mathbf{P}^{(k)}(t)\mathbf{x^0}\| \leq \frac{1}{(k+1)!}\alpha^{k+1}t^{k+1}.$$

This implies that the sequence $\mathbf{P}^{(k)}(t)\mathbf{x^0}$ converges as $k \to +\infty$ to a vector function, say, $\varphi(t, \mathbf{x^0})$, uniformly for all $\mathbf{x^0} \in S^1$, and t in a bounded set.

The remaining step is to show that the limit vector function $\varphi(t, \mathbf{x^0})$ satisfies the integral equation (8.7). To this end, we take the limits of both sides of Eq. (8.8) and interchange the order of limit and integration:

$$\lim_{k\to+\infty}\mathbf{P}^{(k+1)}(t)\mathbf{x^0} = \mathbf{I}\mathbf{x^0} + \int_0^t \lim_{k\to+\infty}\mathbf{A}\mathbf{P}^{(k)}(s)\mathbf{x^0}\,ds.$$

Because of the uniform convergence of the series this process can be justified rigorously to obtain the integral equation

$$\varphi(t, \mathbf{x^0}) = \mathbf{x^0} + \int_0^t \mathbf{A}\varphi(s, \mathbf{x^0})\,ds.$$

Therefore, $\varphi(t, \mathbf{x^0})$ is the solution of Eq. (8.2) satisfying the initial condition $\varphi(0, \mathbf{x^0}) = \mathbf{x^0}$. Since, from Eq. (8.6), $\varphi(t, \mathbf{x^0}) = e^{\mathbf{A}t}\mathbf{x^0}$, the sequence $\mathbf{P}^k(t)$ converges to $e^{\mathbf{A}t}$ as $k \to +\infty$. $\diamondsuit$

It is evident from the lemma above that the principal matrix solution $e^{\mathbf{A}t}$ enjoys many of the properties of the analogous scalar exponential function. Despite these similarities, however, there are also major differences between the scalar and matrix exponential functions, and care must be exercised in computations. For instance, in general, $e^{\mathbf{A}t} \cdot e^{\mathbf{B}t} \neq e^{(\mathbf{A}+\mathbf{B})t}$ for two 2×2 matrices $\mathbf{A}$ and $\mathbf{B}$. Here are two such matrices.

Example 8.6. *Noncommuting matrices:* Consider the matrices

$$\mathbf{A} = \begin{pmatrix} 0 & 1 \\ 0 & 0 \end{pmatrix}, \qquad \mathbf{B} = \begin{pmatrix} 0 & 0 \\ -1 & 0 \end{pmatrix}.$$

Since $\mathbf{A}^n = \mathbf{B}^n = 0$ for $n \geq 2$, the power series expansions of $e^{\mathbf{A}t}$ and $e^{\mathbf{B}t}$ given by Lemma 8.5 consist of the first two terms only. Thus,

$$e^{\mathbf{A}t} = \mathbf{I} + \mathbf{A}t = \begin{pmatrix} 1 & t \\ 0 & 1 \end{pmatrix},$$

$$e^{\mathbf{B}t} = \mathbf{I} + \mathbf{B}t = \begin{pmatrix} 1 & 0 \\ -t & 1 \end{pmatrix},$$

and

$$e^{\mathbf{A}t} \cdot e^{\mathbf{B}t} = \begin{pmatrix} 1 - t^2 & t \\ -t & 1 \end{pmatrix}.$$

The power series of $e^{(\mathbf{A}+\mathbf{B})t}$ is also easy to sum, if we recall that the power series expansions of the scalar functions $\sin t$ and $\cos t$ are given by

$$\sin t = \sum_{n=0}^{+\infty}(-1)^n \frac{t^{2n+1}}{(2n+1)!} = t - \frac{t^3}{3!} + \frac{t^5}{5!} - \frac{t^7}{7!} + \cdots$$

and

$$\cos t = \sum_{n=0}^{+\infty}(-1)^n \frac{t^{2n}}{(2n)!} = 1 - \frac{t^2}{2!} + \frac{t^4}{4!} - \frac{t^6}{6!} + \cdots .$$

Again, using part (iv) of Lemma 8.5 we obtain rather easily that

$$e^{(\mathbf{A}+\mathbf{B})t} = e^{\left(\begin{smallmatrix} 0 & 1 \\ -1 & 0 \end{smallmatrix}\right)t} = \left(\begin{array}{cc} \cos t & \sin t \\ -\sin t & \cos t \end{array}\right),$$

which is different from the matrix $e^{\mathbf{A}t}e^{\mathbf{B}t}$. $\Diamond$

The "anomaly" illustrated in the example above can be avoided in certain special but computationally very useful situations.

Lemma 8.7. *If two matrices* $\mathbf{A}$ *and* $\mathbf{B}$ *commute, that is,* $\mathbf{A}\mathbf{B} = \mathbf{B}\mathbf{A}$, *then* $e^{(\mathbf{A}+\mathbf{B})t} = e^{\mathbf{A}t}e^{\mathbf{B}t}$. $\Diamond$

A proof of this lemma is outlined in the exercises. We will shortly put this fact to good use.

It may appear from the calculations performed in Example 8.6 that one could compute $e^{\mathbf{A}t}$ from its power series expansion given in Lemma 8.5. Unfortunately, this is not so in general, as it is impractical to sum the series except for special coefficient matrices. In the following example, we compute $e^{\mathbf{A}t}$ directly from its power series for three such very special matrices $\mathbf{A}$, and determine the phase portraits of the corresponding linear systems $\dot{\mathbf{x}} = \mathbf{A}\mathbf{x}$. Despite their specialized forms, these systems, known as *canonical systems* or *systems in Jordan Normal Form*, are of central importance in the theory of linear systems. In fact, using a bit of linear algebra, we will show in the next section that the computation of $e^{\mathbf{A}t}$ for any coefficient matrix $\mathbf{A}$ can be reduced, by a linear change of coordinates, to one of these canonical cases.

Example 8.8. *Linear Canonical Systems:* Consider the following three classes of linear systems with coefficient matrices:

$$(i) \ \left(\begin{array}{cc} \lambda_1 & 0 \\ 0 & \lambda_2 \end{array}\right), \qquad (ii) \ \left(\begin{array}{cc} \lambda & 1 \\ 0 & \lambda \end{array}\right), \qquad (iii) \ \left(\begin{array}{cc} \alpha & \beta \\ -\beta & \alpha \end{array}\right),$$

where λ, λ_1, λ_2, α, and $\beta \neq 0$ are real numbers. Let us compute the exponentials of these matrices.

Case (i). Diagonal matrix:

$$\text{If}\quad \mathbf{A} = \begin{pmatrix} \lambda_1 & 0 \\ 0 & \lambda_2 \end{pmatrix}, \quad \text{then}\quad e^{\mathbf{A}t} = \begin{pmatrix} e^{\lambda_1 t} & 0 \\ 0 & e^{\lambda_2 t} \end{pmatrix}.$$

In this case, it is easy to sum the power series if we observe that

$$\mathbf{A}^n = \begin{pmatrix} \lambda_1^n & 0 \\ 0 & \lambda_2^n \end{pmatrix}$$

and recognize the series

$$e^t = \sum_{n=0}^{+\infty} \frac{1}{n!} t^n$$

of the scalar exponential function. Notice that the diagonal matrix $\mathbf{A}$ is the coefficient matrix of Example 7.14 and various possible phase portraits were already given in Figure 7.15.

Case (ii). Triangular matrix:

$$\text{If}\quad \mathbf{A} = \begin{pmatrix} \lambda & 1 \\ 0 & \lambda \end{pmatrix}, \quad \text{then}\quad e^{\mathbf{A}t} = e^{\lambda t} \begin{pmatrix} 1 & t \\ 0 & 1 \end{pmatrix}.$$

To compute $e^{\mathbf{A}t}$ we will utilize Lemma 8.7. We can write the matrix $\mathbf{A}$ as the sum of two commuting matrices:

$$\mathbf{A} = \begin{pmatrix} \lambda & 0 \\ 0 & \lambda \end{pmatrix} + \begin{pmatrix} 0 & 1 \\ 0 & 0 \end{pmatrix}.$$

The diagonal part of the matrix can be exponentiated as above and the exponential of the second matrix was already computed in Example 8.6. So, the desired formula for $e^{\mathbf{A}t}$ follows from Lemma 8.7.

Let us now draw the phase portrait of the corresponding linear system. We will suppose that $\lambda < 0$ and leave the remaining cases $\lambda = 0$ and $\lambda > 0$ to you to practice with. We first observe that the origin is the only equilibrium point. Next, using the explicit formula $\varphi(t, \mathbf{x}^0) = e^{\mathbf{A}t}\mathbf{x}^0$ for solutions we see that all orbits approach the origin as $t \to +\infty$. Furthermore, since $dx_2/dx_1 \to 0$ as $t \to +\infty$, all orbits eventually become tangent to the x_1-axis at the origin. The resulting phase portrait as depicted in Figure 8.1 is often called a (stable) *improper node*.

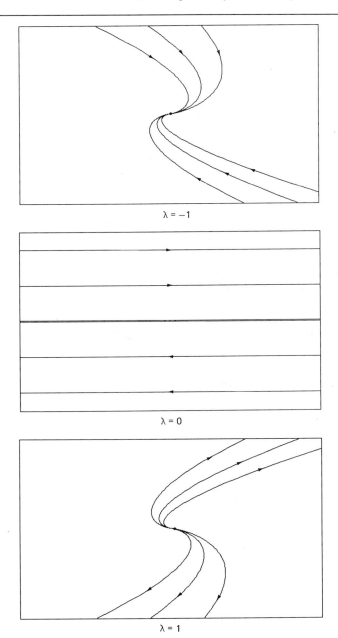

$\lambda = -1$

$\lambda = 0$

$\lambda = 1$

Figure 8.1. *Phase portraits of the linear system* $\dot{x}_1 = \lambda x_1 + x_2$, $\dot{x}_2 = \lambda x_2$ *for* $\lambda = -1$, $\lambda = 0$, *and* $\lambda = 1$.

Case (iii). *"Complex" matrix:*

If $\mathbf{A} = \begin{pmatrix} \alpha & \beta \\ -\beta & \alpha \end{pmatrix}$, then $e^{\mathbf{A}t} = e^{\alpha t} \begin{pmatrix} \cos\beta t & \sin\beta t \\ -\sin\beta t & \cos\beta t \end{pmatrix}$.

Computation of $e^{\mathbf{A}t}$ is similar to the previous case. We write $\mathbf{A}$ as the sum of two commuting matrices

$$\mathbf{A} = \begin{pmatrix} \alpha & 0 \\ 0 & \alpha \end{pmatrix} + \begin{pmatrix} 0 & \beta \\ -\beta & 0 \end{pmatrix}$$

and use Lemma 8.7 together with Example 8.6.

To determine the phase portraits of the corresponding linear systems observe that the components of the vector $e^{\mathbf{A}t}\mathbf{x}^0$ are periodic functions of t which are multiplied by the factor $e^{\alpha t}$. Thus, since we assume that $\beta \neq 0$, when $\alpha < 0$, the solutions spiral towards the unique equilibrium at the origin as $t \to +\infty$. For $\alpha = 0$, all solutions are periodic; and when $\alpha > 0$, the solutions spiral out without bound. We have drawn typical phase portraits in these three cases when $\beta > 0$; see Figure 8.2. You should draw the analogous phase portraits when $\beta < 0$. (*Hint:* examine the vector field as well as the solutions.) The phase portraits of this "complex" case can, of course, be easily determined using polar coordinates. We will explain the reason for the label "complex" in the next section. ◇

Exercises ————————————————————————— ♣♡♠◇

8.1. *An animation sequence:* Draw the phase portraits and indicate the direction of the flows of the following linear systems:

(a) $\dot{\mathbf{x}} = \begin{pmatrix} 0 & 1 \\ -1 & 0 \end{pmatrix} \mathbf{x}$;   (b) $\dot{\mathbf{x}} = \begin{pmatrix} 0 & -1 \\ 1 & 0 \end{pmatrix} \mathbf{x}$;

(c) $\dot{\mathbf{x}} = \begin{pmatrix} -1 & 1 \\ 0 & -1 \end{pmatrix} \mathbf{x}$;   (d) $\dot{\mathbf{x}} = \begin{pmatrix} 0 & 1 \\ 0 & 0 \end{pmatrix} \mathbf{x}$;   (e) $\dot{\mathbf{x}} = \begin{pmatrix} 1 & 1 \\ 0 & 1 \end{pmatrix} \mathbf{x}$.

What is the difference between (a) and (b)? Also, try to visualize the transition from (c) to (d) to (e) as an animated sequence.

8.2. *A proof of Lemma 8.1:* Once the explicit formula in part (*iv*) of Lemma 8.5 is established, the conclusion of Lemma 8.1 is immediate. However, the fact that solutions of a linear system are defined for all time can also be established directly:
  (a) Show that for a given 2×2 matrix $\mathbf{A}$, there exists a $k > 0$ such that $\|\mathbf{A}\mathbf{x}\| \leq k \|\mathbf{x}\|$ for all $\mathbf{x} \in \mathbb{R}^2$.
  (b) Differentiate the norm $\|\mathbf{x}(t)\|^2$ with respect to t and obtain the estimate $\frac{1}{2}\frac{d}{dt}\|\mathbf{x}(t)\|^2 \leq k\|\mathbf{x}(t)\|^2$. Conclude that $\|\mathbf{x}(t)\| \leq \exp(k|t|)$ for all $t \in \mathbb{R}$.

8.3. Prove that if $\mathbf{AB} = \mathbf{BA}$, then $\mathbf{B}e^{\mathbf{A}t} = e^{\mathbf{A}t}\mathbf{B}$.

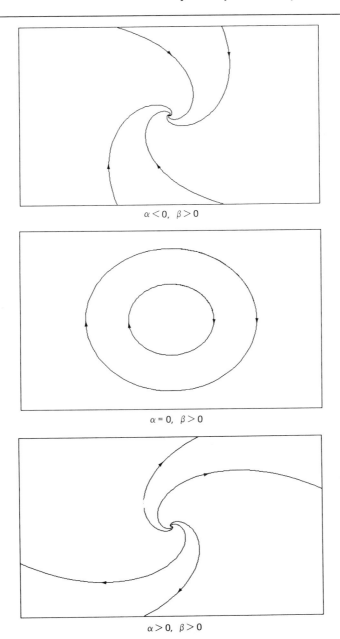

$\alpha < 0, \quad \beta > 0$

$\alpha = 0, \quad \beta > 0$

$\alpha > 0, \quad \beta > 0$

Figure 8.2. *Typical phase portraits of a "complex" linear canonical system.*

Hint: Verify that $\mathbf{B}e^{\mathbf{A}t}$ and $e^{\mathbf{A}t}\mathbf{B}$ are matrix solutions of $\dot{\mathbf{x}} = \mathbf{A}\mathbf{x}$ and use uniqueness.

8.4. Prove that if $\mathbf{AB} = \mathbf{BA}$, then $e^{(\mathbf{A}+\mathbf{B})t} = e^{\mathbf{A}t}e^{\mathbf{B}t}$.

Hint: Using the previous exercise, verify that $e^{(\mathbf{A}+\mathbf{B})t}$ and $e^{\mathbf{A}t}e^{\mathbf{B}t}$ are matrix solutions of the linear system $\dot{\mathbf{x}} = (\mathbf{A}+\mathbf{B})\mathbf{x}$.

8.5. *A square under the flow:* Draw the image of the unit square in the first quadrant under the flow of the linear system $\dot{x}_1 = -x_1$, $\dot{x}_2 = x_2$ at $t = 1$ and also at $t = -1$. What are the areas of the images?

8.2. Reduction to Canonical Forms

In this section, we will show how to put a matrix $\mathbf{A}$ into one of the three canonical forms given in Example 8.8. To accomplish this, we first need to consider how a linear system $\dot{\mathbf{x}} = \mathbf{A}\mathbf{x}$ is affected by a linear change of coordinates.

Let $\mathbf{P}$ be an invertible 2×2 matrix and consider the new variables $\mathbf{y}$ given by $\mathbf{x} = \mathbf{P}\mathbf{y}$, or equivalently, $\mathbf{y} = \mathbf{P}^{-1}\mathbf{x}$. Then, in the new coordinates the linear system $\dot{\mathbf{x}} = \mathbf{A}\mathbf{x}$ becomes

$$\dot{\mathbf{y}} = \mathbf{P}^{-1}\mathbf{A}\mathbf{P}\,\mathbf{y}. \tag{8.9}$$

It is also easy to determine the flow in the new coordinates using Eq. (8.6):

$$\mathbf{x}(t) = \mathbf{P}\mathbf{y}(t) = \mathbf{P}e^{\mathbf{P}^{-1}\mathbf{A}\mathbf{P}t}\,\mathbf{y}^0 = \left(\mathbf{P}e^{\mathbf{P}^{-1}\mathbf{A}\mathbf{P}t}\mathbf{P}^{-1}\right)\mathbf{x}^0,$$

and thus

$$e^{\mathbf{A}t} = \mathbf{P}e^{\mathbf{P}^{-1}\mathbf{A}\mathbf{P}t}\mathbf{P}^{-1}. \tag{8.10}$$

Multiplication by $\mathbf{P}^{-1}$ on the left and by $\mathbf{P}$ on the right results in the convenient formula

$$e^{\mathbf{P}^{-1}\mathbf{A}\mathbf{P}t} = \mathbf{P}^{-1}e^{\mathbf{A}t}\mathbf{P}, \tag{8.11}$$

which should be thought of as an addendum to the list of the properties of the principal matrix solution given in Lemma 8.6.

The main computational utility of linear coordinate transformations lies in formula (8.10). The strategy is to find an appropriate $\mathbf{P}$ so that the new coefficient matrix $\mathbf{P}^{-1}\mathbf{A}\mathbf{P}$ becomes one of the canonical matrices given in Example 8.7. As we saw in that example, it is rather easy to exponentiate such matrices. So, Eq. (8.10) enables us to reduce the computation of $e^{\mathbf{A}t}$ to a couple of matrix multiplications. Let us illustrate this idea on a simple example.

Example 8.9. *Changing to canonical:* Consider the linear system $\dot{\mathbf{x}} = \mathbf{A}\mathbf{x}$ with

$$\mathbf{A} = \begin{pmatrix} 5 & -4 \\ 4 & -5 \end{pmatrix}, \tag{8.12}$$

and the transformation matrix $\mathbf{P}$ and its inverse $\mathbf{P}^{-1}$ given by

$$\mathbf{P} = \begin{pmatrix} 2 & 1 \\ 1 & 2 \end{pmatrix}, \qquad \mathbf{P}^{-1} = \begin{pmatrix} 2/3 & -1/3 \\ -1/3 & 2/3 \end{pmatrix}.$$

Then, in the new coordinates $\mathbf{y} = \mathbf{P}^{-1}\mathbf{x}$, the linear system becomes the canonical system

$$\dot{\mathbf{y}} = \mathbf{P}^{-1}\mathbf{A}\mathbf{P}\,\mathbf{y} = \begin{pmatrix} 3 & 0 \\ 0 & -3 \end{pmatrix}\mathbf{y}.$$

As we saw in Example 8.8, the flow of this canonical system is given by

$$\varphi(t,\,\mathbf{y}^0) = e^{\mathbf{P}^{-1}\mathbf{A}\mathbf{P}t}\,\mathbf{y}^0 = \begin{pmatrix} e^{3t} & 0 \\ 0 & e^{-3t} \end{pmatrix}\mathbf{y}^0.$$

Now, using formula (8.10), we can readily determine the flow in the original coordinates:

$$\varphi(t,\,\mathbf{x}^0) = \mathbf{P} \begin{pmatrix} e^{3t} & 0 \\ 0 & e^{-3t} \end{pmatrix} \mathbf{P}^{-1}\mathbf{x}^0$$

$$= \frac{1}{3} \begin{pmatrix} 4e^{3t} - e^{-3t} & -2e^{3t} + 2e^{-3t} \\ 2e^{3t} - 2e^{-3t} & -e^{3t} + 4e^{-3t} \end{pmatrix} \mathbf{x}^0.$$

In contrast with the flow of the canonical system, this explicit formula does not readily reveal much information about the geometry of the flow; see Figure 8.3. For this reason, in our qualitative study of planar systems, we will usually prefer to transform linear systems into their canonical forms. ◇

Despite the simplicity of these computations, one important question remains: How did we find the coordinate transformation $\mathbf{P}$ that led us to the canonical form? The answer lies in uncovering certain special numbers and vectors associated with the coefficient matrix $\mathbf{A}$.

Definition 8.10. *A (real or complex) number λ is called an eigenvalue of a matrix $\mathbf{A}$ if there is a (real or complex) nonzero vector $\mathbf{v}$ such that*

$$\mathbf{A}\mathbf{v} = \lambda\mathbf{v}.$$

The vector $\mathbf{v}$ is called an eigenvector of $\mathbf{A}$ corresponding to the eigenvalue λ.

For future reference, we should point out that, since $\mathbf{A}$ is a real matrix, a real eigenvalue has a corresponding real eigenvector. A complex eigenvalue, however, has a necessarily complex eigenvector of the form $\mathbf{v}^1 + i\mathbf{v}^2$, where both $\mathbf{v}^1$ and $\mathbf{v}^2$ are real and nonzero vectors.

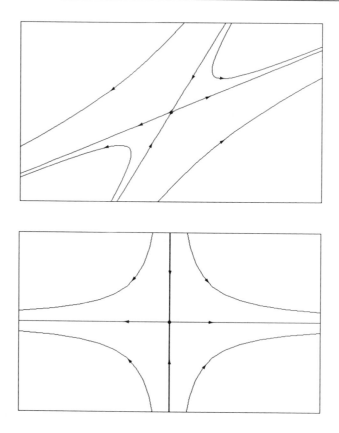

Figure 8.3. *Phase portraits of Eq. (8.12) and its canonical form.*

In computations, we first determine the eigenvalues of **A**. To this end, notice that in the definition above we are requiring the homogeneous system of linear equations

$$(\mathbf{A} - \lambda\mathbf{I})\,\mathbf{v} = \mathbf{0} \tag{8.13}$$

to have a nonzero solution. This is equivalent to the fact that

$$\det(\mathbf{A} - \lambda\mathbf{I}) = 0. \tag{8.14}$$

Therefore, the eigenvalues of **A** are the roots of this quadratic polynomial in λ. The quadratic polynomial in Eq. (8.14) is called the *characteristic polynomial* of **A**. Once the eigenvalues are found, corresponding eigenvectors can be computed easily by solving the system of linear equations (8.13).

Example 8.11. *Continuation of Example 8.9:* With the goal of finding the transformation matrix $\mathbf{P}$, let us compute the eigenvalues and their corresponding eigenvectors of the coefficient matrix $\mathbf{A}$ of Example 8.9. From Eq. (8.14), the characteristic polynomial of $\mathbf{A}$ is

$$\det(\mathbf{A} - \lambda\mathbf{I}) = \det \begin{pmatrix} 5 - \lambda & -4 \\ 4 & -5 - \lambda \end{pmatrix}$$
$$= (5 - \lambda)(-5 - \lambda) + 16 = 0.$$

The roots of this quadratic polynomial in λ are $\lambda_1 = 3$ and $\lambda_2 = -3$ which are the eigenvalues of $\mathbf{A}$. To find the eigenvectors corresponding to the eigenvalue $\lambda_1 = 3$, we solve the system of linear equations (8.13):

$$(\mathbf{A} - 3\mathbf{I})\mathbf{v} = \begin{pmatrix} 2 & -4 \\ 4 & -8 \end{pmatrix} \mathbf{v} = \begin{pmatrix} 0 \\ 0 \end{pmatrix}.$$

It is easy to see that solutions of this system are scalar multiples of the vector

$$\mathbf{v}^1 = \begin{pmatrix} 2 \\ 1 \end{pmatrix}.$$

Similarly, we find that the eigenvalues corresponding to the eigenvalue $\lambda_2 = -3$ are scalar multiples of the vector

$$\mathbf{v}^2 = \begin{pmatrix} 1 \\ 2 \end{pmatrix}.$$

Observe now that these two vectors are linearly independent, and the transformation matrix $\mathbf{P}$ we chose in Example 8.9 has its columns as these eigenvectors. $\Diamond$

Some embellishments of the computations in this example yield a constructive proof of the following general result for transforming an arbitrary 2×2 matrix to a canonical matrix.

Theorem 8.12. *Let $\mathbf{A}$ be a 2×2 matrix with real entries. Then, there exists a real invertible 2×2 matrix $\mathbf{P}$ such that*

$$\mathbf{P}^{-1}\mathbf{A}\mathbf{P} = \mathbf{J},$$

where $\mathbf{J}$ is one of the following three matrices in Jordan Normal Form:

$$(i) \begin{pmatrix} \lambda_1 & 0 \\ 0 & \lambda_2 \end{pmatrix}, \qquad (ii) \begin{pmatrix} \lambda & 1 \\ 0 & \lambda \end{pmatrix}, \qquad (iii) \begin{pmatrix} \alpha & \beta \\ -\beta & \alpha \end{pmatrix}.$$

Here, λ, λ_1, λ_2, α, and $\beta \neq 0$ are real numbers.

Proof. (*i*) *Real distinct eigenvalues:* Let us suppose that **A** has two real and distinct eigenvalues λ_1 and λ_2 with corresponding, necessarily real (why?), eigenvectors $\mathbf{v}^1$ and $\mathbf{v}^2$, that is,

$$\mathbf{A}\mathbf{v}^i = \lambda_i \mathbf{v}^i \quad \text{for } i = 1, 2. \tag{8.15}$$

As a candidate for a transformation matrix **P**, consider the matrix whose columns consist of the two eigenvectors:

$$\mathbf{P} = \left(\mathbf{v}^1 \,|\, \mathbf{v}^2\right).$$

We first establish the invertibility of **P** by exhibiting the linear independence of its columns, the eigenvectors. Suppose that there are constants c_1 and c_2 such that

$$c_1 \mathbf{v}^1 + c_2 \mathbf{v}^2 = \mathbf{0}. \tag{8.16}$$

Then, multiplying both sides of this equation by **A** and using relations (8.15), we obtain

$$c_1 \lambda_1 \mathbf{v}^1 + c_2 \lambda_2 \mathbf{v}^2 = \mathbf{0}. \tag{8.17}$$

Now, we multiply Eq. (8.16) by λ_2 and then subtract the resulting equation from Eq. (8.17) to obtain

$$c_1 \left(\lambda_1 - \lambda_2\right) \mathbf{v}^1 = \mathbf{0}.$$

Since we assumed that $\lambda_1 - \lambda_2 \neq 0$, and $\mathbf{v}^1$ is not the zero vector, this implies that $c_1 = 0$; and now Eq. (8.16) yields $c_2 = 0$. Thus, the eigenvectors $\mathbf{v}^1$ and $\mathbf{v}^2$ corresponding to two distinct eigenvalues are linearly independent.

Now, routine matrix multiplications suffice to exhibit the desired transformation property of **P**:

$$\mathbf{A}\mathbf{P} = \left(\mathbf{A}\mathbf{v}^1 \,|\, \mathbf{A}\mathbf{v}^2\right) = \left(\lambda_1 \mathbf{v}^1 \,|\, \lambda_2 \mathbf{v}^2\right) = \mathbf{P}\mathbf{J},$$

where

$$\mathbf{J} = \begin{pmatrix} \lambda_1 & 0 \\ 0 & \lambda_2 \end{pmatrix}.$$

Since **P** is invertible, if we multiply both sides of this matrix equation by $\mathbf{P}^{-1}$, we obtain the desired similarity

$$\mathbf{P}^{-1}\mathbf{A}\mathbf{P} = \mathbf{J} = \begin{pmatrix} \lambda_1 & 0 \\ 0 & \lambda_2 \end{pmatrix}.$$

(*ii*) *Equal eigenvalues:* There are two cases to consider. First, suppose that λ is a double eigenvalue but there are two corresponding linearly independent eigenvectors $\mathbf{v}^1$ and $\mathbf{v}^2$. Then, as in the case of the distinct eigenvalues, the transformation matrix $\mathbf{P} = (\mathbf{v}^1 \,|\, \mathbf{v}^2)$ yields the similarity

$$\mathbf{P}^{-1}\mathbf{A}\mathbf{P} = \mathbf{J} = \begin{pmatrix} \lambda & 0 \\ 0 & \lambda \end{pmatrix}.$$

As the second possibility, let us suppose that λ is still a double eigenvalue but with only one independent eigenvector. Let $\mathbf{v}^1$ be such an eigenvector. We will show momentarily that in this case there is a second vector $\mathbf{v}^2$ which is linearly independent from $\mathbf{v}^1$ and satisfies the equation

$$(\mathbf{A} - \lambda\mathbf{I})\,\mathbf{v}^2 = \mathbf{v}^1. \tag{8.18}$$

Given this fact, if we take the transformation matrix $\mathbf{P}$ as $\mathbf{P} = (\mathbf{v}^1 \,|\, \mathbf{v}^2)$, then

$$\mathbf{A}\mathbf{P} = (\mathbf{A}\mathbf{v}^1 \,|\, \mathbf{A}\mathbf{v}^2) = (\lambda\mathbf{v}^1 \,|\, \mathbf{v}^1 + \lambda\mathbf{v}^2) = \mathbf{P}\mathbf{J},$$

where

$$\mathbf{J} = \begin{pmatrix} \lambda & 1 \\ 0 & \lambda \end{pmatrix}.$$

Since $\mathbf{P}$ is invertible, if we multiply both sides of this matrix equation by $\mathbf{P}^{-1}$, we obtain the desired similarity

$$\mathbf{P}^{-1}\mathbf{A}\mathbf{P} = \mathbf{J} = \begin{pmatrix} \lambda & 1 \\ 0 & \lambda \end{pmatrix}.$$

Let us now show the existence of $\mathbf{v}^2$ satisfying Eq. (8.18). Since $(\mathbf{A} - \lambda\mathbf{I})$ is not the zero matrix (why?), if there is no such vector $\mathbf{v}^2$, then the range of $(\mathbf{A} - \lambda\mathbf{I})$ must contain a vector $\mathbf{u}^1$ so that $\mathbf{u}^1$ and $\mathbf{v}^1$ are linearly independent. Let $\mathbf{u}^2$ be such that $(\mathbf{A} - \lambda\mathbf{I})\,\mathbf{u}^2 = \mathbf{u}^1$. Consequently, any vector, and in particular $\mathbf{u}^2$, can be written as a linear combination of $\mathbf{u}^1$ and $\mathbf{v}^1$:

$$\mathbf{u}^2 = c_1\mathbf{v}^1 + c_2\mathbf{u}^1.$$

If we substitute this into the equation $(\mathbf{A} - \lambda\mathbf{I})\,\mathbf{u}^2 = \mathbf{u}^1$ and use the fact that $\mathbf{v}^1$ is an eigenvector, we obtain

$$c_2\,(\mathbf{A} - \lambda\mathbf{I})\,\mathbf{u}^1 = \mathbf{u}^1.$$

Since $\mathbf{u}^1$ is not the zero vector, it follows that $c_2 \neq 0$ and thus

$$\left[\mathbf{A} - \left(\lambda + \frac{1}{c_2}\right)\mathbf{I}\right]\mathbf{u}^1 = \mathbf{0}.$$

This equation implies that $\lambda + 1/c_2$ is an eigenvalue of $\mathbf{A}$ which is contrary to our assumption that λ was the only eigenvalue. Consequently, there must exist a vector $\mathbf{v}^2$ satisfying Eq. (8.18) from which the linear independence of $\mathbf{v}^1$ and $\mathbf{v}^2$ is almost immediate.

(*iii*) *Complex eigenvalues:* Suppose that $\lambda = \alpha + i\beta$, with $\beta \neq 0$, is a complex eigenvalue of $\mathbf{A}$ with a corresponding complex eigenvector $\mathbf{v}^1 + i\mathbf{v}^2$, that is,

$$\mathbf{A}\left(\mathbf{v}^1 + i\mathbf{v}^2\right) = (\alpha + i\beta)\left(\mathbf{v}^1 + i\mathbf{v}^2\right),\tag{8.19}$$

where $\mathbf{v}^1$ and $\mathbf{v}^2$ are two nonzero real vectors. Furthermore, these two real vectors are linearly independent. Indeed, suppose that they were linearly dependent. Then there would be two nonzero real constants c_1 and c_2 such that $c_1\mathbf{v}^1 + c_2\mathbf{v}^2 = \mathbf{0}$, or equivalently $\mathbf{v}^1 = (c_2/c_1)\mathbf{v}^2$. Using this in Eq. (8.19) we obtain $\mathbf{A}\mathbf{v}^2 = (\alpha + i\beta)\,\mathbf{v}^2$. Since the left-hand side of this equation is real and the right-hand side is complex, we arrive at a contradiction. So, the vectors $\mathbf{v}^1$ and $\mathbf{v}^2$ must be linearly independent.

As a transformation matrix $\mathbf{P}$, consider the matrix whose columns consist of the real and imaginary parts of the complex eigenvector:

$$\mathbf{P} = \left(\mathbf{v}^1 \,|\, \mathbf{v}^2\right).$$

To exhibit the desired transformation property of $\mathbf{P}$, observe that by equating its real and imaginary parts, complex equation (8.19) is equivalent to the following pair of real equations:

$$\mathbf{A}\mathbf{v}^1 = \alpha\mathbf{v}^1 - \beta\mathbf{v}^2, \qquad \mathbf{A}\mathbf{v}^2 = \beta\mathbf{v}^1 + \alpha\mathbf{v}^2.$$

Now, routine matrix multiplications yield

$$\mathbf{A}\mathbf{P} = \left(\mathbf{A}\mathbf{v}^1 \,|\, \mathbf{A}\mathbf{v}^2\right) = \left(\alpha\mathbf{v}^1 - \beta\mathbf{v}^2 \,|\, \beta\mathbf{v}^1 + \alpha\mathbf{v}^2\right) = \mathbf{P}\mathbf{J},$$

where

$$\mathbf{J} = \begin{pmatrix} \alpha & \beta \\ -\beta & \alpha \end{pmatrix}.$$

Since $\mathbf{P}$ is invertible, if we multiply both sides of this matrix equation by $\mathbf{P}^{-1}$, we obtain the desired similarity

$$\mathbf{P}^{-1}\mathbf{A}\mathbf{P} = \mathbf{J} = \begin{pmatrix} \alpha & \beta \\ -\beta & \alpha \end{pmatrix}.$$

This concludes the proof of the theorem and our somewhat extended excursion into linear algebra. ◇

We remarked in Section 8.1 that the space of solutions of a planar linear system is a two-dimensional vector space. The efforts in the proof of the theorem above easily yield a basis for this vector space. If one is interested in only finding the explicit solution of an initial-value problem for a linear system, then one can simply consider a linear combination of these basis vectors and adjust the coefficients appropriately. Sample computations along these lines are suggested in the exercises. With these remarks it may seem that the theory of linear systems is complete. The next two sections, however, contain interesting surprises.

Exercises

8.6. *Changing coordinates:* Consider the linear system $\dot{x}_1 = x_1$, $\dot{x}_2 = -x_2$. Transform this system to new coordinates $\mathbf{y} = \mathbf{P}^{-1}\mathbf{x}$ for various choices of $\mathbf{P}$ below and sketch the phase portraits in these new coordinate systems:

$$\begin{pmatrix} 2 & 0 \\ 0 & 2 \end{pmatrix}, \quad \begin{pmatrix} 0 & 1 \\ 1 & 0 \end{pmatrix}, \quad \begin{pmatrix} 1 & 1 \\ 1 & -1 \end{pmatrix}, \quad \begin{pmatrix} 1 & 1 \\ 2 & -2 \end{pmatrix}.$$

8.7. Find the Jordan Normal Form and also compute $e^{\mathbf{A}t}$ for each of the following matrices:

$$\begin{pmatrix} 0 & 1 \\ 1 & 0 \end{pmatrix}, \quad \frac{1}{2}\begin{pmatrix} 2 & 1 \\ -1 & 0 \end{pmatrix}, \quad \begin{pmatrix} 0 & 1 \\ -\omega^2 & -\rho \end{pmatrix},$$

where ρ and ω are constants, and you will have to consider their relative magnitudes.

8.8. *Determining fundamental matrix solutions:* In this exercise we outline how to find two linearly independent solutions of $\dot{\mathbf{x}} = \mathbf{A}\mathbf{x}$ using eigenvalues and eigenvectors.

(a) Show that if $\lambda_1 \neq \lambda_2$ are two real eigenvalues of $\mathbf{A}$ and $\mathbf{v}^1$ and $\mathbf{v}^2$ are eigenvectors corresponding to λ_1 and λ_2, respectively, then $e^{\lambda_1 t}\mathbf{v}^1$ and $e^{\lambda_2 t}\mathbf{v}^2$ are two linearly independent solutions of the linear system $\dot{\mathbf{x}} = \mathbf{A}\mathbf{x}$.

(b) Show that if $\lambda_1 = \lambda_2$ are the eigenvalues of $\mathbf{A}$ and there is only one independent eigenvector $\mathbf{v}^1$ corresponding to λ_1 and $\mathbf{v}^2$ is any solution of $(\mathbf{A} - \lambda_1\mathbf{I})\mathbf{v}^2 = \mathbf{v}^1$, then $e^{\lambda_1 t}\mathbf{v}^1$ and $e^{\lambda_1 t}(\mathbf{v}^2 + t\mathbf{v}^1)$ are two linearly independent solutions of the linear system $\dot{\mathbf{x}} = \mathbf{A}\mathbf{x}$. As a specific example, find the solution of the initial-value problem

$$\dot{\mathbf{x}} = \frac{1}{2}\begin{pmatrix} 2 & 1 \\ -1 & 0 \end{pmatrix}\mathbf{x}, \qquad \mathbf{x}(0) = \begin{pmatrix} 1 \\ 0 \end{pmatrix}.$$

(c) If $\lambda = \alpha + i\beta$, with $\beta \neq 0$, is a complex eigenvalue of $\mathbf{A}$ and $\mathbf{v} = \mathbf{v}^1 + i\mathbf{v}^2$ be a corresponding complex eigenvector, then

$$\mathbf{x}^1(t) = e^{\alpha t}\left(\mathbf{v}^1 \cos\beta t - \mathbf{v}^2 \sin\beta t\right),$$
$$\mathbf{x}^2(t) = e^{\alpha t}\left(\mathbf{v}^1 \sin\beta t + \mathbf{v}^2 \cos\beta t\right)$$

are two linearly independent solutions of the linear system $\dot{\mathbf{x}} = \mathbf{A}\mathbf{x}$. As a specific example, find the solution of the initial-value problem

$$\dot{\mathbf{x}} = \begin{pmatrix} 1 & -1 \\ 5 & -3 \end{pmatrix} \mathbf{x}, \qquad \mathbf{x}(0) = \begin{pmatrix} 1 \\ 2 \end{pmatrix}.$$

8.9. *All ellipses:* Show that every orbit of the linear system

$$\dot{\mathbf{x}} = \begin{pmatrix} 0 & -2 \\ 8 & 0 \end{pmatrix} \mathbf{x}$$

is an ellipse. The origin is, of course, a degenerate ellipse. Draw the phase portrait of the system.

8.10. *A second-order equation:* Consider the second-order differential equation $\ddot{y} + \rho\dot{y} + \omega^2 y = 0$, where ρ and ω are real constants. Convert this differential equation to an equivalent planar linear system and find a fundamental matrix solution.

8.11. *On characteristic polynomial:* Show that the characteristic polynomial of a 2×2 matrix $\mathbf{A}$ can be written as

$$\det(\mathbf{A} - \lambda\mathbf{I}) = \lambda^2 - (\operatorname{tr} \mathbf{A})\lambda + \det \mathbf{A},$$

where

$$\operatorname{tr} \mathbf{A} = a_{11} + a_{22} = \lambda_1 + \lambda_2, \qquad \det \mathbf{A} = a_{11}a_{22} - a_{21}a_{12} = \lambda_1\lambda_2,$$

and λ_1 and λ_2 are the roots of the equation $\det(\mathbf{A} - \lambda\mathbf{I}) = 0$.

8.12. *A useful estimate:* As an application of Theorem 8.12 on normal forms, establish the following estimate:
If the eigenvalues of a 2×2 matrix $\mathbf{A}$ have negative real parts, then there are positive constants k and α such that, for all $\mathbf{x} \in \mathbb{R}^2$,

$$\|e^{\mathbf{A}t}\mathbf{x}\| \leq k e^{-\alpha t} \|\mathbf{x}\| \qquad \text{for } t \geq 0.$$

The utility of this result will become apparent in the next section.

8.13. *A "stiff" linear system:* Since the exact solutions of linear systems can readily be written down, they often serve as testing grounds for various numerical algorithms for solving initial-value problems for ordinary differential equations. For example, try to solve the following linear system using Euler, Improved Euler, Runge–Kutta, and various step sizes:

$$\dot{\mathbf{x}} = \begin{pmatrix} 998.0 & 1998.0 \\ -999.0 & -1999.0 \end{pmatrix} \mathbf{x}.$$

Any success? Compute the eigenvalues and the eigenvectors of the system to obtain explicit solutions; compare your theoretical findings with the numerical ones. By the way, the term *stiff* does not really have a precise mathematical definition; it usually means that the solutions are doing interesting things on a very short, as well as on a long, time scale. To capture such a behavior numerically, one must use either a "very small" step size, and wait, or detect the places where the drastic changes are and vary the step size accordingly. How do the magnitudes of the eigenvalues of the linear system above compare? For more information on this linear system, see Forsythe et al. [1977], p. 124.

8.3. Qualitative Equivalence in Linear Systems

The purpose of this section is to investigate the question of qualitative equivalence of planar linear systems in the spirit of Section 2.6. Let us begin with a precise definition of the notion of qualitative equivalence.

Definition 8.13. *Two planar linear systems* $\dot{x} = Ax$ *and* $\dot{x} = Bx$ *are said to be topologically equivalent if there is a homeomorphism* $h : \mathbb{R}^2 \to \mathbb{R}^2$ *of the plane, that is,* h *is continuous with continuous inverse, that maps the orbits of* $\dot{x} = Ax$ *onto the orbits of* $\dot{x} = Bx$ *and preserves the sense of direction of time.*

Since we have a formula for the flows of planar linear systems, it is convenient to recast this definition in a somewhat more quantitative form by mapping one flow to the other, that is,

$$h(e^{At}x) = e^{Bt}h(x) \tag{8.20}$$

for every $t \in \mathbb{R}$ and $x \in \mathbb{R}^2$. A homeomorphism h satisfying Eq. (8.20) is a bit more special than the one required in Definition 8.13; while it maps orbits onto orbits, it also preserves the time parametrizations of the orbits. However, as we shall shortly see, in the case A and B are *hyperbolic*, that is, their eigenvalues have nonzero real parts, the homeomorphism h in Definition 8.13 can indeed be chosen to satisfy Eq. (8.20).

The question of topological equivalence of linear systems is a somewhat difficult one. Therefore, to motivate the introduction of the formal definition of qualitative equivalence above, let us first reexamine the results of the previous section. In the course of transforming matrices into their Jordan Normal Forms we have investigated the question of topological equivalence in a limited context by considering only invertible linear maps as our allowable homeomorphisms. In fact, Eqs. (8.9) and (8.11) imply that if there is an invertible 2×2 matrix P such that $A = P^{-1}BP$, then the flows of $\dot{x} = Ax$ and $\dot{x} = Bx$ are related by $Pe^{At} = e^{Bt}P$. In other words, if matrices A and B are *similar*, also called *conjugate*, then the flows of $\dot{x} = Ax$ and $\dot{x} = Bx$ are *linearly equivalent*. It is easy to verify the converse implication by simply differentiating the equation $\mathbf{P}e^{\mathbf{A}t} = e^{\mathbf{B}t}\mathbf{P}$ and then evaluating it at $t = 0$.

Unfortunately, linear equivalence is a bit too restrictive for comparing the qualitative features of flows of linear systems. For example, the two linear systems $\dot{\mathbf{x}} = -\mathbf{I}$ and $\dot{\mathbf{x}} = -2\mathbf{I}$ should be considered qualitatively equivalent yet they are not linearly equivalent; this follows from the fact that the matrices $-\mathbf{I}$ and $-2\mathbf{I}$ are not similar because they have different eigenvalues, as shown in the exercises.

After linear equivalence, one may naturally ponder about the *differentiable equivalence* of linear systems: For given two matrices $\mathbf{A}$ and $\mathbf{B}$, does there exist a diffeomorphism $h : \mathbb{R}^2 \to \mathbb{R}^2$ satisfying $h(e^{\mathbf{A}t}\mathbf{x}) = e^{\mathbf{B}t}h(\mathbf{x})$? In the case of hyperbolic linear systems, differentiable equivalence offers nothing new, as the following simple lemma attests:

Lemma 8.14. *Two hyperbolic linear systems $\dot{\mathbf{x}} = \mathbf{A}\mathbf{x}$ and $\dot{\mathbf{x}} = \mathbf{B}\mathbf{x}$ are differentiably equivalent if and only if they are linearly equivalent.*

Proof. Linear equivalence, of course, implies differentiable equivalence; therefore, we need to show the converse implication. Suppose that $\widehat{h} : \mathbb{R}^2 \to \mathbb{R}^2$ is a diffeomorphism satisfying Eq. (8.20). Let $\widehat{h}(\mathbf{0}) = \mathbf{c}$. Since $\mathbf{0}$ is an equilibrium point of $\dot{\mathbf{x}} = \mathbf{A}\mathbf{x}$, it follows that $\mathbf{c}$ is an equilibrium point of $\dot{\mathbf{x}} = \mathbf{B}\mathbf{x}$, that is, $\mathbf{B}\mathbf{c} = \mathbf{0}$. Now, consider the diffeomorphism of $\mathbb{R}^2$ consisting of the shift $g(\mathbf{x}) \mapsto \mathbf{x} - \mathbf{c}$. The diffeomorphism g takes orbits of $\dot{\mathbf{x}} = \mathbf{B}\mathbf{x}$ into itself, and the diffeomorphism $h = g \circ \widehat{h}$ takes orbits of $\dot{\mathbf{x}} = \mathbf{A}\mathbf{x}$ to the orbits $\dot{\mathbf{x}} = \mathbf{B}\mathbf{x}$ while leaving the origin fixed, that is, h satisfies Eq. (8.20) and $h(\mathbf{0}) = \mathbf{0}$.

Now, if we differentiate Eq. (8.20) with respect to $\mathbf{x}^0$ and set $\mathbf{x}^0 = \mathbf{0}$, we obtain $\mathbf{H}e^{\mathbf{A}t} = e^{\mathbf{B}t}H$, where $\mathbf{H} = D_{\mathbf{x}}h(\mathbf{0})$ which is a linear map. If we differentiate with respect to t and set $t = 0$, then $\mathbf{H}\mathbf{A} = \mathbf{B}\mathbf{H}$. $\Diamond$

It is evident from the foregoing discussion that we have no choice but face the difficult task of deciding the qualitative equivalence of linear systems using homeomorphisms of the plane.

Let us now state two theorems on the topological classification of planar linear systems and discuss some of their implications. Because of their lengths, we will defer the proofs to the end of the section. Here is the first theorem covering the *hyperbolic* linear systems.

Theorem 8.15. *Suppose that the eigenvalues of two matrices $\mathbf{A}$ and $\mathbf{B}$ have nonzero real parts. Then the two linear systems $\dot{\mathbf{x}} = \mathbf{A}\mathbf{x}$ and $\dot{\mathbf{x}} = \mathbf{B}\mathbf{x}$ are topologically equivalent if and only if $\mathbf{A}$ and $\mathbf{B}$ have the same number of eigenvalues with negative (and hence positive) real parts. Consequently, up to topological equivalence, there are three distinct equivalence classes of hyperbolic planar linear systems with, for example, the following representatives:*

(i) $\begin{pmatrix} -1 & 0 \\ 0 & -1 \end{pmatrix}$: *two negative eigenvalues;*

(ii) $\begin{pmatrix} 1 & 0 \\ 0 & 1 \end{pmatrix}$: *two positive eigenvalues;*

(iii) $\begin{pmatrix} 1 & 0 \\ 0 & -1 \end{pmatrix}$: *one positive and one negative eigenvalue.*

In elementary books on differential equations, most often only the linear equivalence of hyperbolic linear systems is considered and a host of terms, such as node, improper node, spiral, etc., are introduced to label various phase portraits; see Figure 8.4. From the topological viewpoint, however, there are only three cases and they are determined solely by the signs of the real parts of the eigenvalues, as asserted in the theorem above. Notice in particular the striking assertion that a stable spiral and a stable node are topologically equivalent. Consequently, it is preferable to employ the following terminology whose appropriateness will become apparent when we study the qualitative features of nonlinear differential equations near an equilibrium: A linear system whose eigenvalues have negative real parts is called a *hyperbolic sink*; a linear system whose eigenvalues have positive real parts is called a *hyperbolic source*; and, when one eigenvalue is negative and the other is positive the linear system is said to be a *hyperbolic saddle.*

We next present the topological classification of nonhyperbolic linear systems.

Theorem 8.16. *If a coefficient matrix* **A** *has at least one eigenvalue with zero real part, then the planar linear system* $\dot{\mathbf{x}} = \mathbf{A}\mathbf{x}$ *is topologically equivalent to precisely one of the following five linear systems with the indicated coefficient matrices:*

(i) $\begin{pmatrix} 0 & 0 \\ 0 & 0 \end{pmatrix}$: *the zero matrix;*

(ii) $\begin{pmatrix} -1 & 0 \\ 0 & 0 \end{pmatrix}$: *one negative and one zero eigenvalue;*

(iii) $\begin{pmatrix} 1 & 0 \\ 0 & 0 \end{pmatrix}$: *one positive and one zero eigenvalue;*

(iv) $\begin{pmatrix} 0 & 1 \\ 0 & 0 \end{pmatrix}$: *two zero eigenvalues but one eigenvector;*

(v) $\begin{pmatrix} 0 & 1 \\ -1 & 0 \end{pmatrix}$: *two purely imaginary eigenvalues.*

Phase portraits of the representatives of the three hyperbolic and the five nonhyperbolic linear systems are depicted in Figure 8.5.

From a visual inspection of phase portraits of linear systems, the conclusions of the two theorems above are quite plausible, yet formal proofs turn out to be somewhat long and intricate. We now begin this arduous task, which will occupy our attention during the remaining part of this section, with some auxiliary results on quadratic forms.

A real symmetric matrix **C**, that is, $\mathbf{C}^T = \mathbf{C}$ where the superscript T denotes the transpose, is said to be *positive definite* if the quadratic form

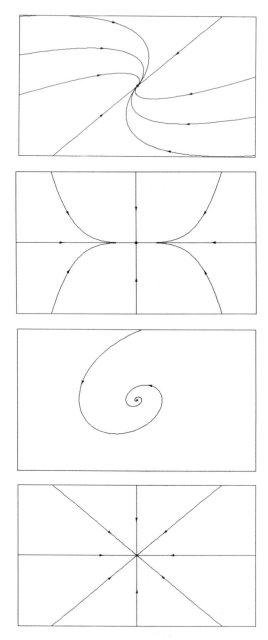

Figure 8.4. (a) *All linear systems whose eigenvalues have negative real parts—sinks—are topologically equivalent* (continued).

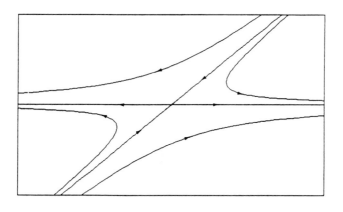

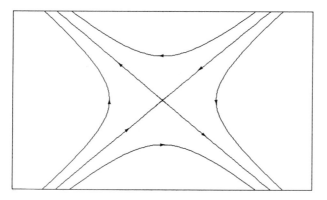

Figure 8.4 Continued. (b) *Linear saddles are topologically equivalent.*

$\mathbf{x}^T\mathbf{C}\mathbf{x} > 0$ for all $\mathbf{x} \neq \mathbf{0}$. An important geometric property of a positive definite quadratic form on the plane is that the *level set* $\{\,\mathbf{x} : \mathbf{x}^T\mathbf{C}\mathbf{x} = k\,\}$, for any $k > 0$, is an ellipse encircling the origin. In the lemma below, we establish the existence of some quadratic forms with additional properties which will play a crucial role in the proof of Theorem 8.15.

Lemma 8.17. *If the eigenvalues of $\mathbf{A}$ have negative real parts, then there exists a positive definite matrix $\mathbf{C}$ such that $\mathbf{A}^T\mathbf{C} + \mathbf{C}\mathbf{A} = -\mathbf{I}$.*

Proof. Without loss of generality, we may assume that the matrix $\mathbf{A}$ is in Jordan Normal Form and consider three cases:

Case (i): $\mathbf{A} = \begin{pmatrix} -\lambda_1 & 0 \\ 0 & -\lambda_2 \end{pmatrix}$, $\lambda_1 > 0$, $\lambda_2 > 0$. Take $\mathbf{C} = \begin{pmatrix} 1/(2\lambda_1) & 0 \\ 0 & 1/(2\lambda_2) \end{pmatrix}$.

Case (ii): $\mathbf{A} = \begin{pmatrix} -\alpha & \beta \\ -\beta & -\alpha \end{pmatrix}$, $\alpha > 0$, and $\beta \neq 0$. Take $\mathbf{C} = 1/(2\alpha)\,\mathbf{I}$.

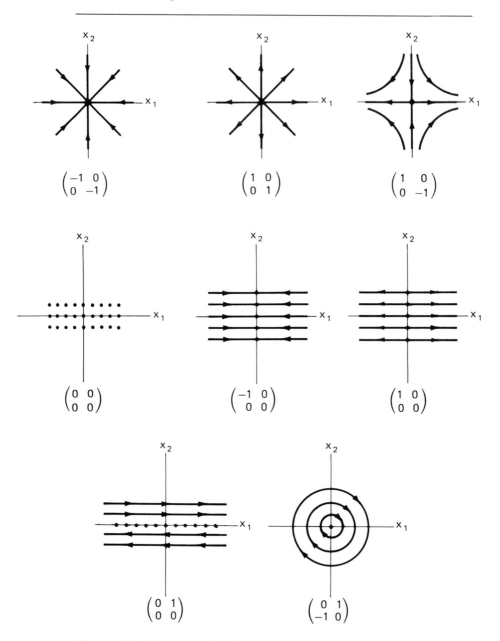

Figure 8.5. Phase portraits of representatives of topological equivalence classes of planar linear systems.

Case (iii): $\mathbf{A} = \begin{pmatrix} -\lambda & 1 \\ 0 & -\lambda \end{pmatrix}$. For any $\varepsilon > 0$, we can assume that $\mathbf{A} = \begin{pmatrix} -\lambda & \varepsilon \\ 0 & -\lambda \end{pmatrix}$ by the transformation of coordinates $\mathbf{x} \mapsto \begin{pmatrix} 1 & 0 \\ 0 & \varepsilon \end{pmatrix} \mathbf{x}$. If we let $\mathbf{D} = 1/(2\lambda)\,\mathbf{I}$, then

$$\mathbf{A}^T \mathbf{D} + \mathbf{D}\mathbf{A} = \begin{pmatrix} -1 & \varepsilon/(2\lambda) \\ \varepsilon/(2\lambda) & -1 \end{pmatrix} \equiv -\mathbf{E}.$$

Now, choose ε so that $1 - [\varepsilon/(2\lambda)]^2 > 0$ and introduce the change of coordinates $\mathbf{x} \mapsto \mathbf{E}^{-1/2}\mathbf{x}$. Then the matrix $\mathbf{C} \equiv \mathbf{E}^{-1/2}\mathbf{D}\mathbf{E}^{-1/2}$ possesses the desired properties. $\Diamond$

We now turn the proof of our first main result of the section.

Proof of Theorem 8.15. Let us first observe that the conditions on the coefficient matrices $\mathbf{A}$ and $\mathbf{B}$ are necessary for topological equivalence. For this purpose, it is not difficult to persuade oneself that if a homeomorphism takes a bounded positive orbit of $\dot{\mathbf{x}} = \mathbf{A}\mathbf{x}$ to a bounded positive orbit of $\dot{\mathbf{x}} = \mathbf{B}\mathbf{x}$, then the homeomorphism also takes the ω-limit set of one positive orbit to that of the other. A similar remark holds for α-limit sets of bounded negative orbits also. Consequently, if $\mathbf{A}$ has both eigenvalues with negative real parts and $\mathbf{A}$ is topologically equivalent to $\mathbf{B}$, then every solution of $\dot{\mathbf{x}} = \mathbf{B}\mathbf{x}$ approaches zero as $t \to +\infty$. Thus, both eigenvalues of $\mathbf{B}$ must have negative real parts. A similar argument holds, as $t \to -\infty$, if both eigenvalues of $\mathbf{A}$ have positive real parts. If $\mathbf{A}$ has one positive and one negative eigenvalue, then there must be one nonequilibrium solution of $\dot{\mathbf{x}} = \mathbf{B}\mathbf{x}$ that approaches zero as $t \to +\infty$ and another solution that approaches zero as $t \to -\infty$. Consequently, $\mathbf{B}$ must have one negative and one positive eigenvalue.

To prove sufficiency, let us begin with *Case (iii)* as it is the simplest. From the previous section, we may assume that if $\mathbf{A}$ and $\mathbf{B}$ have one negative and one positive eigenvalue, then they can be put, by linear change of coordinates, into the following Jordan Normal Forms:

$$\mathbf{A} = \begin{pmatrix} -\lambda_1 & 0 \\ 0 & \lambda_2 \end{pmatrix}, \qquad \mathbf{B} = \begin{pmatrix} -\mu_1 & 0 \\ 0 & \mu_2 \end{pmatrix},$$

where each $\lambda_i > 0$ and $\mu_i > 0$. Now, recall from Section 2.6 that the two linear scalar differential equations

$$\dot{x}_1 = -\lambda_1 x_1, \qquad \dot{x}_1 = -\mu_1 x_1$$

are topologically equivalent with some homeomorphism $h_1 : \mathbb{R} \to \mathbb{R}$. Similarly, the scalar equations

$$\dot{x}_2 = \lambda_2 x_2, \qquad \dot{x}_2 = \mu_2 x_2$$

are topologically equivalent via a homeomorphism $h_2 : \mathbb{R} \to \mathbb{R}$. The flows of the planar linear systems $\dot{\mathbf{x}} = \mathbf{A}\mathbf{x}$ and $\dot{\mathbf{x}} = \mathbf{B}\mathbf{x}$ are then topologically equivalent with the homeomorphism

$$h : \mathbb{R}^2 \to \mathbb{R}^2; \quad h(\mathbf{x}) = (h_1(x_1), h_2(x_2)).$$

We now turn to *Case* (*i*) and construct homeomorphisms to establish the topological equivalence of planar linear systems whose eigenvalues have negative real parts. *Case* (*ii*) follows from *Case* (*i*) by simply replacing t with $-t$. The construction below may appear a bit technical, but the geometric idea behind it is quite simple. For each linear system, find an ellipse encircling the origin such that each orbit, except the equilibrium point at the origin, crosses the ellipse in the same direction and only at one point. Now, map homeomorphically one ellipse to the other, and map an orbit through a point on the first ellipse to the orbit passing through the image point on the second ellipse. Finally, extend such a homeomorphism to include the origin. This idea is reminiscent of polar coordinates where the ellipse is the "angular variable" and the orbits are the "radial variable."

To prove the existence of two ellipses with the desired properties mentioned above, we use Lemma 8.17 and some ideas from the theory of "Liapunov functions." We will delve into the topic of Liapunov functions in Chapter 9. Here, we will be content to point out that if the derivative of a positive definite quadratic function along the solutions of a linear system $\dot{\mathbf{x}} = \mathbf{A}\mathbf{x}$ is negative, except at the origin, then the orbits cross the elliptical level sets inward. There exist two positive definite symmetric matrices $\mathbf{C_A}$ and $\mathbf{C_B}$ such that

$$\mathbf{A}^T\mathbf{C_A} + \mathbf{C_A}\mathbf{A} = -\mathbf{I}, \qquad \mathbf{B}^T\mathbf{C_B} + \mathbf{C_B}\mathbf{B} = -\mathbf{I}.$$

Let $\mathbf{x}(t)$ be a solution of $\dot{\mathbf{x}} = \mathbf{A}\mathbf{x}$, then

$$\frac{d}{dt}\left[\mathbf{x}^T(t)\,\mathbf{C_A}\,\mathbf{x}(t)\right] = \mathbf{x}^T(t)\left[\mathbf{A}^T\mathbf{C_A} + \mathbf{C_A}\mathbf{A}\right]\mathbf{x}(t) = -\mathbf{x}^T(t)\,\mathbf{x}(t).$$

Therefore, any nonequilibrium solution $\mathbf{x}(t)$ of $\dot{\mathbf{x}} = \mathbf{A}\mathbf{x}$ crosses the level sets of $\mathbf{x}^T\mathbf{C_A}\mathbf{x}$ inward. A similar computation shows that any nonequilibrium solution $\mathbf{x}(t)$ of $\dot{\mathbf{x}} = \mathbf{B}\mathbf{x}$ crosses the level sets of $\mathbf{x}^T\mathbf{C_B}\mathbf{x}$ inward also; see Figure 8.6.

It is now easy to define a homeomorphism $h : \mathbb{R}^2 - \mathbf{0} \to \mathbb{R}^2 - \mathbf{0}$ that takes the orbits of $\dot{\mathbf{x}} = \mathbf{A}\mathbf{x}$ to the orbits of $\dot{\mathbf{x}} = \mathbf{B}\mathbf{x}$. Let $\widehat{h} : \{\mathbf{x} : \mathbf{x}^T\mathbf{C_A}\mathbf{x} = 1\} \to \{\mathbf{x} : \mathbf{x}^T\mathbf{C_B}\mathbf{x} = 1\}$ be a given homeomorphism of the two ellipses. For any $\mathbf{x}^0 \neq \mathbf{0}$, there is a unique time $t_{\mathbf{x}^0}$ such that $\mathbf{x}^1 \equiv e^{-\mathbf{A}t_{\mathbf{x}^0}}\mathbf{x}^0$ lies on the ellipse $\mathbf{x}^T\mathbf{C_A}\mathbf{x} = 1$. Now, define $h(\mathbf{x}^0) = e^{\mathbf{B}t_{\mathbf{x}^0}}\widehat{h}(\mathbf{x}^1)$; see Figure 8.7. It is evident, from continuous dependence of solutions on initial data, that h is a homeomorphism; in fact, it is a diffeomorphism if $\widehat{h}$ is chosen to be a diffeomorphism.

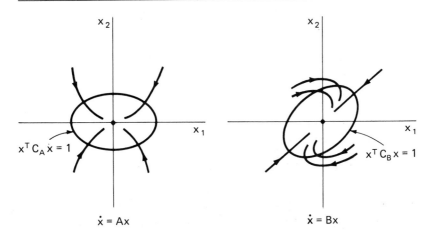

Figure 8.6. *Orbits of a planar linear system whose eigenvalues have negative linear parts cross the level curves of an appropriate positive definite quadratic function, ellipses, inward.*

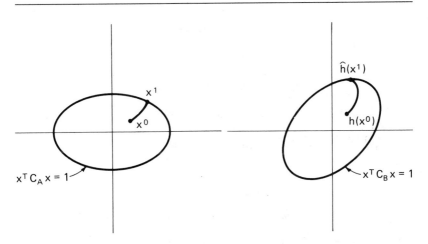

Figure 8.7. *Constructing a homeomorphism between the flows of two planar linear systems whose eigenvalues have negative real parts.*

To extend the domain of the homeomorphism h above to all of $\mathbb{R}^2$, we define $h(\mathbf{0}) = \mathbf{0}$. It remains to show that h is continuous at $\mathbf{0}$. For this purpose, suppose that $0 < \delta < 1$ is given and consider $\mathbf{x}^0 \in \{\, \mathbf{x} : \mathbf{x}^T \mathbf{C_A} \mathbf{x} \leq \delta\,\}$. Then there is $t_0 = t_0(\delta)$ with $t_{\mathbf{x}^0} \geq t_0$, and $t_0(\delta) \to +\infty$ as $\delta \to 0$. Since the set $\{\, \mathbf{x} : \mathbf{x}^T \mathbf{C_B} \mathbf{x} = 1\,\}$ is bounded by a positive constant, say,

M, and $\|e^{\mathbf{B}t}\mathbf{x}\| \leq ke^{-\alpha t}$ for $t \geq 0$ and some positive constants k and α, it follows that

$$\|h(\mathbf{x}^0)\| = \|e^{\mathbf{B}t_{\mathbf{x}^0}}\widehat{h}(e^{-\mathbf{A}t_{\mathbf{x}^0}}\mathbf{x}^0)\| \leq ke^{-\alpha t_0(\delta)}M.$$

Therefore, for any $\varepsilon > 0$, there is $\delta > 0$ such that $\|h(\mathbf{x}^0)\| < \varepsilon$ so long as $(\mathbf{x}^0)^T\mathbf{C_A}\mathbf{x}^0 \leq \delta$. Since $\mathbf{x}^T\mathbf{C_A}\mathbf{x}$ is norm equivalent to the Euclidean norm, h is continuous at the origin. As we saw in an example above, however, h cannot always be made a diffeomorphism at the origin. $\Diamond$

We now give an outline of a proof of the classification of nonhyperbolic planar linear systems.

Proof of Theorem 8.16. It is not difficult to establish that a nonhyperbolic planar linear system is topologically equivalent to one of the five cases by putting the coefficient matrix in Jordan Normal Form and using the ideas from the proof of Theorem 8.15. In the last case it is necessary to reparametrize the periodic orbits. Therefore, it remains to show that no two cases in the statement of the theorem are topologically equivalent. Here are the distinguishing qualitative features of each case that makes it unique:

$\mathbf{A} = \left(\begin{smallmatrix} 0 & 0 \\ 0 & 0 \end{smallmatrix}\right)$: Every orbit is an equilibrium point.

$\mathbf{A} = \left(\begin{smallmatrix} -1 & 0 \\ 0 & 0 \end{smallmatrix}\right)$: The ω-limit sets of all positive orbits are equilibrium points.

$\mathbf{A} = \left(\begin{smallmatrix} 1 & 0 \\ 0 & 0 \end{smallmatrix}\right)$: The α-limit sets of all negative orbits are equilibrium points.

$\mathbf{A} = \left(\begin{smallmatrix} 0 & 1 \\ 0 & 0 \end{smallmatrix}\right)$: All positive and negative nonequilibrium orbits are unbounded.

$\mathbf{A} = \left(\begin{smallmatrix} 0 & 1 \\ -1 & 0 \end{smallmatrix}\right)$: Each nonequilibrium orbit is periodic. $\Diamond$

Exercises _____ ♣♡♠♢

8.14. *Eigenvalues and similarity:* Show that similar matrices have the same eigenvalues.
  Hint: $\det\left(\mathbf{P}^{-1}\mathbf{AP} - \lambda\mathbf{I}\right) = \det\left(\mathbf{P}^{-1}\mathbf{AP} - \mathbf{P}^{-1}\lambda\mathbf{IP}\right) = \det\left(\mathbf{A} - \lambda\mathbf{I}\right).$

8.15. *On topological and linear equivalence:* Show that the following two linear systems have the same eigenvalues and thus are topologically equivalent; however, they are not linearly equivalent:

$$\dot{\mathbf{x}} = \begin{pmatrix} -2 & 0 \\ 0 & -2 \end{pmatrix}\mathbf{x}, \qquad \dot{\mathbf{x}} = \begin{pmatrix} -2 & 1 \\ 0 & -2 \end{pmatrix}\mathbf{x}.$$

Draw the phase portraits of these two linear systems and compare.

8.16. Construct a homeomorphism of $\mathbb{R}^2$ to establish the topological equivalence of the flows of the following two linear systems:

$$\dot{\mathbf{x}} = \begin{pmatrix} -1 & 0 \\ 0 & 2 \end{pmatrix} \mathbf{x}, \qquad \dot{\mathbf{x}} = \begin{pmatrix} -1 & 0 \\ -2 & 1 \end{pmatrix} \mathbf{x}.$$

8.17. *All the same:* Consider the following five coefficient matrices $\mathbf{A}$ of the linear systems $\dot{\mathbf{x}} = \mathbf{A}\mathbf{x}$:

$$\begin{pmatrix} -1 & 0 \\ 0 & -1 \end{pmatrix}, \quad \begin{pmatrix} -1 & 1 \\ 0 & -1 \end{pmatrix}, \quad \begin{pmatrix} -1 & 0 \\ 0 & -2 \end{pmatrix},$$

$$\begin{pmatrix} 0 & 2 \\ -1 & -3 \end{pmatrix}, \quad \begin{pmatrix} -1 & 1 \\ 0 & -2 \end{pmatrix}.$$

(a) For each $\mathbf{A}$ above, determine a symmetric matrix $\mathbf{C_A}$ satisfying the matrix equation $\mathbf{A}^T \mathbf{C_A} + \mathbf{C_A} \mathbf{A} = -\mathbf{I}$.
(b) Draw the level set $\{\, \mathbf{x} : \mathbf{x}^T \mathbf{C_A} \mathbf{x} = 1 \,\}$ for each of your $\mathbf{C_A}$.
(c) Construct homeomorphisms of $\mathbb{R}^2$ so as to establish the topological equivalence of the flows of all five linear systems.

8.4. Bifurcations in Linear Systems

In this section, we first explore the prominence of hyperbolic linear planar systems among the set of all planar linear systems. We observe, as expected, that under small perturbations, hyperbolic linear systems do not undergo any bifurcations. Moreover, we show that "almost all" linear systems are hyperbolic. Then, despite the ubiquity of hyperbolic linear systems, we reason why one must also investigate the possible bifurcations of nonhyperbolic linear systems and determine such bifurcations.

When dealing with questions regarding all linear systems, it is convenient to view a 2×2 matrix as a point in $\mathbb{R}^4$ by, for example, ordering the entries of such a matrix into a four-vector. The lemma below shows that the subset of $\mathbb{R}^4$ consisting of the hyperbolic 2×2 matrices is an open subset of $\mathbb{R}^4$.

Lemma 8.18. *Suppose that $\mathbf{A}$ is hyperbolic. Then there is an open neighborhood U of $\mathbf{A}$ in $\mathbb{R}^4$ such that for any $\mathbf{B} \in U$ the linear system $\dot{\mathbf{x}} = \mathbf{B}\mathbf{x}$ is topologically equivalent to $\dot{\mathbf{x}} = \mathbf{A}\mathbf{x}$.*

Proof. It follows from the quadratic formula that the real parts of the eigenvalues of $\mathbf{A}$ are continuous functions of the entries of $\mathbf{A}$. Now, use Theorem 8.15. $\Diamond$

Of course, the main implication of this theorem in bifurcation theory is that the phase portrait of a hyperbolic linear system does not change topologically if the entries of its coefficient matrix are varied by a small amount.

The next lemma asserts the ubiquity of hyperbolic linear systems; in mathematical terms, they are dense in the set of all linear systems. Let us first recall the definition of dense subset: a subset $\mathcal{H} \subset \mathbb{R}^4$ is *dense* in $\mathbb{R}^4$ if for any $\varepsilon > 0$ and any point $\mathbf{x} \in \mathbb{R}^4$, there is a point $\hat{\mathbf{x}} \in \mathcal{H}$ such that $\|\mathbf{x} - \hat{\mathbf{x}}\| < \varepsilon$.

Lemma 8.19. *Hyperbolic linear systems are dense in the set of all linear systems.*

Proof. Let $\mathbf{A}$ be any nonhyperbolic 2×2 matrix in Jordan Normal Form. Observe that, for any $\varepsilon \neq 0$ and $|\varepsilon|$ small, the matrix $\mathbf{A} + \varepsilon \mathbf{I}$ is hyperbolic. $\diamond$

A subset of $\mathcal{H}$ of $\mathbb{R}^4$ that is both dense and open constitutes almost all of $\mathbb{R}^4$: the density implies that every point in the complement of $\mathcal{H}$ can be well approximated by points of $\mathcal{H}$, and the openness implies that no point in $\mathcal{H}$ can be approximated arbitrarily closely by the points in the complement of $\mathcal{H}$.

A property concerning 2×2 matrices is said to be *generic* if the set of matrices possessing this property contains an open and dense subset of $\mathbb{R}^4$. In particular, hyperbolicity is a generic property.

One perplexing implication of the genericity of hyperbolic linear systems in the set of all linear systems is that under arbitrarily small perturbations a nonhyperbolic system can be made hyperbolic. In applications, a coefficient matrix may be known only approximately and thus it appears quite likely that one should encounter only hyperbolic linear systems in real life. Despite this reasoning, however, there are common situations where one cannot avoid nonhyperbolic systems. For example, the physical system under investigation may possess certain symmetries or conserve energy. Another situation is that we may be given a linear system that contains parameters and that the parameters enter into the system as continuous functions. In this case, it is unavoidable to hit one of the nonhyperbolic systems as we vary the parameters continuously if there are parameter values at which the linear systems belong to two different hyperbolic equivalence classes.

In the discussion of bifurcations of linear systems below, our point of departure will be a bit different than what we have been accustomed to in earlier chapters. Instead of analyzing the possible bifurcations of a given linear system containing parameters, we will insert a minimum number of parameters in any given nonhyperbolic system in such a way that as the parameters are varied we can reach, up to topological equivalence, any linear system in a small neighborhood of the nonhyperbolic linear system.

The resulting parameter-dependent linear system is called an *unfolding* of the nonhyperbolic linear system. Various generalizations of this idea are fundamental in the modern theory of local bifurcations.

Let us proceed by recalling that all of the relevant information regarding the eigenvalues of a 2×2 matrix $\mathbf{A}$ is contained in its characteristic polynomial

$$\det(\mathbf{A} - \lambda \mathbf{I}) = \lambda^2 - (\operatorname{tr} \mathbf{A})\lambda + \det \mathbf{A},$$

where

$$\operatorname{tr} \mathbf{A} = a_{11} + a_{22} = \lambda_1 + \lambda_2, \qquad \det \mathbf{A} = a_{11}a_{22} - a_{21}a_{12} = \lambda_1 \lambda_2, \quad (8.21)$$

and λ_1 and λ_2 are the roots of the equation $\det(\mathbf{A} - \lambda \mathbf{I}) = 0$.

It suffices to investigate the bifurcations of nonhyperbolic systems in Jordan Normal Form. Excluding the zero matrix, there are three such matrices:

$$\begin{pmatrix} \lambda & 0 \\ 0 & 0 \end{pmatrix}, \qquad \begin{pmatrix} 0 & \beta \\ -\beta & 0 \end{pmatrix}, \qquad \begin{pmatrix} 0 & 1 \\ 0 & 0 \end{pmatrix},$$

where $\lambda \neq 0$ and $\beta \neq 0$. Furthermore, without loss of generality, we shall take $\lambda = -1$ and $\beta = 1$.

Case (i): Unfolding $\mathbf{A}_0 = \begin{pmatrix} -1 & 0 \\ 0 & 0 \end{pmatrix}$. It follows from Theorem 8.16 that there is a neighborhood of $\mathbf{A}_0$ in $\mathbb{R}^4$ such that in this neighborhood the matrices that are topologically equivalent to the nonhyperbolic matrix $\mathbf{A}_0$ satisfy the conditions $\operatorname{tr} \mathbf{A} < 0$ and $\det \mathbf{A} = 0$. The set of such matrices is a three-dimensional surface Σ_1 in $\mathbb{R}^4$ containing the matrix $\mathbf{A}_0 = \begin{pmatrix} -1 & 0 \\ 0 & 0 \end{pmatrix}$. To verify that Σ_1 is indeed a three-dimensional surface, consider the determinant as a function $\det : \mathbb{R}^4 \to \mathbb{R}$ and compute that $\frac{\partial}{\partial a_{22}} \det \mathbf{A} \big|_{\mathbf{A}_0} = -1$. Now, the desired geometric conclusion can be deduced as a corollary of the Implicit Function Theorem; see the Submanifold Theorem in the Appendix. Observe that all the matrices on one side of the surface Σ_1 are topologically equivalent.

We would now like to construct a curve of matrices which crosses this surface "transversally" at the point $\mathbf{A}_0$. In other words, we define a matrix $\mathbf{A}(\mu)$ depending on a scalar parameter μ such that $\mathbf{A}(0) = \mathbf{A}_0$ and that as the parameter μ is varied from negative to positive values, one crosses the surface Σ_1 from the side with $\operatorname{tr} \mathbf{A}(\mu) < 0$ and $\det \mathbf{A}(\mu) < 0$ to the side with $\operatorname{tr} \mathbf{A}(\mu) < 0$ and $\det \mathbf{A}(\mu) > 0$. Observe that on the first side, the matrices have two real negative eigenvalues, and on the other side, they have one negative and one positive eigenvalue. One such one-parameter matrix is

$$\mathbf{A}(\mu) = \begin{pmatrix} -1 & 0 \\ 0 & \mu \end{pmatrix}, \tag{8.22}$$

where μ is a small real parameter. For reasons that are self-evident, the nonhyperbolic matrix $\mathbf{A}_0$ is often said to be a *codimension-one singularity*. An oversimplified three-dimensional depiction of this bifurcation is illustrated in Figure 8.8.

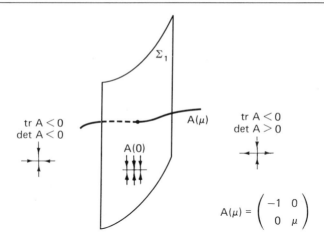

Figure 8.8. *An unfolding of $\left(\begin{smallmatrix} -1 & 0 \\ 0 & 0 \end{smallmatrix}\right)$ depending on one parameter.*

Case (ii): Unfolding $\mathbf{A_0} = \left(\begin{smallmatrix} 0 & 1 \\ -1 & 0 \end{smallmatrix}\right)$. The construction of an unfolding of this nonhyperbolic matrix is similar to that of the previous case if the role of determinant and trace are interchanged. Indeed, the matrices that are topologically equivalent to the nonhyperbolic matrix $\mathbf{A_0} = \left(\begin{smallmatrix} 0 & 1 \\ -1 & 0 \end{smallmatrix}\right)$ above satisfy the conditions tr $\mathbf{A} = 0$ and det $\mathbf{A} > 0$. The set of such matrices in a neighborhood of $\mathbf{A_0}$ is a three-dimensional surface Σ_2 in $\mathbb{R}^4$ containing the matrix $\mathbf{A_0}$. To wit, consider trace as a function tr $: \mathbb{R}^4 \to \mathbb{R}$; compute that $\frac{\partial}{\partial a_{11}}$tr $\mathbf{A}\big|_{\mathbf{A_0}} = 1$.

We would now like to construct a matrix $\mathbf{A}(\mu)$ depending on a scalar parameter μ such that $\mathbf{A}(0) = \mathbf{A_0}$ and that as the parameter μ is varied from negative to positive values, one crosses the surface Σ_2 from the side with tr $\mathbf{A}(\mu) < 0$ and det $\mathbf{A}(\mu) > 0$ to the side with tr $\mathbf{A}(\mu) > 0$ and det $\mathbf{A}(\mu) > 0$. Observe that on the first side, the matrices have eigenvalues with negative real parts, and on the other side, the eigenvalues have positive real parts. One such 1-parameter matrix is

$$\mathbf{A}(\mu) = \begin{pmatrix} \mu & 1 \\ -1 & \mu \end{pmatrix}, \tag{8.23}$$

where μ is a real parameter near zero. The geometry of this codimension-one bifurcation is depicted in Figure 8.9.

Case (iii): Unfolding $\mathbf{A_0} = \left(\begin{smallmatrix} 0 & 1 \\ 0 & 0 \end{smallmatrix}\right)$. The construction of an unfolding of this nonhyperbolic matrix is essentially a combination of the two previous cases above. The matrices that are topologically equivalent to $\mathbf{A_0}$ satisfy the conditions $\mathbf{A} \neq \mathbf{0}$ with det $\mathbf{A} = 0$ and tr $\mathbf{A} = 0$. These conditions

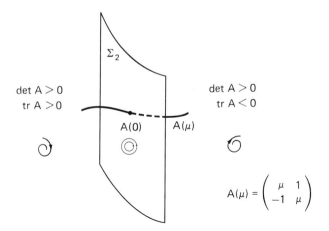

Figure 8.9. *An unfolding of* $\left(\begin{smallmatrix} 0 & 1 \\ -1 & 0 \end{smallmatrix}\right)$ *depending on one parameter.*

define a two-dimensional surface, say, $\mathcal{M}$ in a neighborhood of $\mathbf{A_0}$ in $\mathbb{R}^4$. Notice that this two-dimensional surface $\mathcal{M}$ is the intersection of the three-dimensional surface Σ_1 where det $\mathbf{A} = 0$ and the three-dimensional surface Σ_2 where tr $\mathbf{A} = 0$. Now, our task is to define a two-dimensional surface which intersects the two-dimensional surface $\mathcal{M}$ transversally at the point $\mathbf{A_0}$, that is, at the point $\mathbf{A_0}$, the direct sum of the tangent planes of the two two-dimensional surfaces is $\mathbb{R}^4$. In other words, we need to construct a matrix $\mathbf{A}(\mu_1, \mu_2)$ depending on two parameters such that $\mathbf{A}(0, 0) = \mathbf{A_0}$ and as we vary the parameters we can reach matrices whose eigenvalues are either both negative, one negative and the other positive, have negative real parts, or have positive real parts. Two such two-parameter matrices are

$$\mathbf{A}(\mu_1, \mu_2) = \begin{pmatrix} \mu_1 & 1 \\ -\mu_2 & \mu_1 \end{pmatrix}, \qquad \mathbf{A}(\mu_1, \mu_2) = \begin{pmatrix} 0 & 1 \\ \mu_1 & \mu_2 \end{pmatrix}, \qquad (8.24)$$

where μ_1 and μ_2 are real parameters near zero. For a pictorial summary of this bifurcation, see Figure 8.10.

The nonhyperbolic matrix $\left(\begin{smallmatrix} 0 & 1 \\ 0 & 0 \end{smallmatrix}\right)$ is, in a way, the simplest example of a *codimension-two singularity*. Its second unfolding above will be of paramount importance when we investigate the bifurcations of an equilibrium point of a nonlinear system in Chapter 13.

Exercises _____ ♣ ♡ ♠ ◇

8.18. *Open but not dense:* Show that the set of planar linear systems for which the origin is a sink is an open but not dense subset of $\mathbb{R}^4$.

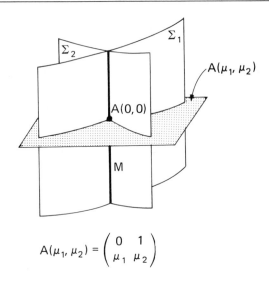

$$A(\mu_1, \mu_2) = \begin{pmatrix} 0 & 1 \\ \mu_1 & \mu_2 \end{pmatrix}$$

Figure 8.10. *An unfolding of* $\left(\begin{smallmatrix} 0 & 1 \\ 0 & 0 \end{smallmatrix}\right)$ *depending on two parameters.*

8.19. *Is it generic?:* Determine if each of the following properties of the coefficient matrices **A** of planar linear systems is generic:
(a) det **A** $\neq 0$;   (b) det **A** $= 1$;   (c) $0 < $ det **A** < 1;
(d) tr **A** $\neq 0$;   (e) tr **A** $= 0$;   (f) tr **A**$^2 \neq 0$;
(g) no real eigenvalues;   (h) both eigenvalues real.

8.20. *Dense open sets are "fat":* Prove that if S_1, ..., S_k are open and dense subsets of $\mathbb{R}^n$, then their intersection $S_1 \cap \cdots \cap S_k$ is also open and dense in $\mathbb{R}^n$.

8.21. For the following matrices, discuss the transition from one topological equivalence class to another as the scalar parameter λ is varied:

(a) $\begin{pmatrix} 1 & \lambda \\ \lambda & \lambda \end{pmatrix}$;   (b) $\begin{pmatrix} 1 & -2\lambda \\ \lambda & -2\lambda + \lambda^2 \end{pmatrix}$.

8.22. Does the matrix

$$\mathbf{A}(\mu_1, \mu_2, \mu_3) = \begin{pmatrix} \mu_1 & 1 + \mu_2 \\ 0 & \mu_3 \end{pmatrix}$$

depending on three real parameters μ_1, μ_2, and μ_3 unfold the codimension-two singularity

$$\mathbf{A}(0, 0, 0) = \begin{pmatrix} 0 & 1 \\ 0 & 0 \end{pmatrix}?$$

Moral: It is not just the number of parameters that counts, it is also where they are inserted.

8.23. *Make a choice:* We gave two unfoldings [Eq. (8.24)] of $\left(\begin{smallmatrix} 0 & 1 \\ 0 & 0 \end{smallmatrix}\right)$ depending on two parameters. Do you prefer one over the other? Argue for or against.

8.24. *Zero matrix:* How many parameters are necessary to unfold the zero matrix?

8.25. *Any bifurcations?* Let $\mathbf{A}(\lambda)$ be a 2×2 real matrix depending on a real parameter λ, for $0 \leq \lambda \leq 1$, with $\mathbf{A}(0) = \mathbf{A}_0$ and $\mathbf{A}(1) = \mathbf{A}_1$. For the following choices of $\mathbf{A}_0$ and $\mathbf{A}_1$, must there be any bifurcations as the parameter is varied?

(a) $\mathbf{A}_0 = \begin{pmatrix} -1 & 0 \\ 0 & -1 \end{pmatrix}, \qquad \mathbf{A}_1 = \begin{pmatrix} -1 & 0 \\ 0 & 1 \end{pmatrix};$

(b) $\mathbf{A}_0 = \begin{pmatrix} -1 & 0 \\ 0 & 1 \end{pmatrix}, \qquad \mathbf{A}_1 = \begin{pmatrix} 1 & 0 \\ 0 & -1 \end{pmatrix};$

(c) $\mathbf{A}_0 = \begin{pmatrix} -1 & 0 \\ 0 & -2 \end{pmatrix}, \qquad \mathbf{A}_1 = \begin{pmatrix} -1 & 1 \\ 0 & -1 \end{pmatrix}.$

8.5. Nonhomogeneous Linear Systems

In this brief section, we will determine an explicit formula for the flow of the nonautonomous differential equation

$$\dot{\mathbf{x}} = \mathbf{A}\mathbf{x} + \mathbf{g}(t), \qquad (8.25)$$

where $\mathbf{g}$ is a given vector-valued C^1 function. An equation of the form (8.25), although its vector field is not a linear function in the precise algebraic sense, is called a *nonhomogeneous linear system.*

As in Example 4.6, with the intention of eliminating the term containing $\mathbf{x}$, we introduce the new variable $\mathbf{y}(t)$ given by

$$\mathbf{y}(t) = e^{-\mathbf{A}t}\mathbf{x}(t). \qquad (8.26)$$

To obtain the differential equation in the new variable, we differentiate both sides of Eq. (8.26) with respect to t:

$$\dot{\mathbf{y}} = -\mathbf{A}e^{-\mathbf{A}t}\mathbf{x} + e^{-\mathbf{A}t}\dot{\mathbf{x}}.$$

Now, if we substitute Eq. (8.25) into the equation above, we get the desired differential equation

$$\dot{\mathbf{y}} = e^{-\mathbf{A}t}\mathbf{g}(t). \qquad (8.27)$$

Suppose that we specify the initial condition $\mathbf{x}(t_0) = \mathbf{x}^0$ for the original equation (8.25). In the new coordinates, this is the same as specifying the initial condition $\mathbf{y}(t_0) = e^{-\mathbf{A}t_0}\mathbf{x}^0$ for Eq. (8.27). To obtain the solution

$\mathbf{y}(t)$ satisfying the initial condition above, we simply integrate both sides of Eq. (8.27) from t_0 to t and rearrange the terms:

$$\mathbf{y}(t) = e^{-\mathbf{A}t_0}\mathbf{x}^0 + \int_{t_0}^t e^{-\mathbf{A}s}\mathbf{g}(s)\,ds.$$

Now, using part (ii) of Lemma 8.5, we can transform this solution back to the original coordinates and obtain

$$\mathbf{x}(t) = e^{\mathbf{A}t}\left[e^{-\mathbf{A}t_0}\mathbf{x}^0 + \int_{t_0}^t e^{-\mathbf{A}s}\mathbf{g}(s)\,ds\right]. \qquad (8.28)$$

This formula is known as the *variation of the constants* formula, likely because, in contrast with the solution of the homogeneous part, the expression inside the brackets has now become a function of t.

The variation of constants formula (8.28), of course, gives the flow of the nonautonomous system (8.25) and, when written in the expanded form

$$\varphi(t, t_0, \mathbf{x}^0) = e^{\mathbf{A}(t-t_0)}\mathbf{x}^0 + e^{\mathbf{A}t}\int_{t_0}^t e^{-\mathbf{A}s}\mathbf{g}(s)\,ds, \qquad (8.29)$$

reveals an important insight. Observe that the first term is the flow of the linear part $\dot{\mathbf{x}} = \mathbf{A}\mathbf{x}$ of Eq. (8.25). It is easy to verify that the second term is the "particular solution" of the full equation (8.25) satisfying the initial condition $\mathbf{x}(t_0) = \mathbf{0}$. We should add in conclusion that our interest in this explicit solution of nonhomogeneous linear systems stems from its utility in certain technical estimates related to questions of stability of nonlinear autonomous systems, as we shall see in the next chapter.

Exercises ————————————————————————

8.26. Find the general solution of the affine system

$$\dot{x}_1 = x_2 + 1$$
$$\dot{x}_2 = -x_1 + 1$$

and sketch its phase portrait.

8.27. Study the effect of the change of variables $\mathbf{x} = \mathbf{P}\mathbf{y}$, where $\mathbf{P}$ is a 2×2 invertible matrix, on the affine system $\dot{\mathbf{x}} = \mathbf{A}\mathbf{x} + \mathbf{g}(t)$. If $\mathbf{A}$ has two real distinct eigenvalues, can the affine system be transformed to a product system?

8.28. *Adjoint Equation:* Suppose that $\mathbf{A}$ is a constant square matrix. The linear system $\dot{\mathbf{y}} = -\mathbf{y}\mathbf{A}$, where $\mathbf{y}$ is a row vector, is called the *adjoint equation* for $\dot{\mathbf{x}} = \mathbf{A}\mathbf{x}$.

   (a) Show that the flow $\psi(t, \mathbf{y}^0)$ of the adjoint equation is given by the formula $\psi(t, \mathbf{y}^0) = \mathbf{y}^0 e^{-\mathbf{A}t}$.

(b) Show that $\psi(t, \mathbf{y}^0)\varphi(t, \mathbf{x}^0) = \mathbf{y}^0\mathbf{x}^0$ for all $t \in \mathbb{R}$, where $\varphi(t, \mathbf{x}^0)$ is the flow of $\dot{\mathbf{x}} = \mathbf{A}\mathbf{x}$.

8.29. *Fredholm's Alternative:* Suppose that $\mathbf{f}(t)$ is a continuous 2π-periodic vector function and consider the differential equation $\dot{\mathbf{x}} = \mathbf{A}\mathbf{x} + \mathbf{f}(t)$.

(a) Show that this equation has a 2π-periodic solution if and only if

$$\int_0^{2\pi} \mathbf{y}(t)\,\mathbf{f}(t)\,dt = 0$$

for all 2π-periodic solutions of the adjoint equation.

(b) If there is a 2π-periodic solution $\hat{\mathbf{y}}(t)$ of the adjoint equation such that $\int_0^{2\pi} \hat{\mathbf{y}}(t)\,\mathbf{f}(t)\,dt \neq 0$, then show that every solution is unbounded.

Hints for (a): Recall from linear algebra that a matrix equation $\mathbf{B}\mathbf{x} = \mathbf{c}$ has a solution if and only if $\mathbf{b}\mathbf{c} = 0$ for all row vectors $\mathbf{b}$ such that $\mathbf{b}\mathbf{B} = \mathbf{0}$.

Use the variation of constants formula to show that $\mathbf{x}^0$ is the initial value of a 2π-periodic solution of $\dot{\mathbf{x}} = \mathbf{A}\mathbf{x} + \mathbf{f}(t)$ if and only if

$$\left(\mathbf{I} - e^{-2\pi\mathbf{A}}\right)\mathbf{x}^0 = -\int_0^{2\pi} e^{-\mathbf{A}s}\mathbf{f}(s)\,ds.$$

Show that $\mathbf{y}(t)$ with $\mathbf{y}(0) = \mathbf{y}^0$ is a 2π-periodic solution of the adjoint equation if and only if $\mathbf{y}^0\left(\mathbf{I} - e^{-2\pi\mathbf{A}}\right) = \mathbf{0}$.

Now use the remarks on linear algebra and the previous exercise on adjoint equation.

Hint for (b): If $\hat{\mathbf{x}}(t)$ is a solution of the equation, compute $d\hat{\mathbf{y}}(t)\hat{\mathbf{x}}(t)/dt$, integrate from 0 to t, and show that $\hat{\mathbf{y}}(t)\hat{\mathbf{x}}(t)$ is unbounded in t.

8.30. *Resonance:* Consider the differential equation $\ddot{x} + x = \cos\omega t$, where ω is a constant. Using Fredholm's alternative show that there is a 2π-periodic solution if and only if $\omega \neq 1$. In the case $\omega = 1$, show that every solution is unbounded.

8.31. If $f(t)$ is a continuous 2π-periodic function, show that the differential equation $\ddot{x} + x = f(t)$ has a 2π-periodic solution if and only if

$$\int_0^{2\pi} f(t)\cos t\,dt = 0, \qquad \int_0^{2\pi} f(t)\sin t\,dt = 0.$$

8.6. Linear Systems with 1-periodic Coefficients

In this section, we present certain rudimentary facts about the solutions of a nonautonomous linear system

$$\dot{\mathbf{x}} = \mathbf{A}(t)\mathbf{x}, \tag{8.30}$$

where $\mathbf{A}(t)$ is a 2×2 matrix whose elements are continuous functions of $t \in \mathbb{R}$. We then examine geometrically the important special case where the entries of $\mathbf{A}(t)$ are 1-periodic functions. Such equations are of considerable practical importance in determining the stability type of periodic orbits of nonlinear planar systems when the periodic orbits are known explicitly, as we shall see in Chapter 12.

   As in Section 8.1, the solution of an initial-value problem for Eq. (8.30) can be obtained from any fundamental matrix solution of Eq. (8.30). The proof of the lemma below is similar to that of Lemma 8.4.

Lemma 8.20. *Properties of fundamental matrix solutions:*
   (i) *If $\mathbf{X}(t)$ is a matrix solution of Eq. (8.30) satisfying $\det \mathbf{X}(t_0) \neq 0$, then $\det \mathbf{X}(t) \neq 0$ for all $t \in \mathbb{R}$, that is, $\mathbf{X}(t)$ is a fundamental matrix solution of Eq. (8.30).*
   (ii) *If $\mathbf{X}(t)$ is a fundamental matrix solution of Eq. (8.30), then the solution $\varphi(t, t_0, \mathbf{x}^0)$ satisfying $\varphi(t_0, t_0, \mathbf{x}^0) = \mathbf{x}^0$ is given by*

$$\varphi(t, t_0, \mathbf{x}^0) = \mathbf{X}(t)\,[\mathbf{X}(t)]^{-1}\mathbf{x}^0.$$

   Next, we state and prove a useful formula, known as *Liouville's Formula*, about a matrix solution of Eq. (8.30).

Lemma 8.21. (Liouville) *If $\mathbf{X}(t)$ is a matrix solution of $\dot{\mathbf{x}} = \mathbf{A}(t)\,\mathbf{x}$, then*

$$\det \mathbf{X}(t) = \det \mathbf{X}(t_0)\,\exp\left\{\int_{t_0}^{t} \operatorname{tr} \mathbf{A}(s)\,ds\right\}, \tag{8.31}$$

where $\operatorname{tr} \mathbf{A}(t) = a_{11}(t) + a_{22}(t)$, the sum of the diagonal entries of $\mathbf{A}(t)$.

Proof. We may assume that $\mathbf{X}(t)$ is a fundamental matrix solution of Eq. (8.30), otherwise the formula is trivially true. Now, let $\mathbf{B}(t)$ be defined as the matrix $[\mathbf{X}(t)]^{-1}\mathbf{A}(t)\,\mathbf{X}(t)$, or, equivalently, $\mathbf{A}(t)\,\mathbf{X}(t) = \mathbf{X}(t)\,\mathbf{B}(t)$. Notice that $\operatorname{tr} \mathbf{B}(t) = \operatorname{tr} \mathbf{A}(t)$.

   For simplicity of notation, let $z(t) \equiv \det \mathbf{X}(t)$. To establish the formula of Liouville, it suffices to show that $z(t)$ satisfies the differential equation

$$\dot{z} = [\operatorname{tr} \mathbf{A}(t)]\,z.$$

Now, let $\mathbf{X}(t) = \left[\mathbf{x^1}(t) \mid \mathbf{x^2}(t)\right]$ and compute:

$$\begin{aligned}
\dot{z} &= \det\left[\dot{\mathbf{x}}^1(t) \mid \mathbf{x^2}(t)\right] + \det\left[\mathbf{x^1}(t) \mid \dot{\mathbf{x}}^2(t)\right] \\
&= \det\left[\mathbf{A}(t)\mathbf{x^1}(t) \mid \mathbf{x^2}(t)\right] + \det\left[\mathbf{x^1}(t) \mid \mathbf{A}(t)\mathbf{x^2}(t)\right] \\
&= \det\left[b_{11}(t)\mathbf{x^1}(t) \mid \mathbf{x^2}(t)\right] + \det\left[\mathbf{x^1}(t) \mid b_{22}(t)\mathbf{x^2}(t)\right] \\
&= \left[b_{11}(t) + b_{22}(t)\right] \det\left[\mathbf{x^1}(t) \mid \mathbf{x^2}(t)\right] \\
&= \left[\operatorname{tr}\mathbf{B}(t)\right] z(t) \\
&= \left[\operatorname{tr}\mathbf{A}(t)\right] z(t). \ \Diamond
\end{aligned}$$

In conjunction with Liouville's Formula, it is useful to remember a geometrical fact about determinants: the determinant of a 2×2 matrix is the area of the parallelogram bounded by the column vectors of the matrix. Thus, the formula of Liouville expresses how areas of the parallelogram defined by a matrix solution change in t. We will use this observation momentarily.

For the remaining part of this section, we assume that the coefficient matrix of Eq. (8.30) is also 1-periodic in t, that is, for all t,

$$\dot{\mathbf{x}} = \mathbf{A}(t)\,\mathbf{x}, \qquad \mathbf{A}(t+1) = \mathbf{A}(t). \tag{8.32}$$

To bring out the geometry in the solutions of Eq. (8.32), it is helpful to extend the notion of the Poincaré map of 1-periodic scalar equations from Section 4.2. For this purpose, we need to sample the solutions of Eq. (8.32) at integer values of t.

Let $\mathbf{X}(t)$ be the principal matrix solution of Eq. (8.32) with $\mathbf{X}(0) = \mathbf{I}$. Then $\mathbf{X}(t+1)$ is also a fundamental matrix solution of Eq. (8.32). Therefore, there is an invertible 2×2 matrix $\mathbf{C}$ with constant entries such that

$$\mathbf{X}(t+1) = \mathbf{X}(t)\,\mathbf{C}. \tag{8.33}$$

Now, since $\mathbf{X}(t)$ is the principal matrix solution of Eq. (8.32), if $\mathbf{x}(t)$ is the solution of Eq. (8.32) with $\mathbf{x}(0) = \mathbf{x^0}$, then

$$\mathbf{x}(t) = \mathbf{X}(t)\mathbf{x^0}$$

and

$$\mathbf{x}(t+1) = \mathbf{X}(t+1)\mathbf{x^0} = \mathbf{X}(t)\mathbf{C}\,\mathbf{x^0}.$$

In words, the action of translating the solution $\mathbf{x}(t)$ by 1 unit of time is the same as translating the initial value $\mathbf{x^0}$ into the new initial value $\mathbf{C}\mathbf{x^0}$. In particular,

$$\mathbf{x}(n) = \mathbf{C}^n\mathbf{x^0} \tag{8.34}$$

for any integer n.

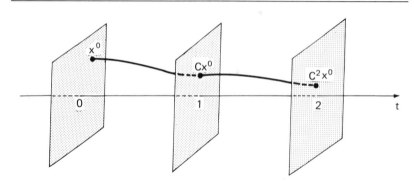

Figure 8.11. *Poincaré map of a 1-periodic linear system.*

The behavior of solutions of Eq. (8.32) are reflected in the dynamics of the iterates of the two-dimensional linear map

$$\mathbf{C} : \mathbb{R}^2 \to \mathbb{R}^2; \qquad \mathbf{x}^0 \mapsto \mathbf{C}\,\mathbf{x}^0,$$

which should be viewed as the *Poincaré map* of the 1-periodic linear system (8.32), see Figure 8.11.

We will undertake a detailed study of planar maps, including, of course, the linear ones, in Chapter 15. For the moment, however, we trust that you will find the few remarks below acceptable. The notions of a fixed point and its stability that we have developed in Chapter 3 for scalar maps are easily generalized to planar maps. This essentially entails replacing scalar quantities with vectors, and absolute values with norms, in several definitions in Chapter 3. With similar replacements, Definitions 4.9 and 4.10 are also readily generalized to yield definitions of stability, asymptotic stability, and instability of a solution of the 1-periodic system (8.32).

A periodic solution of Eq. (8.32) corresponds to a fixed point of the linear planar map (8.34), and the stability type of the periodic solution of Eq. (8.32) is the same as the stability type of the corresponding fixed point of the map (8.34). Notice in particular that the zero solution of Eq. (8.32) corresponds to the fixed point of $\mathbf{C}$ at the origin. We now summarize the main implications of these remarks for the zero solution of Eq. (8.32).

Lemma 8.22. *Let μ_1 and μ_2 be the eigenvalues of the matrix $\mathbf{C}$ as given in Eq. (8.33). Then*
  (i) *If $|\mu_i| < 1$, for $i = 1$, 2, then the zero solution of Eq. (8.32) is asymptotically stable.*
  (ii) *If $|\mu_1| = |\mu_2| = 1$ and $\mu_1 \neq \mu_2$, then the zero solution of Eq. (8.32) is stable.*
 (iii) *If one of the eigenvalues has modulus greater than 1, then the zero solution of Eq. (8.32) is unstable.* ◊

One situation that is not addressed by this lemma is when $|\mu_1| = |\mu_2| = 1$ and $\mu_1 = \mu_2$. Interestingly enough, this difficult case does occur in applications, as we shall see in an example below.

Due to their importance in the theory of 1-periodic linear systems, the eigenvalues of the matrix $\mathbf{C}$ are given their own names which we record here for reference.

Definition 8.23. *The eigenvalues of the matrix $\mathbf{C}$ satisfying Eq. (8.33) are called the characteristic multipliers of the 1-periodic linear system (8.32).*

It is possible to prove considerably more about the structure of linear 1-periodic systems, but we refrain from doing so here. Be warned, however, that unlike the case of linear systems with constant coefficients, the theory of 1-periodic linear systems remains difficult without a complete resolution. To illustrate the richness of this subject, we conclude this section with an example of great prominence.

Example 8.24. *Mathieu's Equation:* The following second-order linear differential equation with 1-periodic coefficients

$$\ddot{y} + (\sigma^2 + \varepsilon \cos 2\pi t)y = 0,$$

where $\sigma \geq 0$ and ε are two real parameters, is known as the linear *Mathieu Equation.*

For computational purposes, it is convenient to convert the second-order equation above to the equivalent 1-periodic linear system

$$\dot{\mathbf{x}} = \mathbf{A}(t)\,\mathbf{x}, \qquad \mathbf{A}(t) = \begin{pmatrix} 0 & 1 \\ -\sigma^2 - \varepsilon \cos 2\pi t & 0 \end{pmatrix}. \qquad (8.35)$$

Let $\mathbf{X}(t)$ be the principal matrix solution of Eq. (8.35) and $\mathbf{C} \equiv \mathbf{X}(1)$. It is easy to discern form Lemma 8.21 that $\det \mathbf{C} = \exp \int_0^1 \mathrm{tr}\,\mathbf{A}(t)\,dt = 1$. Since the determinant is the product of the eigenvalues, we have the important relation below between the eigenvalues of $\mathbf{C}$:

$$\mu_1 \mu_2 = 1. \qquad (8.36)$$

In light of this fact, Lemma 8.22 yields the following theorem for Mathieu's Equation.

Theorem 8.25. *For Mathieu's linear differential equation (8.35):*
  (i) *If μ_1 is not real, then $|\mu_1| = |\mu_2| = 1$ and the zero solution is stable.*
  (ii) *If μ_1 is real and $|\mu_1| < 1$, then $|\mu_2| > 1$ and the zero solution is unstable.* ◇

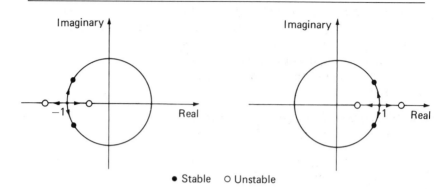

Figure 8.12. *Characteristic multipliers of Mathieu's Equation near bifurcation values.*

Let us now begin the exploration of certain ramifications of this theorem with a geometric observation. From an earlier remark concerning areas and determinants, the relation (8.36) implies that the Poincaré map **C** of Mathieu's equation is an *area-preserving map* of the plane, that is, the area of a region on the plane and the area of its image under the map **C** are the same. Consider now a region of the plane in the form of, say, a square. Part (i) of the theorem above says that in the case of stability a square is rotated around and bent out of shape while its area remains the same. In the case of instability, on the other hand, the image of the square under **C** is elongated in one direction and contracted in the other, as part (ii) implies.

The remaining two possibilities that are not covered by Theorem 8.25 are when $\mu_1 = \mu_2 = 1$ or $\mu_1 = \mu_2 = -1$. In either case, the zero solution will change its stability type if the characteristic multipliers move a little when the parameters are varied—bifurcations. The way they move are depicted in Figure 8.12.

Now, the main problem is to determine which of the two cases in Theorem 8.25 prevail for given values of the parameters σ and ε. To accomplish this, we need to know something about the trace of **C** as a function of the parameters. Unfortunately, no one has succeeded in computing tr **C** for all values of the parameters. In fact, the only case where the trace is known is the trivial case for $\varepsilon = 0$, that is, when the equation of Mathieu is autonomous.

For $\varepsilon = 0$, the eigenvalues of **C** are $e^{i\sigma}$ and $e^{-i\sigma}$, both of which have modulus 1. If $\sigma \neq k\pi$, for $k = 0, 1, 2, \ldots$, then the eigenvalues are not real and thus the zero solution is stable. It is possible to enlarge the domain of stability to bigger subsets of the (σ, ε)-plane than just these pieces of the σ-axis. For example, suppose that $\sigma_0 > 0$ is fixed with $\sigma_0 \neq k\pi$.

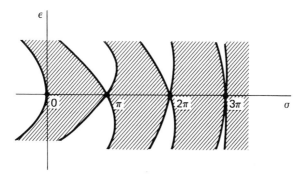

Figure 8.13. *Shaded regions are the zones of stability in the (σ, ε)-plane of the linear Mathieu Equation.*

Then, for each such σ_0, there exists an $\varepsilon_0 > 0$ such that, for $|\varepsilon| < \varepsilon_0$, the eigenvalues of $\mathbf{C}$ remain to be nonreal. This observation follows from the continuous dependence of distinct eigenvalues on the parameters which is a consequence of the quadratic formula. Using sophisticated analysis and numerical computations, it is possible to determine fairly accurately the shapes of the curves on the parameter plane dividing the regions of stability and instability; see Figure 8.13. As a numerical experiment, we have tried in Figure 8.14 to find a point on one of these curves dividing the regions of stability. We fixed $\sigma = 0.6540$, and plotted a solution and its Poincaré map of Mathieu's Equation with initial data near the origin for the values of the parameter $\varepsilon = 0.3900$, $\varepsilon = 0.3950$ and $\varepsilon = 0.4000$. It appears that the point $(0.6540, 0.3995)$ on the (σ, ε)-plane is almost at the boundary of stability. $\Diamond$

We conclude here our study of linear differential equations and next turn to the role of linear theory in the local analysis of nonlinear vector fields.

Exercises ♣ ♡ ♠ ◇

8.32. *Characteristic multipliers of constant coefficients:* Consider a constant matrix $\mathbf{A}$ as a 1-periodic function and find the characteristic multipliers of the system $\dot{\mathbf{x}} = \mathbf{A}\mathbf{x}$.

8.33. *Computing characteristic multipliers:* Find the characteristic multipliers of the following systems:
(a) $\dot{x}_1 = (\sin 2\pi t)\, x_1, \quad \dot{x}_2 = (-1 + \cos 2\pi t)\, x_2$;
(b) $\dot{x}_1 = (\sin t)\, x_1, \quad \dot{x}_2 = (-1 + \cos t)\, x_2$;
(c) $\dot{x}_1 = (\cos 2\pi t)\, x_1, \quad \dot{x}_2 = (\sin 2\pi t)\, x_1 + (-1 + \cos 2\pi t)\, x_2$.

8.34. *Characteristic multipliers are invariants:* Show that the characteristic multipliers are independent of the choice of the fundamental matrix.

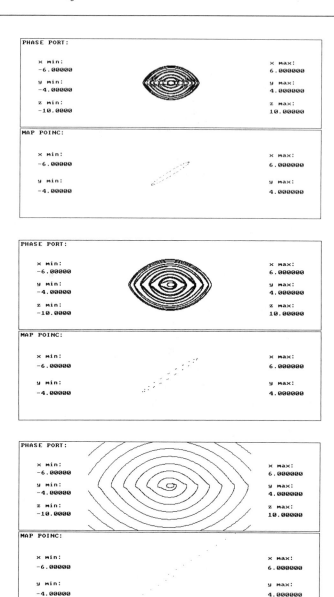

Figure 8.14. *Searching for a point at the boundary of a region of stability of Mathieu's Equation. A solution with initial data (0.2, 0.231) and its Poincaré map for three distinct parameter values $(\sigma, \varepsilon) = (0.6540, 0.3900)$, $(0.6540, 0.3950)$, and $(0.6540, 0.4000)$.*

Hint: If $\mathbf{X}(t)$ and $\widehat{\mathbf{X}}(t)$ are two fundamental matrices, and $\mathbf{X}(t{+}1) = \mathbf{X}(t)\,\mathbf{C}$ and $\widehat{\mathbf{X}}(t + 1) = \widehat{\mathbf{X}}(t)\,\widehat{\mathbf{C}}$, then $\mathbf{C}$ is similar to $\widehat{\mathbf{C}}$.

8.35. *Negative real parts are not enough:* As we saw earlier in this chapter, all solutions of a linear system with constant coefficients eventually go to zero if the eigenvalues of its coefficient matrix have eigenvalues with negative real parts. Unfortunately, this is not so in general for linear 1-periodic systems. Here is an example of this sort from Markus and Yamabe [1960]. Consider the 1-periodic linear system whose coefficient matrix is

$$\mathbf{A}(t) = \begin{pmatrix} -1 + (3/2)\cos^2 t & 1 - (3/2)\cos t \sin t \\ -1 - (3/2)\sin t \cos t & -1 + (3/2)\sin^2 t \end{pmatrix}.$$

(a) Compute the eigenvalues of $\mathbf{A}(t)$ to find $\lambda_{1,2} = \left[-1 \pm i\sqrt{7}\right]/4$. Notice, in particular, that the real parts of the eigenvalues are negative. On the other hand, verify that the vector $(-\cos t,\ \sin t)e^{t/2}$ is a solution of the 1-periodic system above; but this solution is unbounded as $t \to +\infty$.

(b) One of the characteristic multipliers is e^{π}. What is the other multiplier?

(c) If you are brave, you may wish to tackle the somewhat futile exercise of finding a second linearly independent solution of this system.

8.36. *Floquet representation:* If $\mathbf{X}(t)$ is a fundamental matrix solution of a linear system with 1-periodic coefficients (8.32), then there exists a 1-periodic matrix $\mathbf{P}(t)$ and a constant matrix $\mathbf{B}$, with perhaps complex entries such that

$$\mathbf{X}(t) = \mathbf{P}(t)e^{\mathbf{B}t}.$$

Show how to reduce Eq. (8.32) to a linear system with constant coefficients. The catch is that it is very difficult to compute characteristic multipliers. *Hint:* Since $\mathbf{X}(t + 1) = \mathbf{X}(t)\,\mathbf{C}$ and the matrix $\mathbf{C}$ is nonsingular, $\mathbf{C}$ has a logarithm $\mathbf{B}$, that is, $\mathbf{C} = e^{\mathbf{B}}$. Let $\mathbf{P}(t) = \mathbf{X}(t)\,e^{-\mathbf{B}t}$. If $\mathbf{x} = \mathbf{P}(t)\,\mathbf{y}$, then $\dot{\mathbf{y}} = \mathbf{B}\,\mathbf{x}$. When is the matrix $\mathbf{B}$ real?

8.37. *Mathieu on PHASER:* The equation of Mathieu is stored in the library of PHASER under the name *mathieu* in three dimensions. Investigate the zones of stability numerically; in particular, reproduce Figure 8.14.

Bibliographical Notes ⎯⎯⎯⎯⎯⎯⎯⎯⎯⎯⎯⎯⎯⎯⎯⎯⎯⎯⎯⎯⎯⎯⎯⎯

Jordan Normal Form is one of the central results in Linear Algebra, and you can look up the details for higher dimensions in any good book on Linear Algebra. However, most algebraists prefer algebraically closed fields—complex numbers; for the real case and applications to differential equations, see, for example, Hirsch and Smale [1974]. For large systems, one often has to resort to numerics, in which case care is required; see Golub and Wilkinson [1976]. By the way, we never had to put a matrix larger than 4×4 into normal form.

In contrast to the linear classification (Jordan Normal Form) of linear systems, their topological classification is of relatively recent origin; see Arnold [1973] for an exposition, and Landis [1973] and Kuiper [1975] for the complete answer in all dimensions. Applications of these results to control theory of linear systems are contained in Willems [1980].

Genericity is an important consideration in securing general results by leaving out often difficult exceptional cases. Hirsch and Smale [1974] have the linear theory in higher dimensions. In Chapter 13, we will explore this important topic for nonlinear planar systems.

Unfolding, with roots in singularity theory of functions, is one of the key notions in modern bifurcation theory. Unfolding of linear differential equations appears in Arnold [1971]; see also Koçak [1984]. More general settings pertaining to local unfoldings of nonlinear systems are discussed by Arnold [1972 and 1983] and in Chapter 13.

Mathieu's Equation appears in oscillation theory; consult, for example, Hale [1963], McLachlan [1947], and Stoker [1950]. The more prominent Hill's Equation contains Mathieu's Equation as a special case; see Magnus and Winkler [1979].

9

Near
Equilibria

In this chapter, we investigate the stability and instability properties of equilibrium points of planar differential equations. It is evident from our foregoing discussions that the stability type of an equilibrium point of a linear system is determined by the eigenvalues of its coefficient matrix. Analogous to the results in Section 1.3, we prove several theorems to show that, under certain conditions, the stability type of an equilibrium point of a nonlinear planar differential equation is determined by the linear approximation of the vector field in a sufficiently small neighborhood of the equilibrium point. In order to determine how large these "small" neighborhoods can be, we present another, somewhat more geometric, technique—the direct method of Liapunov—for investigating the stability of an equilibrium of a nonlinear system. We continue our presentation with an analysis of some of the finer geometric details of the flows of nonlinear systems in a neighborhood of an equilibrium point of saddle type. Next, we include a discussion of deciding the local equivalence of flows of nonlinear systems from that of their linear approximations. We conclude the chapter with an example illustrating the global dynamics of saddle points—saddle connections.

9.1. Asymptotic Stability from Linearization

In this section, we first give precise definitions of stability, instability, and asymptotic stability of equilibrium points of planar autonomous systems. Then we prove a fundamental result on the asymptotic stability of an equilibrium point from the linearization of a vector field.

Definition 9.1. *An equilibrium point $\bar{x}$ of a planar autonomous system $\dot{x} = f(x)$ is said to be* stable *if, for any given $\epsilon > 0$, there is a $\delta > 0$ (depending only on ϵ) such that, for every x^0 for which $\|x^0 - \bar{x}\| < \delta$, the solution $\varphi(t, x^0)$ of $\dot{x} = f(x)$ through x^0 at $t = 0$ satisfies the inequality $\|\varphi(t, x^0) - \bar{x}\| < \epsilon$ for all $t \geq 0$. The equilibrium $\bar{x}$ is said to be* unstable *if it is not stable, that is, there is an $\eta > 0$ such that, for any $\delta > 0$, there is an x^0 with $\|x^0 - \bar{x}\| < \delta$ and $t_{x^0} > 0$ such that $\|\varphi(t_{x^0}, x^0) - \bar{x}\| = \eta$.*

From the definitions of stability and instability, it is expected that the discussion of instability should, in general, be more difficult than that of stability. As we shall see below, a comparison of the proofs of Theorems 9.5 and 9.7 will attest to this.

Definition 9.2. *An equilibrium point $\bar{x}$ is said to be* asymptotically stable *if it is stable and, in addition, there is an $r > 0$ such that $\|\varphi(t, x^0) - \bar{x}\| \to 0$ as $t \to +\infty$ for all x^0 satisfying $\|x^0 - \bar{x}\| < r$.*

We remarked in Section 1.3 that an equilibrium point of a scalar differential equation is asymptotically stable if any solution starting near the equilibrium point approaches the equilibrium point in forward time. The requirement that the equilibrium point be stable is satisfied automatically. For the asymptotic stability of an equilibrium point of a planar system, however, this additional requirement is essential. There are vector fields on the plane for which all solutions approach an equilibrium point as $t \to +\infty$ and yet the equilibrium point is not stable. We refrain from giving an explicit formula for such a vector field; one possible planar phase portrait with the desired pathology is depicted in Figure 9.1.

With our previous study of linear systems from Chapter 8, it is not a difficult matter to verify the following facts on the asymptotic stability and the instability of equilibria of linear systems:

Theorem 9.3. *If all the eigenvalues of the coefficient matrix $\mathbf{A}$ in the linear system $\dot{x} = \mathbf{A}x$ have negative real parts, then its equilibrium point $\bar{x} = 0$ is asymptotically stable. Moreover, there are positive constants K and α such that*

$$\|e^{\mathbf{A}t}x^0\| \leq Ke^{-\alpha t}\|x^0\| \quad \text{for all } t \geq 0, \ x^0 \in \mathbb{R}^2. \tag{9.1}$$

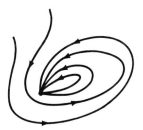

Figure 9.1. *All solutions approach an equilibrium point as $t \to +\infty$ but the equilibrium point is not stable.*

If one of the eigenvalues of the coefficient matrix **A** *has positive real part, then the equilibrium point* $\bar{\mathbf{x}} = \mathbf{0}$ *is unstable.* ◇

It is evident in the definitions above that the stability type of an equilibrium point is a local property. Consequently, as in the case of the scalar equations, it is reasonable to expect that under certain conditions the stability type of $\bar{\mathbf{x}}$ can be determined from the approximation of the vector field **f** with its derivative, which is a linear vector field. For this purpose, let us suppose that $\mathbf{f} = (f_1, f_2)$ is a C^1 function, and let the matrix

$$
Df(\mathbf{x}) = \begin{pmatrix} \dfrac{\partial f_1}{\partial x_1}(\mathbf{x}) & \dfrac{\partial f_1}{\partial x_2}(\mathbf{x}) \\[2ex] \dfrac{\partial f_2}{\partial x_1}(\mathbf{x}) & \dfrac{\partial f_2}{\partial x_2}(\mathbf{x}) \end{pmatrix}
$$

be the *Jacobian matrix of* **f** at the point **x**.

Definition 9.4. *If* $\bar{\mathbf{x}}$ *is an equilibrium point of* $\dot{\mathbf{x}} = \mathbf{f}(\mathbf{x})$, *then the linear differential equation*

$$
\dot{\mathbf{x}} = Df(\bar{\mathbf{x}})\mathbf{x}
$$

is called the linear variational equation *or the* linearization *of the vector field* **f** *at the equilibrium point* $\bar{\mathbf{x}}$.

We now state and prove a fundamental result on the asymptotic stability of an equilibrium point from the linearization of the vector field. The proof may appear a bit technical, but the character of the proof and some of the estimates have wider utility in the stability aspects of differential equations.

Theorem 9.5. *Let* **f** *be a* C^1 *function. If all the eigenvalues of the Jacobian matrix* $Df(\bar{\mathbf{x}})$ *have negative real parts, then the equilibrium point* $\bar{\mathbf{x}}$ *of the differential equation* $\dot{\mathbf{x}} = \mathbf{f}(\mathbf{x})$ *is asymptotically stable.*

Proof. To study the stability properties of $\bar{\mathbf{x}}$, it is convenient to introduce the new variable

$$\mathbf{y}(t) = \mathbf{x}(t) - \bar{\mathbf{x}},$$

so that the equilibrium point $\bar{\mathbf{x}}$ of $\dot{\mathbf{x}} = \mathbf{f}(\mathbf{x})$ corresponds to the equilibrium point $\mathbf{y} = \mathbf{0}$ of the differential equation

$$\dot{\mathbf{y}} = \mathbf{f}(\mathbf{y} + \bar{\mathbf{x}}).$$

Using Taylor's formula in several variables, we can expand the function $\mathbf{f}(\mathbf{y} + \bar{\mathbf{x}})$ about $\bar{\mathbf{x}}$ to obtain

$$\mathbf{f}(\mathbf{y} + \bar{\mathbf{x}}) = \mathbf{f}(\bar{\mathbf{x}}) + D\mathbf{f}(\bar{\mathbf{x}})\mathbf{y} + \mathbf{g}(\mathbf{y}),$$

where the remainder function $\mathbf{g}(\mathbf{y})$ satisfies

$$\mathbf{g}(\mathbf{0}) = \mathbf{0} \quad \text{and} \quad D\mathbf{g}(\mathbf{0}) = \mathbf{0}. \tag{9.2}$$

Therefore, since $\mathbf{f}(\bar{\mathbf{x}}) = \mathbf{0}$, the differential equation $\dot{\mathbf{y}} = \mathbf{f}(\mathbf{y} + \bar{\mathbf{x}})$ can be written in the form

$$\dot{\mathbf{y}} = D\mathbf{f}(\bar{\mathbf{x}})\mathbf{y} + \mathbf{g}(\mathbf{y}). \tag{9.3}$$

We are going to prove the theorem by showing that the solution $\mathbf{y}(t) = \mathbf{0}$ of Eq. (9.3) is asymptotically stable.

For future reference, we note that the properties (9.2) of $\mathbf{g}(\mathbf{y})$ imply that near the origin $\mathbf{g}(\mathbf{y})$ is "small" compared to $\mathbf{y}$. More precisely, it follows from the Mean Value Theorem that, for any $m > 0$, there is an $\varepsilon > 0$ such that

$$\|\mathbf{g}(\mathbf{y})\| \leq m\|\mathbf{y}\| \quad \text{if} \quad \|\mathbf{y}\| < \varepsilon. \tag{9.4}$$

Returning to the differential equation (9.3), let $\mathbf{y}(t)$ be the solution of Eq. (9.3) satisfying the initial condition $\mathbf{y}(0) = \mathbf{y}^0$. If we view $\mathbf{g}(\mathbf{y}(t))$ as a function of t, then, from the variation of the constants formula in Section 8.5, we obtain

$$\mathbf{y}(t) = e^{\mathbf{A}t}\mathbf{y}^0 + \int_0^t e^{\mathbf{A}(t-s)}\mathbf{g}(\mathbf{y}(s))\,ds. \tag{9.5}$$

Although the function $\mathbf{y}(t)$ appears on both sides, we will use this integral equation to estimate $\|\mathbf{y}(t)\|$ in terms of $\|\mathbf{y}^0\|$ as a function of t.

Let us suppose that the constants K and α are given as in Theorem 9.3, $m > 0$ is such that $mK < \alpha$, and $\varepsilon > 0$ is chosen so that Eq. (9.4) is satisfied. Then, from the estimate (9.1) in Theorem 9.3, we have

$$\|\mathbf{y}(t)\| \leq Ke^{-\alpha t}\|\mathbf{y}^0\| + \int_0^t Ke^{-\alpha(t-s)}m\|\mathbf{y}(s)\|\,ds,$$

as long as $\|\mathbf{y}(s)\| \le \varepsilon$ and $0 \le s \le t$. Multiplying both sides of this inequality with $e^{\alpha t}$ yields

$$e^{\alpha t}\|\mathbf{y}(t)\| \le K\|\mathbf{y}^0\| + \int_0^t Kme^{\alpha s}\|\mathbf{y}(s)\|\, ds.$$

If we now apply Gronwall's inequality (see the exercises) to the function $e^{\alpha t}\|\mathbf{y}(t)\|$, we obtain

$$e^{\alpha t}\|\mathbf{y}(t)\| \le K\|\mathbf{y}^0\|e^{Kmt}.$$

Finally, by multiplying both sides of the inequality above by $e^{-\alpha t}$, we obtain the estimate we have been seeking:

$$\|\mathbf{y}(t)\| \le K\|\mathbf{y}^0\|e^{-(\alpha - Km)t} \qquad \text{for } \|\mathbf{y}(t)\| \le \varepsilon. \tag{9.6}$$

To finish the proof, select $\delta > 0$ so that $K\delta < \varepsilon$. If $\|\mathbf{y}^0\| < \delta$, then the inequality (9.6) guarantees that $\|\mathbf{y}(t)\| < \varepsilon$ since $\alpha - Km > 0$. Therefore, the solution $\mathbf{y}(t)$ exists for all $t \ge 0$ and the equilibrium solution $\mathbf{y} = \mathbf{0}$ of Eq. (9.3) is stable. Also from Eq. (9.6), we have $\mathbf{y}(t) \to \mathbf{0}$ as $t \to +\infty$ if $\|\mathbf{y}^0\| < \delta$. Consequently, $\mathbf{y} = \mathbf{0}$ is asymptotically stable and the rate of approach to the equilibrium is exponentially fast. $\Diamond$

Example 9.6. *The damped pendulum:* As an application of the asymptotic stability theorem above, let us consider the equation for the damped pendulum given by

$$\ddot{\theta} + 2a\dot{\theta} + \omega^2 \sin\theta = 0, \tag{9.7}$$

where $a > 0$ reflects friction and $\omega^2 \equiv g/l$. Using the transformation in Example 7.1, it is easy to see that this second-order equation is equivalent to the planar system

$$\begin{aligned} \dot{x}_1 &= x_2 \\ \dot{x}_2 &= -\omega^2 \sin x_1 - 2ax_2. \end{aligned} \tag{9.8}$$

The equilibrium points of this system are $(n\pi, 0)$, where n is any integer. Let us first consider the stability of the equilibrium point at the origin. The Jacobian of the vector field (9.8) is the matrix

$$D\mathbf{f}(\mathbf{x}) = \begin{pmatrix} 0 & 1 \\ -\omega^2 \cos x_1 & -2a \end{pmatrix}.$$

Therefore, the linearization of Eq. (9.8) at the point $\mathbf{x} = (0, 0)$, the linear system $\dot{\mathbf{x}} = D\mathbf{f}(\mathbf{0})\mathbf{x}$, is the following planar linear differential equation:

$$\dot{\mathbf{x}} = \begin{pmatrix} 0 & 1 \\ -\omega^2 & -2a \end{pmatrix} \mathbf{x}. \tag{9.9}$$

The eigenvalues of this linear system can easily be found to be

$$\mu_{1,2} = -a \pm \sqrt{a^2 - \omega^2}.$$

The real parts of both eigenvalues are negative for all $a > 0$ and $\omega > 0$ (why?). Therefore, by Theorem 9.5, the equilibrium point $(0, 0)$ of the non-linear equation (9.8) is asymptotically stable. For comparison, the phase portraits of both the nonlinear equation (9.8) and its linearization (9.9) near the origin have been plotted in Figure 9.2 for several values of the parameters a and ω.

What about the stability of the other equilibrium points? It is easy to see that, when n is an even integer, the linearized equations at the equilibrium points $(n\pi, 0)$ are the same as the system (9.9). So the analysis above applies to these points as well. When n is an odd integer, however, the linearized equations are different. We will consider the stability properties of such points in the next section. $\Diamond$

Exercises ————————————————————————————————

9.1. Using Jordan Normal Forms, provide a proof of Theorem 9.3.

9.2. Show that the zero solution of the differential equations whose vector fields are given below is asymptotically stable:

(a) $\mathbf{f}(\mathbf{x}) = \begin{pmatrix} -x_1 - x_2 - (x_1^2 + x_2^2) \\ x_1 - x_2 + (x_1^2 + x_2^2) \end{pmatrix}$;

(b) $\mathbf{f}(\mathbf{x}) = \begin{pmatrix} \cos x_1 - \sin x_1 - 1 \\ x_1 - x_2 - x_2^2 \end{pmatrix}$.

9.3. *Gronwall's Inequality:* Let K be a nonnegative constant and let f and g be continuous nonnegative functions for $a \le t \le b$ which satisfy

$$f(t) \le K + \int_a^t f(s)\, g(s)\, ds, \qquad a \le t \le b,$$

then

$$f(t) \le K\, e^{\int_a^t g(s)\, ds}, \qquad a \le t \le b.$$

Suggestions: Let $h(t) = K + \int_a^t f(s)\, g(s)\, ds$, and notice that $h(a) = K$. Using calculus, obtain $\dot{h}(t) = f(t)\, g(t) \le h(t)\, g(t)$. Multiply this inequality by $\exp\{-\int_a^t f(s)\, g(s)\, ds\}$ to arrive at

$$\frac{d}{dt}\left[h(t) \exp\left\{ -\int_a^t f(s)\, g(s)\, ds \right\} \right] \le 0.$$

Now integrate from a to t to obtain the desired inequality.

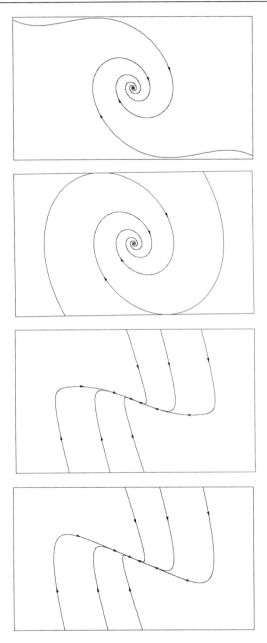

Figure 9.2. *Local phase portraits of the damped nonlinear pendulum and its linearization near the equilibrium point at the origin for the values of the friction coefficients 0.25 and 1.5.*

9.4. *Asymptotic stability under small perturbations:* Suppose that the eigenvalues of the matrix $\mathbf{A}$ have negative real parts and $\mathbf{B}(t)$ is a 1-periodic matrix with $\|\mathbf{B}(t)\,\mathbf{x}\| \leq \delta\,\|\mathbf{x}\|$ for all t and $\mathbf{x}$. Show that if δ is sufficiently small, then all solutions of $\dot{\mathbf{x}} = [\mathbf{A} + \mathbf{B}(t)]\,\mathbf{x}$ approach zero as $t \to +\infty$.

Hint: Use the variations of the constants formula and Gronwall's inequality.

9.2. Instability from Linearization

In this section, we state and prove a fundamental result on instability of an equilibrium point from the linearization of a vector field. The proof is again a bit involved, but the method, especially in the latter half, has important generalizations which we will explore subsequently.

Theorem 9.7. *Let $\mathbf{f}$ be a C^1 function. If at least one of the eigenvalues of the Jacobian matrix $D\mathbf{f}(\bar{\mathbf{x}})$ has positive real part, then the equilibrium point $\bar{\mathbf{x}}$ of the differential equation $\dot{\mathbf{x}} = \mathbf{f}(\mathbf{x})$ is unstable.*

Proof. Using the same notation and transformation as in the proof of Theorem 9.5, it suffices to show that the solution $\mathbf{y}(t) = \mathbf{0}$ of the differential equation

$$\dot{\mathbf{y}} = D\mathbf{f}(\bar{\mathbf{x}})\mathbf{y} + \mathbf{g}(\mathbf{y}) \tag{9.10}$$

is unstable. Depending on the real parts of the eigenvalues of the Jacobian matrix, there are two cases to consider.

Let us suppose first that both eigenvalues of the Jacobian matrix have positive real parts. If we replace t by $-t$ in Eq. (9.10), then we obtain the differential equation

$$\dot{\mathbf{y}} = -\,D\mathbf{f}(\bar{\mathbf{x}})\mathbf{y} - \mathbf{g}(\mathbf{y}). \tag{9.11}$$

Since the eigenvalues of $-D\mathbf{f}(\bar{\mathbf{x}})$ have negative real parts, by Theorem 9.5, the solution $\mathbf{y} = \mathbf{0}$ of Eq. (9.11) is asymptotically stable. In particular, there is an $r > 0$ such that, if $\|\mathbf{y}^0\| = r$, then the solution $\varphi(t, \mathbf{y}^0)$ of Eq. (9.11) approaches $\mathbf{0}$ as $t \to +\infty$.

Now, fix ε and $\mathbf{y}^0$ satisfying $0 < \varepsilon < r$ and $\|\mathbf{y}^0\| = r$. Let $\hat{t}$ be the time, depending on ε and $\mathbf{y}^0$, such that $\|\varphi(\hat{t}, \mathbf{y}^0)\| = \varepsilon$. Notice that $\varphi(-t, \mathbf{y}^0)$ is a solution of Eq. (9.10) and $\varphi(-\hat{t}, \varphi(\hat{t}, \mathbf{y}^0)) = \mathbf{y}^0$. Therefore, the solution of Eq. (9.10) with initial data $\varphi(\hat{t}, \mathbf{y}^0)$, satisfying $\|\varphi(\hat{t}, \mathbf{y}^0)\| = \varepsilon$, reaches the circle $\|\mathbf{y}\| = r$ in a finite amount of time. Since ε can be chosen arbitrarily small, this implies the instability of the equilibrium solution $\mathbf{y} = \mathbf{0}$ of Eq. (9.10).

Let us suppose now, as the second case, that the eigenvalues of the Jacobian matrix are real with $\mu_1 \leq 0 < \mu_2$. By a linear change of coordinates (see Section 8.2) the Jacobian matrix can be put into a diagonal canonical form so that Eq. (9.10) has the form

$$\dot{y}_1 = \mu_1 y_1 + g_1(y_1, y_2)$$
$$\dot{y}_2 = \mu_2 y_2 + g_2(y_1, y_2),$$

(9.12)

where the function $\mathbf{g} = (g_1, g_2)$ still possesses the properties seen in Eq. (9.2).

To prove the instability of the equilibrium at the origin of Eq. (9.12), we will determine a function V and a region near the origin in such a way that the function is increasing along the solutions of Eq. (9.12) in this region. To accomplish this, consider the function

$$V(y_1, y_2) = \tfrac{1}{2}(y_2^2 - y_1^2)$$

and compute its derivative along the solution curves of Eq. (9.12). If $(y_1(t), y_2(t))$ is a solution of Eq. (9.12), and $\|\mathbf{y}(t)\| < \varepsilon$ then using Eq. (9.4) we make the following estimates:

$$\begin{aligned}
\dot{V}(\mathbf{y}(t)) &= \frac{dV}{dt}(y_1(t), y_2(t)) \\
&= y_2 \dot{y}_2 - y_1 \dot{y}_1 \\
&= \mu_2 y_2^2 + y_2 g_2(y_1, y_2) - \mu_1 y_1^2 - y_1 g_1(y_1, y_2) \\
&\geq \mu_2 y_2^2 - m\|\mathbf{y}\| (|y_1| + |y_2|) - \mu_1 y_1^2 \\
&\geq (\mu_2 - m)y_2^2 - 2m|y_1||y_2| - (\mu_1 + m)y_1^2.
\end{aligned}$$

Now, consider the set

$$\Omega \equiv \{ (y_1, y_2) : y_2 > |y_1| \}$$

which is the wedge lying above the lines $y_2 = \pm y_1$, as shown in Figure 9.3. Notice that in the region Ω the level set $V^{-1}(c_2)$ lies above the level set $V^{-1}(c_1)$ if $c_2 > c_1$.

Next, let us choose $\varepsilon > 0$ and m so that Eq. (9.4) is satisfied and $\mu_2 - 4m > 0$. Let $U \equiv \{ (y_1, y_2) : \|\mathbf{y}\| < \varepsilon \}$ be an open neighborhood of the origin. In the region $\Omega \cap U$, we have $V(\mathbf{y}) > 0$ and $\dot{V}(\mathbf{y}) \geq (\mu_2 - 4m)y_2^2 > 0$.

Suppose that $\mathbf{y}^0 \in \Omega \cap U$. Then the solution through $\mathbf{y}^0$ remains in Ω as long as $\mathbf{y}(t) \in U$. To finish the proof, we observe that the solution through $\mathbf{y}^0$ has norm equal to ε for some value of t, that is, the solution must hit the boundary of U. This follows because there is $\delta > 0$ such that $\dot{V}(\mathbf{y}) > \delta$ if $\mathbf{y} \in U$ and $V(\mathbf{y}) > V(\mathbf{y}^0)$. The proof of the instability of the equilibrium point of Eq. (9.12) at the origin is now complete. $\diamond$

Example 9.8. *The damped pendulum continued:* Let us consider the stability type of the equilibrium points $(n\pi, 0)$ of Eq. (9.8) when n is an odd

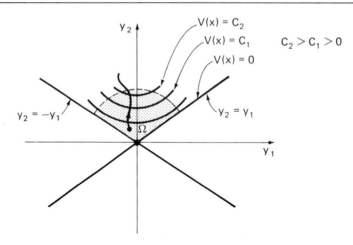

Figure 9.3. *The level sets of V and the region Ω where $\dot{V}(\mathbf{y}) > 0$.*

integer. The linearization of Eq. (9.8) at these equilibrium points is the linear system

$$\dot{\mathbf{x}} = \begin{pmatrix} 0 & 1 \\ \omega^2 & -2a \end{pmatrix} \mathbf{x}. \tag{9.13}$$

The eigenvalues of this linear system can easily be found to be

$$\mu_{1,\,2} = -a \pm \sqrt{a^2 + \omega^2}.$$

Since one eigenvalue is positive (and the other is negative), it follows from Theorem 9.7 that these equilibrium points are unstable.

For comparison, we have plotted in Figure 9.4 the phase portraits of the nonlinear system (9.8) and its linearization (9.13) near one of these unstable equilibrium points. Of course, the linear system is a saddle. Furthermore, the nonlinear system looks much like a saddle also, and the linearization is a good reflection of the local phase portrait of the nonlinear system near the equilibrium. The preservation of a saddle under small perturbations of a linear system is true in general and we will explore this fact further in Section 9.5. ◊

The stability type of an equilibrium point of a nonlinear system cannot always be determined from linearization. It is evident from Theorems 9.5 and 9.7 that such a situation can occur only if some eigenvalue of the linearization has zero real part (and the remaining eigenvalue has negative real part). In this case, we must examine effects of the specific nonlinear terms of the vector field to determine the local dynamics. Indeed, we have already encountered an instance of this difficulty in the saddle-node

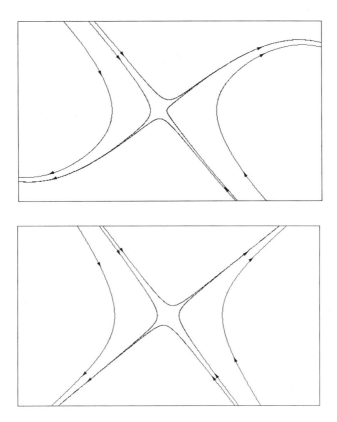

Figure 9.4. *Local phase portraits of the damped nonlinear pendulum and its linearization near one of the unstable equilibria.*

bifurcation given in Example 7.23 where one zero eigenvalue was present. Here is an example with purely imaginary eigenvalues:

Example 9.9. *When linearization does not suffice:* Consider the system of nonlinear differential equations

$$\dot{x}_1 = x_2 + \lambda x_1(x_1^2 + x_2^2)$$
$$\dot{x}_2 = -x_1 + \lambda x_2(x_1^2 + x_2^2), \tag{9.14}$$

where λ is a scalar parameter. These equations are a nonlinear perturbation of the harmonic oscillator (7.4) but the perturbation does not affect the linear terms. For all values of λ, the origin is an equilibrium point and the linearized equation at the origin is, of course, the harmonic oscillator

$$\dot{\mathbf{x}} = \begin{pmatrix} 0 & 1 \\ -1 & 0 \end{pmatrix} \mathbf{x}. \tag{9.15}$$

The eigenvalues of this linear system are $\pm i$, which have zero real parts.

To analyze the behavior of the nonlinear system (9.14), we compute the derivative of the square of the distance of a solution from the origin:

$$\frac{d}{dt}(x_1^2 + x_2^2) = 2\lambda \left(x_1^2 + x_2^2\right)^2.$$

If $\lambda < 0$, then $\|\mathbf{x}(t)\|^2$ approaches zero monotonically as $t \to +\infty$. Thus, the equilibrium at the origin is asymptotically stable. However, if $\lambda > 0$, then all solutions of Eq. (9.14) with initial data $\mathbf{x}^0 \neq \mathbf{0}$ escape to infinity. Therefore, in this case, the origin is unstable. $\Diamond$

Exercises

9.5. *Equilibrium of Van der Pol:* Consider the Lienard form of Van der Pol's oscillator

$$\dot{x}_1 = x_2 - \lambda(x_1^3/3 - x_1)$$
$$\dot{x}_2 = -x_1,$$

where λ is a scalar parameter. Show that the only equilibrium point is at the origin. Determine its stability type as a function of λ.

9.6. Show that the zero solution of the differential equations whose vector fields are given below is unstable:

(a) $\mathbf{f}(\mathbf{x}) = \begin{pmatrix} x_2 \\ x_1 + 2x_2^3 \end{pmatrix}$;

(b) $\mathbf{f}(\mathbf{x}) = \begin{pmatrix} x_1 + 5x_2 + x_1^2 x_2 \\ 5x_1 + x_2 - x_2^3 \end{pmatrix}$;

(c) $\mathbf{f}(\mathbf{x}) = \begin{pmatrix} e^{x_1 + x_2} - 1 \\ \sin(x_1 + x_2) \end{pmatrix}$.

9.7. Find all equilibrium points of the following systems of differential equations and determine their stability properties:

(a) $\dot{x}_1 = 1 - x_1 x_2$, $\dot{x}_2 = x_1 - x_2^3$;

(b) $\dot{x}_1 = 2x_1 - x_1^2 - x_1 x_2$, $\dot{x}_2 = -x_2 + x_1 x_2$;

(c) $\ddot{y} + \dot{y} + y^3 = 0$;

(d) $\dot{x}_1 = \sin(x_1 + x_2)$, $\dot{x}_2 = e^{x_1} - 1$;

(e) $\dot{x}_1 = x_1 - x_1^3 - x_1 x_2^2$, $\dot{x}_2 = 2x_2 - x_2^5 - x_2 x_1^4$.

Try to sketch the phase portraits of the first three equations. Make sure to use the information about the linearized equations at each equilibrium point.

9.8. If the origin is a stable but not asymptotically stable equilibrium point of the planar system $\dot{\mathbf{x}} = \mathbf{f}(\mathbf{x})$, can the origin be a saddle point of the linearized equations?

9.9. *Feedback control:* Consider the equation for the pendulum of length l, mass m, in a viscous medium with friction proportional to the velocity of the pendulum. Suppose now that the objective is to stabilize the pendulum in the vertical position (above its pivot) by a control mechanism which can move the pivot of the pendulum horizontally. Let us assume that θ is the angle from the vertical position measured in the clockwise direction and the restoring force v due to the control mechanism is a linear function of θ and $\dot{\theta}$, that is, $v(\theta, \dot{\theta}) = c_1\theta + c_2\dot{\theta}$. Convince yourself that the differential equation

$$\ddot{\theta} + m\dot{\theta} - \frac{g}{l}\sin\theta - \frac{1}{l}(c_1\theta + c_2\dot{\theta})\cos\theta = 0$$

describes the motion of such a pendulum. Show that the constants c_1 and c_2 can be chosen in such a way so as to make the equilibrium point $(\theta, \dot{\theta}) = (0, 0)$ asymptotically stable.

9.10. *"Pole" placement:* In the problem above, the linearized equations had the form

$$\dot{\mathbf{x}} = \mathbf{A}\mathbf{x} + \mathbf{b}v, \qquad \mathbf{b} = \begin{pmatrix} 0 \\ 1 \end{pmatrix},$$

and the problem was solved by choosing $v = \mathbf{c}^T\mathbf{x} = c_1x_1 + c_2x_2$ so that the eigenvalues of the matrix $\mathbf{A} + \mathbf{b}\mathbf{c}^T$ have negative real parts. Prove the following result:

Theorem: If the matrix $(\mathbf{b} \mid \mathbf{A}\mathbf{b})$ is nonsingular, that is, the linear system above is controllable, then there is a vector $\mathbf{c}$ such that the eigenvalues of $\mathbf{A} + \mathbf{b}\mathbf{c}^T$ have negative real parts.

Hint: There is a vector $\mathbf{c}$ with $\operatorname{tr}(\mathbf{A} + \mathbf{b}\mathbf{c}^T) < 0$ and $\det(\mathbf{A} + \mathbf{b}\mathbf{c}^T) > 0$.

9.3. Liapunov Functions

For an asymptotically stable equilibrium point $\bar{\mathbf{x}}$ of a planar system $\dot{\mathbf{x}} = \mathbf{f}(\mathbf{x})$, it is of considerable practical importance to obtain good estimates of the *basin of attraction* of $\bar{\mathbf{x}}$, that is, the subset of $\mathbb{R}^2$ consisting of the initial data $\mathbf{x}^0$ with the property $\varphi(t, \mathbf{x}^0) \to \bar{\mathbf{x}}$ as $t \to +\infty$. To be sure, it is possible to extract from the proof of Theorem 9.5 an estimate of the basin of attraction of an asymptotically stable equilibrium point. However, this would yield only a crude estimate because in the method of proof we paid no particular attention to any special features that the nonlinear terms in a specific differential equation may possess. In this section, we present a different procedure—the direct method of Liapunov—for studying stability type of an equilibrium point; a procedure which is more geometric, makes explicit use of nonlinear terms, and usually gives a better estimate of the basin of attraction of an asymptotically stable equilibrium point. Variants of this geometric method yield instability results, as well as means to establish the stability type of an equilibrium point in the presence of eigenvalues with zero real parts. As we saw in earlier examples, this latter case cannot be established from linearization. Despite this praise, however, we should point out that the direct method of Liapunov is not without certain serious limitations, as we shall see momentarily.

The main idea behind Liapunov's method is to determine how certain special real-valued functions vary along the solutions of $\dot{\mathbf{x}} = \mathbf{f}(\mathbf{x})$. Let us begin by defining these functions.

Definition 9.10. *Let U be an open subset of $\mathbb{R}^2$ containing the origin. A real-valued C^1 function*

$$V : U \to \mathbb{R}; \quad \mathbf{x} \mapsto V(\mathbf{x})$$

is said to be positive definite on U if
  (i) $V(\mathbf{0}) = 0$;
  (ii) $V(\mathbf{x}) > 0$ *for all $\mathbf{x} \in U$ with $\mathbf{x} \neq \mathbf{0}$.*
A real-valued C^1 function V is said to be negative definite if $-V$ is positive definite.

For instance, the function $V(x_1, x_2) = x_1^2 + x_2^2$ is positive definite on all of $\mathbb{R}^2$, while the function $V(x_1, x_2) = x_1^2 + x_2^2 - x_2^3$ is positive definite on only a sufficiently small strip about the x_1-axis. On the other hand, the functions $V(x_1, x_2) = x_1 + x_2^2$, $V(x_1, x_2) = (x_1 + x_2)^2$, and $V(x_1, x_2) = x_1^2$ are not positive definite in any open neighborhood of the origin.

To gain insight into Liapunov's method, we need to explore some of the geometric aspects of positive definite functions. If V is a positive definite function on U, then V has a minimum at the origin. If this extreme point of V is isolated, then the surface $z = V(\mathbf{x})$ representing the graph of V in $\mathbb{R}^3$ near the origin has the general shape of a parabolic mirror pointing upward. The intersections of this graph with the horizontal plane $z = k$, that is, the *level sets* of V,

$$V^{-1}(k) \equiv \{\, \mathbf{x} \in \mathbb{R}^2 \,:\, V(\mathbf{x}) = k \,\},$$

are closed curves for small $k > 0$. The projection of these level sets onto the (x_1, x_2)-plane results in concentric ovals encircling the origin; see Figure 9.5. These remarks are, of course, obvious in the examples above, and they can be made precise locally for certain classes of functions known as "Morse functions"; namely, those functions $V(\mathbf{x})$ for which the eigenvalues of the *Hessian* matrix $(\partial^2 V(\mathbf{x})/\partial x_i \partial x_j)$ evaluated at local minima are positive. In this case, the Implicit Function Theorem implies that these

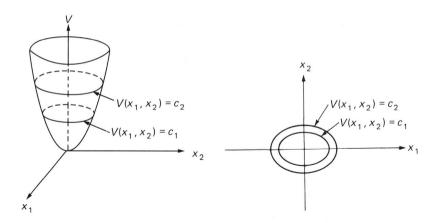

Figure 9.5. *Graph and level curves of a positive definite function V near the origin.*

local minima are isolated. Also, one can show that the level sets of V near the minima are diffeomorphic to circles; see the Appendix. However, a satisfactory characterization of the level sets of an arbitrary positive definite function is not available.

In simple situations, the standard choice of positive definite functions come from homogeneous quadratic polynomials (quadratic forms) in two variables. Here is an elementary test for positive definiteness of such functions:

Lemma 9.11. *A homogeneous quadratic function $V(x_1, x_2) = ax_1^2 + 2bx_1x_2 + cx_2^2$, where a, b, and c are real numbers, is positive definite if and only if $a > 0$ and $ac - b^2 > 0$.*

Proof. We will prove the necessity of the conditions on the coefficients; the sufficiency follows from similar reasoning. Suppose that V is positive definite. Since $V(x_1, 0) > 0$ if $x_1 \neq 0$, we must have $a > 0$. If $x_2 \neq 0$ is fixed, then $V(x_1, x_2) > 0$ for all x_1 and there can be no real zeros x_1 of $V(x_1, x_2)$. Thus the discriminant $4(b^2 - ac)$ of this quadratic function must be negative. $\Diamond$

Now, we would like to determine how the solutions of $\dot{\mathbf{x}} = \mathbf{f}(\mathbf{x})$ cross the level sets of a positive definite function V. If $\mathbf{x}(t)$ is a solution of $\dot{\mathbf{x}} = \mathbf{f}(\mathbf{x})$, then

$$\dot{V}(\mathbf{x}(t)) = \frac{\partial V}{\partial x_1}(\mathbf{x}(t))\,\dot{x}_1(t) + \frac{\partial V}{\partial x_2}(\mathbf{x}(t))\,\dot{x}_2(t). \tag{9.16}$$

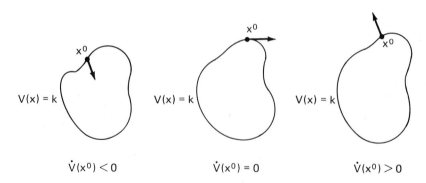

Figure 9.6. *Crossing level curves.*

This expression is simply the inner product of the vector $\mathbf{f}(\mathbf{x})$ with the gradient vector $\nabla V(\mathbf{x})$ of V at the point $\mathbf{x}$:

$$\dot{V}(\mathbf{x}) = \mathbf{f}(\mathbf{x}) \cdot \nabla V(\mathbf{x}) = \|\mathbf{f}(\mathbf{x})\| \cdot \|\nabla V(\mathbf{x})\| \cos\theta, \qquad (9.17)$$

where θ is the angle between $\mathbf{f}(\mathbf{x})$ and $\nabla V(\mathbf{x})$. The gradient vector $\nabla V(\mathbf{x})$ is the outward normal vector to the level curve of V at $\mathbf{x}$. Thus, if $\dot{V}(\mathbf{x}) < 0$, then the angle between $\mathbf{f}(\mathbf{x})$ and $\nabla V(\mathbf{x})$ is obtuse which implies that the orbit through $\mathbf{x}$ is crossing the level curve from the outside to the inside. Similarly, if $\dot{V}(\mathbf{x}) = 0$, then the orbit is tangent to the level curve; if $\dot{V}(\mathbf{x}) > 0$, the orbit is crossing the level curve from the inside to the outside. These three possibilities are shown in Figure 9.6. With these observations, the following basic theorem of Liapunov is quite plausible:

Theorem 9.12. (Liapunov) *Let $\bar{\mathbf{x}} = \mathbf{0}$ be an equilibrium point of $\dot{\mathbf{x}} = \mathbf{f}(\mathbf{x})$ and V be a positive definite C^1 function on a neighborhood U of $\mathbf{0}$.*
   (i) *If $\dot{V}(\mathbf{x}) \le 0$ for $\mathbf{x} \in U - \{\mathbf{0}\}$, then $\mathbf{0}$ is stable.*
   (ii) *If $\dot{V}(\mathbf{x}) < 0$ for $\mathbf{x} \in U - \{\mathbf{0}\}$, then $\mathbf{0}$ is asymptotically stable.*
   (iii) *If $\dot{V}(\mathbf{x}) > 0$ for $\mathbf{x} \in U - \{\mathbf{0}\}$, then $\mathbf{0}$ is unstable.*

Proof. Because of the geometric remarks above, we will give a formal proof of (i) only. Let $\varepsilon > 0$ be sufficiently small so that the neighborhood of the origin consisting of the points with $\|\mathbf{x}\| \le \varepsilon$ is contained in U. Let m be the minimum value of V on the boundary $\|\mathbf{x}\| = \varepsilon$ of this neighborhood. Since V is positive definite and the set $\|\mathbf{x}\| = \varepsilon$ is closed and bounded, we have $m > 0$. Now choose a δ with $0 < \delta \le \varepsilon$ such that $V(\mathbf{x}) < m$ for $\|\mathbf{x}\| \le \delta$. Such a δ always exists because V is continuous with $V(\mathbf{0}) = 0$. If $\|\mathbf{x}^0\| \le \delta$, then the solution $\mathbf{x}(t)$ of $\dot{\mathbf{x}} = \mathbf{f}(\mathbf{x})$ with $\mathbf{x}(0) = \mathbf{x}^0$ satisfies $\|\mathbf{x}(t)\| \le \varepsilon$ for $t \ge 0$ since $\dot{V}(\mathbf{x}(t)) \le 0$ implies that $V(\mathbf{x}(t)) \le V(\mathbf{x}^0)$ for $t \ge 0$. This proves the stability of the equilibrium point at the origin. $\diamond$

Appropriately, a function V implying the stability of an equilibrium point is named after Liapunov:

Definition 9.13. *A positive definite function V on an open neighborhood U of the origin is said to be a Liapunov function for $\dot{\mathbf{x}} = \mathbf{f}(\mathbf{x})$ if $\dot{V}(\mathbf{x}) \leq 0$ for all $\mathbf{x} \in U - \{\mathbf{0}\}$. When $\dot{V}(\mathbf{x}) < 0$ for all $\mathbf{x} \in U - \{\mathbf{0}\}$, the function V is called a strict Liapunov function.*

One of the main difficulties in applying the direct method of Liapunov to specific systems is that it may not at all be easy to determine an appropriate positive definite function. In certain natural systems, however, a pertinent choice, such as the energy function, is usually self-evident.

Example 9.14. *Pendulum:* The linearization of the differential equations of the pendulum (7.1) at the origin has purely imaginary eigenvalues and thus the stability type of the equilibrium point at the origin cannot be deduced from the linear approximation. So, let us try to apply the theorem of Liapunov above. The tempting choice $V(x_1, x_2) = x_1^2 + x_2^2$ as our positive definite function yields $\dot{V}(x_1, x_1) = 2x_1 x_2 + 2x_2(-g/l)\sin x_1$. Since $\dot{V}$ does not have a definite sign in any open set containing the origin, Liapunov's theorem does not seem to be applicable. However, let us consider the total energy of the pendulum

$$V(x_1, x_2) = \tfrac{1}{2}ml^2 x_2^2 + mgl(1 - \cos x_1).$$

Since the mass, length, and the gravitational constants are positive, V is positive definite in a sufficiently small neighborhood of the origin. Moreover, from Eq. (9.16), we have

$$\dot{V}(\mathbf{x}) \equiv 0.$$

Thus, from part (i) of the theorem of Liapunov, the origin is a stable equilibrium point of the planar pendulum (7.2). $\Diamond$

In our next example, we attempt to determine a basin of attraction of an asymptotically stable equilibrium point using a Liapunov function.

Example 9.15. *Basin of attraction:* Consider the second-order differential equation

$$\ddot{z} + 2a\dot{z} + z + z^3 = 0,$$

where a is a constant satisfying $0 < a < 1$, which is equivalent to the system

$$\begin{aligned} \dot{x}_1 &= x_2 \\ \dot{x}_2 &= -x_1 - 2ax_2 - x_1^3. \end{aligned} \tag{9.18}$$

The origin is the only equilibrium point and the eigenvalues of the linearization at the origin are $-a \pm i\beta$, where $\beta = \sqrt{1-a^2}$. Consequently, it follows

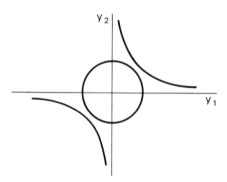

Figure 9.7. *A level set of the function $y_1^2 + y_2^2$ is a circle and that of $\frac{1}{a\beta} y_1^3 y_2$ is a hyperbola.*

from Theorem 9.5 that the origin is asymptotically stable. Let us now try to estimate the basin of attraction of the origin by using a Liapunov function. With the intent of determining a quadratic Liapunov function, we first put the linear part of the vector field into Real Jordan Normal Form. If we use the new variables **y** defined by

$$\mathbf{y} = P\mathbf{x}, \quad P = \begin{pmatrix} 1 & 0 \\ -a & \beta \end{pmatrix}, \quad P^{-1} = \frac{1}{\beta}\begin{pmatrix} \beta & 0 \\ a & 1 \end{pmatrix},$$

then the system (9.18) becomes

$$\dot{y}_1 = -ay_1 + \beta y_2$$
$$\dot{y}_2 = -\beta y_1 - ay_2 - \frac{1}{\beta}y_1^3. \tag{9.19}$$

We now take our Liapunov function as the quadratic function

$$V(y_1, y_2) = \frac{1}{2a}(y_1^2 + y_2^2),$$

then a simple computation yields that

$$\dot{V}(y_1, y_2) = -(y_1^2 + y_2^2) - \frac{1}{a\beta}y_1^3 y_2.$$

Now, the main task is to determine the largest subset of $\mathbb{R}^2$ containing the origin where $-\dot{V}$ is positive definite. This is a difficult task, so let us settle for the largest such disk. Observe that the level sets of the function $y_1^2 + y_2^2$ are circles about the origin and the level sets of the function $\frac{1}{a\beta}y_1^3 y_2$ are similar to hyperbolas; see Figure 9.7.

From the symmetry of $-\dot{V}$, it is evident that the radius r_0 of the largest circle inside which $-\dot{V}$ is positive definite must satisfy

$$r_0^2 - \frac{1}{a\beta}r_0^4 = 0,$$

that is, $r_0 = \sqrt{a\beta}$. Thus, every solution of Eq. (9.19) with initial value $\mathbf{y}^0$ satisfying $\|\mathbf{y}^0\| < r_0$ approaches the origin as $t \to +\infty$. The circle of radius r_0 becomes an ellipse when transformed back to the original variables.

With a little more work, one could obtain a slightly larger subset of the basin of attraction of the origin. However, it is clear that $-\dot{V}$ is not positive definite on all of $\mathbb{R}^2$. One could also try other Liapunov functions to obtain possibly larger subsets of the basin of attraction of the origin. Since it is not clear what to try, the important question remains: How large is the basin of attraction of the origin? We will answer this question after we uncover additional properties of Liapunov functions in the next section. $\diamondsuit$

There is a chapter in the theory of Liapunov functions that is usually referred to as the *converse theorems of Liapunov.* The basic premise of these results is that if an equilibrium point is, for example, asymptotically stable, then there exists an appropriate Liapunov function with the properties listed in (ii) of Theorem 9.12. Although it may be of limited practical use, such a result is of considerable theoretical interest, as we shall see in Chapter 13. In this spirit, let us reprove Theorem 9.5 on asymptotic stability of an equilibrium point from linearization.

Example 9.16. *A Liapunov function for an asymptotically stable equilibrium point:* Let us suppose that the initial transformation of the variables in the proof of Theorem 9.5 have been made and consider the system

$$\dot{\mathbf{x}} = \mathbf{A}\mathbf{x} + \mathbf{g}(\mathbf{x}), \tag{9.20}$$

where $\mathbf{A}$, which is the Jacobian matrix, has eigenvalues with negative real parts, $\mathbf{g}(\mathbf{0}) = \mathbf{0}$ and $D\mathbf{g}(\mathbf{0}) = \mathbf{0}$. We now construct a Liapunov function that implies the asymptotic stability of the origin for Eq. (9.20).

Our Liapunov function will be chosen as a quadratic form $V(\mathbf{x}) = \mathbf{x}^T\mathbf{B}\mathbf{x}$, where the symmetric matrix $\mathbf{B}$ satisfies $\mathbf{A}^T\mathbf{B} + \mathbf{B}\mathbf{A} = -\mathbf{I}$. Lemma 8.17 implies that there is such a V that is positive definite. Then, for Eq. (9.20), we have

$$\dot{V}(\mathbf{x}) = -\mathbf{x}^T\mathbf{x} + 2\mathbf{x}^T\mathbf{B}\mathbf{g}(\mathbf{x}).$$

For any $m > 0$, there is a $\delta > 0$ such that $\|\mathbf{g}(\mathbf{x})\| \leq m\|\mathbf{x}\|$ if $\|\mathbf{x}\| \leq \delta$. Let β be the largest eigenvalue of $\mathbf{B}$, then

$$\dot{V}(\mathbf{x}) \leq -(1 - 2\beta m)\mathbf{x}^T\mathbf{x}.$$

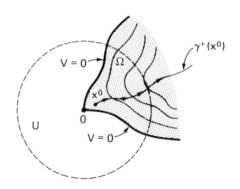

Figure 9.8. *Instability theorem of Četaev.*

If $m < 1/(2\beta)$, then $-\dot{V}$ is positive inside the disk $U = \{\, \mathbf{x} \,:\, \|\mathbf{x}\| < \delta \,\}$. Therefore, the quadratic function V as constructed satisfies the hypotheses of Theorem 9.12 (ii) and thus the origin is an asymptotically stable equilibrium point of Eq. (9.20). $\Diamond$

We conclude this section with an embellishment of the theorem of Liapunov. The instability part of Theorem 9.12 has the deficiency of considering a full neighborhood of the origin and thus is not applicable to equilibria of saddle type. The following theorem of Četaev, which is reminiscent of the latter part of the proof of Theorem 9.7, is a way to remedy this shortcoming:

Theorem 9.17. (*Četaev*) *Let U be a sufficiently small open neighborhood of the origin. If there is an open region Ω and a C^1 function $V : \overline{\Omega} \to \mathbb{R}$ with the properties*
  (i) *the origin is a boundary point of Ω;*
  (ii) *$V(\mathbf{x}) = 0$ for all $\mathbf{x}$ on the boundary points of Ω inside U;*
  (iii) *$V(\mathbf{x}) > 0$ and $\dot{V}(\mathbf{x}) > 0$ for all $\mathbf{x} \in \Omega \cap U$,*
then the origin is an unstable equilibrium point.

Sketch of Proof. We have illustrated in Figure 9.8 a typical situation described by the theorem. From the property (i), there are points in Ω, hence in U, that are arbitrarily close to the origin. From (ii) and (iii), no orbit starting from any one of these points in Ω can cross the boundary of Ω in U. Thus, also from (iii), such orbits must leave the neighborhood U through Ω. Consequently, the origin is an unstable equilibrium point. $\Diamond$

Example 9.18. *Instability with Četaev:* As an application of the theorem of Četaev, let us consider the system of differential equations

$$\dot{x}_1 = x_1^3 + x_2 x_1^2$$
$$\dot{x}_2 = -x_2 + x_1^2,$$

which has an equilibrium point at the origin. Notice that since the eigenvalues of the linearized vector field at the origin are 0 and -1, none of the theorems in the previous two sections is applicable. Therefore, let us consider the function $V(x_1, x_2) = x_1^2/2 - x_2^2/2$ and the open region

$$\Omega = \{ (x_1, x_2) : x_1 > x_2 > -x_1 \}.$$

Observe that $V(\mathbf{x}) > 0$ for $\mathbf{x} \in \Omega$, and $V(\mathbf{x}) = 0$ on the boundary. Next, we compute the derivative of V along the solutions of the differential equations above:

$$\dot{V}(x_1, x_2) = x_1^4 - x_2(x_1^2 - x_1^3) + x_2^2.$$

It is not immediately obvious that we are in the realm of Četaev's theorem. However, in a sufficiently small neighborhood U of the origin, we have the estimate

$$\dot{V}(x_1, x_2) \geq x_1^4 - (1 + \varepsilon)|x_2|x_1^2 + x_2^2,$$

where $\varepsilon \geq 0$ is small. By viewing the right-hand side of this inequality as a quadratic form in $|x_2|$ and x_1^2, it is easy to see that $\dot{V}(\mathbf{x}) > 0$ for $\mathbf{x}$ in a neighborhood of $\mathbf{x} = \mathbf{0}$ and, in particular, for $\mathbf{x} \in \Omega$. Now, the conditions of Theorem 9.17 are satisfied and thus the origin is unstable. $\Diamond$

Exercises

9.11. *Graphics:* Computer graphics is a very useful tool for plotting graphs of functions. You should locate a standard package with 3-D graphics capabilities and plot the graphs of a few functions, for example,

$V(x_1, x_2) = x_1^2 + 4x_2^2;$ $V(x_1, x_2) = x_1^2 + x_1 x_2 + 2x_2^2;$
$V(x_1, x_2) = (x_1 + x + 2)^2;$ $V(x_1, x_2) = \frac{1}{2}x_1^2 + \frac{1}{3}x_1^3 + \frac{1}{2}x_2^2;$
$V(x_1, x_2) = x_1^2 + x_2^2 - x_2^3;$ $V(x_1, x_2) = x_2^2 + (1 - \cos x_1).$

Find the largest neighborhoods of the origin where these functions are positive definite.

9.12. *Odd polynomials:* Show that a homogeneous polynomial of odd degree cannot be positive definite.

Hint: Take $x_2 = ax_1$, and factor out a power of x_1.

9.13. Use Theorem 9.12 to show that, for $c > 0$, every solution of the system $\dot{x}_1 = x_2$, $\dot{x}_2 = -x_1 - cx_2$ approaches the origin as $t \to +\infty$.

Suggestion: Use an appropriate quadratic function; consult Section 8.3.

9.14. *No linear part:* For the three systems below, linearization is of no help in determining the stability type of the equilibrium point at the origin because the Jacobian matrix at the origin is the zero matrix. However, using quadratic functions, determine the stability types of the origin:

(a) $\dot{x}_1 = -x_1^3 + x_1 x_2^2$, $\dot{x}_2 = -2x_1^2 x_2 - x_2^3$;
(b) $\dot{x}_1 = -x_1^3 + 2x_2^3$, $\dot{x}_2 = -2x_1 x_2^2$;
(c) $\dot{x}_1 = x_1^3 - x_2^3$, $\dot{x}_2 = x_1 x_2^2 + 2x_1^2 x_2 + x_2^3$.

9.15. *After Lagrange and Dirichlet:* Consider the conservative system

$$\dot{x}_1 = x_2, \quad \dot{x}_2 = -g(x_1),$$

where the function g is, say, C^1.

(a) Show that each isolated minimum point $\bar{x}_1$ of the potential function $\int_0^{x_1} g(u)\,du$ corresponds to a stable equilibrium point $(\bar{x}_1, 0)$ of the system.
    Help: See Chapter 14.

(b) Give an example of a function g such that $\bar{x}_1$ is not a minimum of the potential function and yet $(\bar{x}_1, 0)$ is a stable equilibrium point.
    Hint: Try a nonanalytic function.

9.16. Suppose that you have the function $V(x_1, x_2) = x_2^2 e^{-x_1}$ defined on the whole (x_1, x_2)-plane and that relative to some planar differential equation $\dot{V}(x_1, x_2) = -x_2^2 V(x_1, x_2)$. Can you conclude anything about the solutions of the original differential equation? If not, what is the trouble?

9.17. Consider the system of equations

$$\dot{x}_1 = x_2 - x_1 f(x_1, x_2)$$
$$\dot{x}_2 = -x_1 - x_2 f(x_1, x_2),$$

where f is a real-valued C^1 function. Notice that the origin is an equilibrium point independent of the specific form of f. Using a quadratic function, show that if $f(x_1, x_2) > 0$ in some open neighborhood of the origin, then the origin is asymptotically stable. What is the stability type of the origin if $f(x_1, x_2) < 0$ in a neighborhood of the origin?

9.18. Show that the origin is an unstable equilibrium point for the system

$$\dot{x}_1 = x_1^3 + x_1 x_2$$
$$\dot{x}_2 = -x_2 + x_2^2 + x_1 x_2 - x_1^3.$$

Hint: Consider the function $V(x_1, x_2) = x_1^4/4 - x_2^2/2$ and show

$$\dot{V}(x_1, x_2) = x_1^6 + x_2 x_1^3(1 + x_1) + x_2^2(1 - x_1 - x_2)$$
$$\geq x_1^6 - \tfrac{1}{2}|x_2|\,|x_1|^3 + \tfrac{1}{2}x_2^2 > 0$$

in a small neighborhood of the origin, except the origin itself.

9.19. *Actual basin of attraction:* Using the function $V(x_1, x_2) = x_1^2/2 + x_2^2/4$, determine the largest ellipse contained in the basin of attraction of the origin of the system

$$\dot{x}_1 = -x_1 + x_2^2$$
$$\dot{x}_2 = -2x_2 + 3x_1^2.$$

Answer: $x_1^2/2 + x_2^2/4 = 1/9$.

Experiment numerically to convince yourself that the actual basin of attraction of the origin is larger than this ellipse. However, it is not the entire plane because there are other equilibrium points.

9.20. *Indirect control:* Suppose that $\psi : \mathbb{R} \to \mathbb{R}$, $\sigma \mapsto \psi(\sigma)$, is a C^1 function satisfying $\psi(0) = 0$, $\sigma\psi(\sigma) > 0$ if $\sigma \neq 0$, and $\int_0^\sigma \psi(s)\, ds \to +\infty$ as $|\sigma| \to +\infty$. For k, c, and ρ positive constants with $k\rho > 0$, show that every solution of the indirect control problem

$$\dot{x} = -kx - \xi, \quad \dot{\xi} = \psi(\sigma), \quad \sigma = cx - \rho\xi,$$

approaches zero as $t \to +\infty$.

The label "indirect control" comes from the fact that in the system above the control variable ξ is not given directly as a function of the state variable x; instead, it is determined indirectly using another differential equation. In certain situations, indirect control turns out to be very efficient; on related matters, see, for example, Lefschetz [1965].

9.4. An Invariance Principle

In this section, we continue our study of Liapunov functions with a more detailed discussion of the basin of attraction of an asymptotically stable equilibrium point. In preparation for later chapters, we will commence our exposition in a setting that is applicable for general limit sets and then specialize to the case of equilibrium points.

Definition 9.19. *A subset U of $\mathbb{R}^2$ is said to be positively invariant [respectively, negatively invariant] under the flow φ of $\dot{\mathbf{x}} = \mathbf{f}(\mathbf{x})$ if, for any $\mathbf{x}^0 \in U$, the positive orbit $\gamma^+(\mathbf{x}^0)$ [respectively, negative orbit $\gamma^-(\mathbf{x}^0)$] through $\mathbf{x}^0$ belongs to U.*

Below, we will be chiefly interested in subsets of the plane that are both positively and negatively invariant.

Definition 9.20. *A subset M of $\mathbb{R}^2$ is said to be invariant under the flow φ of $\dot{\mathbf{x}} = \mathbf{f}(\mathbf{x})$ if, for any $\mathbf{x}^0 \in M$, the orbit $\gamma(\mathbf{x}^0)$ through $\mathbf{x}^0$ belongs to M; or, in consolidated notation, $\varphi(t, M) = M$ for each $t \in \mathbb{R}$.*

Some of the noteworthy examples of invariant sets we have seen are, of course, equilibrium points and periodic orbits. More generally, the lemma below provides some of the important topological properties of the limit set of any positive or negative orbit among which is the fact that the limit set is invariant.

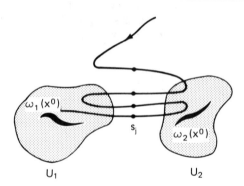

Figure 9.9. *If $\omega(\mathbf{x}^0)$ were not connected.*

Lemma 9.21. *If a positive orbit $\gamma^+(\mathbf{x}^0)$ [respectively, $\gamma^-(\mathbf{x}^0)$] is bounded, then the ω-limit set $\omega(\mathbf{x}^0)$ [respectively, $\alpha(\mathbf{x}^0)$] is a nonempty, compact, and connected invariant set.*

Proof. The fact that $\omega(\mathbf{x}^0)$ is nonempty and compact are relatively easy consequences of the definition of omega-limit set. Let us establish the key property of invariance of $\omega(\mathbf{x}^0)$. If $\mathbf{y}$ is in $\omega(\mathbf{x}^0)$, then there is a sequence $\{t_j\}$, satisfying $t_j \to +\infty$ as $j \to +\infty$, such that $\varphi(t_j, \mathbf{x}^0) \to \mathbf{y}$ as $j \to +\infty$. Consequently, for any fixed t in $(-\infty, +\infty)$, we have $\varphi(t + t_j, \mathbf{x}^0) = \varphi(t, \varphi(t_j, \mathbf{x}^0)) \to \varphi(t, \mathbf{y})$ as $n \to +\infty$ from the continuity of φ. This shows that the orbit through $\mathbf{y}$ belongs to $\omega(\mathbf{x}^0)$ which establishes the invariance of $\omega(\mathbf{x}^0)$.

We now show that $\omega(\mathbf{x}^0)$ is connected. Suppose that $\omega(\mathbf{x}^0)$ is not connected so that there are two nonempty, disjoint, closed sets $\omega_1(\mathbf{x}^0)$ and $\omega_2(\mathbf{x}^0)$ such that $\omega(\mathbf{x}^0) = \omega_1(\mathbf{x}^0) \cup \omega_2(\mathbf{x}^0)$. Then there exist two disjoint open sets U_1 and U_2 with $\omega_1(\mathbf{x}^0) \subset U_1$ and $\omega_2(\mathbf{x}^0) \subset U_2$. Let $\{t_j\}$ be a sequence with $\varphi(t_j, \mathbf{x}^0) \in U_1$, and $\{\tau_j\}$ be another sequence with $\varphi(\tau_j, \mathbf{x}^0) \in U_2$. We can choose these sequences so that $\tau_j < t_j < \tau_{j+1}$. Then there must exist a sequence $\{s_j\}$ satisfying $t_j < s_j < \tau_{j+1}$ such that $\varphi(s_j, \mathbf{x}^0) \notin U_1 \cup U_2$. But there is a limit point of the sequence $\{\varphi(s_j, \mathbf{x}^0)\}$ which is not in $\omega(\mathbf{x}^0)$ and this is a contradiction; see Figure 9.9. $\Diamond$

With the terminology above, we now state the main theorem of this section.

Theorem 9.22. (Invariance Principle) *Let V be a real-valued function and let $U \equiv \{\mathbf{x} \in \mathbb{R}^2 : V(\mathbf{x}) < k\}$, where k is a real number. Suppose further that V is continuous on the closure $\overline{U}$ of U and is C^1 on U with $\dot{V}(\mathbf{x}) \leq 0$ for $\mathbf{x} \in U$. Consider the subset S of $\overline{U}$ defined by*

$$S \equiv \{\mathbf{x} \in \overline{U} : \dot{V}(\mathbf{x}) = 0\}$$

and let M be the largest invariant set in S. Then every positive orbit that starts in U and remains bounded has its ω-limit set in M.

Proof. For $\mathbf{x^0} \in U$, let $\varphi(t, \mathbf{x^0})$ be the solution through $\mathbf{x^0}$ and suppose that it is bounded for $t \geq 0$. Then $\dot{V}(\varphi(t, \mathbf{x^0})) \leq 0$ and thus $V(\varphi(t, \mathbf{x^0})) \leq V(\mathbf{x^0}) \leq k$ for all $t \geq 0$. Consequently, $\varphi(t, \mathbf{x^0}) \in U$ for all $t \geq 0$. Moreover, $V(\varphi(t, \mathbf{x^0})) \to c$, where c is a constant, as $t \to +\infty$. The continuity of V implies that $V(\mathbf{y}) = c$ for any $\mathbf{y} \in \omega(\mathbf{x^0})$. Since $\omega(\mathbf{x^0})$ is invariant, we have $V(\varphi(t, \mathbf{y})) = c$ for all $t \in \mathbb{R}$. Thus $\dot{V}(\varphi(t, \mathbf{y})) = 0$ for all $t \in \mathbb{R}$ and $\omega(\mathbf{x^0}) \subset S$. Now, the invariance of $\omega(\mathbf{x^0})$ implies that $\omega(\mathbf{x^0}) \subset M$. $\diamond$

For practical applications, we state several important consequences of the Invariance Principle:

Corollary 9.23. *If, in addition, every positive orbit is bounded, V is positive definite, and M consists of the origin, $M = \{\mathbf{0}\}$, then the origin is asymptotically stable and all of U belongs to its basin of attraction.* $\diamond$

Corollary 9.24. *If, in addition, every positive orbit is bounded and V is positive definite for $\mathbf{x} \in U - \{\mathbf{0}\}$, then $M = \{\mathbf{0}\}$, that is, the origin is asymptotically stable and all of U belongs to its basin of attraction.* $\diamond$

Theorem 9.25. *If $V : \mathbb{R}^2 \to \mathbb{R}$ is a C^1 function such that $V(\mathbf{x}) \to +\infty$ as $\|\mathbf{x}\| \to +\infty$ and $\dot{V}(\mathbf{x}) \leq 0$, for all $\mathbf{x} \in \mathbb{R}^2$, then every positive orbit is bounded and has its ω-limit set in M, the largest invariant set in $\{\mathbf{x} \in \mathbb{R}^2 : \dot{V}(\mathbf{x}) = 0\}$.* $\diamond$

Example 9.26. *Global orbit structures:* Here, we complete the analyses of two previous examples using the Invariance Principle.

Continuation of Example 9.15: Let us now determine the entire basin of attraction of the asymptotically stable origin. We begin by choosing a different Liapunov function. When $a = 0$, notice that this is a conservative system with the energy function

$$V(x_1, x_2) = \tfrac{1}{2}(x_1^2 + x_2^2) + \tfrac{1}{4}x_1^4,$$

so that the orbits of Eq. (9.18) lie on the level sets of this function; see Figure 9.10. The parameter a, when $a > 0$, causes the system to lose energy. Therefore, it is natural to compute $\dot{V}$ to see how this happens. A simple computation yields

$$\dot{V}(x_1, x_2) = -2ax_2^2.$$

Let us now apply the corollaries of the invariance principle. Since $V(\mathbf{x}) \to +\infty$ as $\|\mathbf{x}\| \to +\infty$, for any k, each of the sets U in the statement of the Invariance Principle is bounded. Moreover, it is clear that the set S is contained in the x_1-axis. Now, we determine the largest invariant set M

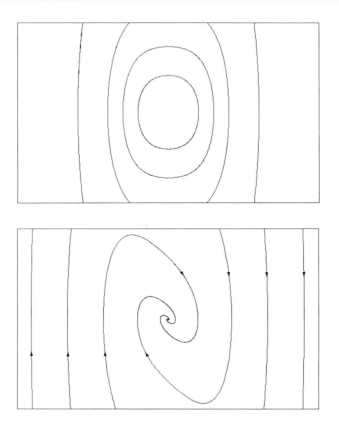

Figure 9.10. *Level sets of the Liapunov function* $V(x_1, x_2) = \frac{1}{2}(x_1^2 + x_2^2) + \frac{1}{4}x_1^4$ *and the phase portrait of Eq. (9.18).*

on the x_1-axis. Since M is invariant, we need only look for solutions of Eq. (9.18) which remain on the x_1-axis for all $t \in \mathbb{R}$. The first equation in Eq. (9.18) implies that $\dot{x}_1 = 0$, or $x_1 = constant$. The second equation implies that $x_1 = 0$. Thus, M consists of the origin in $\mathbb{R}^2$. Consequently, the basin of attraction of the origin is all of $\mathbb{R}^2$; see Figure 9.10.

Continuation of Examples 9.6 and 9.8: Let us consider the damped pendulum equation (9.8). As our Liapunov function, we will use the Hamiltonian

$$V(x_1, x_2) = \frac{1}{2}x_2^2 + (1 - \cos x_1)$$

of the undamped pendulum. A short calculation yields

$$\dot{V}(x_1\, x_2) = -ax_2^2.$$

In order to bound the positive orbits of the damped pendulum, we put its flow on the cylinder $S^1 \times \mathbb{R}$. As such, the flow of the damped pendulum has two equilibria $(0, 0)$ and $(\pi, 0)$ on the cylinder. Recall from the discussion of the flow of the undamped pendulum on the cylinder in Example 7.21 that the level sets of V are closed curves. A positive orbit of the damped pendulum crosses a level curve of V from above to below if $x_2 > 0$, and from below to above if $x_2 < 0$. Consequently, all positive orbits of the damped pendulum on the cylinder are bounded. Now, what we need is an extension of the Invariance Principle to the cylinder; luckily, such an extension is valid but we refrain from presenting the details. Therefore, we conclude that the ω-limit set of each orbit of the damped pendulum on the cylinder is one of the two equilibrium points.

Finally, it is important to observe that the local phase portraits of the undamped and damped penduli are qualitatively the same in a sufficiently small neighborhood of the equilibrium point $(\pi, 0)$. Globally, however, the homoclinic loop of the undamped pendulum is now broken for the damped pendulum. Breaking of homoclinic loops play a significant role in dynamics and we will investigate it further at a later time. ◇

Exercises

9.21. *On unbounded orbits:* The hypothesis of boundedness in the statement of Lemma 9.21 is essential. Draw a picture of an orbit of a planar system whose ω-limit set is empty. Also, draw a picture of an orbit of a planar system whose ω-limit set is nonempty but disconnected. Can you write down specific planar systems with these properties?

9.22. Use Theorem 9.22 to show that, for $c > 0$, every solution of the system $\dot{x}_1 = x_2$, $\dot{x}_2 = -x_1 - cx_2$ approaches the origin as $t \to +\infty$.

9.23. *Consult a classic book:* Decipher the equivalent statement and the proof of Lemma 9.21 as given on page 198 of the classic book of Birkhoff [1927].

9.24. *Where is Van der Pol's periodic orbit?* Consider the Lienard form of Van der Pol's oscillator:

$$\dot{x}_1 = x_2 - \lambda(\tfrac{1}{3}x_1^3 - x_1)$$
$$\dot{x}_2 = -x_1,$$

where $\lambda > 0$. The only equilibrium point is at the origin and it is unstable. We will show in Chapter 12 that this system has a unique periodic orbit, encircling the origin, which attracts all other orbits except the origin. Herein, we determine how far away this periodic orbit must lie from the origin. Observe that reversing time, $t \mapsto -t$, is equivalent to taking $\lambda < 0$. Take $\lambda < 0$ and use the Liapunov function $V(x_1, x_2) = x_1^2/2 + x_2^2/2$ to show that the basin of attraction of the origin contains the interior of the disk $x_1^2 + x_2^2 < 3$. Conclude that when $\lambda > 0$, the periodic orbit must lie in the exterior of this disk.

9.25. *Breaking a homoclinic loop:* Consider the system

$$\dot{x}_1 = x_2$$
$$\dot{x}_2 = -2x_1 - ax_2 - 3x_1^2,$$

where $a > 0$, and explore its dynamics as follows:
1. Show that $(0, 0)$ is asymptotically stable and $(2/3, 0)$ is a saddle point.
2. Draw typical level sets of the function $V(x_1, x_2) = x_2^2/2 + x_1^2 + x_1^3$. This is the energy function of the conservative case when $a = 0$. Notice especially that the level set $V(x_1, x_2) = 0$ contains the origin as well as a loop (a homoclinic loop) which starts and ends at the origin.
3. For every $\mathbf{x}^0 \in \mathbb{R}^2$ for which $\gamma^+(\mathbf{x}^0)$ is bounded, show that $\omega(\mathbf{x}^0)$ is one of the equilibrium points.
4. Estimate the basin of attraction of the origin.
5. Let $W^s(2/3, 0)$ be the set of points $\mathbf{x}^0$ in $\mathbb{R}^2$ with $\omega(\mathbf{x}^0) = (2/3, 0)$. Show that $\|\varphi(t, \mathbf{x}^0)\| \to +\infty$ as $t \to -\infty$. This can be done by contradiction; assume that $\alpha(\mathbf{x}^0)$ is not empty and conclude that $\alpha(\mathbf{x}^0)$ is an equilibrium point.
6. Observe that there are orbits which are unbounded in both directions in time t.
7. Finally, sketch the qualitative phase portrait of the system on the plane.

9.26. *Breaking a heteroclinic loop:* For $a > 0$, discuss the properties of the solutions of the system
$$\dot{x}_1 = x_2$$
$$\dot{x}_2 = -x_1 - 2ax_2 + x_1^3$$

in the same detail as in the previous exercise using the energy function for the case $a = 0$.

9.27. *A difficult problem:* Suppose that $p(t + 1) = p(t) > 0$ for all t. Can you use some of the ideas in this chapter to show that every solution of $\ddot{x} + p(t)\dot{x} + x = 0$ approaches zero as $t \to +\infty$? If not, take some particular $p(t)$ and experiment numerically using PHASER.

9.5. Preservation of a Saddle

We saw earlier in this chapter that if the linearization of a vector field at an equilibrium point is a saddle, then the equilibrium point is unstable. But, what distinguishes a saddle geometrically from other types of unstable equilibria? A close inspection of phase portraits near a saddle, in Figure 9.4, for example, points to the presence of four special orbits that approach the equilibrium point in forward or reverse time. These special orbits, together with the equilibrium point, play a very important role in the qualitative analysis of differential equations. So, we begin the study of these sets with their precise definitions.

Definition 9.27. *Let U be a neighborhood of an equilibrium point $\bar{\mathbf{x}}$. Then the local stable manifold $W^s(\bar{\mathbf{x}}, U)$, and the local unstable manifold $W^u(\bar{\mathbf{x}}, U)$ of $\bar{\mathbf{x}}$ are defined, respectively, to be the following subsets of U:*

$$W^s(\bar{\mathbf{x}}, U) \equiv \{\, \mathbf{x}^0 \in U \ : \ \varphi(t, \mathbf{x}^0) \in U \text{ for } t \geq 0,$$
$$\text{and } \varphi(t, \mathbf{x}^0) \to \bar{\mathbf{x}} \text{ as } t \to +\infty \,\},$$

$$W^u(\bar{\mathbf{x}}, U) \equiv \{\, \mathbf{x}^0 \in U \ : \ \varphi(t, \mathbf{x}^0) \in U \text{ for } t \leq 0,$$
$$\text{and } \varphi(t, \mathbf{x}^0) \to \bar{\mathbf{x}} \text{ as } t \to -\infty \,\}.$$

Let us examine a simple specific nonlinear system and determine its local stable and unstable manifolds.

Example 9.28. *Stable and unstable manifolds:* Consider the nonlinear planar differential equations

$$\dot{x}_1 = -x_1$$
$$\dot{x}_2 = x_2 + x_1^2. \tag{9.21}$$

This system has a unique equilibrium point at the origin, and its linearization at this point is the canonical linear system

$$\dot{\mathbf{x}} = \begin{pmatrix} -1 & 0 \\ 0 & 1 \end{pmatrix} \mathbf{x},$$

which is a saddle. It is self-evident from our previous study of this linear system that the local stable and local unstable manifolds of the origin are, respectively, the x_1- and x_2-axis, including the origin. To determine the corresponding manifolds in the nonlinear system, we compute its flow explicitly. Of course, we have purposely concocted this example so that this is easy to do. In fact, by solving the linear differential equation for x_1 and using it in the second equation, we obtain from the variations of the constants formula

$$x_1(t) = e^{-t} x_1^0$$
$$x_2(t) = e^t \left[x_2^0 + \tfrac{1}{3}(x_1^0)^2 \right] - \tfrac{1}{3} e^{-2t} (x_1^0)^2.$$

Now, it is clear from these formulas that the local unstable and local stable manifolds are, respectively, the x_2-axis and the parabola $x_2 = -(1/3)(x_1^2)$; more formally, for any disk U containing the origin,

$$W^u(\mathbf{0}, U) = \{\, (x_1^0, x_2^0) \in U \ : \ x_1^0 = 0 \,\},$$

$$W^s(\mathbf{0}, U) = \{\, (x_1^0, x_2^0) \in U \ : \ x_2^0 = -\tfrac{1}{3}(x_1^0)^2 \,\}.$$

We have drawn in Figure 9.11 the local stable and local unstable manifolds of Eq. (9.21) as well as those of its linearization at the origin. $\diamondsuit$

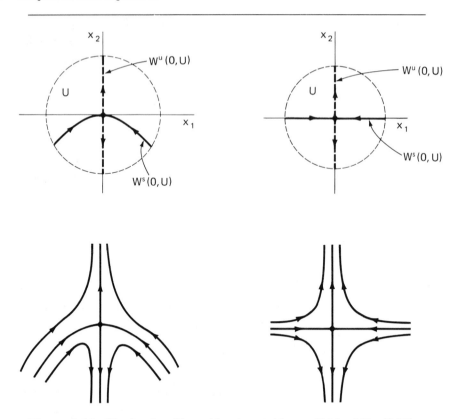

Figure 9.11. *The local stable and local unstable manifolds of Eq.* (9.21) *and its linearization near the origin.*

With the intent of obtaining a general result, let us isolate several noteworthy features of the example above. The local stable [respectively, unstable] manifold of the origin of the nonlinear system is a smooth graph over the local stable [respectively, unstable] manifold of the linearized system. Furthermore, the local stable [respectively, unstable] manifolds of the nonlinear system and its linearization are tangent at the equilibrium point.

If linearization is to reflect the qualitative features of the phase portrait of a nonlinear system near an equilibrium point, then, in general, we expect these observations to be true. The theorem below secures that this is indeed the case. We should emphasize that since linearization is a local process, we can expect success only locally near an equilibrium point. In this sense, the example above is a bit too special; its local stable and unstable manifolds are graphs globally. Later in this section, we will present other examples where this is not the case, and explore this issue further. But, first, let us state our general local theorem.

By an appropriate linear change of coordinates, a general planar differential equation $\dot{\mathbf{x}} = \mathbf{f}(\mathbf{x})$ whose linearization at the origin is a saddle can always be transformed to the "normal form"

$$\dot{x}_1 = \lambda_1 x_1 + g_1(x_1, x_2)$$
$$\dot{x}_2 = \lambda_2 x_2 + g_2(x_1, x_2), \tag{9.22}$$

where $\lambda_1 < 0$, $\lambda_2 > 0$, and the function $\mathbf{g} \equiv (g_1, g_2)$ satisfies $\mathbf{g}(\mathbf{0}) = \mathbf{0}$, $D\mathbf{g}(\mathbf{0}) = \mathbf{0}$. Notice that the local stable and unstable manifolds of the linearization at the origin are, respectively, the x_1- and x_2-axis. Let us now suppose that our differential equation has been put into the normal form (9.22). Then we have the following theorem:

Theorem 9.29. *For the system (9.22), there is a $\delta > 0$ such that, in the neighborhood $U \equiv \{(x_1, x_2) : |x_1| < \delta, |x_2| < \delta\}$ the local stable and local unstable manifolds of the equilibrium point at the origin are given by*

$$W^s(\mathbf{0}, U) = \{(x_1, x_2) : x_2 = h_s(x_1), |x_1| < \delta\},$$

$$W^u(\mathbf{0}, U) = \{(x_1, x_2) : x_1 = h_u(x_2), |x_2| < \delta\},$$

where the functions h_s and h_u are as smooth as the function $\mathbf{g}$ in Eq. (9.22). Furthermore, they satisfy

$$h_s(0) = 0, \quad h_u(0) = 0, \quad \frac{dh_s}{dx_1}(0) = 0, \quad \frac{dh_u}{dx_2}(0) = 0. \tag{9.23}$$

We will not give a proof of this theorem because some of the necessary technical details are beyond the intended scope of our book. Instead, we will devise a constructive method for approximating, say, the function h_s to any desired accuracy in specific differential equations. To be of practical use, such a method, unlike in Example 9.28, should not depend on the explicit knowledge of the flow. We will accomplish this by deriving a scalar differential equation whose solution satisfying $h_s(0) = 0$ is the local stable manifold of the equilibrium point at the origin. If we differentiate with respect to t the defining equation $x_2(t) = h_s(x_1(t))$ given by the theorem above, then we obtain

$$\dot{x}_2 = \frac{dh_s}{dx_1} \dot{x}_1.$$

If we now substitute the expressions for $\dot{x}_1$ and $\dot{x}_2$ given by the differential equation (9.22), and use h_s for the x_2 variable, then the solution of the initial-value problem

$$\frac{dh_s}{dx_1} [\lambda_1 x_1 + g_1(x_1, h_s)] = \lambda_2 h_s + g_2(x_1, h_s), \quad h_s(0) = 0 \tag{9.24}$$

is the local stable manifold of the origin. In an analogous way, we obtain for the local unstable manifold the initial-value problem

$$\frac{dh_u}{dx_2}\left[\lambda_2 x_2 + g_2(h_u, x_2)\right] = \lambda_1 h_u + g_1(h_u, x_2), \quad h_u(0) = 0. \tag{9.25}$$

Admittedly, these differential equations look rather formidable. One can, of course, attempt to solve them numerically. Or, better yet, one can also readily obtain approximate solutions by using Taylor expansions of the functions h_s or h_u near the origin and then equating the coefficients, as illustrated in the example below.

Example 9.30. *Computing local stable and unstable manifolds:* Let us reconsider Example 9.28 in light of the scalar differential equations (9.24) and (9.25). In this case, these equations become

$$\frac{dh_s}{dx_1}\left[-x_1\right] = h_s + x_1^2, \quad h_s(0) = 0, \tag{9.26}$$

$$\frac{dh_u}{dx_2}\left[x_2 + h_u^2\right] = -h_u, \quad h_u(0) = 0, \tag{9.27}$$

respectively. Because of the properties (9.23), the functions h_s and h_u can be expanded in power series of the form

$$h_s(x_1) = \frac{1}{2}a_2 x_1^2 + \frac{1}{3!}a_3 x_1^3 + \cdots,$$

$$h_u(x_2) = \frac{1}{2}b_2 x_2^2 + \frac{1}{3!}b_3 x_2^3 + \cdots,$$

where a_i and b_i are unknown coefficients. We can determine these coefficients by substituting the power series of h_s and h_u into Eqs. (9.26) and (9.27), respectively. Finally, by equating the coefficients we readily see that

$$h_s(x_1) = -\tfrac{1}{3}x_1^2 \quad \text{and} \quad h_u(x_2) = 0,$$

thus recovering the same results we obtained earlier from explicit solutions. ◇

Let us now return to the beginning of this section and try to generalize the concept of local stable and unstable manifolds of an equilibrium point $\bar{x}$. If we do not confine our attention to a local neighborhood U of $\bar{x}$, then we are led to the following definition of two invariant sets:

Definition 9.31. *The global stable manifold* $W^s(\bar{\mathbf{x}})$, *and the global unstable manifold* $W^u(\bar{\mathbf{x}})$ *of an equilibrium point* $\bar{\mathbf{x}}$ *are defined, respectively, to be the following sets:*

$$W^s(\bar{\mathbf{x}}) \equiv \{\mathbf{x}^0 \in \mathbb{R}^2 : \varphi(t, \mathbf{x}^0) \to \bar{\mathbf{x}} \text{ as } t \to +\infty\},$$

$$W^u(\bar{\mathbf{x}}) \equiv \{\mathbf{x}^0 \in \mathbb{R}^2 : \varphi(t, \mathbf{x}^0) \to \bar{\mathbf{x}} \text{ as } t \to -\infty\}.$$

It is clear from this definition that in Example 9.28, we have, in fact, determined the global stable and unstable manifolds of $\bar{\mathbf{x}} = \mathbf{0}$, because in this case $W^s(\bar{\mathbf{x}}, U) = W^s(\bar{\mathbf{x}}) \cap U$ and $W^u(\bar{\mathbf{x}}, U) = W^u(\bar{\mathbf{x}}) \cap U$. Unfortunately, this is not true in general; global stable and unstable manifolds can be very complicated and come arbitrarily close to themselves. This excludes a global analog of Theorem 9.29, as seen in the example below.

Example 9.32. *Global stable and unstable manifolds:* Let us reexamine Example 7.27 at $\lambda = 0$ in light of these remarks. The origin is an equilibrium point of saddle type; thus, by Theorem 9.29, the local stable and unstable manifolds are graphs over the local stable and unstable manifolds, respectively, of the linearized equations. The global stable and global unstable manifolds of the origin, on the other hand, are not graphs because the entire homoclinic orbit is part of both of these manifolds; see Figure 9.12. $\Diamond$

Overlapping of global stable and unstable manifolds usually leads to nonlocal bifurcations. In higher dimensions, this is a cause of complicated dynamical behavior. Despite the difficulties associated with these manifolds, unraveling their geometry is essential in the study of complicated dynamical systems, as we shall see in later chapters.

We conclude this section with several remarks on the definition and computational aspects of global stable and unstable manifolds. It is rather difficult to determine these global manifolds using Definition 9.31 since it appears that one must search through the entire plane for appropriate initial data. An alternative, but equivalent, way is to obtain the global manifolds from the corresponding local ones; to obtain $W^s(\bar{\mathbf{x}})$, let the points in $W^s(\bar{\mathbf{x}}, U)$ flow backward in time, and to obtain $W^u(\bar{\mathbf{x}})$, let the points in $W^u(\bar{\mathbf{x}}, U)$ flow forward in time:

$$W^s(\bar{\mathbf{x}}) \equiv \bigcup_{t \leq 0} \varphi(t, W^s(\bar{\mathbf{x}}, U)),$$

$$W^u(\bar{\mathbf{x}}) \equiv \bigcup_{t \geq 0} \varphi(t, W^u(\bar{\mathbf{x}}, U)).$$

For computational purposes, it is advantageous to observe that, from the uniqueness of solutions, it suffices to follow only the points on the intersection of ∂U, the boundary of U, and the local manifolds:

$$W^s(\bar{\mathbf{x}}) = \bigcup_{t \leq 0} \varphi(t, \partial U \cap W^s(\bar{\mathbf{x}}, U)) \cup W^s(\bar{\mathbf{x}}, U),$$

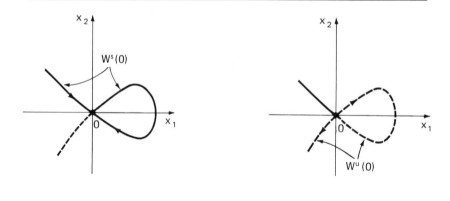

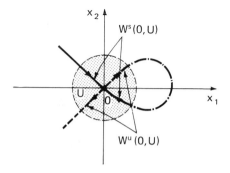

Figure 9.12. *Homoclinic orbit is part of both the global stable and global unstable manifolds of the equilibrium point.*

$$W^u(\bar{\mathbf{x}}) = \bigcup_{t \geq 0} \varphi(t, \, \partial U \cap W^u(\bar{\mathbf{x}}, U)) \cup W^u(\bar{\mathbf{x}}, U).$$

This method, coupled with the approximation procedure for local stable and unstable manifolds we have presented earlier, provides a reasonably practical way to compute the global stable and unstable manifolds.

Exercises ━━━━━━━━━━━━━━━━━━━━━━━━━━━━━ ♣ ♡ ♠ ◇

9.28. Consider the planar systems
 (a) $\dot{x}_1 = -x_1, \quad \dot{x}_2 = x_2 + x_1^3$;
 (b) $\dot{x}_1 = x_2, \quad \dot{x}_2 = x_1 + x_1^3$.
 Show that the origin is a saddle point. Determine the stable and unstable manifolds of the origin for the linearized as well as the nonlinear systems. Sketch and compare the phase portraits of the linearized equations about the origin and the nonlinear equations.

9.29. Find approximations for the local stable and unstable manifolds of all saddle points of the following differential equations:

(a) $\ddot{\theta} + 2\lambda\dot{\theta} + \sin\theta = 0$, where $\lambda > 0$;

(b) $\dot{x}_1 = 1 - x_1 x_2$, $\dot{x}_2 = x_1 - x_2^3$.

9.30. *Travelling Waves of Fisher's Equation:* Let $u(x, t)$ be a real-valued function of the "spacial" scalar variable $x \in \mathbb{R}$ and time, and $r > 0$ be a constant. The partial differential equation

$$\frac{\partial u}{\partial t} = \frac{\partial^2 u}{\partial x^2} + ru(1 - u)$$

has served as a model for interaction between local population growth and global dispersion. A solution $u(x, t)$ of this partial differential equation is said to be a *travelling wave* solution with wave speed $c > 0$ if there is a function $v : \mathbb{R} \to \mathbb{R}$ such that $u(x, t) = v(x - ct)$ for all $x \in \mathbb{R}$ and $t \in \mathbb{R}$. A travelling wave solution must satisfy the second-order ordinary differential equation

$$v'' + cv' + rv(1 - v) = 0,$$

where $v' = dv/ds$ and $s = x - ct$.

Show that, for every $c \geq 2\sqrt{r}$, there is a travelling wave solution satisfying $v(s) \to 1$ as $s \to -\infty$ and $v(s) \to 0$ as $s \to +\infty$, moreover, $v'(s) < 0$.
Suggestion: Discuss the stability properties of the equilibrium points and note special properties of the unstable manifold of the equilibrium point at $(1, 0)$. Then show that there is a triangular region in the (v, v')-plane bounded by the lines $v' = 0$, $v = 1$, and $v' = -\mu v$ (for an appropriate μ) which is positively invariant. For further information, see Kolmogorov, Petrovskii, and Piskunov [1937], and Jones and Sleeman [1983].

9.31. Show that the partial differential equation

$$\frac{\partial u}{\partial t} = \frac{\partial^2 u}{\partial x^2} + u(1 - u)(u - a),$$

where $0 < a < 1/2$, has a travelling wave solution which approaches zero as $t \to -\infty$ and 1 as $t \to +\infty$. Also, show that it is the unique travelling wave solution with these properties.
Suggestion: The ordinary differential equation for the travelling wave will have the equilibrium points $(0, 0)$ and $(1, 0)$ as saddle points. You will need to use the fact that stable and unstable manifolds depend continuously on parameters and also observe how level sets of the conservative system $v'' + u(1 - u)(u - a) = 0$ are crossed.

9.32. *Wazewski's Principle and existence of stable manifold:* In this exercise, we outline a proof the existence of the stable manifold for a hyperbolic equilibrium point with one positive and one negative eigenvalue. The geometric idea behind this proof is a special case of Wazewski's Principle. Consider the system

$$\dot{\mathbf{x}} = \begin{pmatrix} -\lambda_1 & 0 \\ 0 & \lambda_2 \end{pmatrix} \mathbf{x} + \mathbf{g}(\mathbf{x}),$$

where λ_1, λ_2 are positive numbers, and the function **g** satisfies $\mathbf{g}(\mathbf{0}) = \mathbf{0}$ and $D\mathbf{g}(\mathbf{0}) = \mathbf{0}$. Let V be the quadratic function

$$V(\mathbf{x}) = \tfrac{1}{2}\left(-x_1^2/\lambda_1 + x_2^2/\lambda_2\right).$$

1. There are constants $\varepsilon > 0$, $\mu > 0$, and $k > 0$ such that $1 - \mu k > 0$ and, for $\|\mathbf{x}\| < \varepsilon$, we have $\dot{V}(\mathbf{x}) \geq (1 - \mu k)\|\mathbf{x}\|^2$.
2. Plot the level curves of V, and choose a square with its center at the origin so that the square lies inside the disk $\|\mathbf{x}\| < \varepsilon$, as shown in Figure 9.13. Observe that solutions enter the square on the sides e and e', and leave on the other two sides f and f'.

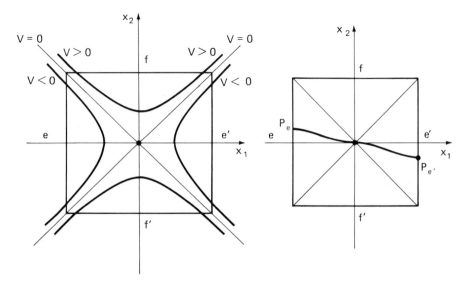

Figure 9.13. *Proving the existence of the stable manifold of a hyperbolic saddle using Wazewski's Principle. The diagonal lines and hyperbolas are the level curves of the quadratic function V.*

3. There is a point p_e on the side e such that the solution $\varphi(t, p_e)$ through p_e remains in the square for all $t \geq 0$. There is also a point $p_{e'}$ on e' with similar properties.
   Hint: Consider two pieces of the side e such that solutions through one piece leave the square via the side f, and the solutions through the other piece leave the square via the side f'. Use continuity with respect to initial data.
4. Show that $\varphi(t, p_e) \to 0$ as $t \to +\infty$.
   Hint: Use the fact $V(\varphi(t, p_e)) \to 0$ as $t \to +\infty$.
5. The point p_e is unique on e. There is also a unique point $p_{e'}$ on e' with similar properties.

Hint: If q_e is any other such point distinct from p_e, then show that $V(p_e - q_e) > 0$ and $\dot{V}(\varphi(t, p_e) - \varphi(t, p_e)) > 0$ for all t. Now, since $\varphi(t, p_e) \to 0$ as $t \to +\infty$, conclude that $\varphi(t, q_e)$ eventually leaves the square.

6. Show that the curve $W = \gamma^+(p_e) \cup \{\,0\,\} \cup \gamma^+(p_{e'})$ is a graph over the x_1-axis.
   Hint: Take smaller squares.

7. Finally, establish that W above is C^1. Therefore, W is the local stable manifold of the origin, that is, $W = W^s_{loc}(0)$.
   Hint: Consider the function $V_\eta(\mathbf{x}) = \frac{1}{2}\left(-x_1^2/\lambda_1 + \eta x_2^2/\lambda_2\right)$ depending on $\eta > 0$, and compute $\dot{V}_\eta$ and let $\eta \to 0$.

9.6. Flow Equivalence Near Hyperbolic Equilibria

We saw previously in this chapter that, in the absence of eigenvalues with zero real part, linearization captures many of the local qualitative features, such as stability type and local stable and unstable manifolds of nonlinear systems near equilibria. Therefore, it is natural to ponder if linearization determines the full orbit structure locally. In this brief section, we state a remarkable theorem which, under certain conditions, provides a positive answer to this difficult question. Let us begin with a precise definition of equivalence of vector fields.

Definition 9.33. *Two planar differential equations* $\dot{\mathbf{x}} = \mathbf{f}(\mathbf{x})$ *and* $\dot{\mathbf{x}} = \mathbf{g}(\mathbf{x})$ *defined on open subsets* U *and* V *of* $\mathbb{R}^2$, *respectively, are said to be* topologically equivalent *if there is a homeomorphism* $\mathbf{h} : U \to V$ *such that* $\mathbf{h}$ *maps the orbits of the vector field* $\mathbf{f}$ *onto the orbits of* $\mathbf{g}$ *and preserves the sense of direction of time.*

In the next definition, we single out certain equilibria about which we can expect, in light of Example 9.9, for instance, to capture the local dynamics of nonlinear systems from linearization.

Definition 9.34. *An equilibrium point* $\bar{\mathbf{x}}$ *of* $\dot{\mathbf{x}} = \mathbf{f}(\mathbf{x})$ *is said to be hyperbolic if all the eigenvalues of the Jacobian matrix* $D\mathbf{f}(\bar{\mathbf{x}})$ *have nonzero real parts.*

Near hyperbolic equilibria linearization tells it all, as the following theorem asserts:

Theorem 9.35. (Grobman–Hartman) *If* $\bar{\mathbf{x}}$ *is a hyperbolic equilibrium point of* $\dot{\mathbf{x}} = \mathbf{f}(\mathbf{x})$, *then there is a neighborhood of* $\bar{\mathbf{x}}$ *in which* $\mathbf{f}$ *is topologically equivalent to the linear vector field* $\dot{\mathbf{x}} = D\mathbf{f}(\bar{\mathbf{x}})\mathbf{x}$. $\Diamond$

Because of the hyperbolicity of the equilibrium point $\bar{\mathbf{x}}$, the homeomorphism $\mathbf{h}$ above can be chosen so as to preserve also the time parametrization of the orbits of the vector fields involved. To paraphrase the theorem of Grobman–Hartman, let $\varphi(t, \mathbf{x}^0)$ be the flow of $\dot{\mathbf{x}} = \mathbf{f}(\mathbf{x})$, and $\psi(t, \mathbf{x}^0)$ be the flow of $\dot{\mathbf{x}} = D\mathbf{f}(\bar{\mathbf{x}})\mathbf{x}$. Then the homeomorphism $\mathbf{h} : U \to \mathbb{R}^2$ can be chosen such that

$$\mathbf{h}\left(\varphi(t, \mathbf{x}^0)\right) = \psi\left(t, \mathbf{h}(\mathbf{x}^0)\right)$$

for all t as long as $\varphi(t, \mathbf{x}^0)$ remains in U.

We have shown in Section 8.3 that any two linear systems with hyperbolic equilibrium points are topologically equivalent if they have an equal number of eigenvalues with positive real parts and an equal number with negative real parts. This result, in conjunction with the theorem of Grobman–Hartman above, provides us a way to determine the topological, or flow, equivalence of two nonlinear vector fields in a sufficiently small neighborhood of a hyperbolic equilibrium point.

Another implication of the theorem of Grobman and Hartman is that the stability type of a hyperbolic equilibrium point is preserved under arbitrary but small nonlinear perturbations. In the next two chapters we will investigate the behavior of nonhyperbolic equilibria under small perturbations and the possible bifurcations.

Exercises ———————————————————————————— ♣ ♡ ♠ ♢

9.33. Find two specific nonlinear planar differential equations with the following properties: each has a unique equilibrium point, the linearized vector fields at these equilibria are topologically equivalent, but the flows of the two nonlinear vector fields on the whole plane are not topologically equivalent.

9.34. Try to prove the theorem of Grobman–Hartman in the special case where $\mathbf{f}(0) = \mathbf{0}$ and the eigenvalues of $D\mathbf{f}(0)$ have negative real parts.
Hint: Borrow ideas from Example 9.16 and the proof of Theorem 8.15.

9.7. Saddle Connections

In the preceding sections we have established that near a hyperbolic saddle point all is well—linearization suffices. From a somewhat global viewpoint, however, there can be complications as we saw in Example 9.32: the unstable manifold wanders out and returns as the stable manifold of the same saddle point. It is also possible that the unstable manifold of one saddle point could become the stable manifold of another saddle point; thus, connecting two saddle points. In this section, we give an example of this important global dynamical phenomenon.

We begin by defining the characteristic feature of a planar flow that we will demonstrate.

Definition 9.36. *A planar differential equation* $\dot{\mathbf{x}} = \mathbf{f}(\mathbf{x})$ *is said to have a saddle connection between two saddle points* $\bar{\mathbf{x}}^1$ *and* $\bar{\mathbf{x}}^2$ *if the intersection of* $W^u(\bar{\mathbf{x}}^1)$ *and* $W^s(\bar{\mathbf{x}}^2)$ *is not empty; equivalently, if there is a point* $\mathbf{p}$ *such that* $\alpha(\mathbf{p}) = \bar{\mathbf{x}}^1$ *and* $\omega(\mathbf{p}) = \bar{\mathbf{x}}^2$.

In the special case when the two saddle points coincide, a saddle connecting orbit is a homoclinic orbit, as in Example 9.32. We now give an example in which a saddle connection occurs between two distinct saddle points—a heteroclinic orbit—and the connection is broken when the vector field is perturbed.

Example 9.37. *Breaking a saddle connection:* Consider the system

$$\begin{aligned} \dot{x}_1 &= \lambda + 2x_1 x_2 \\ \dot{x}_2 &= 1 + x_1^2 - x_2^2 \end{aligned} \tag{9.28}$$

depending on a parameter λ.

Typical phase portraits of this system for three different values of the parameter are depicted in Figure 9.14. For $\lambda = 0$, there are two hyperbolic equilibrium points at $(0, 1)$ and $(0, -1)$, both of which are saddle points. Also, the x_2-axis is invariant, that is, any orbit starting on the axis stays on the axis. In particular, the orbit of any initial point lying on the x_2-axis and in between the two equilibria has its α-limit set as the saddle point $(0, -1)$ and its ω-limit set as the other saddle point $(0, 1)$. Thus, the system (9.28) for $\lambda = 0$ has a heteroclinic saddle connection.

For $|\lambda| \neq 0$, and small, there are still two saddle points, lying on opposite sides of the x_2-axis, but the saddle connection is no longer present. Consequently, the bifurcation that occurs as λ moves away from zero is called breaking a saddle connection. ◇

We will return to saddle connections in Chapter 13, and further explore how they effect the global behavior of planar flows at large.

Exercises ———————————————————————— ♣♡♠◇

9.35. *Breaking saddle connections:* For the following systems show that, for $k = 0$, there are saddle connections, and for $k \neq 0$, there are no saddle connections:
(a) $\dot{x}_1 = x_2, \quad \dot{x}_2 = -\sin x_1 - kx_2$;
(b) $\dot{x}_1 = k\sin x_1 + \sin x_2, \quad \dot{x}_2 = -\sin x_1 + k\sin x_2$.

Bibliographical Notes ———————————————————— ◎◎

The thesis of Liapunov in 1892 was seminal in the development of stability theory. One of his motivations for using positive definite functions was to investigate the stability of nonhyperbolic equilibria. To implement his program, he developed transformation theory, which was the basis for our

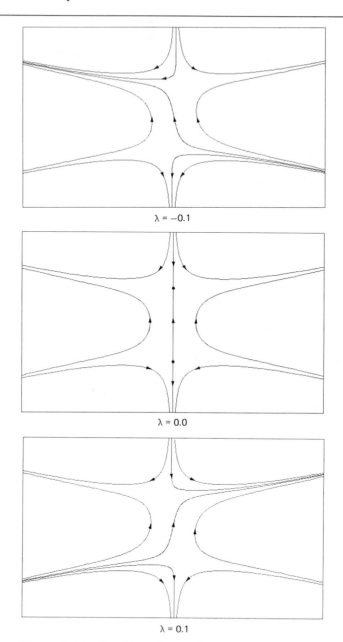

$\lambda = -0.1$

$\lambda = 0.0$

$\lambda = 0.1$

Figure 9.14. *Breaking a heteroclinic saddle connection.*

presentation in Section 5.3, and the converse theorems of stability for special types of vector fields. Lyapunov [1947], with a spelling variation, is the original source; Četaev's instability theorem is from Četaev [1934]. La Salle and Lefschetz [1961] contains an elementary exposition; Yoshizawa [1966] has general converse theorems and historical references.

The Invariance Principle in Section 9.4 is due to La Salle [1960], although similar ideas were developed independently by others; see, for example, Krasovskii [1963]. This principle is used widely in applications and has special prominence in infinite dimensional dynamical systems as explained in Hale [1977], Henry [1981], and Sell [1971].

A nice proof of Theorem 9.29 on the existence of stable and unstable manifolds is in Irwin [1980]; see also Palis and de Melo [1982]. In the exercises, we outlined a procedure for obtaining the existence of the stable manifold of a saddle point by exploiting the fact that, if orbits exit on two opposite sides of a box about the saddle point and enter on the other two sides, then some solution remains in the box. This is a special case of a principle formalized by Wazewski in 1947; precise formulations are in Cesari [1963] and Hartman [1964]. An extension of this idea is the concept of an "isolating block" of an invariant set, the existence of which is equivalent to the property that an invariant set is isolated. Further generalizations yield an index theory which is important in detecting when there are orbits connecting two different invariant sets; see Conley [1978] and Rybakowski [1987].

The linearization theorem near hyperbolic equilibria in Section 9.6 is due to Grobman [1959] and Hartman [1964]. Another proof, which generalizes to infinite dimensions, can be found in Palis and de Melo [1982].

10

In the Presence of a Zero Eigenvalue

 In this chapter, we undertake the study of stability and bifurcations of nonhyperbolic equilibria of a planar system in the case where the linearized vector field has one zero and one negative eigenvalue. Our investigation culminates in the observation that the local dynamics and bifurcations of such a planar system are determined from those of an appropriate scalar differential equation. Analysis of the resulting scalar equation can, of course, be accomplished using the results in Chapters 1 and 2. To provide a geometric view of this reduction from two dimensions to one, we include an exposition of a class of important invariant curves—center manifolds—which capture the asymptotic features of these planar systems.

10.1. Stability

In this section, we show how to determine the stability of a nonhyperbolic equilibrium point of a planar vector field with one zero and one negative eigenvalue. Since linear approximation is of no help in this pursuit, we will have to examine how a particular nonlinear term of a vector field affects the flow near such a nonhyperbolic equilibrium point. Let us begin with a familiar example which encapsulates the essence of the general situation.

Example 10.1. Let $k \geq 1$ be an integer, $a \neq 0$ be a real number and consider the product system

$$\dot{x}_1 = ax_1^k$$
$$\dot{x}_2 = -x_2.$$

Regardless of the values of a and $k > 1$, the eigenvalues of the linearized equation about the equilibrium point at the origin are always 0 and -1. Consequently, to determine the stability type of the origin we need to investigate the effect of the nonlinear term of the vector field. Since $x_2(t) \to 0$ as $t \to +\infty$, the stability properties of the equilibrium point $\bar{x} = 0$ are determined by the first scalar equation $\dot{x}_1 = ax_1^k$. It is now evident that the origin is asymptotically stable if $a < 0$ and k is odd, and unstable otherwise; see Figure 10.1. $\diamondsuit$

We now turn to the general setting. Let $\mathbf{f}$ be a given C^k function, with $k \geq 1$,

$$\mathbf{f} : \mathbb{R}^2 \to \mathbb{R}^2; \quad \mathbf{x} \mapsto \mathbf{f}(\mathbf{x}),$$

satisfying

$$\mathbf{f}(0) = 0, \qquad D\mathbf{f}(0) = 0 \tag{10.1}$$

and consider the planar system of differential equations

$$\dot{x}_1 = f_1(x_1, x_2)$$
$$\dot{x}_2 = -x_2 + f_2(x_1, x_2). \tag{10.2}$$

To bring the linear part of this system to the forefront, let us write it, for a moment, in vector notation:

$$\dot{\mathbf{x}} = \begin{pmatrix} 0 & 0 \\ 0 & -1 \end{pmatrix} \mathbf{x} + \mathbf{f}(\mathbf{x}).$$

Notice that the linear part of the vector field about the equilibrium point at the origin is in Jordan Normal Form with eigenvalues 0 and -1. In applications, the linearization of a vector field with one zero and one negative eigenvalues may not always come in normal form [Eq. (10.2)]; however, such a vector field can always be put into this form with a linear change of coordinates and a rescaling of the independent variable t.

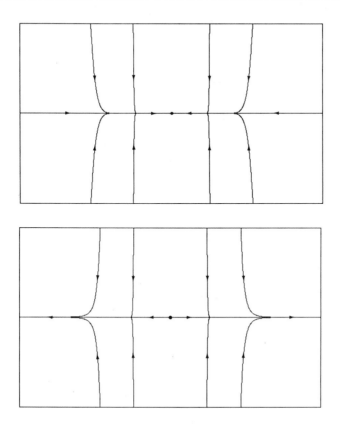

Figure 10.1. *Phase portraits of* $\dot{x}_1 = ax_1^3$, $\dot{x}_2 = -x_2$ *near the origin for* $a = -0.1$ *and* $a = 0.1$.

We begin our investigation of the stability of the origin of Eq. (10.2) with an explanation of the key idea which dates back to Liapunov. If $\mathbf{x}(t) = (x_1(t), x_2(t))$ is a solution of Eq. (10.2) with initial values close to zero, then the variation of $x_1(t)$ occurs more slowly than the variation of $x_2(t)$. Thus, it is reasonable to consider x_1 as constant in the second equation, and consider x_2 in the first equation as the zero $\psi(x_1)$ of $-x_2 + f_2(x_1, x_2)$. As a result, Eq. (10.2) becomes almost like a product system and its dynamics in a neighborhood of the origin should then be determined by the scalar differential equation $\dot{x}_1 = f_1(x_1, \psi(x_1))$.

Let us now make these ideas more precise. Consider the equation

$$\mathcal{F}(x_1, x_2) \equiv -x_2 + f_2(x_1, x_2) = 0.$$

Since $\mathcal{F}(0, 0) = 0$ and $\partial \mathcal{F}(0, 0)/\partial x_2 = -1$, the Implicit Function Theorem implies that there is a constant $\delta > 0$ and a unique C^1 function $\psi : \{ x_1 : |x_1| < \delta \} \to \{ x_2 : |x_2| < \delta \}$ such that

$$-\psi(x_1) + f_2(x_1, \psi(x_1)) = 0, \tag{10.3}$$

$$\psi(0) = 0, \quad \psi'(0) = 0. \tag{10.4}$$

The latter relation in Eq. (10.4) follows from differentiating Eq. (10.3) with respect to x_1 and setting $x_1 = 0$.

The following theorem, which generalizes the observations in Example 10.1, is not difficult to establish, except perhaps the arithmetic of "big O" for which you may wish to consult the Appendix:

Theorem 10.2. *Suppose that* $\mathbf{f} = (f_1, f_2)$ *is a* C^{k+1} *function with*

$$f_1(x_1, \psi(x_1)) = ax_1^k + O(|x_1^{k+1}|) \quad \text{as } x_1 \to 0, \tag{10.5}$$

where $a \neq 0$ *is a real number,* k *is a positive integer, and* $\psi(x_1)$ *is as given in Eq. (10.3). Then the equilibrium point at the origin of the planar system Eq. (10.2) is asymptotically stable if* $a < 0$ *and* k *is an odd integer; otherwise, it is unstable.*

Proof. It is convenient to introduce the new variables $\mathbf{y} = (y_1, y_2)$ defined by

$$x_1 = y_1, \quad x_2 = y_2 + \psi(y_1).$$

In these variables the original system (10.2) becomes

$$\begin{aligned} \dot{y}_1 &= g_1(y_1, y_2) \\ \dot{y}_2 &= -y_2 + g_2(y_1, y_2), \end{aligned} \tag{10.6}$$

where

$$\begin{aligned} g_1(y_1, y_2) &= f_1(y_1, \psi(y_1) + y_2) \\ g_2(y_1, y_2) &= f_2(y_1, \psi(y_1) + y_2) - f_2(y_1, \psi(y_1)) \\ &\quad - \psi'(y_1) f_1(y_1, \psi(y_1) + y_2). \end{aligned}$$

The stability properties of the equilibrium point $\bar{\mathbf{y}} = \mathbf{0}$ of Eq. (10.6) are the same as those of the equilibrium point $\bar{\mathbf{x}} = \mathbf{0}$ of Eq. (10.2).

Since the conclusions of the theorem concern a sufficiently small neighborhood of the origin, we proceed, as you might suspect, to determine the first several terms of the Taylor series of these functions about the origin. Using Eqs. (10.5) and (10.4), we obtain, as $\|\mathbf{y}\| \to 0$,

$$\begin{aligned} g_1(y_1, y_2) &= ay_1^k \left[1 + O\left(\|\mathbf{y}\| \right) \right] + y_2 O\left(\|\mathbf{y}\| \right) \\ g_2(y_1, y_2) &= O\left(|y_1^{k+1}| \right) + y_2 O\left(\|\mathbf{y}\| \right). \end{aligned} \tag{10.7}$$

Let us now consider the function

$$V(y_1, y_2) = -\frac{1}{a(k+1)}\, y_1^{k+1} + \tfrac{1}{2}y_2^2$$

and compute its derivative along the solutions of Eq. (10.6). Utilizing Eq. (10.7), we observe that there is a $\delta > 0$ such that for $\|\mathbf{y}\| < \delta$:

$$\dot{V}(y_1, y_2) = -y_1^{2k}\left[1 + O\left(\|\mathbf{y}\|\right)\right] + y_1^k y_2 O\left(\|\mathbf{y}\|\right) - y_2^2\left[1 + O\left(\|\mathbf{y}\|\right)\right]$$
$$\leq -\tfrac{1}{2}\left[y_1^{2k} - y_1^k y_2 + y_2^2\right].$$

The function $-\dot{V}(y_1, y_2)$ can easily be seen to be positive definite by treating it as a quadratic function in y_1^k and y_2.

Now, the desired conclusions follow rather easily from our earlier results in Chapter 9. If $a < 0$ and k is odd, then $V(y_1, y_2)$ is positive definite; thus, from Theorem 9.14 of Liapunov, $\bar{\mathbf{y}} = \mathbf{0}$ is asymptotically stable. If $a > 0$ and k is odd, then we apply Theorem 9.17 of Četaev to the function $-V$ to conclude the instability of $\bar{\mathbf{y}} = \mathbf{0}$. The remaining case $a \neq 0$ and k even follows from the same theorem. $\Diamond$

Using the Taylor series of the functions involved, it is a routine, but perhaps tedious, task to apply Theorem 10.2 to specific equations.

Example 10.3. Consider the particular instance of Eq. (10.2) given by

$$\begin{aligned}
\dot{x}_1 &= ax_1^3 + x_1 x_2 \\
\dot{x}_2 &= -x_2 + x_2^2 + x_1 x_2 - x_1^3,
\end{aligned} \tag{10.8}$$

where a is a given real constant. The function $\psi(x_1)$ in Eq. (10.3) is a solution of

$$-\psi(x_1) + [\psi(x_1)]^2 + x_1\psi(x_1) - x_1^3 = 0.$$

Substituting a Taylor series for $\psi(x_1)$ and equating the coefficients of like powers of x_1, we obtain $\psi(x_1) = -x_1^3 + O(|x_1^4|)$ and $f_1(x_1, \psi(x_1)) = ax_1^3 - x_1^4 + O(|x_1^5|)$. Now, Theorem 10.2 implies that the equilibrium $\bar{\mathbf{x}} = \mathbf{0}$ of Eq. (10.8) is asymptotically stable if $a < 0$ and unstable if $a \geq 0$.

We have plotted in Figure 10.2 several numerically computed phase portraits of Eq. (10.8). There is an important practical consideration in dealing with a nonhyperbolic equilibrium point which is visible in these apparently incomplete pictures. Unlike the hyperbolic case where approach to an equilibrium point is exponentially fast in t, it can take a very long time to approach a nonhyperbolic asymptotically stable equilibrium point. As a result, it sometimes is disheartening to watch orbits come to a crawl on the computer monitor while investigating bifurcations where many exciting phenomena revolve around nonhyperbolicity. $\Diamond$

We conclude our discussion of stability of equilibria in the presence of a zero eigenvalue with a generalization of Theorem 10.2. A proof of this extension requires considerably more sophisticated ideas, which should explain its absence.

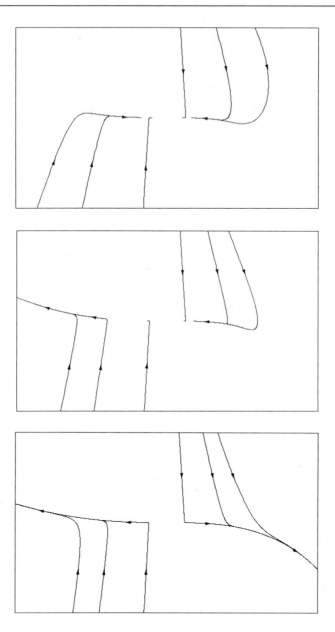

Figure 10.2. *Numerically computed partial phase portraits of Eq. (10.8) for $a = -1.0$, $a = 0.0$, and $a = 1.0$.*

Theorem 10.4. *Let $\psi(x_1)$ be defined as in Eq. (10.3). Then the equilibrium $\bar{\mathbf{x}} = \mathbf{0}$ of Eq. (10.2) is stable [respectively, asymptotically stable, unstable] if and only if the equilibrium $\bar{x}_1 = 0$ of the scalar differential equation*

$$\dot{x}_1 = f_1(x_1, \psi(x_1)) \tag{10.9}$$

is stable [respectively, asymptotically stable, unstable]. $\Diamond$

Exercises ———————————————————————— ♣ ♡ ♠ ◇

10.1. *From Liapunov's thesis:* Consider the following system of differential equations from Liapunov's thesis:

$$\dot{x}_1 = ax_1^2 + bx_1x_2 + cx_2^2$$
$$\dot{x}_2 = -x_2 + \ell x_1^2 + mx_1x_2 + nx_2^2,$$

where a, b, c, ℓ, m, and n are constants. Show that
 1. if $a \neq 0$, then the origin is unstable;
 2. if $a = 0$, $b\ell < 0$, then the origin is asymptotically stable;
 3. if $a = 0$, $b\ell > 0$, then the origin is unstable;
 4. if $a = b = 0$, $c\ell^2 \neq 0$, then the origin is unstable;
 5. if $a = b = c = 0$, then the origin is stable but not asymptotically;
 6. if $a = \ell = 0$, then the origin is stable but not asymptotically.

10.2. *Everything periodic:* Show that the origin of the second-order system $\ddot{y}+y^3 = 0$ is a stable equilibrium point. Use the symmetry of the equation to show that all orbits are indeed periodic.

10.3. Consider the planar system

$$\dot{x}_1 = f_1(x_1, x_2)$$
$$\dot{x}_2 = \mu x_2 + f_2(x_1, x_2),$$

where $\mathbf{f}(\mathbf{0}) = \mathbf{0}$, $D\mathbf{f}(\mathbf{0}) = \mathbf{0}$, and $\mu \neq 0$ is a given constant. Discuss the stability properties of the equilibrium at the origin.
Hint: Rescale the independent variable t, but be aware of the sign of μ when interpreting flows.

10.4. Consider the planar system $\dot{\mathbf{x}} = \mathbf{Ax} + \mathbf{f}(\mathbf{x})$, where $\mathbf{f}(\mathbf{0}) = \mathbf{0}$, $D\mathbf{f}(\mathbf{0}) = \mathbf{0}$, and the matrix $\mathbf{A}$ has one zero and one nonzero eigenvalue. Describe how you would reduce the discussion of the stability of the origin to the case considered in the previous problem.

10.5. As a specific instance of the problem above, determine the stability type of the origin of the system

$$\dot{x}_1 = -x_1 + x_2 + x_1^2 + x_1x_2 + ax_1^3$$
$$\dot{x}_2 = x_1 - x_2 + x_1^2 + bx_1x_2 + x_2^3.$$

10.2. Bifurcations

In this section, we extend the foregoing setting to systems of differential equations that depend on parameters. Our goal is, of course, to investigate the possible bifurcations of nonhyperbolic equilibrium points with one zero and one negative eigenvalue. Initially, we will be content to account for local variations in the number of equilibria and their stability types. Since we are working in the plane, other changes in phase portraits are potentially possible. We will address such issues in the next section.

Let $\mathbf{F}$ be a given C^k function, $k \geq 1$, depending on parameters λ,

$$\mathbf{F} : \mathbb{R}^m \times \mathbb{R}^2 \to \mathbb{R}^2; \quad (\lambda, \mathbf{x}) \mapsto \mathbf{F}(\lambda, \mathbf{x}),$$

satisfying

$$\mathbf{F}(\mathbf{0}, \mathbf{x}) = \mathbf{f}(\mathbf{x}) \quad \text{and} \quad \mathbf{f}(\mathbf{0}) = \mathbf{0}, \quad D\mathbf{f}(\mathbf{0}) = \mathbf{0}. \tag{10.10}$$

Below, we consider the system of differential equations

$$\begin{aligned} \dot{x}_1 &= F_1(\lambda, x_1, x_2) \\ \dot{x}_2 &= -x_2 + F_2(\lambda, x_1, x_2), \end{aligned} \tag{10.11}$$

where $\mathbf{F} = (F_1, F_2)$, and determine the nature of the bifurcations of equilibria near the origin for small values of the parameter λ.

We begin our analysis with a lemma for determining the equilibrium points of the system (10.11) in terms of the zeros of a scalar equation.

Lemma 10.5. *If $\mathbf{F}$ is a C^k function, $k \geq 1$, then there are constants $\lambda_0 > 0$, $\delta > 0$ and a C^k function*

$$\psi : \{\, \lambda \, : \, \|\lambda\| < \lambda_0 \,\} \times \{\, x_1 \, : \, |x_1| < \delta \,\} \to \{\, x_2 \, : \, |x_2| < \delta \,\}$$

satisfying

$$\psi(\mathbf{0}, 0) = 0, \quad \frac{\partial \psi}{\partial x_1}(\mathbf{0}, 0) = 0 \tag{10.12}$$

such that, for each $\bar{\lambda}$, with $\|\bar{\lambda}\| < \lambda_0$, a point $(\bar{x}_1, \bar{x}_2)$, with $|\bar{x}_1| < \delta$ and $|\bar{x}_2| < \delta$, is an equilibrium point of Eq. (10.11) if and only if $\bar{x}_2 = \psi(\bar{\lambda}, \bar{x}_1)$ and $F_1(\bar{\lambda}, \bar{x}_1, \psi(\bar{\lambda}, \bar{x}_1)) = 0$.

Proof. Apply the Implicit Function Theorem to find $x_2 = \psi(\lambda, x_1)$ as the solution of $-x_2 + F_2(\lambda, x_1, x_2) = 0$. $\Diamond$

Definition 10.6. *Let ψ be as in Lemma 10.5. Then the C^k function $G(\lambda, x_1)$ defined by*

$$G(\lambda, x_1) \equiv F_1(\lambda, x_1, \psi(\lambda, x_1)) \tag{10.13}$$

is called the *bifurcation function* and the equation

$$G(\lambda, x_1) = 0 \tag{10.14}$$

is referred to as the *bifurcation equation* of the system Eq. (10.11).

It is now evident that the equilibrium points of the scalar differential equation

$$\dot{x}_1 = G(\lambda, x_1) \tag{10.15}$$

are in one-to-one correspondence with the equilibrium points of the planar system (10.11). It turns out that the stability types of the corresponding equilibria of Eqs. (10.11) and (10.15) agree as well. Before we delve into the stability question, however, let us reexamine one of our stock examples.

Example 10.7. Consider the product system

$$\dot{x}_1 = \lambda + x_1^2$$
$$\dot{x}_2 = -x_2,$$

where λ is a scalar parameter. The equilibrium points of this system are given by $\mathbf{x} = (x_1, x_2)$ with $x_2 = \psi(\lambda, x_1) = 0$ and x_1 satisfying the bifurcation equation $G(\lambda, x_1) = \lambda + x_1^2 = 0$. There are no solutions of the bifurcation equation if $\lambda > 0$, and two solutions $\bar{x}_1 = \pm\sqrt{-\lambda}$ if $\lambda < 0$. The corresponding equilibria of the scalar differential equation $\dot{x}_1 = G(\lambda, x_1)$ are hyperbolic, one of which is asymptotically stable and the other of which is unstable. $\diamondsuit$

In the case where the equilibria of the scalar differential equation (10.15) are hyperbolic, it is not difficult to establish the stability types of the corresponding equilibria of the planar system (10.11).

Theorem 10.8. *Let $\psi(\lambda, x_1)$ be as in Lemma 10.5 and $\bar{\mathbf{x}} = (\bar{x}_1, \psi(\bar{\lambda}, \bar{x}_1))$ is an equilibrium point of Eq. (10.11) with $\|\bar{\lambda}\| < \lambda_0$, $|\bar{x}_1| < \delta$, and $|\psi(\bar{\lambda}, \bar{x}_1)| < \delta$. Then λ_0 and δ can be chosen small enough so that*
 (i) *$\bar{\mathbf{x}}$ is a hyperbolic stable node if $\partial G(\bar{\lambda}, \bar{x}_1)/\partial x_1 < 0$,*
 (ii) *$\bar{\mathbf{x}}$ is a saddle point if $\partial G(\bar{\lambda}, \bar{x}_1)/\partial x_1 > 0$.*

Proof. It is convenient to begin by making the transformation of variables

$$x_1 = y_1, \qquad x_2 = y_2 + \psi(\bar{\lambda}, x_1).$$

In the new coordinates Eq. (10.11) becomes

$$\dot{y}_1 = F_1(\bar{\lambda}, y_1, \psi(\bar{\lambda}, y_1) + y_2)$$
$$\dot{y}_2 = -y_2 + F_2(\bar{\lambda}, y_1, \psi(\bar{\lambda}, y_1) + y_2) - F_2(\bar{\lambda}, y_1, \psi(\bar{\lambda}, y_1))$$
$$- \frac{\partial \psi}{\partial y_1}(\bar{\lambda}, y_1) \, F_1(\bar{\lambda}, y_1, \psi(\bar{\lambda}, y_1) + y_2),$$

and $(\bar{y}_1, \bar{y}_2) = (\bar{x}_1, 0)$ is an equilibrium point. The Jacobian of this vector field evaluated at $(\bar{x}_1, 0)$ is given by the matrix (for the sake of brevity, variables are omitted)

$$
A(\bar{\lambda}, \bar{x}_1) = \begin{pmatrix} \dfrac{\partial G}{\partial y_1} & \dfrac{\partial F_1}{\partial y_2} \\ -\dfrac{\partial \psi}{\partial y_1}\dfrac{\partial G}{\partial y_1} & -1 + \dfrac{\partial F_2}{\partial y_2} - \dfrac{\partial \psi}{\partial y_1}\dfrac{\partial F_1}{\partial y_2} \end{pmatrix}.
$$

To determine the stability type of the equilibrium point $(\bar{x}_1, 0)$ it suffices to discern the signs of the real parts of the eigenvalues of $A(\bar{\lambda}, \bar{x}_1)$. As we have shown in the exercises, the eigenvalues of $A(\bar{\lambda}, \bar{x}_1)$ vary continuously in $\bar{\lambda}$ and $\bar{x}_1$. Since the eigenvalues of $A(0, 0)$ are 0 and -1, for sufficiently small $\bar{\lambda}$ and $\bar{x}_1$, one eigenvalue of $A(\bar{\lambda}, \bar{x}_1)$ is near 0 and the other near -1. We now need to determine the sign of the eigenvalue near 0.

The determinant of the Jacobian $A(\bar{\lambda}, \bar{x}_1)$ is

$$
\det A(\bar{\lambda}, \bar{x}_1) = \frac{\partial G}{\partial y_1}(\bar{\lambda}, \bar{x}_1)\left[-1 + \frac{\partial F_2}{\partial y_2}(\bar{\lambda}, \bar{x}_1, \psi(\bar{\lambda}, \bar{x}_1))\right].
$$

We can choose λ_0 and δ sufficiently small to assure, for example, that $|\partial F_2(\bar{\lambda}, \bar{x}_1, \psi(\bar{\lambda}, \bar{x}_1))/\partial y_2| < 1/2$ for $\|\bar{\lambda}\| < \lambda_0$ and $|\bar{x}_1| < \delta$.

Consequently, since $\det A$ is the product of the eigenvalues of A and one of the eigenvalues is negative, the eigenvalue of A near zero has the same sign as the sign of $\partial G(\bar{\lambda}, \bar{x}_1)/\partial y_1$. Now the desired conclusions follow from the theorems in Sections 9.1 and 9.2. $\Diamond$

As an application of this theorem, we can now state a result that we have alluded to on several previous occasions—saddle-node bifurcation:

Theorem 10.9. *Consider the system* (10.11) *in the case where λ is a scalar parameter. If*

$$
\frac{\partial F_1}{\partial \lambda}(0, 0, 0) \neq 0, \qquad \frac{\partial^2 F_1}{\partial x_1^2}(0, 0, 0) \neq 0,
$$

then there is a saddle-node bifurcation at $\lambda = 0$, that is, when $\lambda \frac{\partial F_1}{\partial \lambda}\frac{\partial^2 F_1}{\partial x_1^2} < 0$, there are two hyperbolic equilibria, one saddle and the other an asymptotically stable node, and no equilibrium when $\lambda \frac{\partial F_1}{\partial \lambda}\frac{\partial^2 F_1}{\partial x_1^2} > 0$.

Proof: Compute the first several terms of the Taylor series of the bifurcation function $G(\lambda, x_1)$, then use the results from Section 2.3. $\Diamond$

To make our exposition of the theory of bifurcation functions complete, we should state a generalization of Theorem 10.8 when the equilibria of Eq. (10.15) are not hyperbolic. It is considerably more difficult to prove this extension, but here is the statement of the general situation.

Theorem 10.10. *Let $\psi(\lambda, x_1)$ be defined as in Lemma 10.5 and, for a given small $\bar{\lambda}$, let $\bar{\mathbf{x}} = (\bar{x}_1, \psi(\bar{\lambda}, \bar{x}_1))$ be an equilibrium point of Eq. (10.11). Then the equilibrium point $\bar{\mathbf{x}}$ is stable [respectively, asymptotically stable, unstable] if and only if the equilibrium $\bar{x}_1$ of the scalar differential equation*

$$\dot{x}_1 = G(\bar{\lambda}, x_1)$$

is stable [respectively, asymptotically stable, unstable]. $\Diamond$

Let us now demonstrate how to use the method of bifurcation function to investigate the dynamics of a somewhat more substantial example of practical interest.

Example 10.11. *Damped pendulum with torque:* Consider the following second-order differential equation describing the motion of a planar pendulum in the presence of damping and constant torque M:

$$\ddot{\theta} + \dot{\theta} + \sin\theta = M. \tag{10.16}$$

If the rod of the pendulum is rigid, then one can interpret the application of the torque as pushing the pendulum with constant force that is perpendicular to the rod. We would like to analyze the possible bifurcations of the equilibrium positions of such a pendulum as a function of the value M of the torque. To utilize the general results above, we have to transform Eq. (10.16) into the normal form (10.11). For this purpose, we begin by converting Eq. (10.16) into the first-order planar system

$$\begin{aligned} \dot{y}_1 &= y_2 \\ \dot{y}_2 &= -y_2 - \sin y_1 + M, \end{aligned} \tag{10.17}$$

where $y_1 = \theta$ and $y_2 = \dot{\theta}$. Since $\sin y_1$ is periodic with period 2π, we confine our discussion to the values of y_1 in the interval $[-\pi, \pi]$.

The equilibrium points of Eq. (10.17) are given by $y_2 = 0$ and $\sin y_1 = M$. For $M > 1$, there are no equilibrium points; when $M = 1$, there is a single equilibrium point at $(\pi/2, 0)$; for $M < 1$, there are two equilibrium points. It appears that the equilibrium point at $(\pi/2, 0)$ undergoes a bifurcation at the parameter value $M = 1$. To analyze the behavior of the orbits of Eq. (10.17) near $M = 1$ and (x_1, x_2) near $(\pi/2, 0)$, let us use the translation of variables

$$M = 1 + \lambda, \quad y_1 = \frac{\pi}{2} + z_1, \quad y_2 = z_2,$$

and transform Eq. (10.17) to the system

$$\begin{aligned} \dot{z}_1 &= z_2 \\ \dot{z}_2 &= -z_2 - \cos z_1 + 1 + \lambda \end{aligned} \tag{10.18}$$

so that for $\lambda = 0$ the origin is an equilibrium point. The coefficient matrix of the linearization of Eq. (10.18) at the origin is

$$\begin{pmatrix} 0 & 1 \\ 0 & -1 \end{pmatrix}.$$

To put this matrix into Jordan Normal Form, observe that its eigenvalues are 0 and -1 with the corresponding eigenvectors $(1, 0)$ and $(-1, 1)$. Therefore, if we make the transformation of variables

$$\mathbf{z} = P\mathbf{x}, \qquad P = \begin{pmatrix} 1 & -1 \\ 0 & 1 \end{pmatrix},$$

then Eq. (10.18) becomes

$$\dot{\mathbf{x}} = \begin{pmatrix} 0 & 0 \\ 0 & -1 \end{pmatrix} \mathbf{x} + P^{-1} \begin{pmatrix} 0 \\ -\cos(x_1 - x_2) + 1 + \lambda \end{pmatrix},$$

or equivalently

$$\begin{aligned} \dot{x}_1 &= -\cos(x_1 - x_2) + 1 + \lambda \\ \dot{x}_2 &= -x_2 - \cos(x_1 - x_2) + 1 + \lambda. \end{aligned} \tag{10.19}$$

The system (10.19) is now in the normal form (10.11) and the hypotheses of Theorem 10.9 are satisfied. Therefore, at $\lambda = 0$ the system (10.19) undergoes a saddle-node bifurcation; when $\lambda < 0$, there are two equilibria, one an asymptotically stable node and the other a saddle, and there are no equilibria when $\lambda > 0$.

For future reference, as well as for practice, let us also compute the bifurcation function for Eq. (10.19). From the second equation, we first need to determine the function $\psi(\lambda, x_1)$ satisfying

$$-\psi(\lambda, x_1) - \cos(x_1 - \psi(\lambda, x_1)) + 1 + \lambda = 0. \tag{10.20}$$

Since we are interested in local bifurcations, it suffices to determine a few terms of the Taylor series of ψ about $(\lambda, x_1) = (0, 0)$. Therefore, consider the power series of ψ given by

$$\psi(\lambda, x_1) = c_{10}\lambda + c_{20}\lambda^2 + c_{11}\lambda x_1 + c_{02}x_1^2 + O((|\lambda| + |x_1|)^3),$$

where the coefficients c_{ij} are to be determined. If we substitute this expansion into Eq. (10.20) and use the Taylor series of the cosine, then a somewhat laborious calculation yields

$$\psi(\lambda, x_1) = \lambda + \tfrac{1}{2}\lambda^2 - \lambda x_1 + \tfrac{1}{2}x_1^2 + O((|\lambda| + |x_1|)^3). \tag{10.21}$$

From the form of the vector field (10.19) and Eq. (10.20) it is clear that the bifurcation function is $G(\lambda, x_1) = \psi(\lambda, x_1)$. Therefore the scalar differential equation (10.15) becomes

$$\dot{x}_1 = \lambda + \tfrac{1}{2}\lambda^2 - \lambda x_1 + \tfrac{1}{2}x_1^2 + O((|\lambda| + |x_1|)^3). \qquad (10.22)$$

By appealing to Section 2.3, you should analyze the bifurcations of equilibria of Eq. (10.22) near the origin, and confirm the presence of a saddle-node bifurcation at $\lambda = 0$. $\Diamond$

Using the bifurcation function we have succeeded in accounting for the bifurcations of equilibria and their stability of the damped pendulum with torque. The complete flow of Eq. (10.19) near the origin, however, still remains to be determined. With this purpose in mind, we now turn to the theory of center manifolds.

Exercises

10.6. Draw the bifurcation diagrams for the equilibrium solutions of the following systems:
   (a) $\dot{x}_1 = -x_2$, $\quad \dot{x}_2 = \lambda x_1 - x_2$;
   (b) $\dot{x}_1 = x_2$, $\quad \dot{x}_2 = -x_2 + x_1^2 - \lambda$;
   (c) $\dot{x}_1 = 3\lambda x_1 - 3\lambda x_2 - x_1^2 - x_2^2$, $\quad \dot{x}_2 = \lambda x_1 - x_2$.

10.7. Analyze the equilibrium points and their stability types of the system

$$\dot{x}_1 = \lambda x_1 - x_1^2 + 2x_1 x_2$$
$$\dot{x}_2 = (\lambda - 1)x_2 + x_2^2 + x_1 x_2,$$

near the origin for small values of the scalar parameter λ.

10.8. Consider the following system depending on two parameters λ and μ:

$$\dot{x}_1 = \mu - x_1^3 - x_1 x_2^2 + x_1 x_2$$
$$\dot{x}_2 = -x_2 + x_2 x_1^2 - x_1^2 + \lambda.$$

Find the approximate bifurcation curves near the origin in the (λ, μ)-plane for the equilibria near $(0, 0)$. Also sketch the representative phase portraits for each region of the parameter space.

10.9. *Odd symmetry:* Suppose λ is a scalar parameter and the functions F_1 and F_2 in Eq. (10.11) satisfy, for $i = 1, 2$,

$$F_i(\lambda, -x_1, -x_2) = -F_i(\lambda, x_1, x_2).$$

   1. Show that $\mathbf{x} = \mathbf{0}$ is always a solution.
   2. Show that the bifurcation function $G(\lambda, x_1)$ is odd in x_1.
   3. If $\partial^2 G(0, 0)/\partial\lambda\partial x_1 \neq 0$ and $\partial^3 G(0, 0)/\partial x_1^3 \neq 0$, show that there is a pitchfork bifurcation at $\lambda = 0$.

4. Suppose that F_1 and F_2 are analytic functions in all variables. Then the bifurcation function $G(\lambda, x_1)$ is also analytic. Using this fact, show that if $\partial^2 G(0, 0)/\partial\lambda\partial x_1 \neq 0$, then either there is a pitchfork bifurcation at $\lambda = 0$, or $G(\lambda, x_1) \equiv 0$ for $\lambda = 0$.

10.10. *Even symmetry:* Suppose that λ is a scalar parameter and the functions F_1 and F_2 in Eq. (10.11) satisfy, for $i = 1, 2$,

$$F_i(\lambda, -x_1, x_2) = F_i(\lambda, x_1, x_2).$$

1. Show that $G(\lambda, x_1)$ is even in x_1.
2. If $\partial G(0, 0)/\partial\lambda \neq 0$ and $\partial^2 G(0, 0)/\partial x_1^2 \neq 0$, show that there is a saddle-node bifurcation at $\lambda = 0$.
3. If $G(\lambda, x_1)$ is analytic and $\partial G(0, 0)/\partial\lambda \neq 0$, show that there is a saddle-node bifurcation at $\lambda = 0$.

10.11. Consider the planar system

$$\dot{x}_1 = f_1(x_1, x_2)$$
$$\dot{x}_2 = \mu x_2 + \nu x_1 + f_2(x_1, x_2),$$

where $\mathbf{f}(0) = 0$, $D\mathbf{f}(0) = 0$, and $\mu \neq 0$ and $\nu \neq 0$ are given constants. Discuss the stability properties of the equilibrium at the origin. First, be daring and compute the bifurcation function without putting the linear part into normal form. Then normalize the linear part and compute the bifurcation function again. How do the two cases compare?

10.12. *Bifurcation from a simple eigenvalue:* In a neighborhood of the origin, obtain those values $\alpha = \alpha^\star(x_1^0)$ such that the following system has an equilibrium point on the line $x_1 = x_1^0$:

$$\dot{x}_1 = \alpha x_1 + f_1(x_1, x_2)$$
$$\dot{x}_2 = -x_2 + f_2(x_1, x_2),$$

where $\mathbf{f}(0) = 0$ and $D\mathbf{f}(0) = 0$. Draw some possible curves $\alpha = \alpha^\star(x_1)$ in the (α, x_1)-plane (the bifurcation curve) and label the stable and unstable equilibria in the usual way.
Hint: Observe that the sign of $\partial G/\partial x_1$ is the sign of $-(\alpha^\star)'(x_1^0)x_1^0$ and use Theorem 10.8.

10.13. Let $A(\lambda)$ be a 2×2 matrix whose entries are continuous functions of parameters $\lambda \in \mathbb{R}^k$. If the eigenvalues of $A(0)$ are distinct, show that the eigenvalues of $A(\lambda)$ are continuous functions of λ in a neighborhood of $\lambda = 0$.
Hint I: (For real eigenvalues) If $\mu_1(\lambda)$ and $\mu_2(\lambda)$ are real eigenvalues of $A(\lambda)$, then $\mu_1(0) \neq \mu_2(0)$ implies that the line $\operatorname{tr} A(0) = \mu_1(0) + \mu_2(0)$ and the hyperbola $\det A(0) = \mu_1(0)\mu_2(0)$ intersect at two points. (Draw their pictures.) For $\lambda \neq 0$ and small, the picture persists.
Hint II: Show that $[\operatorname{tr} A(\lambda)]^2 - 4\det A(\lambda) \neq 0$ at $\lambda = 0$.
Hint III: Let us suppose that $A(\lambda) = A_0 + B(\lambda)$, where $B(0) = 0$ and A_0 is a diagonal matrix of the form $A_0 = \operatorname{diag}(\mu_1^0, \mu_2^0)$. Here is an outline of how

to use the method of bifurcation function to find an eigenvalue μ_1 near μ_1^0 for small $\lambda \neq 0$. A number $\mu = \mu_1^0 + \nu$ is an eigenvalue of $A(\lambda)$ if and only if there is a nonzero vector $\mathbf{x} = (x_1, x_2)$ such that $\left[A(\lambda) - \mu_1^0 I - \nu I\right] \mathbf{x} = \mathbf{0}$, which is equivalent to the pair of equations

$$[-\nu + b_{11}(\lambda)]x_1 + b_{12}(\lambda)x_2 = 0$$
$$b_{21}(\lambda)x_1 + [\mu_2^0 - \mu_1^0 - \nu + b_{22}(\lambda)]x_2 = 0.$$

From the second equation solve for x_2 as a linear function of x_1. Substitute the result into the first equation to obtain the "bifurcation equation." From the bifurcation equation conclude, with the help of the Implicit Function Theorem, that ν can be determined as a function of λ. This method is applicable to $n \times n$ matrices.

10.3. Center Manifolds

In this section, we determine fine structures of flows and bifurcations near an equilibrium point at which the matrix of the linear approximation has one zero and one negative eigenvalue. Our presentation resembles, with certain added complications, that of stable and unstable manifolds near a saddle. In fact, we will see that there is some invariant curve—local center manifold—tangent to the line containing the eigenvectors corresponding to the zero eigenvalue of the linearized vector field. Since the other eigenvalue is negative, all orbits starting near the origin approach this invariant curve. The qualitative behavior of the local flow on the plane can then be determined from the flow of an appropriate scalar differential equation on the center manifold. To fix the main ideas in a simple context, we will first describe the theory of center manifolds for Eq. (10.2). Eventually, we will generalize the setting to the parameter dependent equation (10.11) to study its local bifurcations. Let us begin with a precise definition of local center manifolds for Eq. (10.2).

Definition 10.12. *A C^k curve $W^c(\mathbf{0}, U)$ in a neighborhood U of the origin is said to be a local center manifold for Eq. (10.2) if*

- *$W^c(\mathbf{0}, U)$ is invariant under the flow of Eq. (10.2), that is, if $\mathbf{x}(t)$ is a solution of Eq. (10.2) with the initial value $\mathbf{x}(0) \in W^c(\mathbf{0}, U)$, then $\mathbf{x}(t) \in W^c(\mathbf{0}, U)$ as long as $\mathbf{x}(t) \in U$;*
- *$W^c(\mathbf{0}, U)$ is a graph of a C^k function $h(x_1) = x_2$ and is tangent to the x_1-axis at the origin, that is,*

$$W^c(\mathbf{0}, U) = \{\, (x_1, x_2) \ : \ x_2 = h(x_1), \ (x_1, x_2) \in U \,\},$$

where the function h satisfies

$$h(0) = 0, \quad \frac{\partial h}{\partial x_1}(0) = 0. \tag{10.23}$$

If there is no chance of confusion, we will usually omit the word "local" and the reference to the origin and the neighborhood U. To appreciate some of the subtleties associated with center manifolds, let us reexamine Example 10.1, and inspect Figures 10.1 and 10.2 a bit more closely.

Example 10.13. *Many center manifolds:* Consider the product system

$$\dot{x}_1 = ax_1^3$$
$$\dot{x}_2 = -x_2,$$
(10.24)

where a is a given real number. The origin is an equilibrium point for all values of a and the linearization at the origin is

$$\dot{\mathbf{x}} = \begin{pmatrix} 0 & 0 \\ 0 & -1 \end{pmatrix} \mathbf{x}.$$

The x_1-axis contains the eigenvectors corresponding to the zero eigenvalue and is the center manifold of the linearized system. Since Eq. (10.24) is a product system, it is clear that the graph of $h(x_1) = 0$, which is the x_1-axis, is a center manifold of Eq. (10.24). All orbits approach this center manifold exponentially fast and the flow of the planar system looks essentially like the flow of the scalar differential equation $\dot{x}_1 = ax_1^3$ on the center manifold.

Unlike the stable and unstable manifolds, center manifolds are not always unique. In fact, it is apparent from Figure 10.1 that when $a < 0$ the union of an orbit from the left half-plane and an orbit from the right half-plane, together with the origin, is also a center manifold of Eq. (10.24). More specifically, it is easy to determine that for any two constants c_1 and c_2 the graph of the function

$$h(x_1) = \begin{cases} c_1\, e^{1/(2ax_1^2)} & \text{if } x_1 < 0 \\ 0 & \text{if } x_1 = 0 \\ c_2\, e^{1/(2ax_1^2)} & \text{if } x_1 > 0 \end{cases}$$

yields a center manifold of Eq. (10.24). It is important to notice, however, that on all of these center manifolds the flows are equivalent. Consequently, for the qualitative study of the dynamics it is inconsequential which center manifold we use. Nevertheless, coexistence of many center manifolds is a potential cause for concern as it frequently is troublesome to compute nonunique entities. We will address this issue later in this section. ◊

The example above is admittedly a bit too simple. However, the local phase portrait of the nonproduct system (10.5) plotted in Figure 10.2 also suggests the existence of an attracting center manifold. This is indeed the case for the general system (10.2) as the following theorem asserts:

Theorem 10.14. *Let the vector field* (10.2) *be C^k and consider a sufficiently small neighborhood U of the origin in $\mathbb{R}^2$. Then there exists a local center manifold W^c in U consisting of the graph of a C^k function $h(x_1) = x_2$. Moreover, there are positive constants α and β such that, for any solution $\mathbf{x}(t)$ with initial value $\mathbf{x}(0) \in U$, the estimate*

$$|x_2(t) - h(x_1(t))| \leq \alpha e^{-\beta t}|x_2(0) - h(x_1(0))| \qquad (10.25)$$

holds as long as $\mathbf{x}(t) \in U$. ◊

In lieu of a proof, which is rather technical, here is an important practical consequence of this theorem.

Theorem 10.15. *An equilibrium point of Eq.* (10.2) *in U is stable [respectively, asymptotically stable, unstable] if and only if the corresponding equilibrium point of the scalar differential equation*

$$\dot{x}_1 = f_1(x_1, h(x_1)) \qquad (10.26)$$

on the center manifold defined by h is stable [respectively, asymptotically stable, unstable]. ◊

To utilize the theory of center manifolds in applications, we now show how to compute them in specific equations. For practical purposes, it suffices to have a procedure for the computation of several terms of the Taylor series of a function $h(x_1)$ defining a center manifold. As in the case of the local stable and unstable manifolds, we will accomplish this by deriving an appropriate differential equation. If we differentiate with respect to t the defining equation $x_2(t) = h(x_1(t))$, then we obtain

$$\dot{x}_2 = \frac{\partial h}{\partial x_1}(x_1)\,\dot{x}_1.$$

If we now substitute the expressions for $\dot{x}_1$ and $\dot{x}_2$ given by the differential equation (10.2), then a solution of the partial differential equation

$$-h(x_1) + f_2(x_1, h(x_1)) = \frac{\partial h}{\partial x_1}(x_1)\,f_1(x_1, h(x_1)) \qquad (10.27)$$

subject to the initial values

$$h(0) = 0, \qquad \frac{\partial h}{\partial x_1}(0) = 0$$

yields a center manifold defined by $h(x_1)$. The partial differential equation (10.27) cannot, of course, be solved for h in most cases. Therefore, we opt for a sufficiently accurate approximate solution to facilitate a local analysis of the flow. To accomplish this, we expand $h(x_1)$ into a power series in the variable x_1 as

$$h(x_1) = c_2 x_1^2 + c_3 x_1^3 + O(|x_1|^4), \tag{10.28}$$

where the coefficients c_i are to be determined. We now substitute this series into the partial differential equation (10.27) and determine c_i by equating the coefficients of like terms.

Example 10.16. *Continuation of Example 10.13:* Using the procedure outlined above, we now compute the Taylor series of the center manifold of Eq. (10.24). In this case, the differential equation (10.28) becomes

$$-h(x_1) = \frac{\partial h}{\partial x_1}(x_1) \, a x_1^3. \tag{10.29}$$

If we expand $h(x_1)$ into power series in x_1 as in Eq. (10.28) and then substitute it into Eq. (10.29), we find that all the coefficients are 0. Consequently, $h(x_1) \equiv 0$ is a solution of Eq. (10.29). This, of course, yields the x_1-axis as the center manifold, but what about the other center manifolds of Eq. (10.24)? The resolution of this dilemma lies in the realization that a function does not need to be zero in an interval about the origin even if all its Taylor coefficients vanish. In fact, all the other center manifolds are so "flat" near the origin that all their Taylor coefficients about the origin are 0, as a simple computation shows. ◇

Although center manifolds are not unique, the power series method above always yields a unique Taylor series, if it exists. It turns out that all center manifolds have the same Taylor expansions near the origin; so, our approximation procedure is a good one. Let us make these remarks a bit more precise. For any function

$$g : \mathbb{R} \times \mathbb{R} \to \mathbb{R}; \quad x_1 \mapsto g(x_1)$$

satisfying

$$g(0) = 0, \qquad \frac{\partial g}{\partial x_1}(0) = 0$$

which is C^k, with $k \geq 1$, in a neighborhood of the origin, consider the function $\mathcal{M}(g)(x_1)$ defined by

$$\mathcal{M}(g)(x_1) \equiv \frac{\partial g}{\partial x_1}(x_1) \, f_1(x_1, g(x_1)) + g(x_1) - f_2(x_1, g(x_1)).$$

Then one can prove the following approximation result for any center manifold:

Theorem 10.17. *Let h be a center manifold of Eq. (10.2). Suppose that $\mathcal{M}(g)(x_1) = O(|x_1|^k)$ as $x_1 \to 0$, where $k > 1$. Then, as $x_1 \to 0$,*

$$|h(x_1) - g(x_1)| = O(|x_1|^k).$$

In the comfort of this theorem, let us return to Figure 10.2 and determine the power series of the apparent center manifold resembling a cubic curve.

Example 10.18. *Continuation of Example 10.3:* The power series solution of the partial differential equation (10.27) in this case is given by

$$h(x_1) = -x_1^3 + O(|x_1|^4).$$

Thus the flow on the center manifold is determined by the scalar differential equation

$$\dot{x}_1 = ax_1^3 + x_1 h(x_1) = ax_1^3 - x_1^4 + O(|x_1|^5).$$

It is interesting to notice that the vector field of this scalar differential equation has the same terms up through order four as the ones we have obtained earlier using the method of bifurcation function. ◇

We now generalize the theory of center manifolds to systems of differential equations (10.11) which depend on parameters, and investigate the possible bifurcations near the origin for small λ. This extension may appear formally to consist of insertion of a λ or two into the previous definitions and theorems. For the sake of completeness, we will make such insertions. From a geometric point of view, however, center manifolds become considerably more complicated in the presence of parameters: for each small λ there is a curve, and the collection of these curves form a surface—a center manifold. Here is the precise definition:

Definition 10.19. *A family of C^k curves $W_\lambda^c(0, U)$ in a neighborhood U of the origin is said to be a local center manifold for Eq. (10.11) if*

- *$W_\lambda^c(0, U)$ is invariant under the flow of Eq. (10.11), that is, if $\mathbf{x}(t)$ is a solution of Eq. (10.11) with the initial value $\mathbf{x}(0) \in W_\lambda^c(0, U)$, then $\mathbf{x}(t) \in W_\lambda^c(0, U)$ as long as $\mathbf{x}(t) \in U$;*
- *$W_\lambda^c(0, U)$ is a graph of a C^k-function $h(\lambda, x_1) = x_2$ and, for $\lambda = 0$, is tangent to the x_1-axis at the origin, that is,*

$$W_\lambda^c(0, U) = \{ (x_1, x_2) : x_2 = h(\lambda, x_1), (x_1, x_2) \in U \},$$

where the function h satisfies

$$h(0, 0) = 0, \quad \frac{\partial h}{\partial x_1}(0, 0) = 0. \tag{10.30}$$

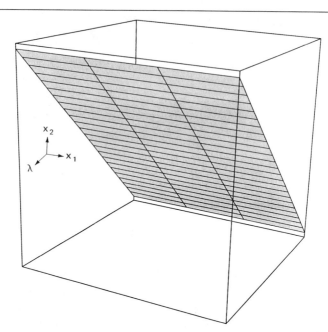

Figure 10.3. *The inclined plane is a center manifold of Example 10.20.*

To appreciate some of the subtleties associated with center manifolds when a vector field depends on parameters, let us examine several examples.

Example 10.20. Consider the product system

$$\dot{x}_1 = -x_1^3$$
$$\dot{x}_2 = -x_2 + \lambda,$$

where λ is a scalar parameter. A center manifold is the graph of $h(\lambda, x_1) = \lambda$ and the flow on the center manifold is given by $\dot{x}_1 = -x_1^3$. On the (x_1, x_2)-plane this center manifold is the family of horizontal lines $x_2 = \lambda$ parametrized by λ. It is more revealing to visualize this center manifold in the (λ, x_1, x_2)-space: it is the inclined plane in Figure 10.3. The intersections of the inclined plane with vertical planes are the family of lines $x_2 = \lambda$. The plane of the center manifold is tangent to the x_1-axis at $\lambda = 0$. $\diamond$

Example 10.21. Consider the linear system

$$\dot{x}_1 = 0$$
$$\dot{x}_2 = \lambda x_1 - x_2,$$

where λ is a scalar parameter. A center manifold is given by $h(\lambda, x_1) = \lambda x_1$, which is a family of rotating lines given by $x_2 = \lambda x_1$ on the (x_1, x_2)-plane. In the (λ, x_1, x_2)-space this center manifold is the "hyperbolic ruled surface" depicted in Figure 10.4. $\diamond$

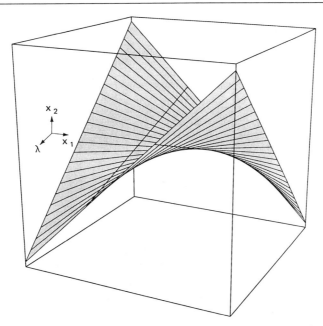

Figure 10.4. *The ruled surface is a center manifold of Example 10.21.*

The existence theorem for an attracting center manifold for Eq. (10.11) reads as follows:

Theorem 10.22. *Let the vector field* (10.11) *be C^k and consider a sufficiently small neighborhood U of the origin in $\mathbb{R}^2$. Then, for $\|\lambda\|$ small, there exists a local center manifold W_λ^c in U consisting of the graph of a C^k function $h(\lambda, x_1) = x_2$. Moreover, for any solution $\mathbf{x}(t)$ with initial value $\mathbf{x}(0) \in U$ there are positive constants α and β such that*

$$|x_2(t) - h(\lambda, x_1(t))| \le \alpha e^{-\beta t}|x_1(0) - h(\lambda, x_1(0))| \tag{10.31}$$

as long as $\mathbf{x}(t) \in U$. ◇

Despite possible nonuniqueness of center manifolds, the attraction estimate (10.31) implies the following property of any center manifold:

Corollary 10.23. *An equilibrium point of Eq.* (10.11) *in U must always be on any center manifold.* ◇

Because W_λ^c is attracting all solutions starting in U, the omega limit set $\omega(\mathbf{x}^0)$ of $\mathbf{x}^0 \in U$ must be in W_λ^c. Since $\omega(\mathbf{x}^0)$ is invariant and W_λ^c is one dimensional, we have the following result:

Corollary 10.24. *The omega limit set $\omega(\mathbf{x}^0)$, if it exists, of any orbit of Eq. (10.11) with initial value $\mathbf{x}^0 \in U$ is an equilibrium point.* $\diamondsuit$

This corollary enables us to "fill in" the phase portraits of Eq. (10.11) near the origin rather easily because it rules out the existence of, for example, periodic orbits. Another practical implication, which is considerably more than a mere corollary, of Theorem 10.22 is the following:

Theorem 10.25. *An equilibrium point of Eq. (10.11) in U is stable [respectively, asymptotically stable, unstable] if and only if the corresponding equilibrium point of the scalar differential equation*

$$\dot{x}_1 = F_1(\lambda,\, x_1,\, h(\lambda, x_1)) \tag{10.32}$$

on the center manifold defined by h is stable [respectively, asymptotically stable, unstable]. $\diamondsuit$

As in the case of Eq. (10.2), the computation of a center manifold for Eq. (10.11) is facilitated through an appropriate partial differential equation for $h(\lambda, x_1)$. If we differentiate with respect to t the defining equation $x_2(t) = h(\lambda, x_1(t))$, then we obtain

$$\dot{x}_2 = \frac{\partial h}{\partial x_1}(\lambda,\, x_1)\, \dot{x}_1.$$

If we now substitute the expressions for $\dot{x}_1$ and $\dot{x}_2$ given by the differential equation (10.11), then a solution of the partial differential equation

$$-h(\lambda,\, x_1) + F_2(\lambda,\, x_1,\, h(\lambda,\, x_1)) = \frac{\partial h}{\partial x_1}(\lambda,\, x_1)\, F_1(\lambda,\, x_1,\, h(\lambda,\, x_1)) \tag{10.33}$$

subject to the initial values

$$h(0,\, 0) = 0, \qquad \frac{\partial h}{\partial x_1}(0,\, 0) = 0$$

yields a center manifold defined by $h(\lambda,\, x_1)$. In search of an approximate solution of Eq. (10.33), we expand $h(\lambda,\, x_1)$ into a power series in the variables λ and x_1 as

$$h(\lambda,\, x_1) = c_{10}\lambda + c_{20}\lambda^2 + c_{11}\lambda x_1 + c_{02}x_1^2 + O((|\lambda| + |x_1|)^3), \tag{10.34}$$

where the coefficients c_{ij} are to be determined. We now substitute this series into the partial differential equation (10.33) and determine c_{ij} by equating the coefficients of like terms.

Although center manifolds are not unique, the computational effectiveness of the power series method above can again be demonstrated. Suppose that

$$g : \mathbb{R}^m \times \mathbb{R} \to \mathbb{R}; \quad (\lambda,\, x_1) \mapsto g(\lambda,\, x_1)$$

satisfying

$$g(0, 0) = 0, \qquad \frac{\partial g}{\partial x_1}(0, 0) = 0$$

is a C^k function, with $k \geq 1$, in a neighborhood of the origin, and consider the function $\mathcal{M}(g)(\lambda, x_1)$ defined by

$$\mathcal{M}(g)(\lambda, x_1) \equiv \frac{\partial g}{\partial x_1}(\lambda, x_1) F_1(\lambda, x_1, g(\lambda, x_1)) + g(\lambda, x_1)$$
$$- F_2(\lambda, x_1, g(\lambda, x_1)).$$

Then one can prove the following approximation result for any center manifold:

Theorem 10.26. *Let $h(\lambda, x_1)$ be a center manifold of Eq. (10.11). Suppose that $\mathcal{M}(g)(\lambda, x_1) = O((|\lambda| + |x_1|)^k)$ as $(\lambda, x_1) \to \mathbf{0}$, where $k > 1$. Then as $(\lambda, x_1) \to \mathbf{0}$,*

$$|h(\lambda, x_1) - g(\lambda, x_1)| = O((|\lambda| + |x_1|)^k).$$

In light of these theoretical and computational remarks, we now complete the analysis of a previous example by determining a center manifold and the flow on it.

Example 10.27. *Damped pendulum with torque continued:* The partial differential equation (10.33) for the system (10.19) is

$$-h(\lambda, x_1) + [-\cos(x_1 + h(\lambda, x_1)) + 1 + \lambda]$$
$$= \frac{\partial h}{\partial x_1}(\lambda, x_1) [-\cos(x_1 + h(\lambda, x_1)) + 1 + \lambda]. \qquad (10.35)$$

As a result of juggling power series there emerges the expansion

$$h(\lambda, x_1) = \lambda - \tfrac{1}{2}\lambda^2 + 2\lambda x_1 - \tfrac{1}{2}x_1^2 + O\left((|\lambda| + |x_1|)^3\right).$$

The scalar differential equation on the center manifold is given by

$$\dot{x}_1 = -\cos(x_1 - h(\lambda, x_1)) + 1 + \lambda$$
$$= \lambda + \tfrac{1}{2}\lambda^2 - \lambda x_1 + \tfrac{1}{2}x_1^2 + O\left((|\lambda| + |x_1|)^3\right).$$

The terms up through order two are the same as the ones we have obtained in Example 10.11 with the method of bifurcation functions. Therefore, using the center manifold theory it is now quite easy to construct the full flow of the system of equations (10.19) on the (x_1, x_2)-plane. To recover the dynamics of the pendulum in the original coordinates (y_1, y_2), all that remains to be done is to undo the effects of shearing by the matrix P and the translation. $\Diamond$

We have presented in this chapter two methods for investigating stability and bifurcations of equilibria in the presence of a zero eigenvalue—bifurcation function and center manifolds. Despite their obvious similarities, they differ in subtle ways. We conclude this chapter with a brief comparison of the two methods.

To understand some of the differences, let us reconsider Eq. (10.2). From Theorem 10.4, we know that the stability properties of the equilibrium point at the origin are the same as the zero solution of

$$\dot{x}_1 = f_1(x_1, \psi(x_1)), \tag{10.36}$$

where the function $\psi(x_1)$ is the solution of

$$-\psi + f_2(x_1, \psi) = 0.$$

From Theorem 10.15, we also know that the same statement is true relative to the equilibria of the equation

$$\dot{x}_1 = f_1(x_1, h(x_1)),$$

where the graph of the function h is a center manifold. The proof of Theorem 10.4 relies heavily on the existence of a center manifold; however, once the result is established, it is better to use Eq. (10.36). Indeed, the function ψ is usually much easier to approximate than h; the computation of ψ involves only f_2, while the computation of h uses both f_1 and f_2. Furthermore, ψ has the theoretical advantage of being uniquely defined and is as smooth in x_1 as f_1 and f_2 (even up to analyticity). Center manifolds do not enjoy these properties.

The theories of bifurcation functions and center manifolds have generalizations for systems of differential equations in $\mathbb{R}^n$ for which the linear approximation at an equilibrium point has k zero and $n - k$ negative eigenvalues. Theorem 10.14 on stability being determined by the flow on a center manifold remains valid in this more general setting. The bifurcation function, on the other hand, is usually insufficient for determining stability if $k > 1$.

Exercises _____ ♣♡♠♢

10.14. *Many center manifolds:* Find *all* center manifolds of the system

$$\dot{x}_1 = x_1^2$$
$$\dot{x}_2 = -x_2.$$

10.15. Draw some center manifolds of the system

$$\dot{x}_1 = \lambda + x_1^2$$
$$\dot{x}_2 = -x_2$$

in the (λ, x_1, x_2)-space. Put $\lambda = 0$ first.

10.16. *No analytic center manifold:* Consider the system

$$\dot{x}_1 = -x_1^3$$
$$\dot{x}_2 = -x_2 + x_1^2.$$

This vector field is, of course, analytic. Show that this system has no analytic center manifold.
Hint: Let $h(x_1) = \sum_{i=2}^{+\infty} c_i x_1^i$ and determine that $c_2 = 1$, $c_i = 0$ for i odd, $c_{i+2} = i c_i$ for i even.

10.17. Show that the equilibrium point at the origin of the system

$$\dot{x}_1 = x_1 x_2 + a x_1^3 + b x_1 x_2^2$$
$$\dot{x}_2 = -x_2 + c x_1^2 + d x_1^2 x_2$$

1. is asymptotically stable if either $a + c < 0$, or $a + c = 0$ and $cd + bc^2 < 0$;
2. is unstable if either $a + c > 0$, or $a + c = 0$ and $cd + bc^2 > 0$.

Do this problem using bifurcation function and a center manifold. Compare the efforts of your computations in the two methods.

10.18. For Eq. (10.2), let $\psi(x_1)$ be the solution of Eq. (10.3), and $h(x_1)$ be a center manifold. Suppose that

$$f_1(x_1, \psi(x_1)) = a x_1^k + O(|x_1|^{k+1}),$$
$$f_1(x_1, h(x_1)) = b x_1^\ell + O(|x_1|^{\ell+1}),$$

with $a \neq 0$ and $b \neq 0$. Show that $k = \ell$ and $a = b$.

10.19. *Rotated pendulum:* Consider a pendulum of mass m and length l hinged to a movable joint and rotated with angular velocity ω about its pivot. The motion of such a pendulum is not planar. However, the projection of the motion of the pendulum on a plane is governed by the differential equation

$$\ddot{\theta} = \omega^2 \cos\theta \sin\theta - \frac{g}{l}\sin\theta - m\dot{\theta},$$

where, as usual, θ is the displacement of the pendulum from its rest position.
1. Show that there is a pitchfork bifurcation at $\omega_0 = \sqrt{g/l}$ and discuss the stability properties of the equilibrium points for all values of ω.
2. If $\omega < \omega_0$, show that every orbit, except the stable manifold $W^s(\pi, 0)$, approaches the origin as $t \to +\infty$.
3. If $\omega > \omega_0$, does every orbit approach an equilibrium point?

10.20. *On the machine:* Plot on the computer, using PHASER, for example, some representative phase portraits of the damped pendulum with torque and rotated pendulum to observe the bifurcations.

Bibliographical Notes

The stability of nonhyperbolic equilibria with one zero eigenvalue was investigated by Liapunov in 1892. His method of reducing a two-dimensional problem to a scalar one has a far-reaching generalization which goes under the name of *Method of Liapunov-Schmidt*. A nice exposition of this important method is given in Hale [1984]; full details are available in Chow and Hale [1982] and Golubitsky and Schaeffer [1985]. For historical reasons, you may also like to see Schmidt [1908].

The center manifold is one of the key ideas in bifurcation theory. Although the idea had been around a long time, the first complete proof of its smoothness in a neighborhood of a nonhyperbolic equilibrium point of an ordinary differential equation appeared in Kelley [1967]. By now, there are many expositions of it; see, for example, Carr [1981]. Center manifolds are important in the case of partial differential equations as well. In fact, one of the most active areas of current research is the existence of finite dimensional manifolds which capture the asymptotic behavior of partial differential equations. For reaction-diffusion equations, the existence of global center manifolds—called inertial manifolds—has been proved. This is important in reducing the dynamics of an infinite dimensional system to those of a finite dimensional one; see Mallet-Paret and Sell [1989].

11

In the Presence of Purely Imaginary Eigenvalues

 In this chapter, we investigate the stability and bifurcations of a nonhyperbolic equilibrium point of a planar differential equation in the case where the linearized vector field has purely imaginary eigenvalues. Using polar coordinates, we capture the dynamics of such a system in the neighborhood of the equilibrium point in terms of the dynamics of an appropriate nonautonomous scalar differential equation with periodic coefficients. For the analysis of this scalar equation, we appeal to results in Chapters 4 and 5. When the vector field is subjected to small perturbations, the original equilibrium point persists, and there can be no new equilibria in the neighborhood. However, if the eigenvalues of the linearized system move away from the imaginary axis, one expects the equilibrium point to change its stability type. This change is typically marked by the appearance of a small periodic orbit encircling the equilibrium point. We present a proof of this celebrated result—the Poincaré–Andronov–Hopf Theorem—and a discussion of the stability of the periodic orbit. We conclude with an exposition of computational procedures for determining bifurcation diagrams of periodic orbits bifurcating from an equilibrium point.

11.1. Stability

As we have seen in Example 9.9, when the eigenvalues of the linearized vector field at an equilibrium point are purely imaginary, the local dynamics about the equilibrium point cannot be determined by the linear approximation. Indeed, depending on the nonlinear terms, the equilibrium can be unstable, stable, or even asymptotically stable. Consequently, we need to investigate the effects of the nonlinear terms in each particular situation. In this section, we show how to carry out such an investigation by reducing the dynamics in the neighborhood of a nonhyperbolic equilibrium point with purely imaginary eigenvalues to the dynamics of a 2π-periodic scalar differential equation.

Let us begin by recalling briefly the dynamics of Example 9.9.

Example 11.1. Consider the planar system

$$
\begin{aligned}
\dot{x}_1 &= x_2 + ax_1(x_1^2 + x_2^2) \\
\dot{x}_2 &= -x_1 + ax_2(x_1^2 + x_2^2),
\end{aligned}
\tag{11.1}
$$

where a is a given real number. Regardless of the value of the constant a, the origin is an equilibrium point and the eigenvalues of the linearization at the origin are $\pm i$. If we introduce polar coordinates (r, θ) defined by

$$
x_1 = r\cos\theta, \qquad x_2 = -r\sin\theta, \tag{11.2}
$$

then Eq. (11.1) becomes

$$
\begin{aligned}
\dot{r} &= ar^3 \\
\dot{\theta} &= 1.
\end{aligned}
\tag{11.3}
$$

This is a rather special product system. Since $\dot{\theta} > 0$, the orbits spiral monotonically in θ around the origin. Therefore, the stability type of the origin of Eq. (11.1) is the same as that of the equilibrium point $r = 0$ of the radial equation $\dot{r} = ar^3$. Of course, $r = 0$ is asymptotically stable if $a < 0$, stable at $a = 0$, and unstable for $a > 0$; see Figure 11.1. $\Diamond$

Unlike the example above, planar systems for which the linearization near an equilibrium point has purely imaginary eigenvalues do not always turn out to be product systems when transformed into polar coordinates. However, with some care, we can still pursue the line of reasoning in this example and reduce the problem to the analysis of a 2π-periodic, rather than autonomous, scalar differential equation. To be specific, let $\mathbf{f}$ be a given C^k function, $k \geq 2$,

$$
\mathbf{f} : \mathbb{R}^2 \to \mathbb{R}^2; \quad \mathbf{x} \mapsto \mathbf{f}(\mathbf{x}),
$$

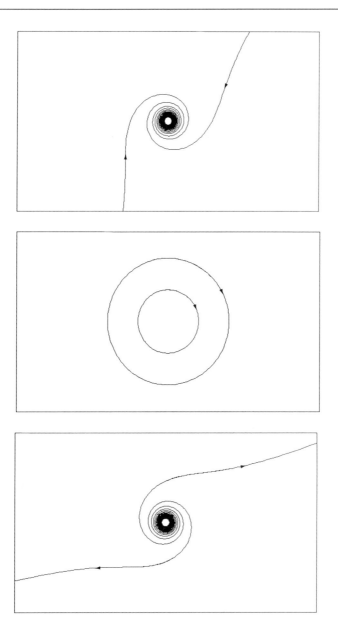

Figure 11.1. For Eq. (11.1), origin is always an equilibrium point with purely imaginary eigenvalues; however, it is asymptotically stable for $a = -0.5$, stable for $a = 0.0$, and unstable for $a = 0.5$.

satisfying

$$f(0) = 0, \qquad \|Df(0)\| < 1, \tag{11.4}$$

where the norm of the Jacobian matrix $\|Df(0)\|$ is a nonnegative real number such that $\|Df(0)x\| \leq \|Df(0)\| \, \|x\|$ for all $x \in \mathbb{R}^2$. We should remark that if we were interested in only stability of equilibria it would be sufficient to prepare the differential equation by appropriate transformations to reduce it to the case $f(0) = 0$, $Df(0) = 0$ so that f contains terms of degree two and higher. We have chosen to impose the lesser restriction on the norm of the derivative of f for the purpose of bifurcation studies which we will undertake in the next section. Now, consider the planar system of differential equations

$$\begin{aligned}
\dot{x}_1 &= x_2 + f_1(x_1, x_2) \\
\dot{x}_2 &= -x_1 + f_2(x_1, x_2).
\end{aligned} \tag{11.5}$$

To bring the linear part of this system to the forefront, let us write it, for a moment, in vector notation:

$$\dot{x} = \begin{pmatrix} 0 & 1 \\ -1 & 0 \end{pmatrix} x + f(x).$$

Notice that the linear part of the vector field about the equilibrium point at the origin is in Jordan Normal Form with eigenvalues $\pm i$. In applications, the linearization of a vector field with purely imaginary eigenvalues may not always come in normal form [Eq. (11.5)]; however, such a vector field can always be put into this form with a linear change of coordinates and a rescaling of the independent variable t.

Since the linear part has rotational symmetry, it is reasonable to introduce polar coordinates (11.2) to investigate the dynamics of the system (11.5) in a sufficiently small neighborhood of the origin. As a result of a short calculation, we see that, in polar coordinates, Eq. (11.5) becomes

$$\begin{aligned}
\dot{r} &= \Re(r, \theta) \\
\dot{\theta} &= 1 + \Theta(r, \theta),
\end{aligned} \tag{11.6}$$

where

$$\Re(r, \theta) = f_1(r\cos\theta, -r\sin\theta)\cos\theta - f_2(r\cos\theta, -r\sin\theta)\sin\theta$$

$$\Theta(r, \theta) = -\frac{1}{r}[f_1(r\cos\theta, -r\sin\theta)\sin\theta + f_2(r\cos\theta, -r\sin\theta)\cos\theta] \tag{11.7}$$

when $r \neq 0$, and at $r = 0$ we define

$$\Theta(0, \theta) = (\sin\theta, \cos\theta)Df(0)\begin{pmatrix} \cos\theta \\ -\sin\theta \end{pmatrix}.$$

Since $\mathbf{f}$ satisfies Eq. (11.4), the function $\Re$ is C^k and Θ is C^{k-1}, and they satisfy

$$\Re(0, 0) = 0, \quad |\Theta(0, \theta)| < 1. \tag{11.8}$$

More importantly, notice that both $\Re(r, \theta)$ and $\Theta(r, \theta)$ are 2π-periodic functions in the variable θ.

Now, we explain how to reduce, when r is sufficiently small, the discussion of the orbits of Eq. (11.5) to the discussion of solutions of a 2π-periodic scalar differential equation. Since $|\Theta(0, \theta)| < 1$ for all θ, and $\Theta(r, \theta)$ is continuous, we can choose $\delta > 0$ such that $1 + \Theta(r, \theta) > 0$ for all θ and $|r| < \delta$. Consequently, in a neighborhood of the origin, we have $\dot\theta > 0$. The pleasant implication of this is that the orbits of Eq. (11.6) spiral monotonically in θ around the origin. Therefore, we can eliminate t in Eq. (11.6) and obtain an equation for r as a function of θ through the differential equation

$$\frac{dr}{d\theta} = \mathcal{R}(r, \theta), \tag{11.9}$$

where

$$\mathcal{R}(r, \theta) = \frac{\Re(r, \theta)}{1 + \Theta(r, \theta)}$$

which is a C^{k-1} function, 2π-periodic, and satisfies

$$\mathcal{R}(0, \theta) = 0. \tag{11.10}$$

The solutions of Eq. (11.9) give the orbits of Eq. (11.5). We can also recover the solutions of Eq. (11.5) as a function of time from the solutions of Eq. (11.9) by following the steps below:

- Fix r_0 and find the solution $r(\theta, r_0)$ of Eq. (11.9) satisfying the initial value $r(0, r_0) = r_0$. The orbit of Eq. (11.5) through the point $\mathbf{x}^0 = (r_0, 0)$ is then given by

$$\gamma(\mathbf{x}^0) = \{ (x_1, x_2) : x_1 = r(\theta, r_0) \cos\theta,$$
$$x_2 = -r(\theta, r_0) \sin\theta, \ 0 \le \theta + \infty \}. \tag{11.11}$$

- Find the solution $\theta(t)$ of the initial-value problem

$$\dot\theta = 1 + \Theta(r(\theta, r_0), \theta), \qquad \theta(0) = 0. \tag{11.12}$$

- The solution $\mathbf{x}(t)$ of Eq. (11.5) through the point $\mathbf{x}^0 = (r_0, 0)$ is then given by

$$x_1(t) = r(\theta(t), r_0) \cos\theta(t)$$
$$x_2(t) = -r(\theta(t), r_0) \sin\theta(t). \tag{11.13}$$

Let us now illustrate, by way of an example, the role of the 2π-periodic scalar differential equation $dr/d\theta$ in the stability analysis of the equilibrium point at the origin of the system (11.5).

Example 11.2. *A damped oscillator:* Consider the second-order differential equation

$$\ddot{y} + y^2\dot{y} + y = 0,$$

or equivalently the system

$$\dot{x}_1 = x_2$$
$$\dot{x}_2 = -x_1 - x_1^2 x_2. \tag{11.14}$$

Using the methods presented in Chapter 9, Liapunov functions, and the Invariance Principle, it can be shown rather easily that the equilibrium point of Eq. (11.14) at the origin is asymptotically stable. Let us establish the same result with the methods presented in this section. In polar coordinates, the scalar equation (11.9) for the system (11.14) is given by

$$\frac{dr}{d\theta} = -[\cos^2\theta \sin^2\theta]r^3 + O(r^4)$$
$$= -\tfrac{1}{8}[1 - \cos(4\theta)]r^3 + O(r^4). \tag{11.15}$$

Let us use the transformation theory from Chapter 5 to make several terms of the Taylor series of the vector field independent of θ. For this purpose, introduce the variable ρ defined by

$$r = \rho + a(\theta)\rho^3, \tag{11.16}$$

where $a(\theta)$ is a 2π-periodic function to be determined. In the new variable, Eq. (11.15) becomes

$$\frac{d\rho}{d\theta} = -\tfrac{1}{8}\left[1 - \cos(4\theta) + 8a'(\theta)\right]\rho^3 + O(\rho^4).$$

To make the coefficient of the ρ^3 term independent of θ, we choose $a'(\theta) = \tfrac{1}{8}\cos(4\theta)$. With this choice of transformation (11.16), we obtain

$$\frac{d\rho}{d\theta} = -\tfrac{1}{8}\rho^3 + O(\rho^4). \tag{11.17}$$

Here, we have abused the "big O" notation a bit, in the sense that the coefficients of the higher-order terms may depend on θ, but this is of no concern. Now, it is evident that $\rho = 0$ is asymptotically stable; hence, from the form of the transformation (11.16), $r = 0$ is also asymptotically stable. With more complicated situations in mind, we should point out that the same conclusion is obtained from the inspection of the Poincaré map Π of Eq. (11.17), which is given by

$$\Pi(\rho_0) = \rho_0 - \tfrac{1}{8}\rho_0^3 + O(\rho_0^4).$$

Despite the asymptotic stability, it is of practical consideration to keep in mind that the approach to the equilibrium is not exponential (slow!) because of nonhyperbolicity; see Figure 11.2. $\Diamond$

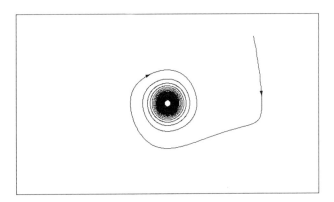

Figure 11.2. *For Eq. (11.14), the origin is an asymptotically stable equilibrium point with purely imaginary eigenvalues.*

The procedure used in the example above can easily be extended to the general case of Eq. (11.5). All that is needed is the transformation theory from Section 5.2 to convert the scalar equation $dr/d\theta$ to the equation $d\rho/d\theta$ for which the lowest order term in the Taylor expansion of the vector field has a nonzero constant coefficient. For Eq. (11.9), it is possible to show that this lowest order term must always be odd (see the exercises). With these remarks, the following result is immediate.

Lemma 11.3. *Suppose that* $\mathbf{f} = (f_1, f_2)$ *is a* C^{2k+2} *function in Eq. (11.5) with the corresponding transformed scalar equation*

$$\frac{d\rho}{d\theta} = a\rho^{2k+1} + o(|\rho^{2k+1}|) \quad \text{as} \quad \rho \to 0,$$

where $a \neq 0$ *is a real number,* k *is a positive integer. Then the equilibrium point at the origin of the planar system (11.5) is asymptotically stable if* $a < 0$; *otherwise it is unstable.* $\Diamond$

In some situations, it may happen that the coefficients of all powers of ρ in the scalar equation $d\rho/d\theta$ are zero, that is, the equation $dr/d\theta$ is formally transformed to $d\rho/d\theta = 0$. In this case, it is quite difficult to decide the stability type of the origin. However, with a further restriction on the function $\mathbf{f}$, we have the following *Center Theorem of Liapunov* which we state without a proof:

Theorem 11.4. *Suppose that* $\mathbf{f}$ *is analytic. Then, for the equilibrium point at the origin of Eq. (11.5), one of the three alternatives holds:*
- *the origin is unstable;*
- *the origin is asymptotically stable;*

- the origin is a *center*, that is, every solution in a neighborhood of the origin is periodic. $\diamondsuit$

Of course, the first two possibilities are covered by Lemma 11.3 where the vector field near the equilibrium point behaves radially as the solutions of $\dot{\rho} = a\rho^{2k+1}$ with $a \neq 0$, for some positive integer k. The third possibility corresponds to the case when $dr/d\theta$ can formally be transformed to zero. For the latter case the analyticity assumption is essential; otherwise, there is another alternative, as the following example demonstrates.

Example 11.5. *Stable but not a center:* Consider in polar coordinates the product system

$$\dot{r} = \begin{cases} 0 & \text{if } r = 0 \\ -e^{-1/r^2} \sin(1/r) & \text{otherwise} \end{cases}$$

$$\dot{\theta} = 1.$$

The Taylor series of $dr/d\theta$ at the origin is identically zero. It is easy to see that the origin is stable; yet, it is not a center. To wit, notice that there are infinitely many concentric periodic orbits encircling the origin with amplitudes, or radii, $(k\pi)^{-1}$, where k is any positive integer. The periodic orbits with amplitudes $[(2k+1)\pi]^{-1}$ are asymptotically stable while the ones with amplitudes $[(2k+2)\pi]^{-1}$ are unstable. Consequently, the origin is stable but not asymptotically stable. $\diamondsuit$

This concludes the stability analysis of an equilibrium point whose linearization has purely imaginary eigenvalues. The method of polar coordinates and reduction to the scalar equation $dr/d\theta$ also provides an effective tool for studying periodic orbits of Eq. (11.5) about the origin. In the remaining part of this section, we study such periodic orbits. The observations below are almost self-evident.

Lemma 11.6. *There is a bounded neighborhood U of the origin in $\mathbb{R}^2$ such that each periodic orbit Γ of Eq. (11.5) lying in U encircles the origin; also, if $\mathbf{x}^0 = (r_0, 0) \in \Gamma$ with $r_0 > 0$, then the solution $r(\theta, r_0)$ of Eq. (11.9) satisfying the initial value $r(0, r_0) = r_0$ is 2π-periodic in θ. Conversely, if $r(\theta, r_0)$ is a 2π-periodic solution of Eq. (11.9), then the orbit $\Gamma(\mathbf{x}^0)$ with $\mathbf{x}^0 = (r_0, 0)$ of the planar system (11.5) is a periodic orbit. The minimal period T of such a Γ is the first value of t for which the solution $\theta(t)$ of Eq. (11.12) satisfies*

$$\theta(T) = 2\pi. \tag{11.18}$$

The reduction of the discussion of the orbits of Eq. (11.5) near the origin to the discussion of solutions of Eq. (11.9) has important consequences for limit sets of orbits of Eq. (11.5). The result below is a special case of the Poincaré–Bendixson Theorem, a fundamental result which we will explore more fully in the next chapter:

Theorem 11.7. *There is a bounded neighborhood U of the origin in $\mathbb{R}^2$ such that if $\mathbf{x}^0 \in U$ and the solution $\varphi(t, \mathbf{x}^0)$ of Eq. (11.5) remains in U for $t \geq 0$ [respectively, $t \leq 0$], then the omega-limit set $\omega(\mathbf{x}^0)$ [respectively, $\alpha(\mathbf{x}^0)$] is either a periodic orbit or the equilibrium point at the origin.*

Proof. Suppose that U is as in Theorem 4.11. If $\varphi(t, \mathbf{x}^0)$ of Eq. (11.5) remains in U for $t \geq 0$ (or $t \leq 0$), then the solution $r(\theta, r_0)$, with $r_0 = \|\mathbf{x}^0\|$, of (11.9) is bounded for $\theta \geq 0$ (or $\theta \leq 0$). Theorem 4.11 implies that $r(\theta, r_0)$ approaches a 2π-periodic solution of Eq. (11.9). Now, the desired conclusions follow from the relations (11.13). $\Diamond$

We now turn to the problem of relating the stability type of a periodic orbit of the planar system (11.5) to that of the corresponding 2π-periodic solution of the scalar equation (11.9). In Chapter 4 we have defined the notions of stability, asymptotic stability of solutions of Eq. (11.9), but, although alluded to it, have not yet agreed on a precise notion of stability of a periodic orbit of a planar autonomous system. By analogy with the definition of stability of an equilibrium point of a planar system, it is tempting to define a periodic orbit as stable if any orbit starting near the periodic orbit stays close to the periodic orbit for all positive time. Unfortunately, this notion turns out to be a bit too restrictive. For example, the equilibrium point at the origin of the planar pendulum is surrounded by concentric periodic orbits; therefore, it is natural to consider any one of these periodic orbits to be stable. However, any two such periodic orbits have different periods and thus at a fixed time they could be at diametrically opposite positions. To overcome this dilemma, we consider a periodic orbit Γ of Eq. (11.5) as a closed curve and neglect its time parametrization. In this context, we define the distance of a point $\widehat{\mathbf{x}} \in \mathbb{R}^2$ to a periodic orbit Γ, denoted by $\operatorname{dist}(\widehat{\mathbf{x}}, \Gamma)$, to as

$$\operatorname{dist}(\widehat{\mathbf{x}}, \Gamma) \equiv \min\{\,\|\widehat{\mathbf{x}} - \mathbf{x}\|, \text{for all } \mathbf{x} \in \Gamma\,\}.$$

Definition 11.8. *A periodic orbit Γ of the planar system (11.5) is said to be orbitally stable if, for any $\varepsilon > 0$, there is a $\delta > 0$ such that $\operatorname{dist}(\mathbf{x}^0, \Gamma) < \delta$ implies that $\operatorname{dist}(\varphi(t, \mathbf{x}^0), \Gamma) < \varepsilon$ for all $t \geq 0$. The periodic orbit Γ is said to be orbitally unstable if it is not orbitally stable.*

Definition 11.9. *A periodic orbit Γ of Eq. (11.5) is said to be orbitally asymptotically stable if it is orbitally stable and, in addition, there is a $b > 0$ such that $\operatorname{dist}(\mathbf{x}^0, \Gamma) < b$ implies that $\operatorname{dist}(\varphi(t, \mathbf{x}^0), \Gamma) \to 0$ as $t \to +\infty$, that is, $\omega(\mathbf{x}^0) \subset \Gamma$.*

Notice that in the definition above we said $\omega(\mathbf{x}^0) \subset \Gamma$. It is a consequence of the Poincaré–Bendixson theorem, to be presented in the following chapter, that $\omega(\mathbf{x}^0) = \Gamma$.

Definition 11.10. *A periodic orbit* Γ *of Eq.* (11.5) *is said to be orbitally asymptotically stable with asymptotic phase if it is orbitally asymptotically stable and, moreover, for any* $\mathbf{x}^0$ *with* dist $(\mathbf{x}^0, \Gamma) < b$ *and* $\mathbf{y}^0 \in \Gamma$ *there exists a real number* ν *such that*

$$\|\varphi(t, \mathbf{x}^0) - \varphi(t + \nu, \mathbf{y}^0)\| \to 0 \quad \text{as } t \to +\infty.$$

With these definitions, in conjunction with our earlier observations in this section, the theorem below is almost immediate:

Theorem 11.11. *Let* Γ *be a periodic orbit of Eq.* (11.5) *and let* $\psi(\theta)$ *be the corresponding* 2π*-periodic solution of Eq.* (11.9). *Then*

- Γ *is orbitally stable* [*respectively, orbitally asymptotically stable*] *if* $\psi(\theta)$ *is stable* [*respectively, asymptotically stable*] *as a solution of Eq.* (11.9),
- Γ *is orbitally asymptotically stable with asymptotic phase if it is orbitally asymptotically stable and, for any solution* $r(\theta, r_0)$ *of Eq.* (11.9) *with* r_0 *near* $\psi(0)$, *there is a real number* ν *such that the solution* $\theta(t)$ *of the initial-value problem*

$$\dot{\theta} = 1 + \Theta(r(\theta, r_0), \theta), \qquad \theta(0) = 0$$

has the property $\theta(t) - t \to \nu$ *as* $t \to +\infty$. $\Diamond$

Exercises _____ ♣ ♡ ♠ ◇

11.1. Consider the second-order equation $\ddot{y} + \dot{y}^3 + y = 0$, or the equivalent system

$$\dot{x}_1 = x_2$$
$$\dot{x}_2 = -x_1 - x_2^3.$$

What is the stability type of the equilibrium point at the origin?

11.2. Discuss the stability properties of the origin for the system

$$\dot{x}_1 = x_2 + x_1 x_2 + a x_1 x_2^2$$
$$\dot{x}_2 = -x_1 - x_1^2 + x_2^2$$

for various values of a.

11.3. Consider the system

$$\dot{x}_1 = x_2,$$
$$\dot{x}_2 = -x_1 + x_1^3.$$

Show that an application of the transformation theory formally takes $dr/d\theta$ to the equation $d\rho/d\theta = 0$. Make sure to exploit the symmetry of the vector field in your computations. Also, show that there is a first integral. Find one and draw the level curves near the origin.

11.4. Discuss the stability and instability of the equilibrium points of the system

$$\dot{x}_1 = -x_2 + x_2^3$$
$$\dot{x}_2 = -x_1 + x_1^3.$$

Note especially the equilibrium point at $(1, 0)$.

11.5. *Always odd power:* Suppose that the transformation theory of Section 5.2 is applied to Eq. (11.9) to obtain $d\rho/d\theta = a\rho^j + o(|\rho|^j)$ with $a \neq 0$. Show that j is an odd integer.
Hint: If j is even, then the origin is unstable for Eq. (11.5). Replacing θ by $-\theta$, one obtains the same result. Show that this is a contradiction.

11.6. C^∞ *vs. analytic:* Find a C^∞, but not analytic, example of Eq. (11.5) so that the transformation theory applied to $dr/d\theta$ leads to $d\rho/d\theta = 0$ yet the origin is unstable.

11.7. *Symmetry and center:* Consider the second-order differential equation $\ddot{x} + \dot{x}^2 + x = 0$, or the equivalent system

$$\dot{x}_1 = x_2$$
$$\dot{x}_2 = -x_1 - x_2^2.$$

(a) Show that there is a center in the neighborhood of the origin.
Hint: Symmetry makes it easy. Observe that if $(x_1(t), x_2(t))$ is a solution, so is $(x_1(-t), -x_2(-t))$.
(b) Show that an application of the transformation theory formally takes $dr/d\theta$ to the equation $d\rho/d\theta = 0$.
(c) Draw the phase portrait of the system on the computer; or, see Diener and Reeb [1986] for the phase portrait of this system as well as many other exciting planar pictures.

11.8. Show that all orbits of the equation $\ddot{y} + y + y^5 = 0$ are periodic. Use symmetry.

11.9. *Two centers:* Show that the system

$$\dot{x}_1 = 2x_1 x_2$$
$$\dot{x}_2 = \tfrac{1}{4} - x_1^2 + x_2^2$$

has two centers.
Hint: Use the change of variable $x_1 \mapsto \tfrac{1}{2} + x_1$ to find one of the centers.

11.10. *Different periods:* Show that every solution of the equation

$$\ddot{y} + \tfrac{1}{2}\left[y^2 + \sqrt{y^4 + \dot{y}^2}\right] y = 0$$

has the form $y(t) = c\sin(ct + d)$, where c and d are constants. Notice the different periods of different periodic solutions. Discuss the stability properties of each solution as well as the stability properties of the orbits on the plane.

11.11. *Hyperbolicity and asymptotic phase:* Let Γ be an orbitally asymptotically stable periodic orbit corresponding to a 2π-periodic solution $\psi(\theta)$ of the scalar equation (11.9). If $\psi(\theta)$ is hyperbolic, show that Γ is orbitally asymptotically stable with asymptotic phase as follows:

1. Without loss of generality, take $\psi(\theta) = 0$, that is, the periodic orbit corresponds to the zero solution of Eq. (11.9).
2. Prove that, for r_0 small, there are positive constants α and k such that $|r(\theta, r_0)| \leq ke^{-\alpha\theta}$ for $\theta \geq 0$.
3. If $\xi(t) = \theta(t) - t$, then $\xi(t) = \int_0^{\theta(t)} \mathcal{R}(r(\theta, r_0), \theta) \frac{dt}{d\theta} \, d\theta$, where $|dt/d\theta| \leq 2$, if r_0 is small.
4. There is a constant K such that $|\mathcal{R}(r(\theta, r_0), \theta) \frac{dt}{d\theta}| \leq Ke^{-\alpha\theta}$, for $\theta \geq 0$.
5. For any $\varepsilon > 0$, there is a $T_0 > 0$ such that $|\xi(t) - \xi(\tau)| < \varepsilon$ if $t, \tau \geq T_0$. Thus, there is a $\nu > 0$ such that $\xi(t) - \nu \to 0$ as $t \to +\infty$.

11.2. Poincaré–Andronov–Hopf Bifurcation

In this section we study bifurcations in the neighborhood of a nonhyperbolic equilibrium point with nonzero purely imaginary eigenvalues. It follows from the Implicit Function Theorem that under small perturbations of the vector field, an equilibrium point persists and no new equilibria are created. However, if the stability type of the equilibrium changes when subjected to perturbations, then this change is usually accompanied with either the appearance or disappearance of a small periodic orbit encircling the equilibrium point. Let us begin with a precise statement of this important bifurcation phenomenon:

Theorem 11.12. (Poincaré–Andronov–Hopf) *Let* $\dot{\mathbf{x}} = A(\lambda)\mathbf{x} + \mathbf{F}(\lambda, \mathbf{x})$ *be a* C^k, *with* $k \geq 3$, *planar vector field depending on a scalar parameter* λ *such that* $\mathbf{F}(\lambda, \mathbf{0}) = \mathbf{0}$ *and* $D_{\mathbf{x}}\mathbf{F}(\lambda, \mathbf{0}) = \mathbf{0}$ *for all sufficiently small* $|\lambda|$. *Assume that the linear part* $A(\lambda)$ *at the origin has the eigenvalues* $\alpha(\lambda) \pm i\beta(\lambda)$ *with* $\alpha(0) = 0$ *and* $\beta(0) \neq 0$. *Furthermore, suppose that the eigenvalues cross the imaginary axis with nonzero speed, that is,*

$$\frac{d\alpha}{d\lambda}(0) \neq 0. \tag{11.19}$$

Then, in any neighborhood U *of the origin in* $\mathbb{R}^2$ *and any given* $\lambda_0 > 0$ *there is a* $\bar{\lambda}$ *with* $|\bar{\lambda}| < \lambda_0$ *such that the differential equation* $\dot{\mathbf{x}} = A(\bar{\lambda})\mathbf{x} + \mathbf{F}(\bar{\lambda}, \mathbf{x})$ *has a nontrivial periodic orbit in* U.

It is remarkable that the essential hypotheses of the theorem concern only the linear part of the vector field. The requirement that the vector field vanish at the origin is inconsequential since it can always be satisfied with a change of variables around an arbitrary equilibrium point. To uncover some of the finer details of this bifurcation, such as the stability of the resulting periodic orbit, one must investigate the effects of the nonlinear terms. This poses nontrivial computational challenges some of which we will address later in this chapter. Before embarking on a proof of the bifurcation theorem above, let us examine several specific examples exhibiting bifurcations of periodic orbits from an equilibrium.

As our first example, let us reconsider a familiar system from Chapter 7 and illustrate some of the possibilities for bifurcation diagrams of periodic orbits near an equilibrium point.

Example 11.13. Consider the rather special planar system of the form

$$\dot{x}_1 = x_2 + \mathcal{F}(\lambda,\, r^2)x_1$$
$$\dot{x}_2 = -x_1 + \mathcal{F}(\lambda,\, r^2)x_2, \tag{11.20}$$

where λ is a scalar parameter, $r^2 = x_1^2 + x_2^2$, and $\mathcal{F}$ satisfies $\mathcal{F}(0,0) = 0$ so that the origin is an isolated equilibrium point. In polar coordinates, this system becomes the product system

$$\dot{r} = \mathcal{F}(\lambda,\, r^2)r$$
$$\dot{\theta} = 1. \tag{11.21}$$

The existence and stability properties of periodic solutions of Eq. (11.20) are the same as the corresponding equilibrium points and their stability types of the scalar differential equation $\dot{r} = \mathcal{F}(\lambda,\, r^2)r$. Indeed, if either $r = 0$ or $\mathcal{F}(\lambda,\, a^2) = 0$, then $(a\cos t,\, -a\sin t)$ is a 2π-periodic solution of Eq. (11.20) with "amplitude" a.

The bifurcation diagram for periodic solutions of Eq. (11.20) is simply a plot of the solutions of $\mathcal{F}(\lambda,\, a^2) = 0$ in the $(\lambda,\, a)$-plane together with the λ-axis. As usual, stable periodic orbits are indicated by solid curves and unstable ones with dashed curves.

Let us now take several specific forms for $\mathcal{F}$ and draw the corresponding bifurcation diagrams.

- For $\mathcal{F}(\lambda,\, r) = \lambda$: we obtain a linear perturbation of the linear harmonic oscillator. There is no nontrivial periodic orbit except at $\lambda = 0$, at which case there is one periodic orbit for each amplitude a. All periodic orbits are orbitally stable. See Figure 11.3a for the bifurcation diagram.
- For $\mathcal{F}(\lambda,\, r) = \lambda - r^2$: there is a unique nontrivial periodic orbit if $\lambda > 0$ for a particular value of a; namely, $a = \sqrt{\lambda}$. The periodic orbit is orbitally asymptotically stable. See Figure 11.3b for the bifurcation

diagram. Because the bifurcation curve emanates from the origin to the right, the bifurcation is called *supercritical*.

- For $\mathcal{F}(\lambda, r) = -(r^2 - c)^2 + c^2 + \lambda$ with $c > 0$ a fixed constant: there are two nontrivial periodic orbits, one orbitally unstable and the other orbitally asymptotically stable, for $-c^2 < \lambda < 0$ with amplitudes $\left[c \pm (\lambda + c^2)^{1/2}\right]^{1/2}$. The two periodic orbits coalesce and disappear as λ decreases through $-c^2$. There is only one periodic orbit for $\lambda > 0$ and it is orbitally asymptotically stable. See Figure 11.3c for the complete bifurcation diagram. Because the bifurcation curve emanates from the origin to the left, the bifurcation is called *subcritical*.

In each of these examples the hypotheses of the Poincaré–Andronov–Hopf bifurcation theorem are satisfied. The existence of a periodic orbit with small amplitude for small λ as asserted by the theorem is evident; however, the stability type of the periodic orbit depends on the nonlinear terms of the vector field. Moreover, as seen in the first and last cases, there can also be additional periodic orbits, possibly with large amplitudes, for a given small $|\lambda|$. Indeed, a glance at the bifurcation diagrams in Figure 11.3 suggests that the bifurcation diagram of the general situation should be a distortion of that of the linear harmonic oscillator but the diagram is still a graph over the a-axis. $\Diamond$

Our second example concerns a variant of the oscillator of Van der Pol. This famous equation exhibits the essential characteristics of the typical manner in which a periodic orbit appears through a Poincaré–Andronov–Hopf bifurcation. More importantly, however, the method of computations is realistic and it points the way to a successful analysis of the general case.

Example 11.14. *Van der Pol's oscillator:* Consider the following second-order differential equation depending on a small scalar parameter λ:

$$\ddot{y} - (2\lambda - y^2)\dot{y} + y = 0,$$

which is equivalent to the planar system

$$\begin{aligned}
\dot{x}_1 &= x_2 \\
\dot{x}_2 &= -x_1 + 2\lambda x_2 - x_1^2 x_2.
\end{aligned} \tag{11.22}$$

The eigenvalues of the linearization of Eq. (11.22) about the equilibrium point at the origin are $\lambda \pm i\sqrt{1 - \lambda^2}$. For $\lambda < 0$, the origin is asymptotically stable because the real parts of the eigenvalues are negative. At $\lambda = 0$, the origin is still asymptotically stable as we have shown in Example 11.2. For $\lambda > 0$, the real parts of the eigenvalues become positive and thus the origin is unstable.

It is clear that the planar system (11.22) satisfies the hypotheses of the Poincaré–Andronov–Hopf bifurcation theorem and thus it must have a

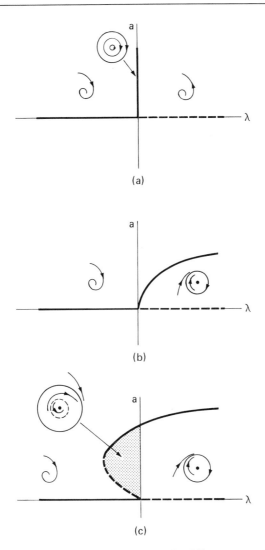

Figure 11.3. *Bifurcation diagrams of Eq. (11.20) for three different functions $\mathcal{F}$: (a) for $\mathcal{F}(\lambda, r^2) = \lambda$ is degenerate; (b) for $\mathcal{F}(\lambda, r^2) = \lambda - r^2$ is supercritical; and (c) for $\mathcal{F}(\lambda, r^2) = -(r^2 - c)^2 + c^2 + \lambda$ is subcritical.*

periodic orbit near the origin for some small values of λ, as seen in Figure 11.4. To gain insight into the dynamics of Eq. (11.22), let us perform some detailed computations which are also indicative of the general situation. Using polar coordinates and the transformation theory presented in the previous section, we will show below that, as the eigenvalues cross the

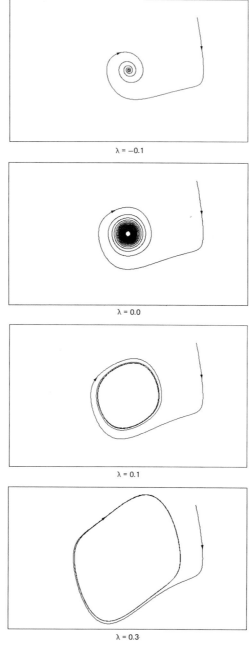

$\lambda = -0.1$

$\lambda = 0.0$

$\lambda = 0.1$

$\lambda = 0.3$

Figure 11.4. *Poincaré–Andronov–Hopf bifurcation in Van der Pol.*

imaginary axis, the origin gives up its stability to a periodic orbit. More specifically, we will demonstrate that, for each small $\lambda > 0$, there is a unique nontrivial periodic orbit Γ_λ near the origin which is orbitally asymptotically stable. Moreover, $\Gamma_\lambda \to \mathbf{0}$ as $\lambda \to 0$.

Let $F_1(\lambda, \mathbf{x}) = 0$ and $F_2(\lambda, \mathbf{x}) = 2\lambda x_2 - x_1^2 x_2$. Then there is a $\lambda_0 > 0$ such that

$$\|D\mathbf{F}(\lambda, \mathbf{0})\| < 1 \quad \text{for} \quad |\lambda| < \lambda_0$$

and thus the conditions (11.4) for reduction to a 2π-periodic scalar equation $dr/d\theta$ are satisfied as long as $|\lambda| < \lambda_0$. Indeed, for the oscillator of Van der Pol, Eq. (11.9) is given by

$$\frac{dr}{d\theta} = \frac{2\lambda r \sin^2 \theta - r^3 \cos^2 \theta \sin^2 \theta}{1 + 2\lambda \sin \theta \cos \theta - r^2 \cos^3 \theta \sin \theta}. \tag{11.23}$$

The next task is to compute a few terms of the Taylor series of Eq. (11.23) in r. For this purpose, it suffices to use the expansion $(1 - z)^{-1} = 1 + z + z^2 + z^3 + O(z^4)$ for $|z| < 1$, and some trigonometric identities. As a result, for $|\lambda|$ and r small we obtain the following as the Taylor series of Eq. (11.23):

$$\frac{dr}{d\theta} = [\lambda - \lambda \cos(2\theta) + O(\lambda^2)]\, r - \tfrac{1}{8}[1 - \cos(4\theta) + O(\lambda)]r^3 + O(r^4). \tag{11.24}$$

We now apply the transformation theory of periodic scalar equations to make the coefficients of the powers of r up through r^3 independent of θ. For the linear terms, if we change to the new variable ρ by putting

$$r = \rho\, e^{-(\lambda/2)\sin(2\theta)},$$

then Eq. (11.24) becomes

$$\frac{d\rho}{d\theta} = [\lambda + O(\lambda^2)]\rho - \tfrac{1}{8}[1 - \cos(4\theta) + O(\lambda)]\rho^3 + O(\rho^4).$$

As for the cubic terms, we let

$$\rho \mapsto \rho + a(\theta)\rho^3, \qquad \frac{da}{d\theta} = \tfrac{1}{8}\cos(4\theta),$$

as in Example 11.2, and arrive at the desired equation

$$\frac{d\rho}{d\theta} = [\lambda + O(\lambda^2)]\rho - [\tfrac{1}{8} + O(\lambda)]\rho^3 + O(\rho^4). \tag{11.25}$$

Notice that the ρ^2 term is not present. The Poincaré map Π of Eq. (11.25) is the map

$$\Pi(\rho_0) = \rho_0 + [\lambda + O(\lambda^2)]\rho_0 - [\tfrac{1}{8} + O(\lambda)]\rho_0^3 + O(\rho_0^4). \tag{11.26}$$

To find the periodic solutions of Eq. (11.25) near the origin, we need to locate the fixed points of its Poincaré map (11.26); equivalently, solve the equation

$$[\lambda + O(\lambda^2)]\rho_0 - [\tfrac{1}{8} + O(\lambda)]\rho_0^3 + O(\rho_0^4) = 0.$$

The solution $\rho_0 = 0$ of this equation corresponds to the equilibrium point at the origin. The other solution satisfies the equation

$$\lambda + O(\lambda^2) - [\tfrac{1}{8} + O(\lambda)]\rho_0^2 + O(\rho_0^3) = 0.$$

This equation can be considered either for λ as a function of ρ_0, or ρ_0 as a function of λ. In either case, this represents a curve—bifurcation curve—in the (λ, ρ_0)-plane which has the property that there are periodic orbits of the equation with approximate amplitude ρ_0 if and only if (λ, ρ_0) lie on this curve. For small positive λ and ρ_0, the approximate bifurcation curve is

$$\lambda = \tfrac{1}{8}\rho_0^2 + O(\rho_0^3) \tag{11.27}$$

or

$$\rho_0 = \sqrt{8}\sqrt{\lambda} + O(\lambda). \tag{11.28}$$

The latter equation (11.28) corresponds to a fixed point of the Poincaré map and the derivative of the Poincaré map at this fixed point is $1 - 2\lambda + O(\lambda^2)$. For small positive λ, this derivative has absolute value less than one which implies that the fixed point, hence the corresponding periodic solution, is asymptotically stable.

The results of these computations on the oscillator of Van der Pol can be summarized most conveniently in the bifurcation diagram depicted in Figure 11.5, where the approximate radius ρ_0 of the periodic orbit is plotted as a function of the parameter λ. The equilibrium point at the origin is included as a periodic orbit as well. The solid curve represents the asymptotically stable periodic orbit and the dashed line the unstable solution. As λ is increased through $\lambda = 0$, the origin gives up its stability to the periodic orbit which grows in amplitude proportional to $\sqrt{\lambda}$. $\Diamond$

We now turn to a proof of the Poincaré–Andronov–Hopf bifurcation theorem. To make the subsequent computations manageable, it is convenient to put the linear part of the vector field in a simpler form. Using a linear change of variables, we may transform the matrix of the linearized vector field at the origin to the matrix

$$\begin{pmatrix} \alpha(\lambda) & \beta(\lambda) \\ -\beta(\lambda) & \alpha(\lambda) \end{pmatrix}.$$

From the assumption $\beta(0) \neq 0$, we can change the time variable so that $\beta(\lambda) = 1$ for $|\lambda|$ small. Also, the assumption $(d\alpha/d\lambda)(0) \neq 0$, in conjunction with the Inverse Function Theorem, implies that there is a one-to-one

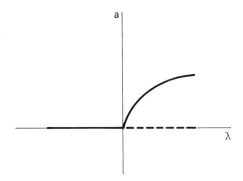

Figure 11.5. *The bifurcation diagram of Van der Pol's oscillator: approximate radius of the periodic orbit, which is asymptotically stable, as a function of the parameter λ.*

correspondence between $\alpha(\lambda)$ and λ. This permits us to use $\alpha(\lambda)$ as the parameter rather than λ. As a result of all these transformations, we may assume that the linear part of the vector field at the equilibrium point is of the form

$$\begin{pmatrix} \lambda & 1 \\ -1 & \lambda \end{pmatrix}.$$

We now assume that the normalization processes above have been performed and confine our attention to the following setting. Let $\mathbf{F}$ be a given C^k function, $k \geq 3$,

$$\mathbf{F} : \mathbb{R} \times \mathbb{R}^2 \to \mathbb{R}^2; \quad (\lambda, \mathbf{x}) \mapsto \mathbf{F}(\lambda, \mathbf{x}),$$

satisfying

$$\mathbf{F}(\lambda, \mathbf{0}) = \mathbf{0}, \qquad D_{\mathbf{x}}\mathbf{F}(\lambda, \mathbf{0}) = \mathbf{0}, \tag{11.29}$$

and consider the system of differential equations

$$\begin{aligned} \dot{x}_1 &= \lambda x_1 + x_2 + F_1(\lambda, x_1, x_2) \\ \dot{x}_2 &= -x_1 + \lambda x_2 + F_2(\lambda, x_1, x_2). \end{aligned} \tag{11.30}$$

For this normalized system, the Poincaré–Andronov–Hopf theorem can be reformulated in the following way, where the variable a should be viewed as the approximate amplitude of a periodic solution:

Theorem 11.15. *For the system* (11.30), *there are constants* $a_0 > 0$, $\lambda_0 > 0$, $\delta_0 > 0$, *and real-valued* C^1 *functions* $\lambda^\star(a)$, $T^\star(a)$ *of a real variable* a, *and a* $T^\star(a)$-*periodic vector-valued function* $\mathbf{x}^\star(t, a)$ *with the following properties: for* $0 \leq a < a_0$,

- $\lambda^\star(0) = 0,\quad T^\star(0) = 2\pi,\quad \|\mathbf{x}^\star(0,\,a)\| = a.$
- The function $\mathbf{x}^\star(t,\,a)$ is a solution of the system (11.30) with the parameter value $\lambda = \lambda^\star(a)$ and its components are given by

$$x_1^\star(t,\,a) = a\cos t + o(|a|),$$
$$x_2^\star(t,\,a) = -a\sin t + o(|a|) \quad \text{as } a \to 0.$$

- For $|\lambda| < \lambda_0$ and $|T - 2\pi| < \delta_0$, every T-periodic solution $\mathbf{x}(t)$ of eq. (11.30) satisfying $\|\mathbf{x}(0)\| = a$ and $\|\mathbf{x}(t)\| < a_0$ must be given by the function $\mathbf{x}^\star(t,\,a)$, except for a possible translation in phase.

Proof. Using the technique of reduction from the previous section, we compute that the 2π-periodic scalar equation (11.9) for the system (11.30) has the form

$$\frac{dr}{d\theta} = \lambda r + \mathcal{P}(\lambda,\,r,\,\theta), \tag{11.31}$$

with

$$\mathcal{P}(\lambda,\,0,\,\theta) = 0, \qquad D_r\mathcal{P}(\lambda,\,0,\,\theta) = 0. \tag{11.32}$$

From Lemma 11.6, we need to investigate the 2π-periodic solutions of Eq. (11.31). If $r(\lambda,\,\theta,\,a)$ is a solution of Eq. (11.31) with initial value $r(\lambda,\,0,\,a) = a$, then $r(\lambda,\,\theta + 2\pi,\,a) = r(\lambda,\,\theta,\,a)$ for all θ if and only if $r(\lambda,\,2\pi,\,a) = a$. From the variation of the constants formula, the solutions of Eq. (11.31) satisfying $r(\lambda,\,0,\,a) = a$ are given by

$$r(\lambda,\,\theta,\,a) = e^{\lambda\theta}a + \int_0^\theta e^{\lambda(\theta-s)}\,\mathcal{P}(\lambda,\,r(\lambda,\,s,\,a),\,s)\,ds \tag{11.33}$$

and thus $r(\lambda,\,2\pi,\,a) = a$ if and only if λ and a satisfy

$$\left(1 - e^{-2\pi\lambda}\right)a + \int_0^{2\pi} e^{-\lambda s}\,\mathcal{P}(\lambda,\,r(\lambda,\,s,\,a),\,s)\,ds = 0. \tag{11.34}$$

Using the Implicit Function Theorem, we will show that the values of a and λ satisfying this equation is a curve in the $(a,\,\lambda)$-plane and that this curve is a graph over the a-axis expressed by λ as a function of a. From Eq. (11.32), $a = 0$ satisfies Eq. (11.34) because the integrand vanishes. This trivial solution corresponds to an equilibrium solution. To find nontrivial periodic solutions, let us consider the function $h(a,\,\lambda)$ given by

$$h(a,\,\lambda) \equiv 1 - e^{-2\pi\lambda} + \frac{1}{a}\int_0^{2\pi} e^{-\lambda s}\,\mathcal{P}(\lambda,\,r(\lambda,\,s,\,a),\,s)\,ds$$

for $a \neq 0$, and define

$$h(0,\,0) = 0.$$

Now, it follows from Eq. (11.32) that h is a C^1 function near $a = \lambda = 0$, and

$$\frac{\partial h}{\partial \lambda}(0, 0) = -2\pi \neq 0.$$

Therefore, from the Implicit Function Theorem, there exists a function $\lambda^\star(a)$ with $\lambda^\star(0) = 0$ such that

$$h(a, \lambda^\star(a)) = 0.$$

Now, with this $\lambda^\star(a)$, the function $r^\star(\theta, a) \equiv r(\lambda^\star(a), \theta, a)$ is a 2π-periodic solution of Eq. (11.30). Consequently, the orbit through the point $\mathbf{x}^0(a) = (a, 0)$ given by

$$\gamma(\mathbf{x}^0(a)) = \{\, (x_1, x_2) \,:\, x_1 = r^\star(\theta, a)\cos\theta,$$
$$x_2 = -r^\star(\theta, a)\sin\theta, \ 0 \le \theta \le 2\pi \,\}$$

is a periodic orbit of the system (11.30).

To obtain the corresponding solution of Eq. (11.30), let $\theta^\star(t, a)$ be the solution of

$$\dot\theta = 1 + \Theta(\lambda^\star(a), r^\star(\theta, a), \theta)$$

satisfying $\theta^\star(0, a) = 0$. Then the minimal period of $\gamma(\mathbf{x}^0(a))$ is determined by the first value $T^\star(a)$ for which

$$\theta^\star(T^\star(a), a) = 2\pi.$$

In particular, we have $T^\star(0) = 2\pi$. If we now define

$$\mathbf{x}^\star(t, a) \equiv \big(r^\star(\theta^\star(t, a), a)\cos\theta^\star(t, a), \, -r^\star(\theta^\star(t, a), a)\sin\theta^\star(t, a)\big),$$

then it is not difficult to see that $\mathbf{x}^\star$ satisfies the conditions of the theorem. $\Diamond$

The stability type of the periodic orbits in the theorem above can be inferred from the derivative of the function $\lambda^\star(a)$ when this derivative is not zero; see Figure 11.6. More specifically, we have the following theorem with a somewhat technical proof:

Theorem 11.16. *Let $\lambda^\star(a)$ be the function given in Theorem 11.15 and let Γ_a be the corresponding periodic orbit of Eq. (11.30). Then, for sufficiently small $a = \bar{a}$, the periodic orbit $\Gamma_{\bar{a}}$ is orbitally asymptotically stable if $d\lambda^\star(\bar{a})/da > 0$, and unstable if $d\lambda^\star(\bar{a})/da < 0$.*

Proof. In this proof we will continue to use the notation developed in the proof of the previous theorem. For fixed $\bar{a}$, let $\bar{\lambda} = \lambda^\star(\bar{a})$. Using the solution

$$r(\bar{\lambda}, \theta, a) = e^{\bar{\lambda}\theta} a + \int_0^\theta e^{\bar{\lambda}(\theta - s)} \, \mathcal{P}(\bar{\lambda}, r(\bar{\lambda}, s, a), s)\, ds \qquad (11.35)$$

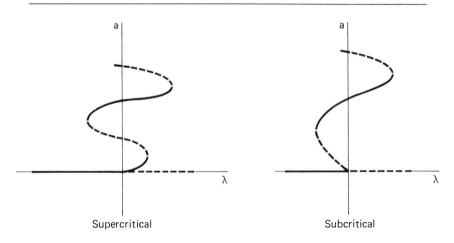

Figure 11.6. *Two typical graphs of the function $\lambda^\star(a)$. Stability type of a periodic orbit $\Gamma_{\bar{a}}$ with amplitude $\bar{a}$ can be inferred from the bifurcation diagram: $\Gamma_{\bar{a}}$ is orbitally asymptotically stable if $d\lambda^\star(\bar{a})/da > 0$, and unstable if $d\lambda^\star(\bar{a})/da < 0$.*

of the scalar differential equation (11.31), we will compute the derivative of the Poincaré map Π of Eq. (11.31) at $a = \bar{a}$. For computational purposes, it is convenient to consider the quantity $q(\bar{a})$ defined by

$$q(\bar{a}) \equiv e^{2\pi\bar{\lambda}} \left[\Pi'(\bar{a}) - 1 \right],$$

so that the stability type of the periodic orbit $\Gamma_{\bar{a}}$ can be inferred from the sign of $q(\bar{a})$. We will now determine this sign by differentiating various expressions.

First, observe that differentiating Eq. (11.35) with respect to a and then evaluating it at $a = \bar{a}$ and $\theta = 2\pi$ results in the following expression for $q(\bar{a})$:

$$q(\bar{a}) = 1 - e^{-2\pi\bar{\lambda}} + \int_0^{2\pi} e^{-\bar{\lambda}s} \frac{\partial P}{\partial r}(\bar{\lambda}, r(\bar{\lambda}, s, \bar{a}), s) \frac{\partial r}{\partial a}(\bar{\lambda}, s, \bar{a}) \, ds. \quad (11.36)$$

Second, using $\lambda = \lambda^\star(a)$, from Eq. (11.34) we have

$$\left(1 - e^{-2\pi\lambda^\star(a)}\right) a + \int_0^{2\pi} e^{-\lambda^\star(a)s} \mathcal{P}(\lambda^\star(a), r(\lambda^\star(a), s, a), s) \, ds = 0.$$

$$(11.37)$$

To make the notation manageable in subsequent computations, it is convenient to rewrite this equation as

$$\left(1 - e^{-2\pi\lambda^\star(a)}\right) a + ag(\lambda^\star(a), a) = 0 \quad (11.38)$$

or

$$1 - e^{-2\pi\lambda^\star(a)} + g(\lambda^\star(a), a) = 0. \tag{11.39}$$

The existence of such a function g follows from Eq. (11.32). Now, differentiating Eq. (11.38) with respect to a and then putting $a = \bar{a}$ yields

$$\left(1 - e^{-2\pi\bar{\lambda}}\right) + \bar{a}\frac{d}{da}\left(1 - e^{-2\pi\lambda^\star(a)}\right)\Big|_{a=\bar{a}}$$

$$+ \frac{d}{da}(ag(\bar{\lambda}, a))\Big|_{a=\bar{a}} + \frac{d}{da}(\bar{a}g(\lambda^\star(a), \bar{a}))\Big|_{a=\bar{a}} = 0.$$

Combining the first and the third "terms" in the equation above in conjunction with Eq. (11.36), and combining the second and the fourth terms in conjunction with Eq. (11.39), we arrive at

$$q(\bar{a}) - \bar{a}\frac{\partial g}{\partial a}(\bar{\lambda}, \bar{a}) = 0.$$

As a result, we obtain the following formula for the derivative of the Poincaré map:

$$\Pi'(\bar{a}) - 1 = \bar{a}e^{-2\pi\bar{\lambda}}\frac{\partial g}{\partial a}(\bar{\lambda}, \bar{a}). \tag{11.40}$$

Finally, differentiate Eq. (11.39) with respect to a and put $a = \bar{a}$:

$$-(\lambda^\star)'(\bar{a})\left[2\pi e^{-2\pi\bar{\lambda}} + \frac{\partial g}{\partial \lambda^\star}(\bar{\lambda}, \bar{a})\right] = \frac{\partial g}{\partial a}(\bar{\lambda}, \bar{a}).$$

Thus, for sufficiently small $\bar{a}$, we have

$$\text{sign}\left[-(\lambda^\star)'(\bar{a})\right] = \text{sign}\,\frac{\partial g}{\partial a}(\bar{\lambda}, \bar{a}). \tag{11.41}$$

Now, from the relations (11.40) and (11.41), the desired conclusion

$$\text{sign}\left[\Pi'(\bar{a}) - 1\right] = \text{sign}\left[-(\lambda^\star)'(\bar{a})\right]$$

is self-evident. $\diamondsuit$

As a reward of this somewhat intricate proof, you should reexamine the earlier bifurcation diagrams in this chapter. In particular, you should differentiate the bifurcation curve (11.27) of the oscillator of Van der Pol.

Exercises ────────────────────────────────── ♣ ♡ ♠ ◇

11.12. *Rayleigh's equation:* The second-order equation $\ddot{y} + \dot{y}^3 - 2\lambda\dot{y} + y = 0$, where λ is a small scalar parameter, arises in the theory of sound and is known as *Rayleigh's equation*. Convert this into a first-order system and investigate the Poincaré–Andronov–Hopf bifurcation about $\lambda = 0$. Determine the stability type of the periodic orbit.

Answer: The approximate bifurcation curve is $\lambda = a^2/8 + O(a^3)$.

11.13. Discuss the periodic orbits of the system

$$\dot{x}_1 = \lambda x_1 + x_2 + x_1 x_2 + a x_1 x_2^2$$
$$\dot{x}_2 = -x_1 + \lambda x_2 - x_1^2 + x_2^2$$

near the origin for $|\lambda|$ small.

11.14. Discuss the Poincaré–Andronov–Hopf bifurcation near the origin for λ near zero of the system

$$\dot{x}_1 = x_1 + x_2$$
$$\dot{x}_2 = (\lambda - 2)x_1 + (\lambda - 1)x_2 - x_1^3 - x_1^2 x_2.$$

In this example, you should first put the linear part for $\lambda = 0$ in Jordan Normal Form.

11.15. *Center in Hamiltonian systems:* Consider a Hamiltonian function of the form

$$H(x_1, x_2) = \frac{\beta}{2}\left(x_1^2 + x_2^2\right) + o\left((|x_1| + |x_2|)^2\right)$$

as $|x_1|, |x_2| \to 0$, and $\beta \neq 0$. Then, the Hamiltonian system

$$\dot{x}_1 = \frac{\partial H}{\partial x_2}$$
$$\dot{x}_2 = -\frac{\partial H}{\partial x_1}$$

has a center at the origin. This is easy to show using the method of Liapunov functions. Establish the existence of a center from the Poincaré–Andronov–Hopf bifurcation theorem.

Hint: Consider the one-parameter system

$$\dot{x}_1 = \lambda \frac{\partial H}{\partial x_1} + \frac{\partial H}{\partial x_2}$$
$$\dot{x}_2 = \lambda \frac{\partial H}{\partial x_2} - \frac{\partial H}{\partial x_1}$$

for λ in a neighborhood of zero. Now verify that this system satisfies the hypotheses of the bifurcation theorem and that there are no periodic orbits except for $\lambda = 0$. For further information on this approach, see Schmidt [1978] and his exposition in Marsden and McCracken [1976].

11.3. Computing Bifurcation Curves

Although the hypotheses of the Poincaré–Andronov–Hopf bifurcation theorem concern only the linear part of the vector field, the shape of the bifurcation curve depends very much on the nonlinear terms. To utilize the results from the previous section, for example, Theorem 11.16, in specific applications, it is important to have constructive procedures for determining these bifurcation curves. In this section we first outline such a procedure which, in effect, is a rephrasing of the transformation theory already presented in Chapter 5. Then we determine the bifurcation diagrams that are most likely to occur in applications.

The main idea behind calculating bifurcation curves is to determine an appropriate transformation of the variable $r(\lambda, \theta)$ so that the 2π-periodic differential equation $dr/d\theta$ is transformed to one with constant coefficients. Once this is accomplished, the approximate form of the Poincaré map, and thus the corresponding bifurcation curve, can be written down rather easily.

Let us suppose that Eq. (11.9) can be formally written as a power series in r:

$$\frac{dr}{d\theta} = \sum_{k=1}^{+\infty} \mathcal{R}_k(\lambda, \theta)\, r^k, \tag{11.42}$$

where the coefficients $\mathcal{R}_k(\lambda, \theta)$ are 2π-periodic functions in θ.

Next, we introduce the new variable ρ defined by

$$r = \rho + \sum_{k=1}^{+\infty} b_k(\lambda, \theta)\, \rho^k, \tag{11.43}$$

where $b_1(0, \theta) = 0$ and $b_k(\lambda, \theta)$ are 2π-periodic functions to be determined. In the new variable the differential equation (11.42) has the form

$$\frac{d\rho}{d\theta} = \sum_{k=1}^{+\infty} \left[c_k(\lambda, \theta) - \frac{\partial b_k}{\partial \theta}(\lambda, \theta) \right] \rho^k.$$

Now, choose the functions b_k so that

$$\frac{\partial b_k}{\partial \theta}(\lambda, \theta) = c_k(\lambda, \theta) - \bar{c}_k(\lambda), \tag{11.44}$$

where $\bar{c}_k$ is the "average" of c_k given by

$$\bar{c}_k(\lambda) = \frac{1}{2\pi} \int_0^{2\pi} c_k(\lambda, s)\, ds.$$

With these choices of b_k, the differential equation (11.42) formally reduces to the following differential equation with constant coefficients:

$$\frac{d\rho}{d\theta} = \sum_{k=1}^{+\infty} \bar{c}_k(\lambda)\, \rho^k. \tag{11.45}$$

Here is an important practical fact about this differential equation:

Lemma 11.17. *The formal power series* (11.44) *contains only the odd powers of ρ, that is,*

$$\frac{d\rho}{d\theta} = \sum_{k=1}^{+\infty} \bar{c}_{2k+1}(\lambda)\,\rho^{2k+1}. \tag{11.46}.$$

Proof. We indicate only an outline. If $r(\lambda, \theta)$ is a solution of Eq. (11.42), then so is $-r(\lambda, \theta + \pi)$. Consequently, the even coefficients $\mathcal{R}_{2k}(\lambda, \theta)$ in Eq. (11.42) have average zero. This implies that the coefficients $c_{2k}(\lambda, \theta)$ in Eq. (11.45) have the same property. $\diamond$

Actual implementation of the computations outlined above can be rather cumbersome in almost any realistic example. Fortunately, there are two reasons for comfort. First, in most applications we need to compute only the first few b_k. Moreover, since we need to know only the approximate shape of bifurcation curves, it is advantageous to make formal expansion in powers of the parameter λ as well and retain only a few of the terms. Second, these tedious computations can be performed on the computer rather routinely with the help of one of many existing symbol manipulation programs, once such a code is developed. With these remarks, you should now recompute the transformation of the variables in Example 11.14, even if you have to resort only to paper and pencil.

In the remaining part of this section we will examine a special case of the Poincaré–Andronov–Hopf bifurcation (Theorem 11.12) by imposing restrictions on the first two terms of the differential equation (11.46). These restrictions are not very stringent and they are met in typical—*generic*—situations.

Consider the following planar differential equation:

$$\dot{\mathbf{x}} = \begin{pmatrix} \alpha(\lambda) & \beta(\lambda) \\ -\beta(\lambda) & \alpha(\lambda) \end{pmatrix} \mathbf{x} + \mathbf{F}(\lambda, \mathbf{x}), \tag{11.47}$$

where λ is a scalar parameter, $\mathbf{F}$ is a C^k function, with $k \geq 4$, satisfying $\mathbf{F}(\lambda, \mathbf{0}) = \mathbf{0}$ and $D_{\mathbf{x}}\mathbf{F}(\lambda, \mathbf{0}) = \mathbf{0}$ for sufficiently small $|\lambda|$. Moreover, assume that $\alpha(0) = 0$ but $\beta(0) \neq 0$. Then using the transformation theory outlined above, one can show that the corresponding scalar differential equation (11.46) with constant coefficients has the form

$$\frac{d\rho}{d\theta} = \left[\frac{\alpha'(0)}{\beta(0)}\lambda + O(\lambda^2)\right]\rho - \left[\frac{c}{\beta(0)} + O(\lambda)\right]\rho^3 + O(\rho^4), \tag{11.48}$$

where $\alpha'(0) = (d\alpha/d\lambda)(0)$ and c is some constant. From the form of the Poincaré map of Eq. (11.48) and Theorem 11.16, the following result on the bifurcation curve $\lambda(a)$, where a is the approximate amplitude, of Eq. (11.47) should be evident:

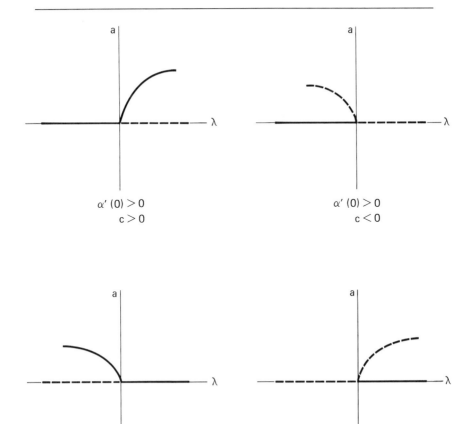

Figure 11.7. *Bifurcation diagrams of periodic orbits in generic Poincaré–Andronov–Hopf bifurcations.*

Theorem 11.18. (Generic Poincaré–Andronov–Hopf Bifurcation) *The bifurcation curve for periodic orbits of Eq. (11.47) with the additional hypotheses $\alpha'(0) \neq 0$ and $c \neq 0$ in Eq. (11.48) is approximately*

$$\lambda(a) = \frac{c}{\alpha'(0)}\, a^2 + O(a^3)$$

as $a \to 0$. The nontrivial periodic orbit is orbitally asymptotically stable if $c\alpha'(0) > 0$, and unstable if $c\alpha'(0) < 0$. $\diamond$

The possible generic bifurcation diagrams resulting from this theorem are depicted in Figure 11.7. It is now instructive to reexamine two of our previous examples in light of these bifurcation diagrams. The bifurcation in the oscillator of Van der Pol in Example 11.14 is generic because $\alpha'(0) = 1$ and $c = 1/8$. The bifurcation in the linear harmonic oscillator of Case I in Example 11.13, however, is not generic since $c = 0$.

We end this section with a remark concerning the coefficient c of the cubic term in Eq. (11.48). One can, of course, use the transformation theory above to derive a general explicit formula for c in terms of the coefficients of the vector field (11.47). Because of its importance in applications, there exist several variants of such a formula. Here is a popular one:

$$
\begin{aligned}
\text{sign of } c = \text{sign of } \quad & (F_1)_{x_1 x_1 x_1} + (F_1)_{x_1 x_2 x_2} + (F_2)_{x_1 x_1 x_2} + (F_2)_{x_2 x_2 x_2} \\
& - \tfrac{1}{\beta(0)} \Big\{ (F_1)_{x_1 x_2} \left[(F_1)_{x_1 x_1} + (F_1)_{x_2 x_2} \right] \\
& \quad - (F_2)_{x_1 x_2} \left[(F_2)_{x_1 x_1} + (F_2)_{x_2 x_2} \right] \\
& \quad - (F_1)_{x_1 x_1} (F_2)_{x_1 x_1} + (F_1)_{x_2 x_2} (F_2)_{x_2 x_2} \Big\},
\end{aligned}
$$

where, for example, $(F_1)_{x_1 x_2}$ denotes $(\partial^2 F_1 / \partial x_1 \partial x_2)(0, \mathbf{0})$. To practice your partial differentiation, after putting the oscillator of Van der Pol into the form (11.47), you may wish to compute the sign of its cubic term with this formula.

Exercises

11.16. *FitzHugh Neuron Model:* In 1961, Fitzhugh proposed the following system as a model of nerve impulse transmission:

$$
\begin{aligned}
\dot{x}_1 &= -x_1(x_1 - a)(x_1 - 1) - x_2 + qH(t) \\
\dot{x}_2 &= b(x_1 - \gamma x_2),
\end{aligned}
$$

where $0 < a < 1$, $\gamma > 0$, $b > 0$, and q are constants; $H(t)$ is the Heaviside function $H(t) = 0$ for $t < 0$ and $H(t) = 1$ for $t \geq 0$. In this model, q represents the stimulus, x_1 represents the response (instantaneous turning on of sodium permeability through the nerve membrane), and x_2 represents a recovery variable (turning on of potassium permeability). Suppose that $\gamma \leq 3(a^2 - a + 1)^{-1}$ and show that there is a critical value q_0 of q such that a Poincaré–Andronov–Hopf bifurcation occurs at q_0. Discuss the stability of the resulting periodic orbit. For further information on this important model, see FitzHugh [1961] and Jones and Sleeman [1983].

11.17. *Chemical instabilities and sustained oscillations:* Consider the chemical reaction

$$
\begin{aligned}
A &\to X \\
B + X &\to Y + D \\
2X + Y &\to 3X \\
X &\to E,
\end{aligned}
$$

where the initial and the final concentrations of the chemicals A, B, D, and E are constant, and the concentrations of X and Y satisfy the differential equations

$$\dot{X} = a - (b+1)X + X^2Y$$
$$\dot{Y} = bX - X^2Y$$

for some positive constants a and b. If you have not studied chemistry, just begin with this system of differential equations and explore their dynamics.

1. Show that $(a, b/a)$ is an equilibrium point which is stable for $a^2 + 1 > b$ and unstable for $a^2 + 1 < b$.
2. Fixing a and using b as a parameter, show that the conditions of the Poincaré–Andronov–Hopf bifurcation theorem are satisfied at $b = a^2 + 1$ for the differential equations $x_1 = X - a$ and $x_2 = Y - b/a$.
3. For $b = a^2 + 1$, determine the stability properties of the solution $(x_1, x_2) = (0, 0)$.
4. Find the approximate formula of the bifurcation curve for the periodic orbits.

For further information on this problem, see Lefever and Nicholis [1971].

11.18. *A predator-prey model:* Consider the system

$$\dot{x}_1 = rx_1\left(1 - \frac{x_1}{k}\right) - \frac{\beta x_1 x_2}{\alpha + x_1}$$
$$\dot{x}_2 = sx_2\left(1 - \frac{x_2}{\nu x_1}\right),$$

where all the parameters are positive. In this model, the predator x_2 becomes satiated during periods of abundance and whose carrying capacity is proportional to the amount of prey, x_1, available. For more information on these equations, see May [1973].

(i) Show that there is exactly one equilibrium point $\bar{x}$ at which the predator and prey coexist, that is, both coordinates of the equilibrium point are positive.
(ii) Show that there is a value of the parameter s at which the equilibrium point $\bar{x}$ undergoes a Poincaré–Andronov–Hopf bifurcation.
(iii) If you try to determine the stability properties of the resulting periodic orbit, be aware. The computations are very lengthy; this may be a good place to explore symbolic manipulations on the machine.

11.19. *A generic bifurcation:* Show that in the system below there is a generic Poincaré–Andronov–Hopf bifurcation from the equilibrium point $(1, 1)$ at $\lambda = 0$:

$$\dot{x}_1 = x_1\left[1 + \tfrac{1}{4}\lambda^2 - \tfrac{1}{4}(x_1 - 1 - \lambda)^2 - x_2\right]$$
$$\dot{x}_2 = x_2(-1 + x_1).$$

Determine whether the bifurcation is subcritical or supercritical.

11.20. *A nongeneric bifurcation:* Consider the differential equation, in polar coordinates, given by

$$\dot{r} = r(-2\lambda^2 + 3\lambda r^2 - r^4)$$
$$\dot{\theta} = 1.$$

Show that the conditions of the generic Poincaré–Andronov–Hopf bifurcation theorem are not satisfied and there are two periodic orbits for each $\lambda > 0$.

11.21. *Damping with one sign change:* Notice that the damping coefficient in Van der Pol's equation changes sign twice. Discuss the Poincaré–Andronov–Hopf bifurcation for the second-order equation

$$\ddot{x} + (\lambda + x)\dot{x} + x + x^2 = 0,$$

where the damping coefficient changes sign only once. For a more general discussion of such equations, see Obi [1954].

Bibliographical Notes ————————————————————

The stability of an equilibrium point whose linearization has purely imaginary eigenvalues was investigated extensively by Liapunov. In fact, he used polar coordinates and transformation theory as given in the text, and obtained Theorem 11.4 by showing that the formal series in the transformation was convergent.

As the label suggests, Theorem 11.12 was independently discovered by Andronov [1929] and Hopf [1943]. Hopf remarks in his paper that the theorem must have been known to Poincaré [1892] because in his writings Poincaré has discussions on the effect of the change of stability of an equilibrium point in the creation of a limit cycle (an isolated periodic orbit). It is not always easy to give due and proper credit to individuals, especially in naming theorems. Over repeated objections, Theorem 11.12 is often to referred in the mathematical literature as simply the Hopf bifurcation. In the interest of fairness, we have deviated from the usual practice and used all three names, however cumbersome this may be.

The number of mathematical sources on the Poincaré–Andronov–Hopf bifurcation is large. In addition to the two classic papers mentioned above, there are two books devoted entirely to this bifurcation: Hassard et al. [1980], which addresses computational issues, and Marsden and McCracken [1976], which contains an English translation of Hopf's paper.

There are more general methods for investigating the local bifurcations of periodic orbits than the one we have presented in this chapter. From the theoretical as well the computational viewpoint, perhaps the most important of them is a reduction to the zeros of a bifurcation function, using an abstract variant of the Method of Liapunov–Schmidt. The setting of this procedure is necessarily infinite dimensional and requires rather sophisticated analysis beyond the intended scope of our book. If you are familiar with Banach Spaces and the Implicit Function Theorem in such spaces, we urge you to consult Chow and Hale [1982], Golubitsky and Schaeffer [1985], and the references therein for this important application of the Method of Liapunov–Schmidt to the study of bifurcations of periodic orbits.

Poincaré–Andronov–Hopf bifurcation has appropriate generalizations for partial differential equations which can be found in, for example, Henry [1981], Kielhofer [1979], or Marsden and McCracken [1976].

12

Periodic Orbits

 After equilibrium points, the most interesting solutions of planar differential equations are periodic orbits. In fact, we have seen in the previous chapter the birth of periodic orbits when a nonhyperbolic equilibrium point undergoes a Poincaré–Andronov–Hopf bifurcation. There can also be periodic orbits far away from equilibrium points. The detection of such orbits is very difficult. In 1900, as part of problem sixteen of his famous list, Hilbert posed the following question: What is the number of (isolated) periodic orbits of a general polynomial system of differential equations on the plane? The problem remains unsolved even for the case where the components of the planar vector field are quadratic polynomials. Our inability to solve this basic problem exemplifies, in a striking way, the limited scope of our knowledge of periodic orbits. Despite the somber note, in this chapter we first present several basic theorems on the presence or absence of periodic orbits of planar systems. We then investigate the stability and local bifurcations of periodic orbits in terms of Poincaré maps. As an important application of these ideas, we establish the existence of a globally attracting periodic orbit of the oscillator of Van der Pol. We conclude the chapter with an example illustrating how a periodic orbit can bifurcate from a homoclinic loop.

12.1. The Poincaré–Bendixson Theorem

We begin this section by stating a fundamental theorem on the possible α- and ω-limit sets of autonomous planar differential equations. For the definitions of limit sets, you may refer back to Chapter 7.

Theorem 12.1. *Suppose that* $\dot{\mathbf{x}} = \mathbf{f}(\mathbf{x})$ *is a planar system with a finite number of equilibrium points. If the positive orbit* $\gamma^+(\mathbf{x}^0)$ *of* $\mathbf{x}^0$ *is bounded, then one of the following is true:*

- *The* ω-*limit set* $\omega(\mathbf{x}^0)$ *is a single point* $\bar{\mathbf{x}}$ *which is an equilibrium point and* $\varphi(t, \mathbf{x}^0) \to \bar{\mathbf{x}}$ *as* $t \to +\infty$.
- $\omega(\mathbf{x}^0)$ *is a periodic orbit* Γ *and either* $\gamma^+(\mathbf{x}^0) = \omega(\mathbf{x}^0) = \Gamma$ *or else* $\gamma^+(\mathbf{x}^0)$ *spirals with increasing time toward* Γ *on one side of* Γ.
- $\omega(\mathbf{x}^0)$ *consists of equilibrium points and orbits whose* α- *and* ω-*limit sets are the equilibrium points.* $\Diamond$

The three possibilities above hold for $\alpha(\mathbf{x}^0)$ when $\gamma^-(\mathbf{x}^0)$ is bounded. We will refrain from presenting a proof of this theorem. However, we should point out that any such proof relies on another fact known as the *Jordan Curve Theorem*, a result that is "geometrically obvious" but notoriously difficult to prove. Since we will use this fact tacitly throughout this chapter, here is its statement:

Theorem 12.2. (Jordan Curve Theorem) *A closed curve in* $\mathbb{R}^2$ *which does not intersect itself separates* $\mathbb{R}^2$ *into two connected components, one bounded, which is called the interior of the curve, and the other unbounded, which is called the exterior of the curve.* $\Diamond$

After this short diversion, let us return to Theorem 12.1 on limit sets. We have already encountered the occurrence of the first two cases in specific planar differential equations. Because of its common occurrence in theory and in applications, the latter possibility of the second case is given a special name:

Definition 12.3. *A periodic orbit* Γ *is called a* limit cycle *if there are two points in* $\mathbb{R}^2$, *one in the interior of* Γ *and the other in the exterior, such that the* α- *or* ω-*limit sets of the orbits through these points is the periodic orbit* Γ.

We will devote a great deal of attention to limit cycles in this chapter. First, however, let us exhibit the occurrence of a limit set described in the third case of Theorem 12.1 in a specific planar system.

Example 12.4. *Homoclinic loop as a limit set:* Consider the planar system

$$\dot{x}_1 = 2x_2$$
$$\dot{x}_2 = 2x_1 - 3x_1^2 + \lambda x_2(x_1^3 - x_1^2 + x_2^2), \tag{12.1}$$

where λ is a scalar parameter. There are two equilibrium points; the one at $(0, 0)$ is a saddle for all values λ, and the other at $(2/3, 0)$ is a source (unstable) if $\lambda < 0$ and a sink (stable) when $\lambda > 0$. Notice that at $\lambda = 0$, these equations essentially coincide with the Hamiltonian system—the fish—whose Hamiltonian is given in Eq. (7.34). Indeed, the perturbation terms are chosen so that the homoclinic loop of the fish remains intact under the perturbation. To see this, we compute the derivative of the function

$$V(x_1, x_2) = x_1^3 - x_1^2 + x_2^2$$

along the solutions of Eq. (12.1) to obtain

$$\dot{V}(x_1, x_2) = 2\lambda x_2^2 (x_1^3 - x_1^2 + x_2^2).$$

Thus, for all values of the parameter λ, the curve $x_1^3 - x_1^2 + x_2^2 = 0$ is invariant for the flow; this curve is, of course, the homoclinic loop containing the origin.

Let us now examine the flow inside the homoclinic loop. Using the Invariance Principle, it is not difficult to show the following: when $\lambda < 0$, the ω-limit set, and when $\lambda > 0$, the α-limit set of any point inside the homoclinic loop, except the equilibrium point $(2/3, 0)$, is part of the homoclinic loop. With a little more work, using some of the ideas to be presented in Section 12.3, one can show that these limit sets are, in fact, the entire homoclinic loop. Typical phase portraits of Eq. (12.1) depicting these situations are shown in Figure 12.1. If you try to locate the homoclinic loop as a limit set of an orbit on the computer, for any finite time the "computed limit set" looks like a periodic orbit. However, if you watch carefully, you will notice that the orbit moves ever so slowly as it passes closer to the origin, and it takes a long time to make one round about the loop. $\Diamond$

Let us now turn our attention to periodic orbits and isolate an important special case of Theorem 12.1 which suggests a way to determine the existence of a nontrivial periodic orbit that is not necessarily close to an equilibrium point.

Theorem 12.5. (Poincaré–Bendixson) *If $\omega(\mathbf{x}^0)$ is a bounded set which contains no equilibrium point, then $\omega(\mathbf{x}^0)$ is a periodic orbit.* $\Diamond$

In order to use the Poincaré–Bendixson theorem to exhibit the existence of a nontrivial periodic orbit, one could attempt to construct an open bounded set $\mathcal{D}$ in $\mathbb{R}^2$ which contains no equilibrium point and such that any solution that begins in $\mathcal{D}$ remains in $\mathcal{D}$ for all $t \geq 0$, that is, $\mathcal{D}$ is an open and bounded positively invariant set. Next, for any $\mathbf{x}^0 \in \mathcal{D}$, one also shows that $\omega(\mathbf{x}^0)$ contains no points on the boundary of $\mathcal{D}$. Then, since $\mathcal{D}$ contains no equilibrium points, $\omega(\mathbf{x}^0)$ must be a periodic orbit. Let us illustrate these remarks on a concocted example.

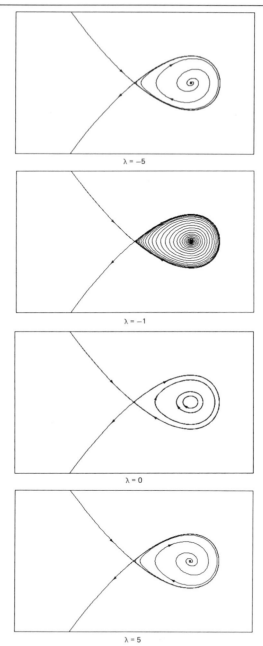

$\lambda = -5$

$\lambda = -1$

$\lambda = 0$

$\lambda = 5$

Figure 12.1. *Phase portraits of Eq. (12.1). Notice that the homoclinic loop is an ω-limit set when $\lambda < 0$ and an α-limit set when $\lambda > 0$.*

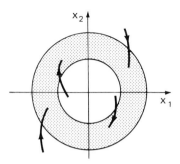

Figure 12.2. *Partial phase portrait of Eq.* (12.2) *near the boundary of the annulus* $\frac{1}{2} < x_1^2 + x_2^2 < 1$.

Example 12.6. *A positively invariant set:* Consider the system

$$\dot{x}_1 = x_2$$
$$\dot{x}_2 = -x_1 + x_2(1 - x_1^2 - 2x_2^2). \tag{12.2}$$

Observe that the origin is the only equilibrium point of Eq. (12.2). Therefore, we will attempt to construct an annular region $\mathcal{D}$ with the desired properties mentioned above. To accomplish this, we compute the derivative of the function $V(x_1, x_2) = (x_1^2 + x_2^2)/2$ along the solutions of Eq. (12.2):

$$\dot{V}(x_1, x_2) = x_2^2(1 - x_1^2 - 2x_2^2).$$

Since $\dot{V}(x_1, x_2) \geq 0$ for $x_1^2 + x_2^2 < \frac{1}{2}$, and $\dot{V}(x_1, x_2) \leq 0$ for $x_1^2 + x_2^2 > 1$, any solution which starts in the annulus $\frac{1}{2} < x_1^2 + x_2^2 < 1$ remains in this annulus for all $t \geq 0$. Since the origin is not in the closure of this annulus, from the Poincaré–Bendixson theorem, there exists at least one periodic orbit of Eq. (12.2) in the annulus; see Figure 12.2. ◇

In general, there may be several difficulties in utilizing the theorem of Poincaré–Bendixson to locate periodic orbits of specific differential equations. First, unlike the example above, it is usually a nontrivial task to construct a region $\mathcal{D}$ with the desired properties. Second, to determine the number of periodic orbits in $\mathcal{D}$ one often has to uncover special properties of the differential equation.

One particularly important case is when we can ascertain that there is at most one periodic orbit in $\mathcal{D}$, then there will be exactly one, say Γ, and $\omega(\mathbf{x}^0) = \Gamma$ for all $\mathbf{x}^0 \in \mathcal{D}$. Since $\gamma^+(\mathbf{x}^0)$ spirals with increasing time toward Γ on one side of Γ and Γ belongs to the interior of $\mathcal{D}$, it follows that points $\mathbf{x}^0$ on both sides of Γ have their positive orbits approaching Γ. Thus, Γ is asymptotically stable. Let us now illustrate some of these remarks on the oscillator of Van der Pol, $\ddot{y} - (\lambda - y^2)\dot{y} + y = 0$. In Section 11.2, we have shown for $\lambda > 0$ and small that there is a periodic orbit of approximate radius $2\sqrt{\lambda}$. We now would like to discuss the case for any positive λ. To accomplish this, it is advantageous and justifiable to scale y as $\sqrt{\lambda}y$ to arrive at the equivalent equation $\ddot{y} - \lambda(1 - y^2)\dot{y} + y = 0$. In the analysis below, it will be more convenient to convert, in the usual way, this second-order equation to an equivalent first-order planar system.

Theorem 12.7. *The equation of Van der Pol*

$$\begin{aligned}
\dot{x}_1 &= x_2 \\
\dot{x}_2 &= -x_1 + \lambda(1 - x_1^2)x_2
\end{aligned} \tag{12.3}$$

has a nontrivial periodic orbit for all values of the scalar parameter λ.

Proof. We will consider the case $\lambda > 0$. When $\lambda = 0$, the equation becomes the linear harmonic oscillator. The case $\lambda < 0$ can be reduced to the first case by reversing time. To show the existence of a periodic orbit we will invoke the Poincaré–Bendixson theorem. For this purpose, we will construct a positively invariant region bounded by a closed curve encircling the origin. Since the origin is the only equilibrium point and the eigenvalues of the linearized equations at the origin have positive real parts, no orbit, except the origin itself, can have the origin as its ω-limit set. This implies that there must be a periodic orbit inside the positively invariant region.

We now begin the construction of a simple (without self-intersections) closed curve $\mathcal{K}$ which will form the outer boundary of a positively invariant region. The idea for constructing the curve $\mathcal{K}$ is to begin at a point A on the negative x_2-axis and use the special properties of the vector field to obtain a curve lying in the left half-plane which intersects the positive x_2-axis at a point E and such that the angle between the tangent vector to this curve and the vector field (12.3) is in the interval $(0, \pi)$. Since Eq. (12.3) is symmetric with respect to the origin, we can also define the reflection of this curve through the origin and obtain points A' and E'. If $A' > E'$, then the curve $AEA'E'A$ will be a suitable curve $\mathcal{K}$.

The precise construction of $\mathcal{K}$ consists of piecing together various curve segments. First, we draw an auxiliary curve $\mathcal{Q}$,

$$\mathcal{Q}(x_1, x_2) \equiv -x_1 + \lambda(1 - x_1^2)x_2 = 0, \tag{12.4}$$

which has three components and their asymptotes given by $x_1 = \pm 1$ and $x_2 = 0$; see Figure 12.3. The component of this curve with asymptotes $x_2 = 0$ and $x_1 = -1$ is crossed from left to right by the vector field (12.3).

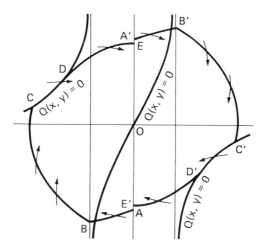

Figure 12.3. *Constructing a positively invariant region for the oscillator of Van der Pol.*

To construct the first piece of our curve $\mathcal{K}$, we take a point $A = (0, x_2^0)$ on the negative x_2-axis sufficiently far away from the origin, and follow the orbit of the system

$$\dot{x}_1 = x_2$$
$$\dot{x}_2 = \lambda(1 - x_1^2)x_2 \tag{12.5}$$

passing through the point A. An easy integration shows that this orbit intersects the line $x_1 = -1$ at the point $B = (-1, -2\lambda/3 + x_2^0)$. Along the arc AB we have

$$\frac{-x_1 + \lambda(1 - x_1^2)x_2}{x_2} - \frac{\lambda(1 - x_1^2)x_2}{x_2} = -\frac{x_1}{x_2} < 0;$$

thus, the orbits of Eq. (12.3) cross AB from right to left.

Next, we follow the orbit of the differential equation

$$\dot{x}_1 = x_2$$
$$\dot{x}_2 = -x_1 \tag{12.6}$$

emanating from B until the orbit, which is a circular arc with its center at the origin, hits the component of the curve (12.4) in the upper left quadrant. Let C denote the point of intersection; such a point always exists if the point A is taken sufficiently far from the origin. Along the curve BC we have

$$\frac{-x_1 + \lambda(1 - x_1^2)x_2}{x_2} + \frac{x_1}{x_2} = \lambda(1 - x_1^2) < 0;$$

thus, the orbits of Eq. (12.3) cross BC from left to right.

For the next piece of our curve $\mathcal{K}$, we first study the points of tangency of an orbit of the differential equation

$$
\begin{aligned}
\dot{x}_1 &= x_2 \\
\dot{x}_2 &= -x_1 + \lambda x_2
\end{aligned}
\tag{12.7}
$$

with the component of the curve (12.4) on the upper left quadrant. By implicitly differentiating Eq. (12.4), it is not difficult to compute that the first coordinate x_1^1 of such a point $D = (x_1^1,\ x_2^1)$ satisfies the equation

$$
1 + (1 - \lambda^2)x^2 + 2\lambda^2 x^4 - \lambda^2 x^6 = 0.
\tag{12.8}
$$

When $x = -1$, the left-hand side of Eq. (12.8) is positive, and when $|x|$ is sufficiently large the left-hand side is negative; hence, there exists a solution x_1^1 of Eq. (12.8). Among the solutions of Eq. (12.8) we take the one nearest to -1. When the point A is sufficiently far from the origin, the point D lies to the right of C. It is clear that the orbits of Eq. (12.3) are crossing the segment of the curve (12.4) between C and D from left to right.

To continue our curve $\mathcal{K}$, we follow the orbit of Eq. (12.7) starting from point D until it hits the x_2-axis at a point which we will denote by E. Since

$$
\frac{-x_1 + \lambda(1 - x_1^2)x_2}{x_2} - \frac{-x_1 + \lambda x_2}{x_2} = -\lambda x_1^2 < 0,
$$

the orbits of Eq. (12.3) cross the curve DE from left to right.

The first half of our curve is now constructed and it is $ABCDE$. To construct the other half, we observe that Eq. (12.3) is symmetric with respect to the origin. Let $A'B'C'D'E'$ be the reflection of $ABCDE$ with respect to the origin. Since D, hence E, is fixed, we may ensure that A' lies above E by taking A far from the origin. Also, observe that orbits of Eq. (12.3) cross the curve segment EA' from left to right. From the symmetry, it is now clear that the region encircled by the closed curve $\mathcal{K} = ABCDEA'B'C'D'E'A$ is positively invariant for the flow of the oscillator of Van der Pol.

In the proof above, we used a point of tangency D of the curve $\mathcal{Q}$ and the vector field (12.7). The proof could have been completed by choosing any other point D on the same component of $\mathcal{Q}$. However, the choice in the proof is optimal in the sense that it allows one to construct the smallest region containing periodic orbits. More precisely, taking E as fixed by the construction above, one can vary A so that either the point C also becomes a point of tangency for $\mathcal{Q}$ and the vector field (12.6) or the point $E = -A$. This is the optimal $\mathcal{K}$ by the method of construction given in the proof. $\diamondsuit$

After developing a little more mathematics, we will prove in the next section that the equation of Van der Pol has in fact a unique nontrivial periodic orbit which is stable when $\lambda > 0$, and unstable when $\lambda < 0$.

As is evident in the proof above, searching for periodic orbits in specific differential equations can be rather cumbersome. It is best to avoid undertaking such a task if the equations do not have a periodic orbit in the first place. Here is a useful result to know for ruling out the existence of periodic orbits in certain instances.

Theorem 12.8. (Bendixson's Criterion) *Let D be a simply connected open subset of $\mathbb{R}^2$ (a region with no "holes," the interior of a disk, for example). If $\operatorname{div} \mathbf{f} \equiv \partial f_1/\partial x_1 + \partial f_2/\partial x_2$ is of constant sign and not identically zero in D, then $\dot{\mathbf{x}} = \mathbf{f}(\mathbf{x})$ has no periodic orbit lying entirely in the region D.*

Proof. Let us suppose that there is a periodic orbit Γ lying in D. We will show that this assumption leads to a contradiction. The interior S of Γ is simply connected; hence, from Green's theorem we have

$$\oint_\Gamma (f_1 \, dx_2 - f_2 \, dx_1) = \iint_S \left(\frac{\partial f_1}{\partial x_1} + \frac{\partial f_2}{\partial x_2} \right) dx_1 \, dx_2.$$

On any orbit, in particular, on Γ, we have $f_1 \, dx_2 - f_2 \, dx_1 = 0$. Therefore, the integral on the left has value zero. The integral on the right, however, cannot be zero because the integrand has constant sign and not zero in S. ◇

An easy but important generalization of the criterion of Bendixson is the following theorem:

Theorem 12.9. (Dulac's Criterion) *Let $D \subseteq \mathbb{R}^2$ be a simply connected open set and $B(x_1, x_2)$ be a real-valued C^1 function in D. If the function $\operatorname{div} B\mathbf{f} = \partial(Bf_1)/\partial x_1 + \partial(Bf_2)/\partial x_2$ is of constant sign and not identically zero in D, then $\dot{\mathbf{x}} = \mathbf{f}(\mathbf{x})$ has no periodic orbit lying entirely in the region D.* ◇

The function B is called a *Dulac function*. This theorem reverts to Bendixson in the special case when $B(x_1, x_2) \equiv 1$. There is no general method for determining an appropriate Dulac function for a given system of planar differential equations, much like the difficulty of finding an "integrating factor."

Example 12.10. *No periodic orbit:* Consider the following quadratic system on the plane:

$$\dot{x}_1 = -x_2 + x_1^2 - x_1 x_2$$
$$\dot{x}_2 = x_1 + x_1 x_2.$$

Since $f_1(x_1, x_2) = -x_2 + x_1^2 - x_1 x_2$ and $f_2(x_1, x_2) = x_1 + x_1 x_2$, we have

$$\frac{\partial f_1}{\partial x_1} + \frac{\partial f_2}{\partial x_2} = 3x_1 - x_2.$$

Hence, Bendixson's Criterion is insufficient to exclude the existence of a periodic orbit. However, if we use the mysterious Dulac function given by

$$B(x_1, x_2) = (1 + x_2)^{-3}(-1 - x_1)^{-1},$$

then we have

$$\frac{\partial B f_1}{\partial x_1} + \frac{\partial B f_2}{\partial x_2} = x_1^2 (1 + x_2)^{-3}(-1 - x_1)^{-2}.$$

Again, this expression is not of one sign, but do not despair. Notice that the line $x_2 = -1$ is invariant under the flow, and the vector field crosses the line $x_1 = -1$ in the same direction. Thus if there is a periodic orbit it must lie entirely in one of the four regions separated by these two lines. However, the function div $B\mathbf{f}$ keeps one sign on any one of these four regions. Consequently, from Dulac's Criterion, our quadratic system has no periodic orbits. ◇

We end this section with the statement of a geometric theorem on the symbiotic relationship of a periodic orbit with an equilibrium point.

Theorem 12.11. *Let Γ be a periodic orbit enclosing an open set U on which the vector field is defined. Then U contains an equilibrium point.* ◇

Exercises ──────────────────────────────────── ♣♡♠◇

12.1. Determine the phase portraits of the system

$$\dot{x}_1 = x_2$$
$$\dot{x}_2 = x_1 - 2x_1^3 + \lambda x_2 (x_1^4 - x_1^2 + x_2^2),$$

for all values of the scalar parameter λ. What are the possible limit sets? *Hint:* Use $V(x_1, x_2) = x_1^4 - x_1^2 + x_2^2$ and the invariance principle.

12.2. *A nonconvex periodic orbit:* Show that, when λ is sufficiently large, say, $\lambda > 10$, a periodic orbit of Van der Pol's equation is not a convex closed curve.
Hint: At the points $(-1, x_2)$ with $x_2 > 0$ show that $dx_2/dx_1 > 0$ and $d^2 x_2/dx_1^2 > 0$.

12.3. *No periodic orbit:* Show that the system of equations

$$\dot{x}_1 = x_1 + x_2^2 + x_1^3$$
$$\dot{x}_2 = -x_1 + x_2 + x_2 x_1^2$$

has no nontrivial periodic orbit.

12.4. *No periodic orbit:* Show that the system of equations

$$\dot{x}_1 = x_1 - x_1 x_2^2 + x_2^3$$
$$\dot{x}_2 = 3x_2 - x_2 x_1^2 + x_1^3$$

has no nontrivial periodic orbit in the region $x_1^2 + x_2^2 \le 4$.

12.5. Draw the phase portrait of the system of equations

$$\dot{x}_1 = (2 - x_1 - 2x_2)x_1$$
$$\dot{x}_2 = (2 - 2x_1 - x_2)x_2$$

on the first quadrant of the (x_1, x_2)-plane.
Suggestions: First, find the equilibrium points and determine their stability types. Then, investigate the stable and unstable manifolds of the equilibria. Notice that the lines $x_1 = x_2$, $x_1 = 0$, and $x_2 = 0$ are invariant. Next, show that there are no periodic orbits using Theorem 12.11. Finally, apply Theorem 12.1. What is the vector field doing for large x_1 and x_2?

12.2. Stability of Periodic Orbits

In this section we show how the behavior of solutions near a periodic orbit can be investigated in terms of the dynamics of a scalar monotone map—the Poincaré map. We also derive a formula for the derivative of the Poincaré map and use it to complete the analysis of the oscillator of Van der Pol.

The construction of the Poincaré map of a planar system near a periodic orbit is similar to the one presented in Chapter 4 for 1-periodic scalar equations. However, some precautions must be taken to insure that the map is well-defined. To be specific, let $\varphi(t, \mathbf{p})$ be a periodic solution with minimal period T of the differential equation $\dot{\mathbf{x}} = \mathbf{f}(\mathbf{x})$ and denote the corresponding periodic orbit by Γ. Now, choose a vector $\mathbf{v} \in \mathbb{R}^2$ so that $\mathbf{v}$ and the tangent vector $\mathbf{f}(\mathbf{p})$ to Γ at $\mathbf{p}$ are linearly independent. Let L_ε be the line segment defined by

$$L_\varepsilon = \{\, \mathbf{x} \in \mathbb{R}^2 \ : \ \mathbf{x} = \mathbf{p} + a\mathbf{v}, \ \ 0 \le |a| \le \varepsilon \,\}.$$

The line segment L_ε is called a *transversal section* to the periodic orbit Γ at the point $\mathbf{p}$.

Next, we define a map on a subset of L_ε induced by the flow. Let us choose ε so small that L_ε intersects the periodic orbit Γ only at the point $\mathbf{p}$, and that all orbits crossing L_ε do so in the same direction; see Figure 12.4. The latter requirement can be met using the Flow Box Theorem. Since $\varphi(T, \mathbf{p}) = \mathbf{p}$ and solutions depend continuously on initial values, there is $\delta > 0$ such that, if $\mathbf{x}^0 \in L_\delta$, then there is a first time $T(\mathbf{x}^0) > 0$, depending on $\mathbf{x}^0$, at which $\varphi(T(\mathbf{x}^0), \mathbf{x}^0) \in L_\varepsilon$.

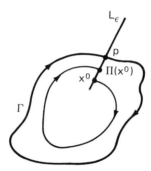

Figure 12.4. *Local transversal section L_ε to the periodic orbit Γ at the point **p**, and the Poincaré map.*

Definition 12.12. *The Poincaré map or first return map* Π *near a periodic orbit* Γ *is defined to be the map*

$$\Pi : L_\delta \to L_\varepsilon; \quad \mathbf{x}^0 \mapsto \varphi(T(\mathbf{x}^0), \mathbf{x}^0).$$

The points on the transversal section L_ε have a natural ordering: two points $\mathbf{x}^0 = \mathbf{p} + a_0 \mathbf{v}$ and $\mathbf{x}^1 = \mathbf{p} + a_1 \mathbf{v}$ satisfy $\mathbf{x}^0 \geq \mathbf{x}^1$ if and only if $a_0 \geq a_1$. Using this ordering, the Poincaré map Π is said to be *monotone* if $\mathbf{x}^0 \geq \mathbf{x}^1$ on L_δ implies $\Pi(\mathbf{x}^0) \geq \Pi(\mathbf{x}^1)$.

Some of the most basic properties of the Poincaré map are listed in the theorem below:

Theorem 12.13. *The Poincaré map possesses the following properties:*
(i) *The Poincaré map* Π *near the periodic orbit* Γ *is a monotone* C^1 *map.*
(ii) *The orbit* $\gamma(\mathbf{x}^0)$ *of a point* $\mathbf{x}^0 \in L_\delta$ *is a periodic orbit if and only if* $\mathbf{x}^0$ *is a fixed point of the Poincaré map, that is,* $\Pi(\mathbf{x}^0) = \mathbf{x}^0$.
(iii) *The periodic orbit* Γ, *with* $\mathbf{p} \in \Gamma$, *is orbitally asymptotically stable if* $\Pi'(\mathbf{p}) < 1$, *and unstable if* $\Pi'(\mathbf{p}) > 1$.

Proof. We indicate a graphic reason for the monotonicity of Π. The remaining assertions are geometrically self-evident from our earlier studies of scalar maps; therefore, we refrain from giving formal details.

Let $\mathbf{x}^0$ and $\mathbf{x}^1$ be on L_δ with $\mathbf{x}^0 \geq \mathbf{x}^1$. Consider the simple closed curve $\mathcal{C}_{\mathbf{x}^1}$ consisting of the part of the orbit $\gamma(\mathbf{x}^1)$ between the points $\mathbf{x}^1$ and $\Pi(\mathbf{x}^1)$ together with the line segment on L_ε between $\mathbf{x}^1$ and $\Pi(\mathbf{x}^1)$; see Figure 12.5. Then, from the Jordan Curve Theorem, $\mathcal{C}_{\mathbf{x}^1}$ has an interior and an exterior. Since the orbits cross L_ε in the same direction, and the orbits cannot intersect each other, we have $\Pi(\mathbf{x}^0) \geq \Pi(\mathbf{x}^1)$ which implies the monotonicity of the Poincaré map. $\Diamond$

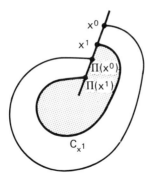

Figure 12.5. *Monotonicity of the Poincaré map.*

Definition 12.14. *The periodic orbit* Γ *through the point* $\mathbf{p}$ *is said to be hyperbolic if* $\mathbf{p}$ *is a hyperbolic fixed point of the Poincaré map* Π, *that is,* $\Pi'(\mathbf{p}_0) \neq 1$.

The somewhat formal setup above is not much use for stability analysis of periodic orbits unless one can compute the derivative of the Poincaré map. We now present a formula for such a computation if the corresponding periodic solution is explicitly known.

Theorem 12.15. *Let* $\varphi(t, \mathbf{p})$ *be a* T-*periodic solution through* $\mathbf{p}$ *of the planar differential equation* $\dot{\mathbf{x}} = \mathbf{f}(\mathbf{x})$. *Then*

$$\Pi'(\mathbf{p}) = \exp\left\{ \int_0^T \left(\frac{\partial f_1}{\partial x_1} + \frac{\partial f_2}{\partial x_2} \right) (\varphi(t, \mathbf{p})) \, dt \right\}. \tag{12.9}$$

Proof. Consider the T-periodic linear variational equation of $\dot{\mathbf{x}} = \mathbf{f}(\mathbf{x})$ about the T-periodic solution $\varphi(t, \mathbf{p})$:

$$\dot{\mathbf{x}} = D\mathbf{f}(\varphi(t, \mathbf{p}))\mathbf{x}. \tag{12.10}$$

Recall from Lemma 8.21, the Liouville's formula, that the fundamental matrix solution $\mathbf{X}(t)$ with $\mathbf{X}(0) = \mathbf{I}$ of the T-periodic linear system (12.10) satisfies the relation

$$\det \mathbf{X}(T) = \exp\left\{ \int_0^T \operatorname{tr} D\mathbf{f}(\varphi(t, \mathbf{p})) \, dt \right\}.$$

We will establish the formula by showing that $\Pi'(\mathbf{p}) = \det \mathbf{X}(T)$.

To this end, we determine the matrix representation of $\mathbf{X}(T)$ in the basis $\{ \mathbf{f}(\mathbf{p}), \mathbf{p} \}$ of $\mathbb{R}^2$. Therefore, we need to compute the vectors $\mathbf{X}(T)\mathbf{f}(\mathbf{p})$ and $\mathbf{X}(T)\mathbf{v}$. Notice that the function $\dot{\varphi}(t, \mathbf{p})$ is a solution of Eq. (12.10) and $\dot{\varphi}(t, \mathbf{p}) = \dot{\varphi}(t + T, \mathbf{p})$. Since $\dot{\varphi}(0, \mathbf{p}) = \dot{\varphi}(T, \mathbf{p}) = \mathbf{f}(\mathbf{p})$ and $\dot{\varphi}(t, \mathbf{p}) = \mathbf{X}(t)\mathbf{f}(\mathbf{p})$, we have

$$\mathbf{X}(T)\mathbf{f}(\mathbf{p}) = \mathbf{f}(\mathbf{p}). \tag{12.11}$$

We next compute $\mathbf{X}(T)\mathbf{v}$. If we differentiate the Poincaré map

$$\Pi(\mathbf{p} + a\mathbf{v}) = \varphi\big(T(\mathbf{p} + a\mathbf{v}), (\mathbf{p} + a\mathbf{v})\big), \qquad |a| < \delta,$$

with respect to a and put $a = 0$, we obtain

$$\Pi'(\mathbf{p})\mathbf{v} = \frac{\partial T}{\partial a}(\mathbf{p})\,\mathbf{f}(\mathbf{p}) + \mathbf{X}(T)\mathbf{v}. \tag{12.12}$$

From Eqs. (12.11) and (12.12), it follows that the matrix of $\mathbf{X}(T)$ in the basis $\{\,\mathbf{f}(\mathbf{p}),\,\mathbf{p}\,\}$ is

$$\mathbf{X}(T) = \begin{pmatrix} 1 & -\partial T(\mathbf{p})/\partial a \\ 0 & \Pi'(\mathbf{p}) \end{pmatrix}.$$

It is now clear that $\Pi'(\mathbf{p}) = \det\mathbf{X}(T)$. $\Diamond$

Let us now illustrate some of the properties of the Poincaré map on one of our standard examples.

Example 12.16. *An explicit Poincaré map:* Consider the familiar planar system

$$\begin{aligned} \dot{x}_1 &= -x_2 + x_1(1 - x_1^2 - x_2^2) \\ \dot{x}_2 &= x_1 + x_2(1 - x_1^2 - x_2^2) \end{aligned} \tag{12.13}$$

and its periodic solution $(x_1(t), x_2(t)) = (\cos t, \sin t)$. Let us determine the Poincaré map of Eq. (12.13) about the periodic orbit and determine the stability type of the periodic orbit by differentiating the Poincaré map. We take as our transversal section

$$L = \{\,(x_1, x_2) \in \mathbb{R}^2 \colon x_1 > 0,\ x_2 = 0\,\}.$$

If we transform everything into polar coordinates $x_1 = r\cos\theta$ and $x_2 = r\sin\theta$, then Eq. (12.13) becomes

$$\begin{aligned} \dot{r} &= r(1 - r^2) \\ \dot{\theta} &= 1, \end{aligned} \tag{12.14}$$

the periodic orbit is given by $r(t) = 1$, and the transversal section L becomes

$$L = \{\,(r, \theta) \in \mathbb{R}^+ \times S^1 \colon r > 0,\ \theta = 0\,\}.$$

Since the system (12.14) is a product system, it is easy to obtain the general solution:

$$\varphi(t, r_0, \theta_0) = \left(\left[1 + \left(r_0^{-2} - 1 \right) e^{-2t} \right]^{-1/2}, \, t + \theta_0 \right).$$

The return time for any point on L is 2π. Thus the Poincaré map is given by

$$\Pi(r_0) = \left[1 + \left(r_0^{-2} - 1 \right) e^{-4\pi} \right]^{-1/2}.$$

This map has a fixed point at $r_0 = 1$, which, of course corresponds to the periodic orbit. The derivative of the Poincaré map at this fixed point is easily seen to be

$$\Pi'(1) = \left. \frac{d\Pi}{dr_0} \right|_{r_0=1} = e^{-4\pi} < 1.$$

Consequently, the fixed point $r_0 = 1$, hence the corresponding periodic orbit, is hyperbolic and, moreover, orbitally asymptotically stable. The same conclusion can also be obtained from the formula in Theorem 12.15 without the explicit knowledge of the Poincaré map. You may wish to perform the integration in Eq. (12.9) to confirm the computations above. $\Diamond$

In certain situations it is possible to put the formula (12.9) in Theorem 12.15 to good use even if the periodic orbit is not explicitly known. Here is one such successful case:

Theorem 12.17. *A nontrivial periodic orbit of the oscillator (12.3) of Van der Pol with $\lambda > 0$ is hyperbolic and orbitally asymptotically stable.*

Proof. Let Γ be a nontrivial periodic orbit with period T and let $\mathbf{x}(t)$ be the corresponding solution. From Theorem 12.13 and the formula (12.9), we need to show that the derivative of the Poincaré map satisfies

$$\Pi'(\mathbf{x}(0)) = \exp\left\{ \int_0^T \lambda \left(1 - [x_1(t)]^2 \right) dt \right\} < 1.$$

We will accomplish this by proving that the value of the integral is negative. Consider the function

$$V(x_1, x_2) = \tfrac{1}{2}(x_1^2 + x_2^2)$$

and compute $\dot{V}$ along the solutions of Eq. (12.3):

$$\dot{V}(x_1, x_2) = \lambda(1 - x_1^2)x_2^2 = -2\lambda(1 - x_1^2) \left[\tfrac{1}{2}x_1^2 - V(x_1, x_2) \right]. \qquad (12.15)$$

The function V assumes a minimum value on the periodic orbit Γ at some point, say, $\mathbf{x}(\bar{t})$. Thus, $\dot{V}(\mathbf{x}(\bar{t})) = 0$ and either $x_2(\bar{t}) = \dot{x}_1(\bar{t}) = 0$ or $x_1(\bar{t}) = \pm 1$.

We now show that the first case is impossible. If $x_2(\bar{t}) = 0$, then $\dot{x}_2(\bar{t}) \neq 0$ for otherwise we would have $x_1(\bar{t}) = 0$; uniqueness of the solutions would then imply that $\mathbf{x}(t) = 0$ for all t which is impossible. Therefore, if $\dot{x}_1(\bar{t}) = x_2(\bar{t}) = 0$, then $\ddot{x}_1(\bar{t}) = \dot{x}_2(\bar{t}) \neq 0$ and thus $x_1(t)$ has either a maximum or a minimum at $t = \bar{t}$. Consequently, the function $1 - [x_1(t)]^2$ has a fixed sign for t near $\bar{t}$; hence, $\dot{V}(\mathbf{x}(t))$ has constant sign for t near $\bar{t}$. Thus, $V(\mathbf{x}(t))$ is strictly monotone for t near $\bar{t}$, which contradicts the fact that it has a minimum at $t = \bar{t}$.

From the argument above, the fact that $\dot{V}(\mathbf{x}(\bar{t})) = 0$ implies that $x_2(\bar{t}) \neq 0$ and $x_1(\bar{t}) = \pm 1$. Therefore, $V(\mathbf{x}(t)) > 1/2$ for all t.

Now, a simple rearrangement of Eq. (12.15) yields the formula

$$-\frac{\dot{V}(\mathbf{x}(t))}{V(\mathbf{x}(t)) - \frac{1}{2}} + 2\lambda\left(1 - [x_1(t)]^2\right) = -\frac{\lambda\left(1 - [x_1(t)]^2\right)^2}{V(\mathbf{x}(t)) - \frac{1}{2}}.$$

Integrating both sides of this equation over the periodic orbit Γ results in

$$2\int_0^T \lambda\left(1 - [x_1(t)]^2\right)\,dt = -\lambda\int_0^T \frac{\left(1 - [x_1(t)]^2\right)^2}{V(\mathbf{x}(t)) - \frac{1}{2}}\,dt < 0$$

because the integral of the first term on the left vanishes, the parameter λ and the integrand on the right are both positive. $\diamond$

This seemingly local result has an easy but very important consequence for the global dynamics of the oscillator of Van der Pol.

Theorem 12.18. (Uniqueness of limit cycle in Van der Pol) *For $\lambda > 0$, the oscillator (12.3) of Van der Pol has a stable limit cycle to which every nonequilibrium solution tends with increasing time.*

Proof. We have shown in the previous theorem that every periodic orbit of Van der Pol's oscillator is orbitally asymptotically stable. Let us suppose that there are two such periodic orbits, say Γ_1 and Γ_2. Since, from Theorem 12.11, every periodic orbit must contain an equilibrium point in its interior and the origin is the only equilibrium point of the differential equation (12.3), one of the periodic orbits must be in the interior of the other. Now, let $\mathbf{x}^0$ be a point in between Γ_1 and Γ_2 such that its omega-limit set $\omega(\mathbf{x}^0) = \Gamma_1$. The alpha-limit set $\alpha(\mathbf{x}^0)$ of $\mathbf{x}^0$ cannot be Γ_2. Therefore, there must be an unstable periodic orbit between Γ_1 and Γ_2, which contradicts the fact that all periodic orbits are orbitally asymptotically stable. $\diamond$

We end this section with an example of a planar system possessing a nonhyperbolic periodic orbit.

Example 12.19. Consider the planar system

$$\begin{aligned}
\dot{x}_1 &= -x_2 + x_1(1 - x_1^2 - x_2^2)^2 \\
\dot{x}_2 &= x_1 + x_2(1 - x_1^2 - x_2^2)^2.
\end{aligned} \tag{12.16}$$

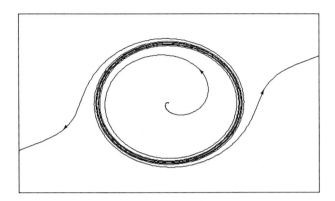

Figure 12.6. *Phase portrait of Eq. (12.16). The unit circle is an unstable and nonhyperbolic periodic orbit.*

and its periodic solution $\big(x_1(t),\ x_2(t)\big) = \big(\cos t,\ \sin t\big)$. It is easy to see, using formula (12.9) in Theorem 12.15, that the periodic orbit is nonhyperbolic. The global phase portrait of Eq. (12.16) can easily be determined by transforming the system to polar coordinates $x_1 = r\cos\theta$ and $x_2 = r\sin\theta$:

$$\dot{r} = r(1 - r^2)^2$$
$$\dot{\theta} = 1.$$

The periodic orbit is unstable, as seen in Figure 12.6. $\diamondsuit$

Bifurcations of nonhyperbolic periodic orbits will be the subject of the next section where we will also investigate an interesting perturbation of the example above.

Exercises ━━━━━━━━━━━━━━━━━━━━━━━━━━━━━

12.6. *Linearizing about a periodic orbit:* Verify the asymptotic stability of the periodic orbit of Example 12.16 without using the Poincaré map: Linearize the equations about the periodic orbit and study the resulting 2π-periodic linear system.

12.7. *Rigorous numerics:* Demonstrate that the planar system

$$\dot{x}_1 = x_2$$
$$\dot{x}_2 = \alpha \left(1 - x_1^2 - x_2^2\right) x_2 - x_1$$

has an orbitally asymptotically stable periodic orbit for the parameter value $\alpha = 0.05$. This system is used in Franke and Selgrade [1979] as a test case of a *rigorous* numerical procedure for locating an orbitally asymptotically stable periodic orbit of a planar system.

12.8. *Oscillators with hyperbolic periodic orbit:* Using reasoning similar to the one in the proof of Theorem 12.18, establish the following more general result: Consider the second-order equation $\ddot{x} + f(x)\dot{x} + g(x) = 0$, where f and g, are, say, C^1. Suppose that

(i) $f(x) < 0$ for $x_1 < x < x_2$, and $f(x) > 0$ for $x < x_1$ or $x > x_2$, where $x_1 < 0 < x_2$;

(ii) $xg(x) > 0$ for $x \neq 0$;

(iii) $G(x_1) = G(x_2)$, where $G(x) = \int_0^x g(x)\,dx$.

Show that every periodic orbit is hyperbolic with $\Pi'(\mathbf{p}) < 1$. Thus, if there is a periodic orbit, it is unique. For assistance, see Coppel [1965], p. 86. Under mild additional assumptions it can be proved that a periodic orbit actually exists, see Levinson and Smith [1942].

12.9. *Hyperbolicity and finiteness:* Consider a planar differential equation $\dot{\mathbf{x}} = \mathbf{f}(\mathbf{x})$ with the property that the vector field $\mathbf{f}$ is transversal to the unit disk D and that D is positively invariant. Suppose further that all equilibrium points and all periodic orbits are hyperbolic and there is no orbit connecting saddles. Prove that the number of periodic orbits in D is finite using the following suggested steps:

1. There are only a finite number of equilibria in D.

2. If there were infinitely many periodic orbits, then there must be a nested sequence of them: (i) $\gamma_1 \supset \gamma_2 \supset \ldots$, or (ii) $\gamma_1 \subset \gamma_2 \subset \ldots$.

3. Let $\widehat{\gamma}_j = \gamma_j \cup (\text{interior of} \gamma_j)$ and suppose that (i) is satisfied and define $S = \cap_j \widehat{\gamma}_j$. Show that the boundary of S is invariant.

4. S is either a closed orbit or contains an equilibrium point. Show that this leads to a contradiction.

5. How do you handle case (ii)?

12.3. Local Bifurcations of Periodic Orbits

Let Γ_0 be a periodic orbit of $\dot{\mathbf{x}} = \mathbf{f}(\mathbf{x})$. Consider the perturbed differential equation

$$\dot{\mathbf{x}} = \mathbf{F}(\lambda, \mathbf{x}), \qquad (12.17)$$

where

$$\mathbf{F} : \mathbb{R}^k \times \mathbb{R}^2 \to \mathbb{R}^2; \quad (\lambda, \mathbf{x}) \mapsto F(\lambda, \mathbf{x})$$

satisfying $\mathbf{F}(0, \mathbf{x}) = \mathbf{f}(\mathbf{x})$. In this section we investigate the behavior of solutions of Eq. (12.17) in a neighborhood of the periodic orbit Γ_0 for small values of the parameter λ near $\lambda = 0$.

On the theoretical level, it is quite easy to study the local bifurcations of Eq. (12.17) near the periodic orbit Γ_0 in terms of its Poincaré map. Indeed, let a local transversal section L_ϵ to the periodic orbit Γ_0 be defined as in the previous section. There are $\lambda_0 > 0$ and $\delta > 0$ such that for $0 \leq |\lambda| < \lambda_0$, and $\mathbf{x}^0 \in L_\delta$, there is a first time $T(\lambda, \mathbf{x}^0) > 0$ such that the solution $\varphi(\lambda, t, \mathbf{x}^0)$ of Eq. (12.17) satisfies $\varphi(\lambda, T(\lambda, \mathbf{x}^0), \mathbf{x}^0) \in L_\varepsilon$. There-fore, we define the Poincaré map depending on parameters as $\Pi(\lambda, \mathbf{x}^0) = \varphi(\lambda, T(\lambda, \mathbf{x}^0), \mathbf{x}^0)$ mapping L_δ into L_ε. The Poincaré map $\Pi(\lambda, \mathbf{x}^0)$ will be monotone for the same reason that $\Pi(0, \mathbf{x}^0)$ was in Theorem 12.13. Of course, periodic orbits near Γ_0 correspond to fixed points of $\Pi(\lambda, \mathbf{x}^0)$.

Now, the general results in Section 3.3 on fixed points of monotone maps can be applied to the Poincaré map $\Pi(\lambda, \mathbf{x}^0)$. For example, if Γ_0 is hyperbolic, then for each λ with $|\lambda|$ small, there is a unique periodic orbit Γ_λ near Γ_0 and Γ_λ is also hyperbolic.

When Γ_0 is nonhyperbolic, the bifurcations near the periodic orbit Γ_0 are determined from the bifurcations of the Poincaré map $\Pi(\lambda, \mathbf{x}^0)$. The example below corresponds to a saddle-node bifurcation of the nonhyper-bolic fixed point of the Poincaré map at, for instance, $\mathbf{x}^0 = (1, 0)$.

Example 12.20. *A saddle-node bifurcation of periodic orbits:* Consider the planar system

$$\dot{x}_1 = -x_1 \sin \lambda - x_2 \cos \lambda + (1 - x_1^2 - x_2^2)^2 (x_1 \cos \lambda - x_2 \sin \lambda)$$
$$\dot{x}_2 = x_1 \cos \lambda - x_2 \sin \lambda + (1 - x_1^2 - x_2^2)^2 (x_1 \sin \lambda + x_2 \cos \lambda), \tag{12.18}$$

where λ is a real small parameter. This one-parameter system is the rota-tion of the vector field of Example 12.16 through an angle λ.

In polar coordinates $x_1 = r \cos \theta$ and $x_2 = r \sin \theta$, the system (12.18) has the form

$$\dot{r} = r \left[(1 - r^2)^2 \cos \lambda - \sin \lambda \right]$$
$$\dot{\theta} = (1 - r^2)^2 \sin \lambda + \cos \lambda. \tag{12.19}$$

Since the first equation is independent of θ, it is easy to see that, in the radial direction, the system above undergoes a saddle-node bifurcation as the parameter λ passes through zero. Indeed, if $\lambda > 0$, and is sufficiently close to zero, then the system (12.19) has two periodic orbits; an unstable periodic orbit

$$x_1^2 + x_2^2 = 1 + \sqrt{\tan \lambda},$$

which is a circle of radius greater than 1, and a stable periodic orbit

$$x_1^2 + x_2^2 = 1 - \sqrt{\tan \lambda},$$

which is a circle of radius less than 1. For the parameter value $\lambda = 0$, this, of course, is Example 12.9 which has a single nonhyperbolic (unstable) periodic orbit at $x_1^2 + x_2^2 = 1$. If $\lambda < 0$, then system (12.19) has no periodic orbits because $\dot{r} > 0$ and all the solutions, except the origin, go to infinity as $t \to +\infty$; see Figure 12.7.

The system (12.19) undergoes other bifurcations at the parameter val-ues, for example, $\lambda = \pi/4$ or $\lambda = \pi/2$ which you might like to investigate. If you need assistance, consult Andronov et al. [1973], pp. 212–217. $\Diamond$

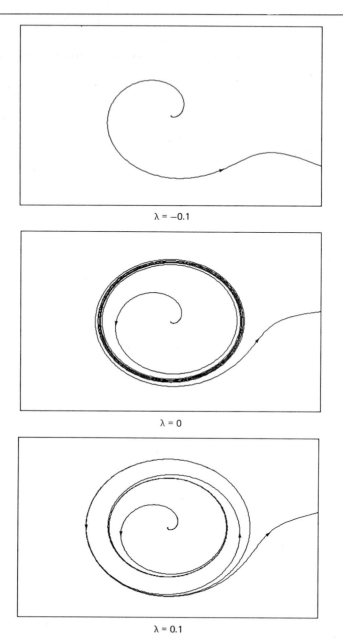

$\lambda = -0.1$

$\lambda = 0$

$\lambda = 0.1$

Figure 12.7. *A saddle-node bifurcation of periodic orbits of Eq.* (12.19).

Exercises ———————————————————————— ♣ ♡ ♠ ◇

12.10. *A quadratic system with two limit cycles:* Consider the quadratic planar system depending on a parameter λ:

$$\dot{x}_1 = P(x_1, x_2)\cos\lambda - Q(x_1, x_2)\sin\lambda$$
$$\dot{x}_2 = P(x_1, x_2)\sin\lambda + Q(x_1, x_2)\cos\lambda,$$

where $P(x_1, x_2)$ and $Q(x_1, x_2)$ are given by

$$P(x_1, x_2) = 169(x_1 - 1)^2 - 16(x_2 - 1)^2 - 153,$$
$$Q(x_1, x_2) = 144(x_1 - 1)^2 - 9(x_2 - 1)^2 - 135.$$

Notice that the parameter λ rotates the vector field. For $\lambda = 0.8$, there is a stable limit cycle surrounding the equilibrium point $(0, 0)$, and an unstable limit cycle around the equilibrium point $(2, 2)$. Locate these limit cycles using a computer. To find the unstable one, you should run the solutions backward with a negative step size. This system is contained in the library of PHASER under the name *hilbert2*.

12.4. A Homoclinic Bifurcation

In the search for periodic orbits, it is important to understand how periodic orbits can be created or destroyed through bifurcations. The most famous instance of this is, of course, the Poincaré–Andronov–Hopf bifurcation. In this brief section, which consists of one example, we will illustrate how a periodic orbit can bifurcate from or be absorbed by a homoclinic loop.

Example 12.21. *Periodic orbit from a homoclinic loop:* Consider the planar system

$$\dot{x}_1 = 2x_2$$
$$\dot{x}_2 = 2x_1 - 3x_1^2 - x_2(x_1^3 - x_1^2 + x_2^2 - c),$$
$$\tag{12.20}$$

where c is a scalar parameter. Notice that when $c = 0$, these equations reduce to the special case of Example 12.4 with $\lambda = -1$: there is a homoclinic loop through the origin and attracts from within, as seen in Figure 12.8b.

We now analyze the phase portrait of this system for small nonzero values of the parameter c near zero. For all values of c, there are two equilibrium points; the one at $(0, 0)$ is always a saddle, and the other at $(2/3, 0)$ is unstable (source) when $c > -4/27$.

To analyze the details of the phase portraits of Eq. (12.20) further, we consider the function

$$V(x_1, x_2) = x_1^3 - x_1^2 + x_2^2$$

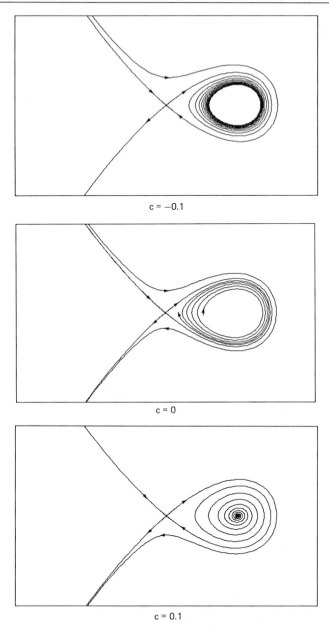

c = −0.1

c = 0

c = 0.1

Figure 12.8. *A homoclinic bifurcation in Example 12.21 near* $c = 0$. *When* $c < 0$, *there is a periodic orbit; at* $c = 0$, *it becomes the homoclinic loop; for* $c > 0$, *the homoclinic loop is broken and there is no periodic orbit nearby.*

and compute its derivative along the solutions of Eq. (12.20) to obtain

$$\dot{V}(x_1, x_2) = -2x_2(x_1^3 - x_1^2 + x_2^2 - c).$$

For $-4/27 < c < 0$, using the Invariance Principle, one can see that there is an orbitally asymptotically stable periodic orbit lying on the curve $x_1^3 - x_1^2 + x_2^2 - c = 0$, with $x_1 > 0$. As $c \to 0$, the periodic orbit approaches the homoclinic loop through the origin. For $c > 0$, the homoclinic loop is broken and also there is no periodic orbit. This sequence of bifurcations is illustrated in Figure 12.8. ◇

When a homoclinic loop is broken, the birth of a (unique) periodic orbit can be established under fairly general assumptions; however, we refrain here from such a formulation. Homoclinic loops play a significant role in bifurcation theory and we will encounter them again in Chapter 13 when we consider planar flows at large.

Exercises ♣♡♠◇

12.11. Show that in Example 12.21 the equilibrium point at $(2/3, 0)$ undergoes a Poincaré–Andronov–Hopf bifurcation at the parameter value $c = -4/27$. Observe that as the parameter c is decreased from 0 towards $-4/27$, the periodic orbit springing off the homoclinic loop gradually becomes smaller and finally disappears into the equilibrium point at $(2/3, 0)$.

12.12. Consider the planar system

$$\dot{x}_1 = x_2$$
$$\dot{x}_2 = x_1 - x_1^2 + \lambda x_2 + a x_1 x_2.$$

Observe that the origin is a saddle point. Fix the value of a at a positive value. Then experiment on PHASER to convince yourself that there is a negative value of λ at which the system has a homoclinic loop. Now, change λ a little to break the homoclinic loop, and search for the unique periodic orbit in the proximity of the original homoclinic loop.

Bibliographical Notes

The details of the central result of Poincaré and Bendixson can be found in many sources; see, for example, Hale [1980], Hartman [1964], and Hirsch and Smale [1974]. A proof of the Jordan Curve Theorem is in Newman [1953]. A generalization of the Poincaré–Bendixson theory for two-dimensional surfaces other than the plane are given in Schwartz [1963].

A comprehensive reference on the search for limit cycles, Dulac functions, etc., on the plane is Yeh [1986]. The proof of the existence of a limit cycle of the oscillator of Van der Pol is from Ye [1986] and that of its hyperbolicity is from Coppel [1965].

The sixteenth problem of Hilbert has a fascinating history and remains one of the few unanswered questions of his list. Curiously, number sixteen is absent in the collection by Browder [1976] on the progress of Hilbert's problems. The first major result on the sixteenth problem was the "theorem" of Dulac [1923] asserting the finiteness of the total number of limit cycles of a general polynomial planar vector field. Later, a gap in his "proof" was discovered; it turned out to be nonrectifiable. Recently, the theorem of Dulac has been proved by Bamón [1987] for quadratic polynomials, with major contributions by Il'yaschenko. For general polynomials, there are announcements of Dulac's theorem in Ecalle et al. [1987] and Il'yaschenko [1990]. A bound for the number of limit cycles, however, still appears to be distant, even for quadratic polynomials, although there have been false proofs and claims. The general feeling is that there can be at most four limit cycles in the case of quadratics, and there is such an example by Wang and also by Shi [1980] stored in the library of PHASER under the name *hilbert4*. For easy reading on quadratic vector fields, consult Chicone and Tian [1982].

One important question regarding the bifurcations of a homoclinic loop is the number of resulting periodic orbits. Under fairly general assumptions, most notably the requirement that the trace of the linearization at the saddle point be different from zero, one can establish the uniqueness of the bifurcating periodic orbit. If the trace condition is not met, it is possible to obtain any number of periodic orbits. Bifurcation of periodic orbits of planar systems is covered in great detail in Andronov et al. [1973], including results on bifurcations of homoclinic loops. Do not miss especially the last chapter of this book, where many specific examples are presented. Some of these topics and the role of planar bifurcations in higher dimensions are contained in Chow and Hale [1982].

13

All Planar Things Considered

 In the numerous chapters that have come before, we have encountered many bifurcations, such as saddle-node for equilibria and periodic orbits, Poincaré–Andronov–Hopf, and breaking homoclinic loops and saddle connections. It is natural to ponder when, if ever, we will stop adding to the list and produce a complete catalog of all possible bifurcations. In this chapter, we indeed provide such a list for "generic" bifurcations of planar vector fields depending on one parameter. However, due to the overwhelming difficulty of the subject matter, our exposition, while precise, is devoid of verifications. To circumvent certain technical complications, we confine our attention to a closed and bounded region of the plane, and in such a region characterize the structurally stable vector fields. To motivate this confinement, we then make a short digression to describe a class of vector fields whose dynamics are naturally confined to a bounded region—dissipative systems. Next, we explore the geometry of sets of mildly structurally unstable vector fields—first-order structural instability. By determining the sets of such vector fields forming hypersurfaces in the set of all vector fields, we arrive at a list of one-parameter "generic" bifurcations. You will undoubtedly notice that some of the familiar bifurcations are absent from the list. We provide an explanation for this as well, in terms of symmetries. We end the chapter with a glimpse into the intricate bifurcations of two-parameter vector fields.

13.1. Structurally Stable Vector Fields

On many previous occasions we have considered properties of vector fields that are preserved under small perturbations of the vector field. Here, we finally cast, in the spirit of Section 2.6, this important topic of qualitative dynamics in a general setting, and give a complete characterization of those planar vector fields for which the orbit structure remains qualitatively unchanged under small perturbations—structurally stable vector fields.

Our first-order task is to introduce a suitable distance on the space of planar vector fields so as to make the notion of "small perturbation" of a vector field precise. Because of the unboundedness of the plane, this turns out to be more difficult than it first appears. To circumvent this and other difficulties, we will restrict our comparison of vector fields to some *compact*—closed and bounded—subset of the plane.

Let $\mathcal{D}$ be a compact subset of $\mathbb{R}^2$ with a smooth boundary and let $\mathcal{X}^k(\mathcal{D})$ denote the C^k vector fields defined on $\mathcal{D}$ and pointing inwards at the boundary points of $\mathcal{D}$. To specify a neighborhood of a vector field in $\mathcal{X}^k(\mathcal{D})$, one fixes a norm on $\mathbb{R}^2$ and defines the C^0 *distance* of two vector fields $\mathbf{f}$ and $\mathbf{g}$ in $\mathcal{X}^k(\mathcal{D})$ to be

$$\|\mathbf{f} - \mathbf{g}\|_0 = \sup_{\mathbf{x} \in \mathcal{D}} \big\{ \|\mathbf{f}(\mathbf{x}) - \mathbf{g}(\mathbf{x})\| \big\}.$$

However, the C^0 distance yields neighborhoods that are 'too large.' For instance, two vector fields that are C^0 close may not have the same number of hyperbolic equilibria; see Figure 13.1. To avoid this undesirable situation, we introduce the C^1 *distance* by requiring the functions as well as their derivatives to be close at all points of $\mathcal{D}$:

$$\|\mathbf{f} - \mathbf{g}\|_1 = \sup_{\mathbf{x} \in \mathcal{D}} \big\{ \|\mathbf{f}(\mathbf{x}) - \mathbf{g}(\mathbf{x})\|, \ \|D\mathbf{f}(\mathbf{x}) - D\mathbf{g}(\mathbf{x})\| \big\}.$$

In this definition, we view the derivatives as linear functions and use any norm on $\mathbb{R}^4$. With the C^1 distance, we define the δ *neighborhood* of $\mathbf{f}$ to be the set of all vector fields $\mathbf{g}$ in $\mathcal{X}^k(\mathcal{D})$, with $k \geq 1$, satisfying $\|\mathbf{f} - \mathbf{g}\|_1 < \delta$. The resulting topology on $\mathcal{X}^k(\mathcal{D})$ is called the C^1 *topology*.

We also will have occasion to employ C^r distances by imposing closeness conditions on higher-order derivatives,

$$\|\mathbf{f} - \mathbf{g}\|_r = \sup_{\mathbf{x} \in \mathcal{D}; \ 0 \leq i \leq r} \big\{ \|D^i\mathbf{f}(\mathbf{x}) - D^i\mathbf{g}(\mathbf{x})\| \big\},$$

and use the C^r topology on the set of vector fields $\mathcal{X}^k(\mathcal{D})$; we omit further details.

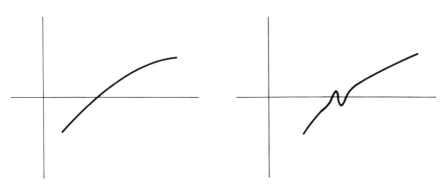

Figure 13.1. *Two (scalar) functions that are close in the C^0 topology but not in the C^1 topology may have different number of zeros.*

To summarize our notation, in this chapter $\mathcal{X}^k(\mathcal{D})$ will denote the space of C^k vector fields, $k \geq 1$, defined on a compact subset $\mathcal{D}$ of the plane and pointing inwards at the boundary points of $\mathcal{D}$. The topology on the infinite dimensional space $\mathcal{X}^k(\mathcal{D})$ will be at least the C^1 topology.

These technical conventions attended, we now return to dynamics.

Definition 13.1. *Two vector fields* **f** *and* **g** *are said to be topologically equivalent if there is a homeomorphism* $h : \mathcal{D} \to \mathcal{D}$ *such that* **h** *maps the orbits of* **f** *onto the orbits of* **g** *and preserves the sense of direction of time.*

Definition 13.2. *A vector field* $\mathbf{f} \in \mathcal{X}^k(\mathcal{D})$*, with* $k \geq 1$*, is structurally stable if there is a neighborhood* $N_{\mathbf{f}}$ *of* **f** *in* $\mathcal{X}^k(\mathcal{D})$ *such that any* $\mathbf{g} \in N_{\mathbf{f}}$ *is topologically equivalent to* **f***.*

This definition is admittedly a bit amalgamated; we consider two vector fields to be "near" if their values as well as their derivatives are close, while settling to compare their phase portraits up to only topological equivalence without reference to differentiability. As we pointed out in Section 8.3, topological equivalence and not differentiable equivalence is the natural concept even when comparing linear systems. It is a remarkable fact that these perplexing choices of comparisons result in a complete characterization of the structurally stable vector fields in a bounded region of the plane. Before we state the result, let us recall several definitions from the past.

Definition 13.3. *An equilibrium point* $\bar{\mathbf{x}}$ *of a vector field* **f** *is called hyperbolic if the linear vector field* $D\mathbf{f}(\bar{\mathbf{x}})$ *has eigenvalues with nonzero real parts. In the case when one eigenvalue is positive and the other negative the equilibrium point* $\bar{\mathbf{x}}$ *is called a saddle point.*

Definition 13.4. *A periodic orbit* $\gamma(\mathbf{x}^0)$ *is called* hyperbolic *if the derivative of the Poincaré map satisfies* $\Pi'(\mathbf{x}^0) \neq 1$.

Definition 13.5. *An orbit is called a* saddle connection *if its* α- *and* ω-*limit sets are saddle points. A saddle connection is called a* heteroclinic *orbit if its* α- *and* ω-*limit sets are distinct saddle points, and* homoclinic *if the limit sets are the same saddle point.*

With these dynamical concepts at our disposal, we can now state the main theorem on structural stability for planar vector fields pointing into a compact region $\mathcal{D}$ of the plane.

Theorem 13.6. *(Structurally stable vector fields) A* C^k *vector field* $\mathbf{f} \in \mathcal{X}^k(\mathcal{D})$, *with* $k \geq 1$, *is structurally stable on* $\mathcal{D}$ *if and only if* $\mathbf{f}$ *has the following properties:*
  (i) *all equilibrium points are hyperbolic;*
  (ii) *all periodic orbits are hyperbolic;*
  (iii) *there are no saddle connections.* $\diamondsuit$

The necessity of these conditions for structural stability are self-evident from our previous examples in the study of local bifurcations. The sufficiency, however, is difficult to establish and thus we say no more about it.

Despite the mathematical elegance of the theorem above, it is not, in general, easy to determine if a specific planar vector field is structurally stable. For example, it is difficult to verify if a vector field points into a compact region, and periodic orbits are nearly impossible to locate. Nevertheless, here is one success story based on our earlier work.

Example 13.7. *Van der Pol is structurally stable:* In the previous chapter, while proving the existence of a periodic orbit of the equation of Van der Pol, we constructed a compact set, say, $\mathcal{D}$, into which all orbits eventually entered. The boundary of this set had a couple of corners, but it is possible to modify the set a bit to smooth out the boundary while ensuring that the flow still points inward at the boundary. Inside this set there is an equilibrium point and a periodic orbit both of which we showed to be hyperbolic. Moreover, there is no saddle connection because the ω-limit set of any orbit except the origin is the periodic orbit. Now that we have verified all the hypotheses of the theorem above, we conclude that the equation of Van der Pol is structurally stable within the class $\mathcal{X}^k(\mathcal{D})$, the set of C^k vector fields pointing inward into $\mathcal{D}$ and endowed with the C^k topology. $\diamondsuit$

Despite the difficulties posed by specific vector fields, the general situation is less of a concern as "most" planar vector fields in $\mathcal{X}^k(\mathcal{D})$ turn out to be structurally stable.

Theorem 13.8. (Genericity of structural stability) *The subset of $\mathcal{X}^k(\mathcal{D})$ consisting of the structurally stable vector fields is open and dense in $\mathcal{X}^k(\mathcal{D})$, that is, structural stability is a generic property.* $\diamondsuit$

In light of this theorem, it appears to be practically convenient to ignore structurally unstable planar vector fields because an arbitrarily small perturbation will usually turn a structurally unstable vector field into a structurally stable one. This is a reasonable strategy for a single planar vector field; if the vector field depends on a parameter, however, the difficulties abound, as we shall see later in this chapter. Before delving into the depths of bifurcations, we now make a short diversion and explore the boundary condition we have elected to impose on our vector fields.

Exercises _____

13.1. *A structurally stable system:* Show that the planar system

$$\dot{x}_1 = 2x_1 + x_2 - x_1(x_1^2 + x_2^2)$$
$$\dot{x}_2 = -x_2 - x_2(x_1^2 + x_2^2)$$

is structurally stable by verifying all the hypotheses of Theorem 13.6.
Hints: The vector field is symmetric with respect to the reflection through origin, and points inward on a large enough disk. There are no periodic orbits because there is an invariant line through the origin. Further help is available in Andronov et al. [1973], p. 190. Also, you might like to plot the phase portrait of this planar system using PHASER.

13.2. Determine the C^0 norm of the vector field in the previous problem on the disk you have chosen. Can you compute its C^1 norm?

13.3. *Why compact?* If you want to know how to introduce a topology on the set of vector fields defined on a noncompact set, look up "*Whitney topology*" in, for example, Hirsch [1976].

13.4. Sketch a flow on a bounded domain such that all equilibria are hyperbolic, yet there are an infinite number of hyperbolic periodic orbits.
Suggestion: Make the periodic orbits accumulate on an invariant set which consists of equilibrium points and orbits connecting them.

13.5. *Structural stability and finiteness:* If $\mathbf{f} \in \mathcal{X}^k$ is structurally stable then $\mathbf{f}$ may have only a finite number of equilibrium points and periodic orbits.
Suggestion: You must exclude, among other things, that the situation in the previous exercise does not occur.

13.2. Dissipative Systems

In the setting of the previous section we considered the dynamics of vector fields only on a compact domain and pointing inwards into the domain. This is not an innocuous constraint; for example, the grand theorem on structural stability does not apply even to a linear saddle which of course is structurally stable. However, there is a class of systems of considerable practical interest—dissipative systems—that falls naturally into the setting of the previous section. In this section, we simply point out several facts on dissipative systems. Let us begin with their definition.

Definition 13.9. *A planar differential equation* $\dot{\mathbf{x}} = \mathbf{f}(\mathbf{x})$ *is said to be dissipative if there is a bounded subset B of $\mathbb{R}^2$ such that, for any $\mathbf{x}^0 \in \mathbb{R}^2$, there is a time t_0, which depends on $\mathbf{x}^0$ and B, so that the solution $\varphi(t, \mathbf{x}^0)$ through $\mathbf{x}^0$ satisfies $\varphi(t, \mathbf{x}^0) \in B$ for $t \geq t_0$.*

This definition can be rephrased by saying that a system is dissipative if infinity is a source; every orbit moves a certain distance from infinity and remains away. For our purposes in this chapter, one of the important properties of dissipative systems is the following:

Theorem 13.10. *If $\dot{\mathbf{x}} = \mathbf{f}(\mathbf{x})$ is dissipative, then there exists a set $\mathcal{D}$ diffeomorphic to a disk with the property that on the boundary of $\mathcal{D}$ the vector field points inside $\mathcal{D}$.* $\Diamond$

We will momentarily say more about this theorem. First, however, we define the omega-limit set of a set.

Definition 13.11. *The omega-limit set $\omega(U)$ of a subset U of $\mathbb{R}^2$ under the flow of $\dot{\mathbf{x}} = \mathbf{f}(\mathbf{x})$ is defined to be the set*

$$\omega(U) = \bigcap_{\tau \geq 0} \overline{\gamma^+(\varphi(\tau, U))},$$

where the overline is the closure of the positive orbits of $\dot{\mathbf{x}} = \mathbf{f}(\mathbf{x})$ through all $\mathbf{x}^0 \in U$.

Asymptotic behavior of a dissipative system can be localized to the study of a flow on a subset of the plane with many desirable properties.

Definition 13.12. *A set $\mathcal{A}$ is said to be a global attractor of a dissipative system if $\mathcal{A}$ is compact, connected, invariant, and $\omega(U) \subset \mathcal{A}$ for every bounded set U.*

Theorem 13.13. *A dissipative system has a unique global attractor.* $\Diamond$

Using the attractor, one can show the existence of an appropriate global Liapunov function with the following properties: The Liapunov function vanishes on the attractor, the level curves surround the attractor, and

Figure 13.2. *Global attractor of the dissipative equation $\ddot{x} + \dot{x} - x + x^3 = 0$.*

the vector field points inside the level curves except the zero level curve. This is how a proof of Theorem 13.10 goes; however, the existence of the Liapunov function is just that and one does not generally expect to be able to construct it.

These discouraging words aside, let us now consider some specific examples of dissipative systems. We can, of course, recall Van der Pol again, in which case the global attractor is the limit cycle together with its interior. We now offer another example of a dissipative system.

Example 13.14. *The global attractor of a damped conservative system:* Consider the second-order equation $\ddot{x} + \dot{x} - x + x^3 = 0$, or the equivalent first-order system

$$\dot{x}_1 = x_2$$
$$\dot{x}_2 = -x_2 + x_1 - x_1^3.$$

We will show that the global attractor of this system, as seen in Figure 13.2, is the unstable manifold of the origin together with the equilibrium points $(-1, 0)$ and $(1, 0)$:

$$\mathcal{A} = W^u(0, 0) \cup \{(-1, 0)\} \cup \{(1, 0)\}.$$

We first establish the existence of the global attractor by using the Invariance Principle, Theorem 9.25, and the Liapunov function

$$V(\mathbf{x}) = -\tfrac{1}{2}x_1^2 + \tfrac{1}{4}x_1^4 + \tfrac{1}{2}x_2^2.$$

Since $\dot{V}(\mathbf{x}) = -x_2^2 \leq 0$ and the level curve $V^{-1}(c)$ for c large is a closed curve, each orbit is bounded. From the Invariance Principle, for any $\mathbf{x}^0 \in \mathbb{R}^2$, the omega-limit set $\omega(\mathbf{x}^0)$ is one of the three equilibria $(-1, 0)$, $(0, 0)$, or $(1, 0)$. The system is therefore dissipative and there exists a global attractor.

We now determine the structure of the attractor. The equilibrium points $(-1, 0)$ and $(1, 0)$ are asymptotically stable while $(0, 0)$ is a saddle point. The set $W^u(0, 0) \cup \{(-1, 0)\} \cup \{(1, 0)\}$ is a compact invariant set and thus must belong to the global attractor. To establish that this set is indeed the entire global attractor, we need to show that the alpha-limit set $\alpha(\mathbf{x}^0)$ for any point $\mathbf{x}^0 \in \mathcal{A}$ is an equilibrium point.

If $\mathbf{y} \in \alpha(\mathbf{x}^0)$, then there is a sequence $t_n \to -\infty$ as $n \to +\infty$, with $t_{n-1} \geq 1 + t_n$, so that $\varphi(t_n, \mathbf{x}^0) \to \mathbf{y}$ as $n \to +\infty$. Since $V(\varphi(\tau, \mathbf{x}^0)) \leq V(\varphi(t, \mathbf{x}^0))$ for $\tau \geq t$, we have

$$V(\varphi(t_{n-1}, \mathbf{x}^0)) \leq V(\varphi(t_n + t, \mathbf{x}^0)) \leq V(\varphi(t_n, \mathbf{x}^0))$$

for all n and $t \in [0, 1]$. Taking the limit as $t \to +\infty$, we have $V(\varphi(t, \mathbf{y})) = V(\mathbf{y})$ for $t \in [0, 1]$ and, consequently, for $t \in \mathbb{R}$. Thus, $\dot{V}(\varphi(t, \mathbf{y})) = 0$ for $t \in \mathbb{R}$ and $\mathbf{y}$ is an equilibrium point. $\Diamond$

This concludes our brief diversion into dissipative systems and we now return to our main task in this chapter, the generic bifurcations of one-parameter planar vector fields.

Exercises ─────────────────────────────────

13.6. *Changing attractors:* Convert each second-order equation below to a first-order system and determine if it is dissipative. Plot their global attractors using PHASER as you vary the parameters λ or μ. Make sure to use initial conditions very near the saddle points:
   (a) $\ddot{x} + \lambda\dot{x} - x + x^3 = 0$;
   (b) $\ddot{x} + \dot{x} + \lambda x + x^3 = 0$;
   (c) $\ddot{x} + \dot{x} + \mu + \lambda x + x^3 = 0$;
   (d) $\ddot{x} - \dot{x}(\lambda - x^2) + x = 0$.

13.7. *On limit set of a set:* For any subset U of $\mathbb{R}^2$, one can define another limit set $\hat{\omega}(U)$ of U as follows:

$$\hat{\omega}(U) = \bigcup_{\mathbf{x}^0 \in U} \omega(\mathbf{x}^0).$$

Using an example from the previous exercise, show that $\omega(U) \neq \hat{\omega}(U)$.

13.3. One-parameter Generic Bifurcations

We saw earlier in this chapter that the structurally stable vector fields in $\mathcal{X}^k(\mathcal{D})$ form an open and dense subset of this infinite dimensional space. For the purposes of tracking down bifurcations, however, we need to understand how the structurally unstable vector fields sit in the space of vector fields $\mathcal{X}^k(\mathcal{D})$. Of course, the variety of structurally unstable vector fields is endless and thus the possible bifurcations are numerous. However, in the case of one-parameter vector fields the situation is not hopeless if we content ourselves with identifying the most likely—generic—bifurcations. In this section, we finally produce a catalog of such bifurcations of planar vector fields depending on one parameter.

To motivate our search for generic bifurcations, let us begin our exposition with a simple analogy: in three dimensions, for example, a curve cannot avoid intersecting a surface while going from one side to the other; a curve, on the other hand, is not likely to intersect another curve or a point while moving around in three dimensions. A one-parameter family of vector fields in $\mathcal{X}^k(\mathcal{D})$ defines a curve in this infinite dimensional space of vector fields. Suppose that there is a hypersurface—codimension-one submanifold—consisting of structurally unstable vector fields and that structurally stable vector fields on the two sides of the hypersurface are topologically different. While varying the parameter, if the curve of vector fields moves from one side of the hypersurface to the other, then the curve cannot avoid hitting the hypersurface, but in all likelihood it would miss subsets of smaller dimensions.

Let us proceed by making the notion of a codimension-one submanifold of the infinite dimensional space $\mathcal{X}^k(\mathcal{D})$ precise.

Definition 13.15. *A subset $\mathcal{S}$ of $\mathcal{X}^k(\mathcal{D})$ is said to be a C^r submanifold of codimension one if there is an open subset U of $\mathcal{X}^k(\mathcal{D})$ and a C^r function $H : U \to \mathbb{R}$, for some $r \geq 1$, such that $\mathcal{S} = \{\, \mathbf{f} \in U \,:\, H(\mathbf{f}) = 0 \,\}$ and that $DH(\mathbf{f}) \neq \mathbf{0}$ for all $\mathbf{f} \in U$.*

With the intent of identifying certain codimension-one submanifolds consisting of structurally unstable vector fields, we will single out various subsets of $\mathcal{X}^k(\mathcal{D})$. In particular, we would like a vector field on one of these submanifolds to be structurally stable with respect to the perturbations along the submanifold, while the vector field is structurally unstable with respect to perturbations in $\mathcal{X}^k(\mathcal{D})$.

Definition 13.16. *A structurally unstable vector field $\mathbf{f}$ is said to be first-order structurally unstable if there is a neighborhood of $\mathbf{f}$ in the subset of structurally unstable vector fields with the induced topology, such that every structurally unstable vector field in this neighborhood is topologically equivalent to $\mathbf{f}$.*

We will identify codimension-one submanifolds of $\mathcal{X}^k(\mathcal{D})$ consisting of first-order structurally unstable vector fields by violating the conditions for structural stability as listed in Theorem 13.6 in as mild a way as possible. In the case of equilibria, we consider two types of nonhyperbolic equilibria. The first kind is when zero is a simple eigenvalue of the linearization and the scalar vector field on the center manifold, or the bifurcation function, starts with a quadratic term. This kind of nonhyperbolic equilibria was the subject of Chapter 10; see, in particular, Definition 10.6, Theorem 10.10, and Theorem 10.15.

Definition 13.17. *A nonhyperbolic equilibrium point $\bar{\mathbf{x}}$ of $\mathbf{f}$ is called an elementary saddle node if zero is a simple eigenvalue of $D\mathbf{f}(\bar{\mathbf{x}})$ and the* scalar vector field on a center manifold has the form

$$\dot{x}_1 = ax_1^2 + O(x_1^3) \quad \text{as} \quad x_1 \to 0, \qquad a \neq 0.$$

The second kind of nonhyperbolic equilibrium point is the one with purely imaginary eigenvalues. Following the setting of Chapter 11, in polar coordinates, we will assume that the radial equation $d\rho/dt$ in normal form starts with a cubic term; see, in particular, Lemma 11.3. Recall that the quadratic term is necessarily zero.

Definition 13.18. *A nonhyperbolic equilibrium point* $\bar{\mathbf{x}}$ *of* $\mathbf{f}$ *is called an elementary composed focus if the eigenvalues of* $D\mathbf{f}(\bar{\mathbf{x}})$ *are purely imaginary and in polar coordinates the radial equation in normal form is given by*

$$\frac{d\rho}{dt} = a\rho^3 + O(\rho^4) \quad \text{as} \quad \rho \to 0, \qquad a \neq 0.$$

These two kinds of mildly nonhyperbolic equilibrium points are referred to as *quasi-hyperbolic equilibria*. Next, we consider mildly nonhyperbolic periodic orbits.

Definition 13.19. *A periodic orbit* $\gamma(\mathbf{x}^0)$ *is called quasi-hyperbolic if the derivatives of the Poincaré map satisfy* $\Pi'(\mathbf{x}^0) = 1$ *and* $\Pi''(\mathbf{x}^0) \neq 1$.

Homoclinic loops must also satisfy a nondegeneracy condition.

Definition 13.20. *Let* $\bar{\mathbf{x}}$ *be a saddle point with trace* $D\mathbf{f}(\bar{\mathbf{x}}) \neq 0$. *An orbit for which both the* α- *and* ω-*limit sets are* $\bar{\mathbf{x}}$ *is called an elementary homoclinic loop.*

We need one final definition for the exclusion of certain global complications.

Definition 13.21. *If in each neighborhood of a point* $\mathbf{p}$ *there is a point* $\mathbf{q}$ *such that* $\omega(\mathbf{p}) \neq \omega(\mathbf{q})$, *or* $\alpha(\mathbf{p}) \neq \alpha(\mathbf{q})$, *then the orbit through* $\mathbf{p}$ *is called a separatrix.*

With these definitions, we now produce a complete, albeit long, list of first-order structurally unstable planar vector fields.

Theorem 13.22. (First-order structurally unstable vector fields) *A vector field in* $\mathcal{X}^k(\mathcal{D})$, *with* $k \geq 4$, *is first-order structurally unstable if and only if it belongs to one of the sets below:*

$\mathcal{S}_1$: *The set of vector fields which have only one elementary saddle node as nonhyperbolic equilibrium point, have all other equilibrium points and periodic orbits hyperbolic, and do not have any saddle connections. Moreover, none of the separatrices of the saddle node may go to a saddle point, and no two separatrices of the saddle node are continuations of each other (see Figure 13.3).*

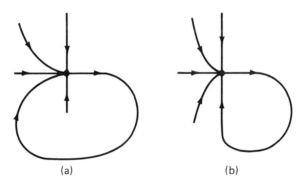

Figure 13.3. *Partial phase portraits of* (a) *first-order structurally unstable, and* (b) *not first-order structurally unstable vector field.*

$\mathcal{S}_2$: *The set of vector fields which have only one elementary composed focus as nonhyperbolic equilibrium point, have all other equilibrium points and periodic orbits hyperbolic, and do not have any saddle connections.*

$\mathcal{S}_3$: *The set of vector fields which have only one quasi-hyperbolic periodic orbit, have all other periodic orbits and equilibrium points hyperbolic, and do not have any saddle connections. Moreover, there may not be two saddle point separatrices going to the quasi-hyperbolic periodic orbit, one for $t \to -\infty$ and the other for $t \to +\infty$ (see Figure 13.4).*

$\mathcal{S}_4$: *The set of vector fields which have an elementary homoclinic loop as the only saddle connection, and have all equilibrium points and periodic orbits hyperbolic. Moreover, a separatrix of a saddle point may not go to the elementary homoclinic loop for $t \to -\infty$ or $t \to +\infty$.*

$\mathcal{S}_5$: *The set of vector fields which have only one saddle connection from one saddle point to another, and have all equilibrium points and periodic orbits hyperbolic.* $\Diamond$

The collection of the sets of first-order structurally unstable vector fields described in the theorem above forms a codimension-one submanifold of $\mathcal{X}^k(\mathcal{D})$.

Theorem 13.23. *The set $\bigcup_{i=1}^5 \mathcal{S}_i$ is a codimension-one C^{k-1} submanifold of $\mathcal{X}^k(\mathcal{D})$, with $k \geq 4$, and it is open in the set of all structurally unstable vector fields (in the induced topology).* $\Diamond$

To uncover the bifurcations of first-order structurally unstable vector fields lying on one of these codimension-one submanifolds, we need to describe a neighborhood of $\mathbf{f} \in \mathcal{S}_i$ in the set of vector fields $\mathcal{X}^k(\mathcal{D})$. Since $\mathcal{S}_i$ is a codimension-one submanifold of $\mathcal{X}^k(\mathcal{D})$, a vector field $\mathbf{f} \in \mathcal{S}_i$ has a neighborhood $W_i \subset \mathcal{X}^k(\mathcal{D})$ consisting of three disjoint sets, $W_i = U_i \cup S_i \cup V_i$,

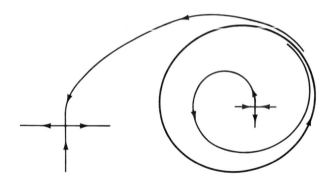

Figure 13.4. *Partial phase portrait of a vector field that is not first-order structurally unstable. When the nonhyperbolic periodic orbit disappears, a saddle connection appears.*

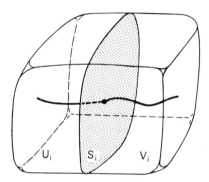

Figure 13.5. *Neighborhood of a first-order structurally unstable vector field.*

such that U_i and V_i are open subsets of $\mathcal{X}^k(\mathcal{D})$ containing structurally stable vector fields, and S_i is an open subset of $\mathcal{S}_i$. Moreover, the vector fields in each component of W_i are topologically equivalent. We will momentarily describe the dynamics of vector fields in each component of a neighborhood W_i. First, however, let us explain the consequences of such knowledge for bifurcations.

A one-parameter vector field is a curve in $\mathcal{X}^k(\mathcal{D})$. Let us suppose that such a curve of vector fields crosses one of the codimension-one submanifolds $\mathcal{S}_i$ transversally for some value of the parameter; see Figure 13.5. Here, transversal crossing means that at the point of contact the curve is not tangent to the codimension-one submanifold; this is the typical situation. Now, the knowledge of the flows of the vector fields in each neighborhood W_i yields a characterization of the typical bifurcations of first-order structurally unstable vector fields. We should emphasize that while the list of bifurcations is for local variations of the parameter near a bifurcation value, the consideration of the flows are global in $\mathcal{D}$.

We now proceed with the enumeration of neighborhoods of first-order structurally unstable vector fields. In each description, we only indicate the changes near the nonhyperbolic equilibria or periodic orbits and saddle connections; see Figure 13.6.

Theorem 13.24. (All codimension-one bifurcations of first-order structurally unstable vector fields) *Each first-order structurally unstable vector field* $\mathbf{f} \in \mathcal{S}_i$ *has a neighborhood* $W_i = U_i \cup S_i \cup V_i$ *such that, in each component, vector fields have the following dynamics:*

- *Near $\mathcal{S}_1$: There are two cases to consider. (1) Saddle-node bifurcation of an equilibrium: A vector field $\mathbf{f} \in \mathcal{S}_1$ has an elementary saddle node at an equilibrium point $\bar{\mathbf{x}}$ and no separatrix of the saddle node forms a loop; there are no equilibrium points of $\mathbf{f}$ near $\bar{\mathbf{x}}$ if $\mathbf{f} \in U_1$; there is a hyperbolic saddle and a hyperbolic node near $\bar{\mathbf{x}}$ if $\mathbf{f} \in V_1$.*
  (2) Saddle-node bifurcation on a loop: A vector field $\mathbf{f} \in \mathcal{S}_1$ has an elementary saddle node at an equilibrium point $\bar{\mathbf{x}}$ and a separatrix of the saddle-node forms a loop; there are no equilibrium points of $\mathbf{f}$ near $\bar{\mathbf{x}}$ if $\mathbf{f} \in U_1$, no separatrix loop but a hyperbolic periodic orbit near its neighborhood; there is a hyperbolic saddle and a hyperbolic node near $\bar{\mathbf{x}}$ and a connecting orbit from the saddle to the node near the original saddle-node loop if $\mathbf{f} \in V_1$.
- *Near $\mathcal{S}_2$: Generic Poincaré–Andronov–Hopf bifurcation: A vector field $\mathbf{f} \in \mathcal{S}_2$ has an elementary composed focus at an equilibrium point $\bar{\mathbf{x}}$; there is a hyperbolic equilibrium point but no periodic orbit of $\mathbf{f}$ near $\bar{\mathbf{x}}$ if $\mathbf{f} \in U_2$; there is a hyperbolic equilibrium surrounded by a hyperbolic periodic orbit near $\bar{\mathbf{x}}$ if $\mathbf{f} \in V_2$.*
- *Near $\mathcal{S}_3$: Saddle-node bifurcation of a periodic orbit: A vector field $\mathbf{f} \in \mathcal{S}_3$ has a quasi-hyperbolic periodic orbit γ which is stable from one side and unstable from the other; there is no periodic orbit of $\mathbf{f}$ near γ if $\mathbf{f} \in U_3$; there are two hyperbolic periodic orbits, one orbitally asymptotically stable and the other unstable, near γ if $\mathbf{f} \in V_3$.*
- *Near $\mathcal{S}_4$: Elementary homoclinic loop bifurcation: A vector field $\mathbf{f} \in \mathcal{S}_4$ has an elementary homoclinic loop Γ at an equilibrium point $\bar{\mathbf{x}}$; there is no periodic orbit of $\mathbf{f}$ near Γ if $\mathbf{f} \in U_4$; there is a unique hyperbolic periodic orbit near Γ if $\mathbf{f} \in V_4$.*
- *Near $\mathcal{S}_5$: Breaking a saddle connection: A vector field $\mathbf{f} \in \mathcal{S}_5$ has a saddle connection; there is no saddle connection if $\mathbf{f} \in U_5$ or in V_5. $\Diamond$*

We hope that the list in the theorem above and the illustrations in Figure 13.6 make our efforts of almost a dozen chapters a bit more meaningful.

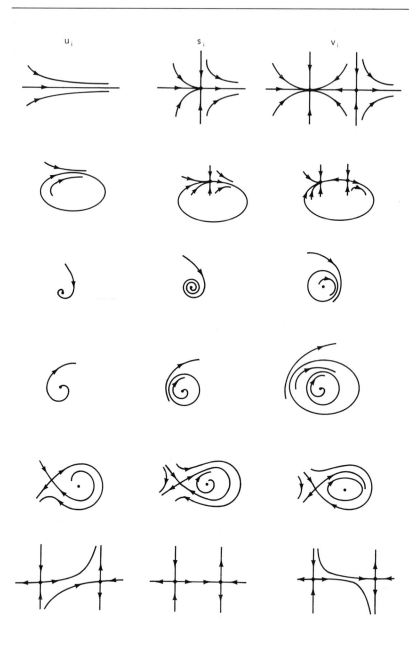

Figure 13.6. Generic bifurcations of first-order structurally unstable planar vector fields depending on one parameter. Only partial phase portraits near orbits undergoing changes are shown.

Yet, some of the one-parameter bifurcations that we have elaborated on at length are not contained in the list. We now turn to a brief explanation of their absence.

Exercises

13.8. *Saddle node on a loop:* Consider the system

$$\dot{x}_1 = x_1(1 - x_1^2 - x_2^2) + x_2(1 + \lambda + x_1)$$
$$\dot{x}_2 = -x_1(1 + \lambda + x_1) + x_2(1 - x_1^2 - x_2^2).$$

In polar coordinates $x_1 = r\cos\theta$ and $x_2 = -r\sin\theta$, show that the system becomes

$$\dot{r} = r(1 - r^2)$$
$$\dot{\theta} = 1 + \lambda + r\cos\theta.$$

Now, verify the following dynamics as the parameter λ is varied: The circle $r = 1$ is always invariant under the flow and all solutions except the origin approach this circle in forward time. On the circle $r = 1$, there are two equilibria for $\lambda < 0$, and $|\lambda|$ small, corresponding to the zeros of $1 + \lambda + \cos\theta = 0$. Observe that one of the equilibria is a saddle and the other is a node. For $\lambda > 0$, there are no equilibria and the circle $r = 1$ is a periodic orbit.

13.4. Bifurcations in the Presence of Symmetry

The concept of structural stability depends very much on the class of allowable perturbations. A vector field that is structurally unstable with respect to arbitrary perturbations chosen from the general set of vector fields $\mathcal{X}^k(\mathcal{D})$ may remain topologically unchanged if the allowable perturbations are confined to a smaller set of vector fields. Even if a vector field is structurally unstable, when it is subjected to special perturbations, the vector field may undergo bifurcations other than the ones we have enumerated earlier. In this section, we give several simple instances of these special, yet practically important, considerations.

Example 13.25. *Saddle connection preserved:* Let us consider the planar vector field

$$\dot{x}_1 = 2x_1 x_2$$
$$\dot{x}_2 = 1 + x_1^2 - x_2^2 \tag{13.1}$$

from Example 9.37. As we saw there, this system has two saddle points $(0, 1)$ and $(0, -1)$. Also, there is a saddle connection between them, part of the x_2-axis, so that the system (13.1) is structurally unstable. For example, the perturbation

$$\dot{x}_1 = \lambda + 2x_1 x_2$$
$$\dot{x}_2 = 1 + x_1^2 - x_2^2 \tag{13.2}$$

has no saddle connection for any nonzero value of the parameter λ.

Now, let us observe that the system (13.1) is symmetric with respect to the x_2-axis. In particular, the x_2-axis is invariant. To preserve this symmetry, let us allow perturbations of which the first component is odd in x_1, and the second even in x_2. With such perturbations, it is impossible to break the saddle connection. For example, the one-parameter perturbation

$$\dot{x}_1 = \lambda x_1 + 2x_1 x_2$$
$$\dot{x}_2 = 1 + x_1^2 - x_2^2,$$

for $|\lambda|$ sufficiently small, leaves the topological type of Eq. (13.1) unchanged.

Of course, the perturbation (13.2) does not posses the symmetries of the original system (13.1) above. $\Diamond$

Example 13.26. *Transcritical bifurcation reconsidered:* The vector field

$$\dot{x}_1 = x_1^2$$
$$\dot{x}_2 = -x_2$$

has an equilibrium point at the origin which is quasi-hyperbolic of saddle-node type. As we saw in the previous section, this system belongs to a codimension-one submanifold of type $\mathcal{S}_1$ of Theorem 13.3.

Now, let us consider the one-parameter perturbations of this vector field in such a way that for all parameter values the origin remains an equilibrium point with no restriction on its stability type. The one-parameter vector field

$$\dot{x}_1 = \lambda x_1 + x_1^2$$
$$\dot{x}_2 = -x_2,$$
\hfill (13.3)

for instance, satisfies this requirement. At $\lambda = 0$, we have the original vector field. For nonzero λ there are two hyperbolic equilibrium points; the origin is asymptotically stable and the other unstable when $\lambda < 0$, the origin is unstable and the other is asymptotically stable when $\lambda > 0$. This, of course, is a transcritical bifurcation, not a saddle-node bifurcation.

We infer now that evidently the vector field (13.3) does not cross the codimension-one submanifold $\mathcal{S}_1$ transversally as the parameter λ is varied, and thus the resulting bifurcation is not one of the generic codimension-one bifurcations; see Figure 13.7. $\Diamond$

Restricting the set of allowable perturbations is an important practical consideration in many contexts. For instance, a vector field may possess certain symmetry properties, such as being conservative or gradient, that must simply be preserved when perturbed. Before we investigate the dynamics and bifurcations of such vector fields in the following chapter, we take a short diversion and examine a couple of planar vector fields depending on two parameters.

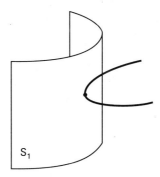

Figure 13.7. *A one-parameter vector field undergoing a transcritical bifurcation meets the codimension-one submanifold $\mathcal{S}_1$ nontransversally.*

Exercises _____ ♣ ♡ ♠ ◇

13.9. Consider the planar vector field

$$\dot{x}_1 = \lambda x_1 + 2x_1 x_2 + x_1^3$$
$$\dot{x}_2 = \tfrac{1}{4} - x_1^2 - x_2^2$$

depending on a parameter λ. For $\lambda = 0$, locate a saddle connection. Can you break this saddle connection for any value of λ?

13.10. *Pitchfork is odd:* Verify that the standard example

$$\dot{x}_1 = \lambda x_1 - x_1^3$$
$$\dot{x}_2 = -x_2$$

exhibiting a pitchfork bifurcation has odd symmetry, that is, the vector field satisfies $\mathbf{f}(-\mathbf{x}) = -\mathbf{f}(\mathbf{x})$.

13.5. Local Two-parameter Bifurcations

To classify generic bifurcations of planar vector fields depending on two parameters, one could attempt to generalize the topological ideas presented in Section 13.3. One could first identify certain codimension-two submanifolds, consisting of structurally unstable vector fields, in the space of vector fields $\mathcal{X}^k(\mathcal{D})$. Then by uncovering the changes in the dynamics of two-parameter vector fields crossing these submanifolds transversally, one could arrive at a catalog of generic two-parameter bifurcations. This is a difficult program to implement and the results are presently fragmentary.

Rather than pursuing such a general setting, the goal of this modest section is to introduce several prominent two-parameter examples of planar vector fields.

The vector fields in the examples below concern the dynamics in the neighborhood of an equilibrium point of planar vector fields whose linearization has two zero eigenvalues but one eigenvector, that is, the Jordan Normal Form of the linearized vector field has the form

$$\mathbf{A} = \begin{pmatrix} 0 & 1 \\ 0 & 0 \end{pmatrix}.$$

As we saw earlier in Section 8.4, this matrix represents a codimension-two singularity in the space of all planar linear vector fields and an unfolding of this matrix is

$$\mathbf{A}(\lambda_1, \lambda_2) = \begin{pmatrix} 0 & 1 \\ \lambda_1 & \lambda_2 \end{pmatrix},$$

where λ_1 and λ_2 are two parameters near zero. Any linear system in a sufficiently small neighborhood of the matrix $\mathbf{A}$ is topologically equivalent to $\mathbf{A}(\lambda_1, \lambda_2)$ for some values of the parameters. Below, we want to generalize this idea of unfolding to nonlinear vector fields in the neighborhood of an equilibrium point whose linear part is the matrix $\mathbf{A}$.

Example 13.27. *Unfolding a double zero eigenvalue after Bogdanov and Takens:* Consider a planar vector field depending on two parameters such that the vector field has an equilibrium point for some fixed values of the parameters. Also, suppose that this equilibrium point is nonhyperbolic with two zero eigenvalues but one eigenvector. Moreover, we assume that our vector field is "in general position" in the set of all planar vector fields with the above properties. This last condition is a bit too technical to state here precisely; it is a certain generic transversality condition met by almost all vector fields, and is obtained from the theory of normal forms.

Now, the remarkable fact is that any vector field possessing the properties above is locally topologically equivalent to one of the two two-parameter vector fields

$$\begin{aligned} \dot{x}_1 &= x_2 \\ \dot{x}_2 &= \lambda_1 + \lambda_2 x_1 + x_1^2 \pm x_1 x_2, \end{aligned} \tag{13.4}$$

depending on the $\pm$ sign. More specifically, there are a change of parameters (with nonzero Jacobian at the fixed values of the parameters) and a homeomorphism (depending continuously on the parameters) of a sufficiently small neighborhood of the equilibrium point which maps the orbits of any planar vector field with the properties above to the orbits of Eq. (13.4) in the neighborhood of the origin while preserving the sense of directions of the orbits. In this sense, the vector field (13.4) is said to be an unfolding of the nonhyperbolic singularity $\mathbf{A}$.

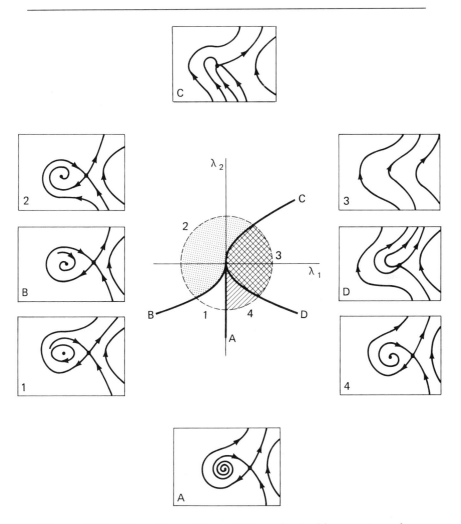

Figure 13.8. *Bifurcations of the unfolding of a double zero eigenvalue.*

It is rather difficult to establish the claims above. Therefore, we content ourselves with only a pictorial summary of the bifurcation diagram of the two-parameter family (13.4) for the case of the $+$ sign as shown in Figure 13.8; the case of the $-$ sign is similar. A careful examination of this bifurcation diagram reveals the inclusion of some of the generic one-parameter bifurcations such as the saddle-node and Poincaré–Andronov–Hopf. It is particularly noteworthy that even though we confined the topological equivalence to a neighborhood of an equilibrium point, we are invariably led to consider a global bifurcation such as the breaking of a homoclinic loop.

Consequently, the determination of the phase portraits of Eq. (13.4) is also nontrivial. In particular, to establish the uniqueness of the resulting limit cycle, one must resort to such things as Abelian integrals. $\Diamond$

There are two other important restricted unfoldings of the nonhyperbolic singularity **A** in smaller sets of vector fields possessing special symmetries.

Example 13.28. *Unfolding a double zero eigenvalue with the equilibrium point fixed:* Consider the following two-parameter nonlinear vector fields:

$$\dot{x}_1 = x_2$$
$$\dot{x}_2 = \lambda_1 x_1 + \lambda_2 x_2 + x_1^2 \pm x_2^2, \tag{13.5}$$

depending on the $\pm$ sign. Notice that the linear part of the vector field (13.5) is the unfolding of **A** in the set of linear vector fields that we have determined in Section 8.4. The distinguishing feature of the vector field (13.5) is that the origin is an equilibrium point for all values of the parameters. This two-parameter vector field has a similar bifurcation diagram to that of the previous example, except that there is no saddle node-bifurcation. Further information on Eq. (13.5) with the $+$ sign is contained in the exercises. $\Diamond$

Example 13.29. *Unfolding a double zero eigenvalue with odd symmetry:* Consider the two-parameter vector fields

$$\dot{x}_1 = x_2$$
$$\dot{x}_2 = \lambda_1 x_1 + \lambda_2 x_2 \pm x_1^3 - x_1^2 x_2, \tag{13.6}$$

depending on the $\pm$ sign. Again, the linear part of this vector field is the unfolding of **A** in the set of linear vector fields. This time, however, the vector field has the odd symmetry, that is, $\mathbf{f}(\lambda_1, \lambda_2, -\mathbf{x}) = -\mathbf{f}(\lambda_1, \lambda_2, \mathbf{x})$. Consequently, the nonlinear terms are third degree. We refrain from saying more about this system except to invite you to ponder about the rather complicated bifurcation diagram for the $-$ sign depicted in Figure 13.9. $\Diamond$

Here we conclude Chapter 13. Despite the title of this chapter, we continue with the prevailing theme of vector fields with special properties and turn next to the dynamics and bifurcations of conservative and gradient systems.

Exercises _____

13.11. *On the Bogdanov–Takens example:*
  (a) Draw the phase portrait of Eq. (13.4) with $+$ sign for the parameter values $(\lambda_1, \lambda_2) = (0, 0)$.

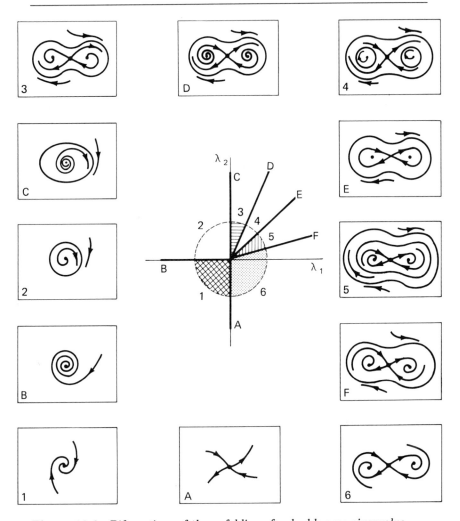

Figure 13.9. *Bifurcations of the unfolding of a double zero eigenvalue with odd symmetry.*

(b) Calculate the bifurcation curves on which the planar system undergoes saddle-node and Poincaré–Andronov–Hopf bifurcations.

(c) This system is stored in the library of PHASER under the name *dzero1*. Experiment on the machine and try to detect the bifurcations numerically.

13.12. *Scaled double zero eigenvalue with origin fixed:* It not easy to see the dynamical features of the system (13.5) in computer simulations. To facilitate such an exploration, as well as for a theoretical proof, it is convenient to

scale the equations to obtain

$$\dot{x}_1 = x_2$$
$$\dot{x}_2 = x_1 - x_1^2 + (b + dx_1)x_2,$$

where b and d are two parameters. For the choice of the scaling, see Chow and Hale [1982], p. 445. Use the initial data $(1.5, 0)$, $(1.3, 0)$, $(1.1, 0)$, and $(-0.1, 0.25)$ and plot the four orbits for the following parameter values:
fix $b = -0.1$, and vary d through 0.1, 0.103, 0.105, and 0.115;
fix $b = 1$, and vary d through 1, 1.101, 1.105, 1.1, and 1.15;
fix $b = 3$, and vary d through 3, 3.05, 3.17, 3.23, 3.255, and 3.26.
Assistance: These equations are stored in the library of PHASER under the name *dzero2*.

13.13. *Scaled double zero eigenvalue with odd symmetry:* For numerical investigations, it is convenient to scale Eq. (13.6) to obtain

$$\dot{x}_1 = x_2$$
$$\dot{x}_2 = x_1 - x_1^3 + (b + dx_1^2)x_2,$$

where b and d are two parameters. Use the initial data $(0.1, 0)$ and $(1.5, 0)$ forward in time, and $(2, 0)$ backward in time, with the following parameter values: fix $b = -0.1$, and vary d through 0.05, 0.1, 0.105, 0.109, 0.125, 0.126, 0.135, and 0.139.
Assistance: These equations are stored in the library of PHASER under the name *dzero3*.

13.14. *Poincaré–Andronov–Hopf bifurcation:* If an equilibrium point has purely imaginary eigenvalues but is not an elementary composed focus, then the radial equation in normal form must begin with terms of degree of at least five. Suppose that

$$\frac{d\rho}{dt} = a\rho^5 + O(\rho^6) \quad \text{as } \rho \to 0, \qquad a \neq 0$$

and show that the following properties hold near the origin:
  (i) Any C^k perturbation, with $k \geq 6$, of the vector field can have at most two periodic orbits.
  (ii) Give a specific perturbation for which there are exactly two periodic orbits.

Bibliographical Notes

Detailed descriptions of norms, topologies, etc., on function spaces are given in, for example, Irwin [1980].

The notion of structural stability was developed by Andronov and Pontrjagin [1937]. Their definition went as follows: $\mathbf{f}$ is structurally stable if, for any $\varepsilon > 0$ there is a $\delta > 0$ such that $\|\mathbf{f} - \mathbf{g}\|_1 < \delta$ implies that there exists an ε homeomorphism $h : \mathcal{D} \rightarrow \mathcal{D}$ which maps the orbits of $\mathbf{f}$ to the orbits of $\mathbf{g}$ while preserving the sense of direction of time. By the way, an ε homeomorphism moves no point more than ε. Peixoto [1959] showed that this definition of structural stability is equivalent to the one given in the text. With this result, the openness of the set of structurally stable systems became self-evident.

The statement of the theorem on the characterization of structurally stable vector fields on a disk was first given by Andronov and Pontrjagin [1937]. The appropriate generalization of this result to orientable compact two-manifolds and the density of structurally stable systems are due to Peixoto [1962]; see also Palis and de Melo [1982]. In dimensions greater than two, structurally stable systems are not dense, as shown by Smale [1966].

A general exposition of dissipative systems is given in the recent monograph by Hale [1988]; see also Temam [1988].

The restriction to a disk is a nontrivial assumption. In applications, it is not easy to establish that a given system is dissipative. In the case of a nondissipative system, there can be bifurcations that can only be observed by the global considerations of the vector field on the entire plane. Such bifurcations are called *bifurcations at infinity.* For polynomial vector fields on the plane, bifurcations at infinity are studied by Sotomayor [1985] and Sotomayor and Paterlini [1987].

The study of dynamics and bifurcations in the presence of symmetry is currently an exciting area of our subject. A comprehensive exposition is available in Golubitsky and Schaeffer [1985] and Golubitsky, Stewart, and Schaeffer [1988].

The general ideas on the unfolding of nonlinear vector fields near an equilibrium point are discussed by Arnold [1983]. Example 13.27 was investigated by Bogdanov [1981] and Takens [1974]. The proof of the versality of the unfolding is quite difficult; full details are in Bogdanov [1981]. Example 13.28 is analyzed in Carr [1981] and Chow and Hale [1982]. The system with odd symmetry in Example 13.29 is studied in Takens [1974], with further essential computations in Carr [1981].

14

Conservative and Gradient Systems

 In this chapter, we investigate the dynamics of two classes of vector fields with special characteristics—conservative and gradient. Both types of vector fields have the common property that they are defined in terms of functions; however, their flows are completely different. While periodic and homoclinic orbits may be omnipresent in conservative systems, the limit sets of orbits of gradient systems are necessarily part of the set of equilibria. We first uncover certain basic relations between the phase portraits of these systems and the geometry of underlying functions. Then we identify subsets of desirable "generic" functions. The vector fields of generic functions are structurally stable in the restricted sense that they are insensitive to small perturbations of the underlying functions. Analysis in the generic situations is made possible by the fact that the flows of both types of vector fields are essentially determined by the unstable manifolds of the saddle points. We also illustrate typical one-parameter bifurcations of conservative and gradient systems in nongeneric cases. Of course, the setting for the bifurcation theory of these systems has the important restriction that change of parameters preserve the conservative or gradient character of vector fields.

14.1. Second-order Conservative Systems

Here, we return to the theme of Section 7.4 and further discuss some of the basic properties of second-order conservative systems. Such a conservative system is defined in terms of its potential function, a real-valued function of a real variable. We first point out how to determine certain features of the phase portrait of a conservative system from its potential function. Then we identify a class of potential functions—generic potentials—for which the phase portraits are determined by the behavior of the unstable manifolds of the saddle points. Finally, we show that the phase portrait of a conservative system with a generic potential does not change qualitatively under small perturbations of its potential function.

For a given C^1 function $v : \mathbb{R} \to \mathbb{R}$, the second-order differential equation

$$\ddot{z} + v(z) = 0$$

or the equivalent system

$$\begin{aligned} \dot{x}_1 &= x_2 \\ \dot{x}_2 &= -v(x_1) \end{aligned} \qquad (14.1)$$

is said to be a *second-order conservative system*. Indeed, the function

$$E(x_1, x_2) = \tfrac{1}{2}x_2^2 + V(x_1),$$

where

$$V(x_1) = \int_0^{x_1} v(s)\, ds$$

is a conserved quantity, or first integral of Eq. (14.1). That is, E is constant on the orbits of Eq. (14.1) because, for each solution $(x_1(t), x_2(t))$ of Eq. (14.1), we have $\dot{E}(x_1(t), x_2(t)) = 0$ for all t. The conserved quantity E is called the *total energy* of the system (14.1); the term $x_2^2/2$ is called the *kinetic energy* and the function V is the *potential energy*.

As we saw in Section 7.4, the shapes of the orbits of the conservative system (14.1) can be determined from the knowledge of the level sets of its energy function. Indeed, the orbit through $\mathbf{x^0}$ with $E(\mathbf{x^0}) = E_0$ lies on the level curve

$$E^{-1}(E_0) = \{\, (x_1, x_2) : x_2 = \pm[2(E_0 - V(x_1)]^{\frac{1}{2}}, \; V(x_1) \le E_0\} \qquad (14.2)$$

of the energy function. A simple, but practically important, observation about the level curves of an energy function is that they are symmetric with respect to the x_1-axis.

It is preferable to extract as much information as possible from the scalar potential function, rather than the energy function. With this intent, we proceed with an examination of the role of the extreme values of a potential function.

Definition 14.1. *A point $\bar{x}_1$ is called a* critical point *of the C^1 function $V : \mathbb{R} \to \mathbb{R}$ if $V'(\bar{x}_1) = 0$. The value of V at a critical point is called a* critical value. *A critical point $\bar{x}_1$ is called* nondegenerate *if $V''(\bar{x}_1) \neq 0$.*

It is evident that equilibrium points of the conservative system (14.1) lie on the x_1-axis and that a point $(\bar{x}_1, 0)$ is an equilibrium point if and only if $\bar{x}_1$ is a critical point of the potential function. The equilibria corresponding to nondegenerate critical points are isolated. Moreover, their types can easily be determined: isolated maxima correspond to saddle points and isolated minima correspond to centers.

Lemma 14.2. *Suppose that $\bar{x}_1$ is a critical point of a potential function V so that $\bar{x} = (\bar{x}_1, 0)$ is an equilibrium point of the conservative vector field (14.1). Then*
  (i) *$\bar{x}$ is a saddle point if $V''(\bar{x}_1) < 0$;*
  (ii) *$\bar{x}$ is a center if $V''(\bar{x}_1) > 0$.*

Proof. It is no loss in generality to assume that $\bar{x}_1 = 0$. If $V''(0) < 0$, then the matrix of the linear variational equation about the origin has nonzero eigenvalues, one being positive and the other negative. Thus, $\mathbf{0}$ is a saddle point. To establish the second part, we observe that the Taylor Expansion of the energy function about the origin is of the form

$$E(x_1, x_2) = \tfrac{1}{2}x_2^2 + \tfrac{1}{2}x_1^2 V''(0)\big[1 + O(x_1)\big]$$

as $\mathbf{x} \to \mathbf{0}$. If $V''(0) > 0$, then the level curves of E near the origin are ellipses. Since the origin is an isolated equilibrium point, each orbit near the origin is a closed curve, that is, the origin is a center. ◇

Unstable manifolds of saddle points play a key role in the shapes of phase portraits of conservative systems. Here are some facts on such unstable manifolds.

Lemma 14.3. *Suppose $\bar{x} = (\bar{x}_1, 0)$ is a saddle point of Eq. (14.1) with stable and unstable manifolds $W^s(\bar{x})$ and $W^u(\bar{x})$, respectively. Then*
  (i) *the reflection of $W^s(\bar{x})$ through the x_1-axis is $W^u(\bar{x})$;*
  (ii) *if $\gamma^+(\mathbf{y})$ for $\mathbf{y} \in W^u(\bar{x})$ is bounded and $V(z) \neq V(\bar{x}_1)$ for all saddle points $(z, 0)$ different from $\bar{x}$, then $\gamma(\mathbf{y})$ is a homoclinic orbit with $\alpha(\mathbf{y}) = \omega(\mathbf{y}) = \bar{x}$.*

Proof. To prove the first part of the lemma, we observe that the value of the energy on $W^s(\bar{x})$ and $W^u(\bar{x})$ is equal to $E(\bar{x})$. Also, the direction of the motion along orbits is reversed as the orbit crosses the x_1-axis. This is sufficient to obtain the conclusion in the first part.

For the second part, we notice that the Poincaré–Bendixson Theorem implies that the ω-limit set of $\mathbf{y} \in W^u(\bar{x})$ cannot be a periodic orbit and therefore must contain equilibrium points and orbits connecting them. The hypothesis that no two saddle points correspond to the same potential energy value implies that the ω-limit set of $\mathbf{y}$ is $\bar{\mathbf{x}}$. $\diamond$

With the intent of identifying a class of well-behaved conservative systems, we now present several examples and indicate how to determine their phase portraits from their potential functions.

Example 14.4. *The Fish:* In Example 7.27, we discussed the system

$$\dot{x}_1 = x_2$$
$$\dot{x}_2 = x_1 - x_1^2.$$

This system is conservative with the potential function

$$V(x_1) = -\tfrac{1}{2}x_1^2 + \tfrac{1}{3}x_1^3.$$

The critical point at $x_1 = 1$ of the potential function is a nondegenerate minimum, and thus the point $(1, 0)$ is a center for the corresponding flow. The point $x_1 = 0$ is a nondegenerate maximum of the potential function and thus the corresponding equilibrium point $(0, 0)$ of the flow is a saddle point. Moreover, using the formula for the level curves of the energy function, we observe that the right piece of the unstable manifold of the saddle point at the origin is a homoclinic loop encircling the center. Since there are no other equilibria inside the loop, all orbits inside the loop are periodic. We have drawn in Figure 14.1 the graph of the potential function as well as the phase portrait of the Fish. $\diamond$

Example 14.5. *Duffing's equation:* Consider the potential function

$$V(x_1) = -\tfrac{1}{2}x_1^2 + \tfrac{1}{4}x_1^4$$

and the corresponding conservative system

$$\dot{x}_1 = x_2$$
$$\dot{x}_2 = x_1 - x_1^3.$$

The potential function has nondegenerate minima at -1 and 1; thus the vector field has centers at $(-1, 0)$ and $(1, 0)$. Also, there is a nondegenerate maximum at the origin and thus a saddle point at $(0, 0)$. Moreover, using the formula (14.2) for the level curves of the total energy, we observe that the unstable manifold of the saddle point is a figure eight. The full flow is shown in Figure 14.2: inside each loop of the figure eight there is a center, and outside the figure eight, all orbits are periodic. $\diamond$

We now present a pair of examples to illustrate the decisive role of the unstable manifolds of saddle points in the flows of conservative systems.

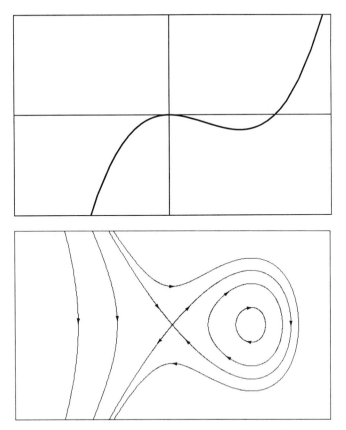

Figure 14.1. *The potential function* $V(x_1) = -x_1^2/2 + x_1^3/3$ *and phase portrait of the fish in Example 14.4.*

Example 14.6. Consider the potential function

$$V(x_1) = \tfrac{1}{3}x_1^2 + \tfrac{1}{9}x_1^3 - \tfrac{1}{4}x_1^4$$

and the corresponding conservative system

$$\dot{x}_1 = x_2$$
$$\dot{x}_2 = -x_1(1 - x_1)(\tfrac{2}{3} + x_1).$$

The vector field has a center at $(0, 0)$ corresponding to the nondegenerate minimum of the potential function at 0. Also, there are two saddle points, $(-\tfrac{2}{3}, 0)$ and $(1, 0)$, corresponding to the nondegenerate critical points at $-\tfrac{2}{3}$ and 1. To complete the phase portrait, we need to look at the unstable manifolds of the saddle points. Since $V(1) > V(-\tfrac{2}{3})$, the unstable manifold

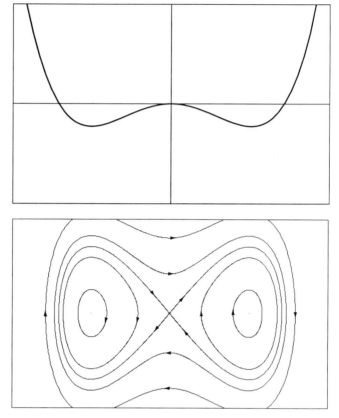

Figure 14.2. *The potential function $V(x_1) = -x_1^2/2 + x_1^4/4$ and phase portrait of the equation of Duffing in Example 14.5.*

of the saddle point $(1, 0)$ is above the unstable manifold of $(-\frac{2}{3}, 0)$ on the upper plane. Also, the right piece of the unstable manifold of $(-\frac{2}{3}, 0)$ is a homoclinic loop encircling the center. Now, the rest of the phase portrait can readily be filled in, as shown in Figure 14.3. ◇

Example 14.7. Consider the potential function

$$V(x_1) = \tfrac{2}{3}x_1^2 - \tfrac{1}{9}x_1^3 - \tfrac{1}{4}x_1^4$$

and the corresponding conservative system

$$\dot{x}_1 = x_2$$
$$\dot{x}_2 = -x_1(1 - x_1)(\tfrac{4}{3} + x_1).$$

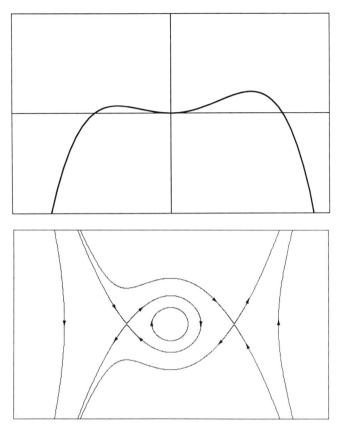

Figure 14.3. *The potential function $V(x_1) = \frac{1}{3}x_1^2 + \frac{1}{9}x_1^3 - \frac{1}{4}x_1^4$ and phase portrait of Example 14.6.*

As in the previous example, the potential function has two nondegenerate maxima, one at $-\frac{4}{3}$ and the other at 1, and one nondegenerate minimum at 0. Therefore, local phase portraits of the two conservative systems are similar. Globally, however, they are quite different. This time, the unstable manifold of the saddle point $(1, 0)$ sits below that of $(-\frac{4}{3}, 0)$ because $V(1) < V(-\frac{4}{3})$. We have drawn in Figure 14.4 the complete phase portrait of this system and its potential function. We will return to this and the previous example in the following section on bifurcations of second-order conservative systems. $\Diamond$

The behavior of a potential function at infinity can have profound effects on the flow of the corresponding conservative system, as the following example illustrates.

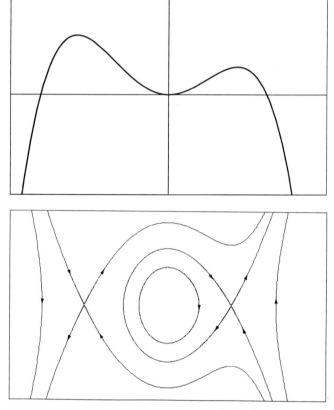

Figure 14.4. *The potential function $V(x_1) = \frac{2}{3}x_1^2 - \frac{1}{9}x_1^3 - \frac{1}{4}x_1^4$ and phase portrait of Example 14.7.*

Example 14.8. *Bounded potential:* Consider the potential function

$$V(x_1) = -x_1 e^{-x_1}. \tag{14.3}$$

The only critical point of this function is a nondegenerate minimum at $x_1 = 1$, thus the corresponding conservative system

$$\dot{x}_1 = x_2$$
$$\dot{x}_2 = (1 - x_1)e^{-x_1}$$

has a center at $(1, 0)$. As long as the value of the potential is negative, the orbits are periodic. When the potential is positive, however, each orbit is unbounded, as shown in Figure 14.5.

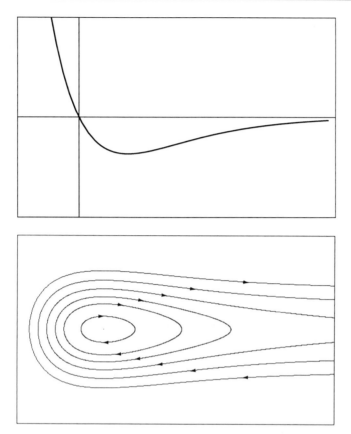

Figure 14.5. *The bounded potential function $V(x_1) = -x_1 e^{-x_1}$ and phase portrait of Example 14.8.*

The importance of the potential (14.3) becomes self-evident when compared with the potential $V(x_1) = \frac{1}{2}x_1^2$ of the linear harmonic oscillator. Both potential functions have a single nondegenerate minimum, but the phase portraits of the corresponding conservative systems are not equivalent. The source of the difficulty is the boundedness of the potential (14.3) as $x \to +\infty$. $\Diamond$

There are certain noteworthy common characteristics of the foregoing potential functions which facilitate the analysis of the corresponding conservative vector fields. The most apparent feature is the nondegeneracy of critical points for ease of local determination of phase portraits. At the global level, there are three key elements. First, the number of critical points is finite. Second, the critical values of any one of these potential functions are distinct; the importance of this observation will become self-evident in the following section. Last, we need to avoid the difficulty associated with boundedness in one or both directions, as exhibited by the potential function (14.3), and prefer that they be unbounded for $x_1 \to +\infty$ and $x_1 \to -\infty$. The flow of a conservative system with a potential function possessing these properties can be constructed from the knowledge of the unstable manifolds of the saddle points.

With these remarks, we now identify a class of desirable potential functions for which the unstable manifolds of the saddles determine the conservative flow.

Definition 14.9. *A potential function V is called generic if it satisfies the following conditions:*
  (i) *there are finitely many critical points of V;*
 (ii) *each critical point of V is nondegenerate, that is, $V''(\bar{x}_1) \neq 0$ for all critical points $\bar{x}_1$;*
(iii) *no two maximum values of V are equal;*
 (iv) *$|V(x_1)| \to +\infty$ as $|x_1| \to +\infty$, that is, V is unbounded for both $x_1 \to +\infty$ and $x_1 \to -\infty$.*

Let us now explain what makes generic potentials so desirable and justify our slight abuse of the word generic in this context. For this purpose, we need to consider the space of all C^2 potential functions. However, for the sake of precision and brevity, we shall now consider C^2 functions defined on a compact interval. This restriction is necessitated by the difficulty of introducing a reasonable topology on the space of functions defined on unbounded sets. On a compact interval, the requirement (iv) in Definition 14.9 becomes irrelevant.

Theorem 14.10. *Let $C^2(\mathcal{I})$ be the set of real-valued C^2 functions defined on a compact interval $\mathcal{I}$ and having no critical points at the end points of the interval. Also, endow this function space with the C^2 topology. Then a given generic potential in $C^2(\mathcal{I})$ has a neighborhood in $C^2(\mathcal{I})$ such that the conservative vector field of any potential function in this neighborhood is topologically equivalent to the vector field of the given generic potential function. Moreover, the subset of generic potentials in $C^2(\mathcal{I})$ is open and dense.* $\lozenge$

We refrain from giving a formal proof of this theorem; however, here are some of the basic ingredients. The first part of this result follows essentially from the Implicit Function Theorem. A small C^2 perturbation leaves the number and the type of nondegenerate critical points of the original generic potential function unchanged. Moreover, the inequality of the maximum values remains unaffected. Finally, you must convince yourselves that the relative positions of the unstable manifolds of the saddle points are also preserved under perturbations. For the second part of the theorem on the density of generic potentials one resorts to "Sard's Theorem;" see the Appendix

It is important to contrast the theorem above with the structural stability results in the previous chapter. To be sure, a conservative system with a generic potential is not always structurally stable because it may have a saddle connection in the form of a homoclinic loop. However, if we allow only conservative perturbations of a conservative system whose potential is generic with finitely many critical points, then such a system remains qualitatively intact. This is structural stability in a limited way, only in the confines of conservative systems.

With these remarks, we conclude our study of generic potentials and consider the possibilities that arise when we encounter a nongeneric potential.

Exercises —————————————————————

14.1. Determine the potential functions of the second-order conservative systems below. From these functions, construct the phase portraits:
(a) $\ddot{x} + x + x^3 = 0$;   (b) $\ddot{x} + x - x^3 = 0$;
(c) $\ddot{x} + x - x^2 = 0$;   (d) $\ddot{x} + x(1 - x)(0.1 - x) = 0$.

14.2. Prove that nondegenerate critical points of a potential function are isolated.

14.3. *A minimum for the potential does not imply a center:* Find a potential function $V(x_1)$ that has a minimum at $\bar{x}_1$ and yet there is no neighborhood of $(\bar{x}_1, 0)$ in which all orbits of Eq. (14.1) are periodic. Can this happen if we require the function $V(x_1)$ to be analytic?

14.4. *A maximum for the potential does not imply instability:* Find a potential function $V(x_1)$ that has a maximum at $\bar{x}_1$ and yet the equilibrium point $(\bar{x}_1, 0)$ is stable. Can this happen if we require the function $V(x_1)$ to be analytic?

14.5. *Unstable manifolds say it all:* Formulate and prove a result to the effect that "the unstable manifolds of the saddle points determine the structure of the flow of a generic potential function." Even if you are apprehensive about this difficult and somewhat vague problem, use the fact to draw phase portraits from generic potentials.

14.6. *Hamiltonian flows preserve area:* The result of this problem explains the absence of sinks or sources in the phase portraits of Hamiltonian, hence conservative, systems. Recall from Section 7.4 that, for a given C^1 function $H : \mathbb{R}^2 \to \mathbb{R}$, the planar system

$$\dot{q} = \frac{\partial H(q, p)}{\partial p}$$

$$\dot{p} = -\frac{\partial H(q, p)}{\partial q}$$

is called a *Hamiltonian system* with the Hamiltonian function H.

(a) Show that the total energy of a second-order conservative system is a Hamiltonian function.

(b) Let D_0 be a region, say, with a smooth boundary, in the plane and consider the image of D_0 under the flow of a planar differential equation. That is, consider the set $D(t) = \{\, \varphi(t, \mathbf{x}^0) : \mathbf{x}^0 \in D_0 \,\}$, where $\varphi(t, \mathbf{x}^0)$ is the solution of the equation $\dot{\mathbf{x}} = \mathbf{f}(\mathbf{x})$ through $\mathbf{x}^0$. If $A(t)$ is the area of $D(t)$, prove that

$$\dot{A}(t) = \int_{D(t)} \operatorname{div} \mathbf{f}(\mathbf{x})\, d\mathbf{x}.$$

Suggestion: Use the fact that

$$A(t) = \int_{D_0} \det \frac{\partial \varphi(t, \mathbf{x})}{\partial \mathbf{x}}\, d\mathbf{x}.$$

(c) Prove now that the flow of a Hamiltonian system preserves area.

14.7. Find a Hamiltonian for the system

$$\dot{x}_1 = -\cos x_1 \sin x_2, \qquad \dot{x}_2 = \sin x_1 \cos x_2$$

and draw the flow.

14.8. *Kepler motion:* In an appropriate coordinate system, the motion of a planet around the sun (considered as fixed) with the attractive force being proportional to the inverse square of the distance $|z|$ of the planet from the sun is given by the solution of the second-order conservative system with the potential function $-|z|^{-1}$ for $z \neq 0$. Show that the orbits are closed curves if the total energy is negative. You can also show that they are ellipses by considering the second-order differential equation obtained by the change of variables $u = |z|^{-1}$.

14.9. *Neutral damping:* The Hamiltonian character of some differential equations may not be self-evident. Here is such an equation.
   (i) Show that the second-order equation

$$\ddot{z} + \dot{z}^2 + g(z) = 0$$

may be transformed to Hamiltonian form by letting $q = z$ and $p = \dot{q}e^{2q}$. *Answer:* The Hamiltonian function is

$$H(q, p) = \tfrac{1}{2}p^2 e^{-2q} + \int_0^q g(z)e^{2z}\, dz.$$

   (ii) Draw the phase portrait for $g(z) = z$.

14.2. Bifurcations in Conservative Systems

In the previous section, we saw that the qualitative structure of the flow of a conservative system with a generic potential function does not change under small perturbations of its potential function. Here, we illustrate with several examples some of the likely bifurcations when nongeneric potentials are perturbed. In these examples, we selectively violate the conditions for genericity as listed in Definition 14.9.

Example 14.11. *Potential with a degenerate critical point:* Consider the potential function

$$V(\lambda,\, x_1) = -\lambda x_1 + \tfrac{1}{3} x_1^3$$

depending on a scalar parameter λ. For $\lambda = 0$, the point $x_1 = 0$ is a degenerate critical point of $V(0,\, x_1)$ and thus the potential is not generic. Let us now discuss how the flow of the corresponding conservative system

$$\dot{x}_1 = x_2$$
$$\dot{x}_2 = \lambda - x_1^2$$

changes as the parameter λ is varied from negative to positive values.

$\lambda < 0$: The potential function is generic. There are two critical points of the potential function both of which are nondegenerate. The minimum at $\sqrt{|\lambda|}$ corresponds to a center; the maximum at $-\sqrt{|\lambda|}$ corresponds to a saddle point and it has a homoclinic loop encircling the center. The phase portrait is shown in Figure 14.6a.

$\lambda = 0$: As we have observed above, the potential function is not generic. The critical point at 0 is degenerate and corresponds to an unstable equilibrium point and its stable and unstable sets form a cusp. These sets as well as the complete flow are depicted in Figure 14.6b.

$\lambda > 0$: Again, the potential function is generic. Indeed, there are no critical points of the potential function and thus no equilibrium points of the flow. The complete flow is shown in Figure 14.6c.

In summary, as λ is varied from negative to positive values, the nondegenerate maximum and minimum of the potential function merge into a degenerate critical point and disappear. In the phase portrait, the homoclinic loop shrinks to a point and disappears. The bifurcation at $\lambda = 0$ is called a *saddle-center bifurcation*. $\Diamond$

Example 14.12. *Potential with equal maxima:* Consider the potential function

$$V(\lambda,\, x_1) = \tfrac{1}{2}(1+\lambda)x_1^2 - \tfrac{1}{3}\lambda x_1^3 - \tfrac{1}{4}x_1^4$$

depending on a scalar parameter λ in the range $-1 < \lambda < 1$.

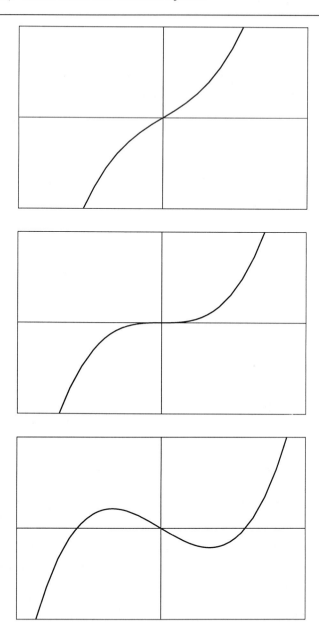

Figure 14.6. A saddle-center bifurcation in the potential $V(\lambda, x_1) = -\lambda x_1 + x_1^3/3$ with a degenerate critical point for $\lambda = 0$. The corresponding phase portraits are on the following page.

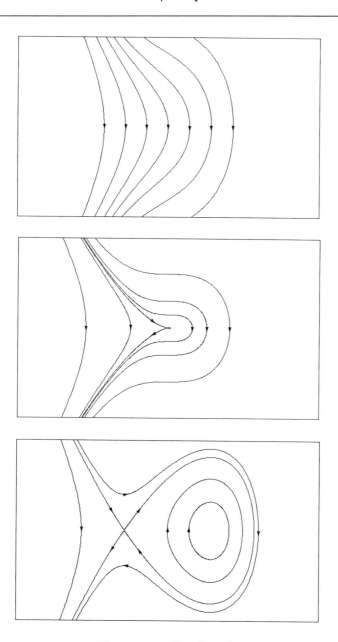

Figure 14.6 Continued.

The critical points 0, 1, and $-1 - \lambda$ of $V(\lambda, x_1)$ are nondegenerate; the points 1 and $-1 - \lambda$ are maxima and the point 0 is a minimum. Moreover, the potential function tends to infinity as $|x| \to +\infty$ for each λ. However, for $\lambda = 0$, the potential function $V(0, x_1)$ is not generic because $V(0, 1) = V(0, -1)$. On the other hand, we see that for $\lambda \neq 0$, the potential function is generic with $V(\lambda, 1) > V(\lambda, -1 - \lambda)$ for $\lambda < 0$, and $V(\lambda, 1) < V(\lambda, -1 - \lambda)$ for $\lambda > 0$. Using this information, we determine that the flow of the corresponding conservative system

$$\dot{x}_1 = x_2$$
$$\dot{x}_2 = -x_1(1 - x_1)(1 + \lambda + x_1)$$

changes with λ as depicted in Figure 14.7.

There is clearly a bifurcation in the dynamics of the equation at $\lambda = 0$. As $\lambda \to 0$ from negative values, the homoclinic orbit defined by the unstable manifold of the saddle point at $(-1 - \lambda, 0)$ becomes larger, coinciding eventually with parts of the stable and unstable manifold of the saddle point $(-1, 0)$ to form heteroclinic orbits between the two saddle points. For $\lambda \to 0$ from positive values, a phenomenon similar to that for negative λ occurs but with the role of the saddle points interchanged. ◇

The bifurcations that we have encountered in the two preceding examples represent typical situations for one-parameter families of potential functions. In fact, with only one parameter to vary, one does not expect more than two maximal values of the potential to coincide, or to have more than two nondegenerate critical points merge to a degenerate critical point. In lieu of making this statement precise, we now present a potential function depending on two parameters.

Example 14.13. *A two-parameter potential:* Consider the potential function

$$V(\lambda, \mu, x_1) = \lambda x_1 + \tfrac{1}{2}\mu x_1^2 - \tfrac{1}{4}x_1^4$$

depending on two real parameters λ and μ. We now analyze the bifurcations of the corresponding conservative vector field

$$\dot{x}_1 = x_2$$
$$\dot{x}_2 = -\lambda - \mu x_1 + x_1^3$$

for small values of the parameters near zero.

Since the critical points of V are the zeros of a cubic polynomial, there can be at most three critical points. In fact, we showed in Chapter 2 that the bifurcation curve in the (λ, μ)-space for the critical points of V is the cusp $4\mu^3 = 27\lambda^2$, as shown in Figure 14.8.

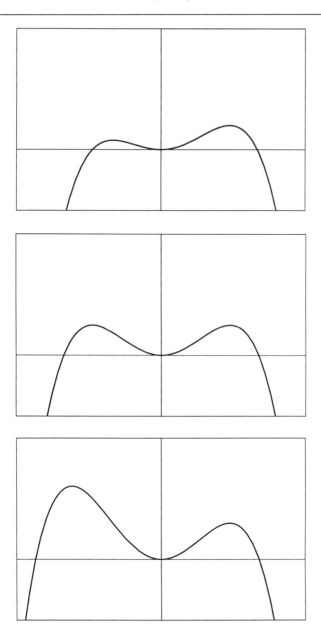

Figure 14.7. *Formation of the heteroclinic orbits in the bifurcation of the potential* $V(\lambda, x_1) = \frac{1}{2}(1 + \lambda)x_1^2 - \frac{1}{3}\lambda x_1^3 - \frac{1}{4}x_1^4$ *with equal maxima. The corresponding phase portraits are on the following page.*

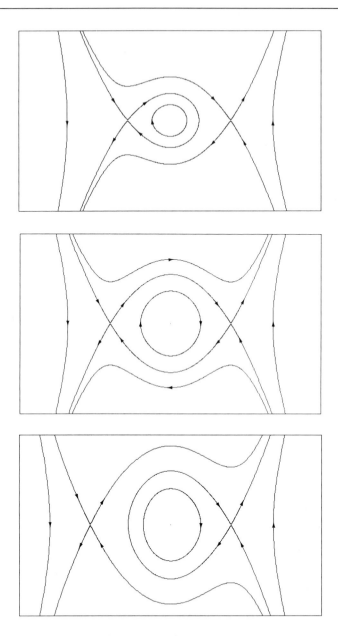

Figure 14.7 Continued.

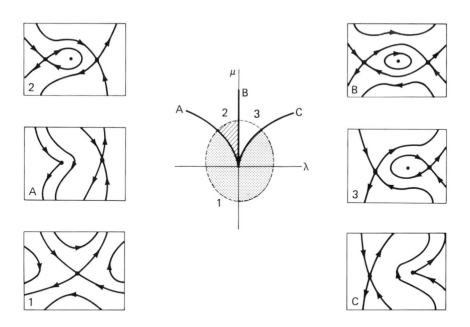

Figure 14.8. *Bifurcations of the two-parameter potential* $V(\lambda, \mu, x_1) = \lambda x_1 + \frac{1}{2}\mu x_1^2 - \frac{1}{4}x_1^4$ *in Example 14.13.*

If (λ, μ) lies below this cusp, there is only one critical point of V which is a nondegenerate maximum and the flow is just a hyperbolic saddle. For $(\lambda, \mu) = (0, 0)$, the only critical point of V is $x_1 = 0$ and it is degenerate; the flow has only an unstable equilibrium point.

For the parameter values on the cusp, the potential function has two critical points, one of which is a nondegenerate maximum, and the other of which is degenerate. As the parameter values cross into the cusp, the degenerate critical point splits into a nondegenerate maximum and a nondegenerate minimum, as in Example 14.11. Inside the cusp, the critical points are always two nondegenerate maxima and a nondegenerate minimum, but the flows are not equivalent for all parameter values. Indeed, there are parameter values at which the two maxima have the same maximum values; such parameter values are exactly the positive μ-axis. As parameter values cross the positive μ-axis, the homoclinic orbit turns into two heteroclinic orbits and then back to a homoclinic orbit, as in Example 14.12. ◇

This concludes our study of dynamics and bifurcations of second-order conservative systems. We now turn our attention to gradient systems, another class of vector fields defined in terms of a function yet with different dynamics.

Exercises ———————————————————————————— ♣ ♡ ♠ ◇

14.10. *Perturbations of bounded potentials:* For λ a real parameter, discuss the bifurcations in the flows for the following potential functions which are bounded in one or both directions:

(i) $V(x_1) = \lambda - x_1 e^{-x_1}$;

(ii) $V(x_1) = 1 - \lambda - \cos x_1$.

14.11. *Small perturbations at infinity may change the flow:* Given a potential function $V_0(z)$ which approaches ∞ as $|z| \to \infty$ and $V_0'(z) > 0$ for $|z|$ large, show that there is a perturbation $V_\lambda(z)$ such that $V_\lambda(z) \to \infty$ as $|z| \to \infty$ and also has two critical points which approach ∞ as $\lambda \to 0$.

14.12. *Rotating pendulum:* Consider a pendulum of mass m and length l constrained to oscillate in a plane rotating with angular velocity ω about a vertical line. If u denotes the angular deviation of the pendulum from the vertical and I is the moment of inertia, then

$$I\ddot{u} - m\omega^2 l^2 \sin u \cos u + mgl \sin u = 0.$$

By changing the time scale, this is equivalent to

$$\ddot{u} - (\cos u - \lambda) \sin u = 0,$$

where $\lambda = g/(\omega^2 l)$. Discuss the flows for each $\lambda > 0$ paying particular attention to the bifurcations in the flow.

Help: Consult Hale [1980], p. 178.

14.13. Show that the following equations are Hamiltonian and discuss the bifurcations in the flow for nonnegative values of the parameters:

(i) $\dot{x}_1 = x_1(a - bx_2)$, $\qquad \dot{x}_2 = -x_2(c - dx_1)$.

   Hint: Let $x_1 = e^q$ and $x_2 = e^p$.

(ii) $\dot{x}_1 = x_1(1 - x_1)(a - bx_2)$, $\qquad \dot{x}_2 = -x_2(1 - x_2)(c - dx_1)$.

   Hint: Let $x_1 = e^q/(1 + e^q)$ and $x_2 = e^p/(1 + e^p)$.

14.14. Discuss the bifurcations in the flows of the following equations depending on two parameters, λ and μ:

(i) $\dot{x}_1 = x_2$, $\qquad \dot{x}_2 = \lambda + \mu x_1 - x_1^3$;

(ii) $\dot{x}_1 = x_2$, $\qquad \dot{x}_2 = -(x_1 - \mu)(1 - x_1)(1 + \lambda + x_1)$.

14.3. Gradient Vector Fields

In this section, we investigate a class of planar vector fields that are gradients of functions. What makes these gradient vector fields worthy of a section of their own is the simplicity of their asymptotic dynamics: the α- and ω-limit sets of bounded orbits belong to the set of equilibria. Following a short summary of rudimentary facts, we establish this result as a consequence of the observation that the defining function is nonincreasing along the solutions of a gradient system and the Invariance Principle from Chapter 9. We then explore the dynamics and bifurcations of several specific gradient systems, some of which are old favorites. Using these examples as a guide, we identify a subclass of structurally stable gradient systems that are also generic in the space of gradient vector fields.

Definition 14.14. *If $F : \mathbb{R}^2 \to \mathbb{R}$ is a C^2 function, the gradient vector field is*

$$-\nabla F(\mathbf{x}) \equiv - \left(\frac{\partial}{\partial x_1} F(\mathbf{x}), \ \frac{\partial}{\partial x_2} F(\mathbf{x}) \right)$$

and the corresponding gradient system of differential equations is

$$\dot{\mathbf{x}} = -\nabla F(\mathbf{x}). \tag{14.4}$$

The points at which the gradient of F vanishes correspond to the equilibria of the gradient system (14.4). To underline the significance of such points, we introduce the following definition:

Definition 14.15. *A point $\bar{\mathbf{x}}$ is said to be a critical point of F if $\nabla F(\mathbf{x}) = 0$. A critical point $\bar{\mathbf{x}}$ is called nondegenerate if the eigenvalues of the Hessian at $\bar{\mathbf{x}}$, the matrix of the second partial derivatives*

$$\begin{pmatrix} \dfrac{\partial^2 F}{\partial x_1 \partial x_1}(\bar{\mathbf{x}}) & \dfrac{\partial^2 F}{\partial x_1 \partial x_2}(\bar{\mathbf{x}}) \\[2ex] \dfrac{\partial^2 F}{\partial x_2 \partial x_1}(\bar{\mathbf{x}}) & \dfrac{\partial^2 F}{\partial x_2 \partial x_2}(\bar{\mathbf{x}}) \end{pmatrix},$$

are nonzero.

In the case of nondegenerate critical points, precise information on the dynamics of the corresponding equilibria can be obtained from the local geometry of the graph of the function F.

Lemma 14.16. *An equilibrium point of a gradient system (14.4) is hyperbolic if and only if the corresponding critical point of F is nondegenerate. If $\bar{\mathbf{x}}$ is a hyperbolic equilibrium of (14.4), then*

- *$\bar{\mathbf{x}}$ is an unstable node if and only if F has an isolated maximum at $\bar{\mathbf{x}}$;*
- *$\bar{\mathbf{x}}$ is asymptotically stable if and only if F has an isolated minimum at $\bar{\mathbf{x}}$;*
- *$\bar{\mathbf{x}}$ is a saddle point if and only if F has a saddle at $\bar{\mathbf{x}}$.* ◇

The key observation for the verification of this lemma is that the matrix of the linear variational equation about an equilibrium point of the gradient system (14.4) is the Hessian matrix of F evaluated at that point. The eigenvalues of the Hessian matrix are real because it is a symmetric matrix. Consequently, nondegenerate critical points correspond to hyperbolic equilibrium points.

The most remarkable aspect of the dynamics of gradient systems is that equilibrium points are the only possible limit sets.

Theorem 14.17. *If $\gamma^+(\mathbf{x}^0)$ is a bounded positive orbit of a gradient system* (14.4), *then the ω-limit set $\omega(\mathbf{x}^0)$ belongs to the set of equilibria of Eq.* (14.4). *If the equilibrium points are isolated, then such an ω-limit set is a single equilibrium point. If $F(\mathbf{x}) \to +\infty$ as $\|\mathbf{x}\| \to +\infty$, then every positive orbit of Eq.* (14.4) *is bounded. Similarly, if $\gamma^-(\mathbf{x}^0)$ is a bounded negative orbit of a gradient system* (14.4), *then the α-limit set $\alpha(\mathbf{x}^0)$ belongs to the set of equilibria of Eq.* (14.4). *If the equilibrium points are isolated, then such an α-limit set is a single equilibrium point. If $F(\mathbf{x}) \to -\infty$ as $\|\mathbf{x}\| \to +\infty$, then every negative orbit of Eq.* (14.4) *is bounded.*

Proof. We indicate the proof for positive orbits. If $\mathbf{x}(t)$ is a bounded solution of Eq. (14.4) for $t \geq 0$, then the derivative of F along such a solution satisfies

$$\frac{d}{dt}F(\mathbf{x}(t)) = -|\nabla F(\mathbf{x}(t))|^2 \leq 0.$$

Now, the Invariance Principle of Section 9.4 implies that $\omega(\mathbf{x}(0))$ belongs to the set of equilibria. If the equilibria are isolated, then $\omega(\mathbf{x}(0))$ is a single point because the limit set is connected. The statement on the boundedness of positive orbits is a consequence of the fact that $F(\mathbf{x}(t)) \leq F(\mathbf{x}(0))$. $\Diamond$

This theorem and its proof have several noteworthy implications. The first corollary is that the solutions of a gradient vector field cross the level sets of the function F orthogonally and inward, except at critical points. Thus, from the geometry of the level sets of F, one can infer much about the phase portrait of a gradient system. The second implication, which is the important one in the sequel, is that a gradient system cannot have any periodic or homoclinic orbits.

With these general facts at our disposal, it is now time to investigate the dynamics of specific examples of planar gradient systems. Before proceeding with the more substantial examples below, you may wish to reconsider the linear gradient system in Example 7.14.

Example 14.18. *Reaction-diffusion:* The planar differential equations

$$\dot{x}_1 = 2(x_2 - x_1) + x_1(1 - x_1^2)$$
$$\dot{x}_2 = -2(x_2 - x_1) + x_2(1 - x_2^2)$$

arise as a model of a reaction-diffusion system. This is a gradient system for the function

$$F(x_1, x_2) = \tfrac{1}{4}(x_1^4 + x_2^4 + 2x_1^2 + 2x_2^2 - 8x_1x_2),$$

whose graph is as shown in Figure 14.9a. The critical points of F are $(0, 0)$, $(1, 1)$, and $(-1, -1)$, all of which are nondegenerate. Therefore,

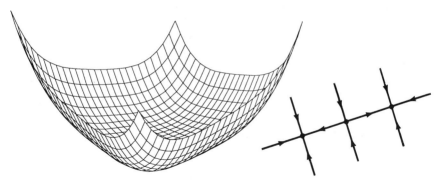

Figure 14.9. *The function F and the corresponding gradient flow of the reaction-diffusion gradient system in Example 14.18.*

these points are hyperbolic equilibria of the gradient system. It is easy to verify that the origin is a saddle point with a one-dimensional unstable manifold whereas the other equilibria are asymptotically stable. Furthermore, all positive orbits are bounded because $F(\mathbf{x}) \to +\infty$ as $\|\mathbf{x}\| \to +\infty$. Consequently, it follows from Theorem 14.17 that the ω-limit set of any orbit must be one of the three equilibrium points. Using the fact that the line $x_1 = x_2$ is invariant, we see that the complete flow is the one shown in Figure 14.9b. $\Diamond$

Example 14.19. *Vibrating membranes:* In the study of the vibrations of rectangular membranes, an application of the method of Liapunov–Schmidt leads to the system of equations

$$\dot{x}_1 = -x_1^3 - bx_1x_2^2 + x_1$$
$$\dot{x}_2 = -x_2^3 - bx_1^2x_2 + x_2,$$

where b is a real parameter. This is a gradient system for the function

$$F(x_1,\, x_2) = \frac{x_1^4}{4} + \frac{b}{2}x_1^2x_2^2 + \frac{x_2^4}{4} - \frac{x_1^2}{2} - \frac{x_2^2}{2},$$

whose graph is given in Figure 14.10a.

Let us now outline the construction of the phase portrait of this gradient system for the parameter range $0 < b < 1$. To locate the equilibria, we have drawn in Figure 14.10b the zero set of the first component of the vector field with dashed curves, and the zero set of the second component with solid curves. There are nine equilibrium points corresponding to the intersections of these two cubic curves. All equilibria are nondegenerate and labeled with SN, S, and UN to designate, respectively, stable node,

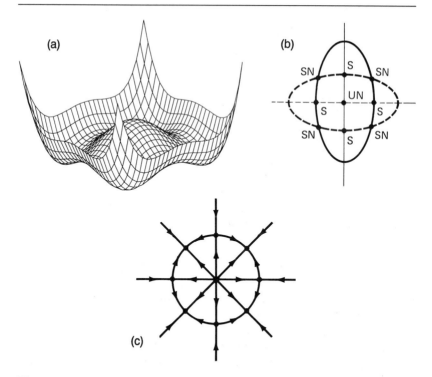

Figure 14.10. *On vibrating membranes in Example 14.19: (a) The graph of F; (b) locating the equilibria; and (c) the gradient flow.*

saddle point, and unstable node. Of course, this information is obtained by computing the eigenvalues of the Hessian of F at the equilibria.

All the positive orbits of this gradient system are bounded because $F(\mathbf{x}) \to +\infty$ as $\|\mathbf{x}\| \to +\infty$. It thus follows from Theorem 14.17 that every solution approaches one of the nine equilibria as $t \to +\infty$. Finally, we observe from the symmetries of the vector field that the x_1-axis, x_2-axis as well as the lines $x_1 = x_2$ and $x_1 = -x_2$ are invariant under the flow. With this information and the stability properties of the equilibria, we conclude that the flow of the gradient system has the qualitative structure as shown in Figure 14.10c. $\Diamond$

In order to identify a class of gradient vector fields for which the qualitative structure of the flow will not change under small perturbations of the defining function, we must understand the typical types of bifurcations that may occur. The first possibility is, of course, the bifurcation of a nonhyperbolic equilibrium point of a gradient system corresponding to a degenerate critical point of F. Since the eigenvalues of the linear variational equation must be real, the saddle-node bifurcation will be typical if only one of the eigenvalues is zero. We now reproduce an example of such a bifurcation from Chapter 7.

Example 14.20. *Saddle-node bifurcation:* Let us reconsider the one-parameter product system

$$\dot{x}_1 = -\lambda + x_1^2$$
$$\dot{x}_2 = -x_2$$

from Example 7.23. As we saw there, the origin undergoes a saddle-node bifurcation at $\lambda = 0$. All we need to add here is the fact that this system is a gradient system for the function

$$F(\lambda, \mathbf{x}) = \lambda x_1 - \tfrac{1}{3}x_1^3 + \tfrac{1}{2}x_2^2,$$

and that, for $\lambda = 0$, the origin is a degenerate critical point of F where the Hessian has a simple zero eigenvalue. $\diamond$

As we have indicated above, a gradient system cannot have periodic or homoclinic orbits; thus, no need to be concerned with their bifurcations. However, the possibility of a heteroclinic saddle connection is real, as we can cite Example 9.17.

Example 14.21. *Breaking a heteroclinic saddle connection:* Let us reconsider the one-parameter system

$$\dot{x}_1 = \lambda + 2x_1 x_2$$
$$\dot{x}_2 = 1 + x_1^2 - x_2^2$$

from Example 9.17. As we saw there, for all parameter values near zero there are always two hyperbolic saddle points. However, for $\lambda = 0$, there is a heteroclinic saddle connection between these two saddle points, but it is broken for any nonzero value of the parameter. Now, the relevance of this example in the context of the present section is the fact that this is a gradient system with the function

$$F(\lambda, \mathbf{x}) = -\lambda x_1 - x_2 + \tfrac{1}{3}x_2^3 - x_1^2 x_2.$$

Therefore, a heteroclinic saddle connection can be broken with a perturbation that does not change the gradient character of the vector field. $\diamond$

We now turn to the task of identifying a subset of gradient systems whose flows do not change under small perturbations of the defining functions. With the foregoing examples in mind, we proceed as follows:

Definition 14.22. *A gradient vector field* $-\nabla F$ *is said to be generic if*
 (i) *F has a finite number of critical points;*
 (ii) *each critical point of F is nondegenerate;*
 (iii) *there are no saddle connections;*
 (iv) $|F(\mathbf{x})| \to +\infty$ *as* $\|\mathbf{x}\| \to +\infty$.

This definition invites a comparison with the definition of generic conservative systems. Unlike Definition 14.9, the definition of a generic gradient system is not strictly in terms of the underlying function F but involves a global property of the vector field itself. This is not an entirely satisfactory situation with no easy resolution.

As in the case of conservative systems, we will now restrict our considerations to compact subsets of the plane to state the role of generic gradient vector fields in the set of all gradient vector fields.

Theorem 14.23. *Let* $\mathcal{D}$ *be a compact subset of* $\mathbb{R}^2$ *with a smooth boundary and* $\mathcal{G}^2(\mathcal{D})$ *be the real-valued* C^2 *functions on* $\mathcal{D}$ *with no critical points on the boundary. Also, endow this function space with the* C^2 *topology. Then, a given function in* $\mathcal{G}^2(\mathcal{D})$ *with generic gradient vector field has a neighborhood in* $\mathcal{G}^2(\mathcal{D})$ *such that the gradient vector field of any function in this neighborhood is topologically equivalent to the given generic gradient vector field. Moreover, the subset of functions in* $\mathcal{G}^2(\mathcal{D})$ *with generic gradient vector fields is open and dense.* $\diamond$

It may be evident from the considerations in Section 13.1 that generic gradient vector fields are structurally stable at large, save the boundary conditions, not just in the set of gradient vector fields. The density, however, stands only in the set of gradient vector fields.

We conclude here our study of dynamics and bifurcations of planar vector fields. However, we will remain on the plane for one more chapter and explore the intriguing world of planar maps.

Exercises ──────────────────────────────────── ♣ ♡ ♠ ◇

14.15. *Linear gradient systems:* Show that a linear system $\dot{\mathbf{x}} = \mathbf{A}\,\mathbf{x}$ is gradient if $\mathbf{A}$ is a symmetric matrix.

14.16. *Always gradient:* Show that any product system

$$\dot{x}_1 = f_1(x_1), \qquad \dot{x}_2 = f_2(x_2)$$

is a gradient system.

14.17. *Many degenerate critical points:* Consider the function $F(x_1, x_2) = x_1^2 x_2^2$. Find the critical points of F and show that they are all degenerate. Draw the graph of the function. Finally, construct the phase portrait of the corresponding gradient vector field.

14.18. Show that each of the systems below is a gradient system, determine the values of the parameters for which the vector field is generic, and discuss the bifurcations in the flows:

(i) $\dot{x}_1 = x_1 + \beta x_2$, $\dot{x}_2 = x_2 + \beta x_1$ for $\beta \in \mathbb{R}$.

(ii) $\dot{x}_1 = \mu(x_2 - x_1) + x_1(1 - x_1^2)$, $\dot{x}_2 = -\mu(x_2 - x_1) + x_2(1 - x_2^2)$ for $\mu > 0$.

14.19. Analyze the flow of the gradient system

$$\dot{x}_1 = -x_1^3 - bx_1x_2^2 + x_1, \qquad \dot{x}_2 = -x_2^3 - bx_1^2x_2 + x_2$$

coming from vibrating membranes for the parameter values $b > 1$.

14.20. *Global attractors:* Identify the global attractors of two examples from the text, the reaction-diffusion and vibrating membrane.

14.21. *No homoclinic orbits:* Show that a gradient system on the plane cannot have a homoclinic orbit.

14.22. *A nongradient system:* Consider the damped conservative system

$$\dot{x}_1 = x_2, \qquad \dot{x}_2 = x_1 - x_1^3 - x_2.$$

(a) Show that the ω-limit set of every solution exists and is an equilibrium point.

(b) Show that this system cannot be a gradient system.
Hint: Compute the eigenvalues of the linear variational equation at the equilibrium point $(1, 0)$.

14.23. *Unfolding a function and its gradient vector field:* There is a theory of unfolding a function which is one of the cornerstone ideas of catastrophe theory. For example, the universal unfolding of the function $F(x_1, x_2) = \frac{1}{3}(x_1^3 + x_2^3)$ is the three-parameter family of functions

$$F(\lambda_1, \lambda_2, \lambda_3, x_1, x_2) = \tfrac{1}{3}(x_1^3 + x_2^3) - \lambda_1 x_1 x_2 - \lambda_2 x_1 - \lambda_3 x_2.$$

Thom calls the catastrophe, the set in the three-dimensional parameter space for which F changes its number of preimages, associated with this function the *hyperbolic umbilic.* Explore the phase portrait of the corresponding gradient vector field. One could also unfold this gradient vector field in the set of gradient vector fields. Comparison of the two unfoldings raises interesting issues, as discussed in the second reference below.
References: General facts about catastrophe theory including the hyperbolic umbilic can be found in, for example, Poston and Stewart [1978]. The gradient vector field above is discussed by Guckenheimer in Peixoto [1973].

14.24. *Unfolding the elliptic umbilic:* Another entry in Thom's famous list of seven elementary catastrophes is the unfolding of the elliptic umbilic

$$F(\lambda_1, \lambda_2, \lambda_3, x_1, x_2) = \tfrac{1}{6}x_2^3 - \tfrac{1}{2}x_1^2x_2 + \lambda_1 x_1 + \lambda_2 x_2 + \lambda_3(x_1^2 + x_2^2).$$

Explore the phase portrait of the gradient vector field of this function.
Help: Information about the elliptic umbilic is contained in Poston and
Stewart [1978]. The gradient vector field of the elliptic umbilic is stored in
the library of PHASER under the name *gradient*.

Bibliographical Notes ⎯⎯⎯⎯⎯⎯⎯⎯⎯⎯⎯⎯⎯⎯⎯⎯⎯ ◎〜◎

Critical points of functions played a prominent role throughout this chapter. A deep study of nondegenerate critical points of a real-valued function
is the subject of Morse Theory; see the Appendix and the standard reference Milnor [1963]. The statement of the Theorem of Sard is given in
the Appendix; consult Milnor [1965] or Smith [1983] for further details. A
relevant application of Sard's Theorem for our purposes is on page 37 of
Milnor [1963].

In the case where the elements of the set of equilibria of a gradient
system are not isolated, it is possible that the ω-limit set of a bounded
orbit is a continuum of equilibria; such an example is on page 14 of Palis
and de Melo [1982]. There are several applications where the equilibrium
points are not isolated and yet the ω-limit set of a bounded orbit is a
single point; see, for example, Aulbach [1984], Hale and Massatt [1982],
and Henry [1981].

Gradient systems have diverse uses. They play an important role in
catastrophe theory; see, for example, Thom [1969], Poston and Stewart
[1978], and Zeeman [1977]. In differential topology, especially in Morse
theory, one flows along gradient vector fields to take one manifold to another, as described in Milnor [1963]. Similar ideas are used in Smale [1961
and 1961a] to affirm the Poincaré conjecture in higher dimensions. In
numerical analysis, computing methods under the names "conjugate gradient" or "steepest descent" essentially consist of flowing along gradient
vector fields; see, for example, Conte and deBoor [1972]. Because of the
fact that bounded solutions approach an equilibrium, computations yield
convergent results.

The example of a vibrating membrane is studied in Chow, Hale, and
Mallet-Paret [1976]. A good reference for reaction-diffusion equations is
Fife [1979].

It is evident that the dynamics of a gradient system are essentially
determined by the equilibria and the possible connecting orbits between
any pair of equilibria. This observation can be made precise and practical
by resorting to simple combinatorics. One associates the vertices of a graph
with equilibria and the edges with connecting orbits. Such graphs are used
to classify all two-dimensional gradient flows, as explained by Peixoto in
Peixoto [1973] and Hale [1977].

It is not possible to characterize all structurally stable systems in dimensions greater than two. However, there is a nice result in Smale [1961]

and Palis and Smale [1970] for gradient vector fields on a compact manifold of any dimension: gradient systems with only hyperbolic equilibria and transversal intersection of stable and unstable manifolds are structurally stable, and structurally stable gradient systems are open and dense in the set of all gradient systems.

15

Planar
Maps

After about a dozen chapters on differential equations, we return here to the theme of Chapter 3 and explore, this time, some of the basic dynamics and bifurcations of planar maps. Our motives for delving into planar maps are akin to the ones for studying scalar maps; namely, as numerical approximations of solutions of differential equations or as Poincaré maps. We begin our exposition with an introduction to the dynamics of linear planar maps. Then, following a section on linearization, we turn to numerical analysis and give examples of planar maps arising from "one-step" approximations of planar differential equations or from "two-step" approximations of scalar differential equations. Afterwards, we undertake, as usual, a detailed study of bifurcations of fixed points, including the Poincaré–Andronov–Hopf bifurcation for maps. The final part of the chapter is devoted to a synopsis of area-preserving maps, an important class arising from classical mechanics and possessing a rich history. The subject of planar maps is a vast one that is also mathematically rather sophisticated. Yet, many planar maps with innocuous appearances continue to defy satisfactory mathematical analysis. Indeed, the purpose of this modest, albeit long, chapter is to acquaint you with several famous planar maps and encourage you to explore their dynamics on the computer; for further mathematical nourishment, we will refer you to other sources.

15.1. Linear Maps

For a given function $\mathbf{f} : \mathbb{R}^2 \to \mathbb{R}^2$, consider the first-order planar difference equation

$$\mathbf{x}^{n+1} = \mathbf{f}(\mathbf{x}^n), \tag{15.1}$$

which is an iteration under the map $\mathbf{f}$. To avoid drowning in sub or superscripts, as well as to bring the function $\mathbf{f}$ to the forefront, it is often convenient to write such a difference equation as

$$\mathbf{x} \mapsto \mathbf{f}(\mathbf{x}).$$

In this section, after several brief general remarks, we explore the geometry of the orbits of maps in the case $\mathbf{f}$ is a linear function.

Most of the necessary notation and many of the concepts from the theory of scalar maps as expounded in Chapter 3 are easily generalized to planar maps. Since it is quite likely that you have studied that chapter a long while ago, let us rapidly record several of these generalizations. For instance, the *positive orbit* γ^+ of a point $\mathbf{x}^0$ in $\mathbb{R}^2$ is the sequence of images of $\mathbf{x}^0$ under the successive compositions of the map $\mathbf{f}$:

$$\gamma^+(\mathbf{x}^0) = \{\, \mathbf{x}^0, \mathbf{f}(\mathbf{x}^0), \ldots, \mathbf{f}^n(\mathbf{x}^0), \ldots \,\}.$$

If the map $\mathbf{f}$ is invertible, we use the notation $\mathbf{f}^{-n}$ to denote the n-fold composition of $\mathbf{f}^{-1}$ with itself, and define the *negative orbit* γ^- of $\mathbf{x}^0$ to be

$$\gamma^-(\mathbf{x}^0) = \{\, \mathbf{x}^0, \mathbf{f}^{-1}(\mathbf{x}^0), \ldots, \mathbf{f}^{-n}(\mathbf{x}^0), \ldots \,\}.$$

When both the positive and negative orbits exist, the *orbit* γ of $\mathbf{x}^0$ is the union of the two: $\gamma(\mathbf{x}^0) = \gamma^+(\mathbf{x}^0) \cup \gamma^-(\mathbf{x}^0)$.

The most notable positive orbit of a map is one consisting of a single point that is fixed under all iterates of the map.

Definition 15.1. *A point $\bar{\mathbf{x}} \in \mathbb{R}^2$ is called a fixed point of $\mathbf{f}$ if $\mathbf{f}(\bar{\mathbf{x}}) = \bar{\mathbf{x}}$. A fixed point $\bar{\mathbf{x}}$ of $\mathbf{f}$ is said to be stable if, for any $\varepsilon > 0$, there is a $\delta > 0$ such that, for every $\mathbf{x}^0$ for which $\|\mathbf{x}^0 - \bar{\mathbf{x}}\| < \delta$, the iterates of $\mathbf{x}^0$ satisfy $\|\mathbf{f}^n(\mathbf{x}^0) - \bar{\mathbf{x}}\| < \varepsilon$ for all $n \geq 0$. A fixed point $\bar{\mathbf{x}}$ is said to be unstable if it is not stable. A fixed point is said to be asymptotically stable if it is stable and, in addition, there is an $r > 0$ such that $\mathbf{f}^n(\mathbf{x}^0) \to \bar{\mathbf{x}}$ as $n \to +\infty$ for all $\mathbf{x}^0$ satisfying $\|\mathbf{x}^0 - \bar{\mathbf{x}}\| < r$.*

Orbits that are fixed points of some iterate of a map play a prominent role in the dynamics of planar maps.

Definition 15.2. *A point* $\mathbf{x}^{\star} \in \mathbb{R}^2$ *is called a periodic point of minimal period n if $\mathbf{f}^n(\mathbf{x}^{\star}) = \mathbf{x}^{\star}$ and n is the least such positive integer. The set of all iterates of a periodic point is called a periodic orbit.*

The notions of stability, asymptotic stability, and instability for periodic orbits are immediate by considering the corresponding fixed points of the appropriate power of the map.

Definition 15.3. *A point $\mathbf{y}$ is called an ω-limit point of the positive orbit $\gamma^+(\mathbf{x}^0)$ of $\mathbf{x}^0$ if there is a sequence of positive integers n_i such that $n_i \to +\infty$ and $\mathbf{f}^{n_i}(\mathbf{x}^0) \to \mathbf{y}$ as $i \to +\infty$. The ω-limit set $\omega(\mathbf{x}^0)$ of $\gamma^+(\mathbf{x}^0)$ is the set of all its ω-limit points. In the case $\mathbf{f}$ is invertible, the α-limit set of $\gamma^-(\mathbf{x}^0)$ is defined similarly by taking n_i to be negative integers.*

Definition 15.4. *A set M in $\mathbb{R}^2$ is said to be invariant under the map $\mathbf{f}$ if $\mathbf{f}(M) = M$, that is, for any $\mathbf{x}$ in M we have $\mathbf{f}(\mathbf{x}) \in M$ and there is a point $\mathbf{y}$ in M such that $\mathbf{f}(\mathbf{y}) = \mathbf{x}$.*

After these generalities, we now turn to the main topic of this section—planar linear maps. A linear map on $\mathbb{R}^2$, in a basis, is given by

$$\mathbf{x} \mapsto \mathbf{A}\,\mathbf{x} \tag{15.2}$$

for some 2×2 matrix $\mathbf{A}$. The positive orbit of $\mathbf{x}^0 \in \mathbb{R}^2$ is the sequence of images of $\mathbf{x}^0$ under the positive powers of the coefficient matrix:

$$\gamma^+(\mathbf{x}^0) = \{\,\mathbf{x}^0,\, \mathbf{A}\mathbf{x}^0,\, \ldots,\, \mathbf{A}^n\mathbf{x}^0,\, \ldots\,\}.$$

As in the case of planar linear differential equations, we can use the Jordan Normal Form of a coefficient matrix $\mathbf{A}$ to compute the orbits of the linear map (15.2). To accomplish this, let us determine the effect of a linear transformation on the orbits of Eq. (15.2): if P is an invertible 2×2 matrix, then

$$(\mathbf{P}^{-1}\mathbf{A}\,\mathbf{P})^n = \mathbf{P}^{-1}\mathbf{A}^n\mathbf{P} \tag{15.3}$$

for any positive integer n. Consequently, we choose a transformation matrix $\mathbf{P}$ so that $\mathbf{P}^{-1}\mathbf{A}\mathbf{P}$ is in Jordan Normal Form, and compute. As we saw in Chapter 8, it is easy to compute the powers of a matrix in Normal Form. Let us now analyze the dynamics of several examples of linear maps and plot their phase portraits.

Example 15.5. *A hyperbolic sink:* Consider the following coefficient matrix in Jordan Normal Form:

$$\mathbf{A} = \begin{pmatrix} 0.9 & 0 \\ 0 & 0.8 \end{pmatrix}.$$

We first need to find the fixed points of the linear map $\mathbf{x} \mapsto \mathbf{A}\mathbf{x}$, that is, determine the solutions of the linear system $(\mathbf{A} - \mathbf{I})\bar{\mathbf{x}} = \mathbf{0}$. Since $\mathbf{A} - \mathbf{I}$ is

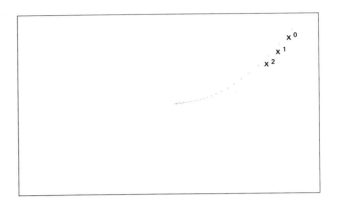

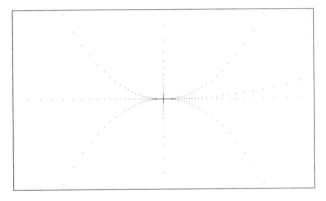

Figure 15.1. *A single orbit, and the phase portrait of the hyperbolic sink in Example 15.5.*

invertible, the origin $\bar{\mathbf{x}} = \mathbf{0}$ is the only fixed point of the linear map. The powers of $\mathbf{A}$ are given by

$$\mathbf{A}^n = \begin{pmatrix} (0.9)^n & 0 \\ 0 & (0.8)^n \end{pmatrix}.$$

It is evident that $\mathbf{A}^n$ approaches the zero matrix as $n \to +\infty$. Thus, the origin is an asymptotically stable fixed point, as seen in Figure 15.1. To infer how a positive orbit approaches the origin, notice that the eigenvalues of $\mathbf{A}$ are 0.9 and 0.8 with corresponding eigenvectors $\mathbf{v}^1 = (1, 0)$ and $\mathbf{v}^2 = (0, 1)$, respectively. For any initial value $\mathbf{x}^0 = (x_1^0, x_2^0)$, we have

$$\mathbf{A}^n \mathbf{x}^0 = (0.9)^n x_1^0 \mathbf{v}^1 + (0.8)^n x_2^0 \mathbf{v}^2.$$

Consequently, the positive orbits approach the origin faster in the direction of $\mathbf{v}^2$ than in the direction of $\mathbf{v}^1$. $\diamond$

When examining static pictures of phase portraits of planar maps you should keep in mind that an orbit is just a sequence of discrete points and not a continuous connected curve. As a result, it could at times be difficult to interpret certain phase portraits. Here is an example of this sort.

Example 15.6. *A hyperbolic sink with reflection:* Consider the linear map with the coefficient matrix

$$\mathbf{A} = \begin{pmatrix} 0.9 & 0 \\ 0 & -0.8 \end{pmatrix}.$$

Following the notations and the computations in the previous example, for any initial vector $\mathbf{x}^0 = (x_1^0, x_2^0)$, we have

$$\mathbf{A}^n \mathbf{x}^0 = (0.9)^n x_1^0 \mathbf{v}^1 + (-1)^n (0.8)^n x_2^0 \mathbf{v}^2.$$

Again, $\mathbf{A}^n \mathbf{x}^0 \to 0$ as $n \to +\infty$ for every $\mathbf{x}^0$. However, due to the presence of the negative eigenvalue, the positive orbit through $\mathbf{x}^0$ jumps back and forth across the x_1-axis; see Figure 15.2. If we try to fit smooth curves through the points on an orbit, there would usually be a piece above the x_1-axis and another one below, the two forming a cusp at the origin. $\diamond$

Example 15.7. *A hyperbolic source:* Consider the coefficient matrix

$$\mathbf{A} = \begin{pmatrix} 1.1 & 0 \\ 0 & 1.2 \end{pmatrix},$$

whose eigenvalues are greater than one. For any initial value $\mathbf{x}^0 = (x_1^0, x_2^0)$, we have

$$\mathbf{A}^n \mathbf{x}^0 = (1.1)^n x_1^0 \mathbf{v}^1 + (1.2)^n x_2^0 \mathbf{v}^2.$$

It is evident that the origin is an unstable fixed point; see Figure 15.3. Furthermore, by considering the iterates of the inverse of this map, $\mathbf{A}^{-n}$, it is easy to deduce that the α-limit set of any point is the origin. $\diamond$

Example 15.8. *A hyperbolic saddle:* Consider the coefficient matrix

$$\mathbf{A} = \begin{pmatrix} 1.1 & 0 \\ 0 & 0.9 \end{pmatrix},$$

with one eigenvalue greater and the other larger than one in absolute value. Since for any initial vector $\mathbf{x}^0 = (x_1^0, x_2^0)$, we have

$$\mathbf{A}^n \mathbf{x}^0 = (1.1)^n x_1^0 \mathbf{v}^1 + (0.9)^n x_2^0 \mathbf{v}^2,$$

the origin is unstable. However, unlike a source, $\omega(0, x_2^0) = (0, 0)$ and $\alpha(x_1^0, 0) = (0, 0)$; see Figure 15.4. $\diamond$

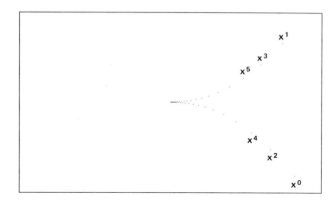

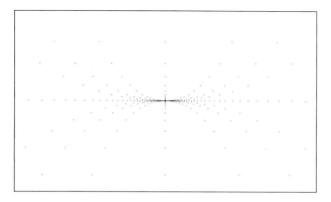

Figure 15.2. *A single orbit, and the phase portrait of the hyperbolic sink with reflection in Example 15.6.*

Example 15.9. *A nonhyperbolic linear map:* Consider the linear system

$$\mathbf{A} = \begin{pmatrix} 0.9 & 0 \\ 0 & 1 \end{pmatrix},$$

whose eigenvalues are 0.9 and 1. Observe now that, in addition to the origin, every point on the x_2-axis is a fixed point. Moreover, since the iterates of an initial vector $\mathbf{x}^0 = (x_1^0, x_2^0)$ are

$$\mathbf{A}^n \mathbf{x}^0 = (0.9)^n x_1^0 \, \mathbf{v}^1 + x_2^0 \, \mathbf{v}^2,$$

we have $\omega(x_1^0, x_2^0) = (0, x_2^0)$. It is evident that the origin is a stable fixed point; see Figure 15.5. ◇

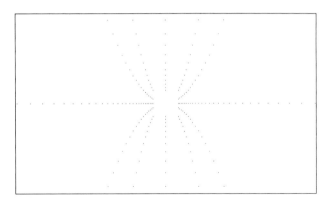

Figure 15.3. *Phase portrait of the hyperbolic source in Example 15.7.*

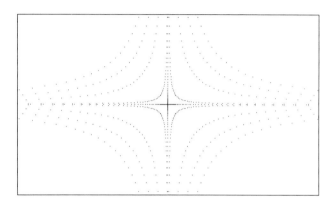

Figure 15.4. *Phase portrait of the hyperbolic saddle in Example 15.8.*

Example 15.10. *A nonhyperbolic linear map with reflection:* Consider the linear system

$$\mathbf{A} = \begin{pmatrix} 0.9 & 0 \\ 0 & -1 \end{pmatrix},$$

whose eigenvalues are 0.9 and -1. The origin is the only fixed point. The iterates of an initial vector $\mathbf{x^0} = (x_1^0,\, x_2^0)$ are

$$\mathbf{A}^n \mathbf{x^0} = (0.9)^n x_1^0 \, \mathbf{v^1} + (-1)^n x_2^0 \, \mathbf{v^2}.$$

It is clear that the origin is stable. Observe, moreover, that every point on the x_2-axis is a periodic point of minimal period 2, and the ω-limit set of any initial point $(x_1^0,\, x_2^0)$ is the periodic orbit consisting of the points $(0,\, x_2^0)$ and $(0,\, -x_2^0)$; see Figure 15.6. $\diamondsuit$

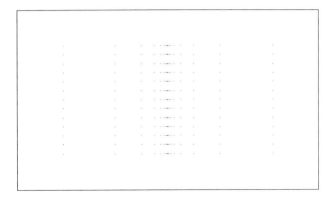

Figure 15.5. *Phase portrait of the nonhyperbolic linear map in Example 15.9.*

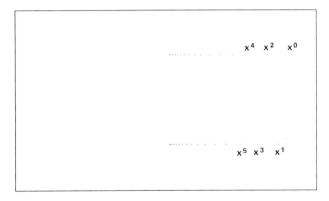

Figure 15.6. *A single orbit of the nonhyperbolic linear map with reflection in Example 15.10.*

Example 15.11. *Complex eigenvalues:* Let us consider the following coefficient matrix with complex eigenvalues:

$$\mathbf{A} = \begin{pmatrix} \alpha & \beta \\ -\beta & \alpha \end{pmatrix}, \qquad \beta \neq 0.$$

The eigenvalues of $\mathbf{A}$ are $\alpha \pm i\beta$. If we write these complex numbers in their polar representation

$$\alpha = \lambda \cos\omega, \qquad \beta = \lambda \sin\omega,$$

where $\lambda = \sqrt{\alpha^2 + \beta^2}$ and $-\pi < \omega \leq \pi$, the coefficient matrix becomes

$$\mathbf{A} = \lambda \begin{pmatrix} \cos\omega & \sin\omega \\ -\sin\omega & \cos\omega \end{pmatrix}. \tag{15.4}$$

Now, it is evident that the action of $\mathbf{A}^n$ on a vector $\mathbf{x}$ is to rotate the vector by angle $n\omega$ and then multiply it by λ^n. Consequently, we can readily infer the asymptotic fates of orbits. Indeed, if $\lambda < 1$, the origin is asymptotically stable; when $\lambda > 1$, the origin is unstable. The case $\lambda = 1$ is particularly noteworthy: the orbit through $\mathbf{x}^0$ lies on the circle with radius $\|\mathbf{x}^0\|$ centered about the origin. The flow on such a circle, however, depends on ω. As we saw in Chapter 6, each orbit on the circle is periodic if $\omega/2\pi$ is rational, and dense if $\omega/2\pi$ is irrational. Typical orbits for each of the three cases above are plotted in Figure 15.7.

As λ passes through 1, the linear map (15.4) undergoes a "vertical" bifurcation in the following sense: for $\lambda \neq 1$, there are no closed curves invariant under the map (15.4), except the trivial one consisting of a single point, and, for $\lambda = 1$, there is a one-parameter family of closed invariant curves. In the nonlinear case, the counterpart of this is the Poincaré–Andronov–Hopf bifurcation which we will explore later in this chapter. $\diamond$

It is evident by now from the examples above that the stability type of the fixed point at the origin of a linear map is, in most cases, determined by the moduli, not the real parts, of the eigenvalues of the coefficient matrix. Before we state our general result to this end, let us recall that *modulus*, also referred to as absolute value, of a complex number $\alpha + i\beta$ is defined to be $|\alpha + i\beta| = \sqrt{\alpha^2 + \beta^2}$.

Theorem 15.12. *The fixed point $\bar{\mathbf{x}} = \mathbf{0}$ of a linear map $\mathbf{x} \mapsto \mathbf{A}\mathbf{x}$ is asymptotically stable if and only if the eigenvalues of $\mathbf{A}$ have moduli less than one. If at least one eigenvalue of $\mathbf{A}$ has modulus greater than one, then the fixed point is unstable.* $\diamond$

The class of linear planar maps covered by this theorem play a significant role in local analysis. Therefore, we now define the following terminology that we have already been using:

Definition 15.13. *A linear planar map $\mathbf{x} \mapsto \mathbf{A}\mathbf{x}$ is called hyperbolic if the eigenvalues of $\mathbf{A}$ have moduli different than one.*

The stability type of a nonhyperbolic fixed point of even a linear map cannot always be determined from only its eigenvalues, as indicated in the exercises. Nonhyperbolic fixed points will be the subject of bifurcation theory a bit later in the chapter. For the moment, we turn to a simpler question and point out the role of hyperbolicity in the stability analysis of nonlinear maps near fixed points.

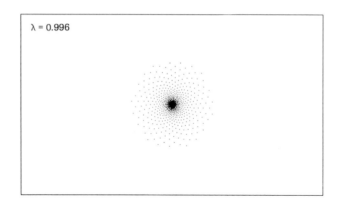

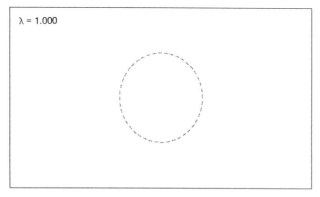

Figure 15.7. *A single* (!) *orbit of the linear map* (15.4) *with complex eigenvalues for* $\lambda = 0.996$, $\lambda = 1.000$, *and* $\lambda = 1.001$.

Exercises _____ ♣ ♡ ♠ ◇

15.1. Describe the dynamics of the linear maps with the coefficient matrices below, and identify each as hyperbolic or not:

a. $\begin{pmatrix} 0.2 & 0 \\ 0 & 2 \end{pmatrix}$; b. $\begin{pmatrix} 0.2 & 0 \\ 0 & -2 \end{pmatrix}$; c. $\begin{pmatrix} 0.2 & 0 \\ 0 & -0.3 \end{pmatrix}$;

d. $\begin{pmatrix} 2 & 1 \\ 1 & 1 \end{pmatrix}$; e. $\begin{pmatrix} 1/\sqrt{2} & 1/\sqrt{2} \\ -1/\sqrt{2} & 1/\sqrt{2} \end{pmatrix}$; f. $\begin{pmatrix} -1 & 0 \\ 0 & 1 \end{pmatrix}$.

Aid: Two-dimensional linear maps are stored in the library of PHASER under the name *dislin2d*, if you wish to plot the phase portraits of linear maps.

15.2. *Stability type of a nonhyperbolic linear map:* In case you got the impression in the text that eigenvalue 1 implies stability, consider the linear map

$$\mathbf{x} \longmapsto \begin{pmatrix} 1 & \lambda \\ 0 & 1 \end{pmatrix} \mathbf{x}.$$

Notice that for all values of the parameter λ, the eigenvalues of the coefficient matrix are 1. At $\lambda = 0$, it is the identity map so that the origin is a stable fixed point. Try to determine the phase portrait of this linear map for nonzero, both positive and negative, values of the parameter λ. Is the origin still stable for nonzero λ? Do the same problem for the nonhyperbolic linear map

$$\mathbf{x} \longmapsto \begin{pmatrix} -1 & \lambda \\ 0 & -1 \end{pmatrix} \mathbf{x}.$$

The difficulty with these examples is the fact that 1, or -1, is not a simple root of the minimal polynomial; for more information, see, for example, La Salle in Hale [1977].

15.3. Show that the ω-limit set of a point is invariant.

15.4. *First integrals for maps:* The orbits of maps are sets of discrete points. However, it is often advantageous to know the equations of the curves on which positive orbits may lie. Such curves are usually the level curves of a function—first integral—that is constant along orbits (see Section 7.4). The notion of a first integral for maps is defined as follows:
A function $H : \mathbb{R}^2 \to \mathbb{R}$ is called a *first integral* for a map $\mathbf{f}$ if

$$(H \circ \mathbf{f})(\mathbf{x}) = H(\mathbf{x})$$

for all $\mathbf{x}$ in the domain of the map.
Show that for the linear map

$$\mathbf{f}(\mathbf{x}) = \begin{pmatrix} \cos \omega & \sin \omega \\ -\sin \omega & \cos \omega \end{pmatrix} \mathbf{x}$$

the function $H(x_1, x_2) = x_1^2 + x_2^2$ is a first integral.

15.5. *Topological types of linear maps:* Following the ideas in Section 8.3, define the notion of topological equivalence of planar maps. Then, try to list as many topologically distinct linear planar maps as you can. Keep in mind that under homeomorphisms a periodic orbit goes to a periodic orbit with the same period. Be warned that there are many more topological types of linear maps than the ones given for linear planar differential equations.

References: The topological classification of linear maps in the plane and the higher dimensions turns out to be a bit more complicated than that of linear differential equations. For a nice exposition on this and related matters, see Robbin [1972] and Kuiper and Robbin [1973].

15.2. Near Fixed Points

In this section we rephrase several selected results from Sections 9.1, 9.2, and 9.4, and cast them in the setting of maps and their fixed points. Included in the list are the determination of stability type of a fixed point from linearization, and the geometry of the stable and unstable manifolds of a saddle point. To expedite the exposition, we have elected to omit all verifications. As you read this section you should keep in mind that everything said holds for periodic orbits as well by simply considering an appropriate power of a map.

Definition 15.14. *If $\bar{\mathbf{x}}$ is a fixed point of a C^1 map $\mathbf{x} \mapsto \mathbf{f}(\mathbf{x})$, then the linear map*

$$\mathbf{x} \mapsto D\mathbf{f}(\bar{\mathbf{x}})\,\mathbf{x}, \tag{15.5}$$

where $D\mathbf{f}(\bar{\mathbf{x}})$ is the Jacobian matrix

$$D\mathbf{f}(\bar{\mathbf{x}}) = \begin{pmatrix} \dfrac{\partial f_1}{\partial x_1}(\bar{\mathbf{x}}) & \dfrac{\partial f_1}{\partial x_2}(\bar{\mathbf{x}}) \\[2mm] \dfrac{\partial f_2}{\partial x_1}(\bar{\mathbf{x}}) & \dfrac{\partial f_2}{\partial x_2}(\bar{\mathbf{x}}) \end{pmatrix},$$

is called the linearization of the map $\mathbf{f}$ at the fixed point $\bar{\mathbf{x}}$.

We single out certain fixed points with well-behaved dynamics.

Definition 15.15. *A fixed point $\bar{\mathbf{x}}$ of $\mathbf{x} \mapsto \mathbf{f}(\mathbf{x})$ is said to be hyperbolic if the linear map (15.5) is hyperbolic, that is, if the Jacobian matrix $D\mathbf{f}(\bar{\mathbf{x}})$ at $\bar{\mathbf{x}}$ has no eigenvalues with modulus one.*

It is easy to determine the stability type of a hyperbolic fixed point from the linearization of the map at the fixed point:

Theorem 15.16. *Let* **f** *be a* C^1 *function with a fixed point* $\bar{\mathbf{x}}$.

(i) *If all the eigenvalues of the Jacobian matrix* $D\mathbf{f}(\bar{\mathbf{x}})$ *have moduli less than one, then the fixed point* $\bar{\mathbf{x}}$ *is asymptotically stable.*

(ii) *If at least one of the eigenvalues of* $D\mathbf{f}(\bar{\mathbf{x}})$ *has modulus greater than one, then* $\bar{\mathbf{x}}$ *is unstable.* $\Diamond$

We now apply the theorem above to an important specific map, to which we will return in a later section for a deeper investigation.

Example 15.17. *Delayed logistic map:* Consider the second-order scalar difference equation

$$y_{n+1} = \lambda y_n (1 - y_{n-1}), \tag{15.6}$$

where λ is a positive real parameter. This is the discrete logistic model of a single population, except that the nonlinear term regulating the population growth contains a time delay of one generation.

To elucidate the geometry of the dynamics of the second-order difference equation (15.6), it is convenient to convert it to an equivalent first-order planar map. To this end, if we let

$$x_1^n = y_{n-1}, \qquad x_2^n = y_n,$$

then we obtain the equivalent first-order system

$$x_1^{n+1} = x_2^n$$
$$x_2^{n+1} = \lambda x_2^n (1 - x_1^n).$$

To avoid the overabundant superscripts, we will prefer to write this system of difference equations as the iteration of the map

$$\begin{pmatrix} x_1 \\ x_2 \end{pmatrix} \mapsto \begin{pmatrix} x_2 \\ \lambda x_2 (1 - x_1) \end{pmatrix}, \tag{15.7}$$

which is called the *delayed logistic map*.

The map (15.7) has two fixed points, one at $(0, 0)$ and the other at $(1 - 1/\lambda, 1 - 1/\lambda)$. The Jacobian matrix at the origin is

$$\begin{pmatrix} 0 & 1 \\ 0 & \lambda \end{pmatrix},$$

which has the eigenvalues 0 and λ. Consequently, the origin is asymptotically stable if $0 < \lambda < 1$, and unstable if $\lambda > 1$.

The Jacobian matrix at the other fixed point $(1 - 1/\lambda, 1 - 1/\lambda)$ is

$$\begin{pmatrix} 0 & 1 \\ 1 - \lambda & 1 \end{pmatrix},$$

which has the eigenvalues

$$\mu(\lambda) = \tfrac{1}{2}\left(1 + \sqrt{5 - 4\lambda}\right), \qquad \bar{\mu}(\lambda) = \tfrac{1}{2}\left(1 - \sqrt{5 - 4\lambda}\right).$$

For $1 < \lambda < 2$, the eigenvalues have moduli less than one, and thus the fixed point is asymptotically stable; see Figure 15.8.

At $\lambda = 2$, the eigenvalues at the fixed point $(1 - 1/\lambda, 1 - 1/\lambda) = (1/2, 1/2)$ have moduli one (on the unit circle in the complex plane) and thus the fixed point is nonhyperbolic. Consequently, the stability of the fixed point cannot be determined from linearization and depends on the nonlinear terms.

Observe that when $\lambda = 2$ the eigenvalues are sixth roots of unity:

$$\tfrac{1}{2}\left(1 \pm i\sqrt{3}\right) = e^{\pm i\pi/3}. \tag{15.8}$$

We will point out later in this chapter the significance of this seemingly irrelevant observation on the bifurcations of the nonhyperbolic fixed point $(1 - 1/\lambda,\ 1 - 1/\lambda)$ as the parameter λ is varied near $\lambda = 2$. $\Diamond$

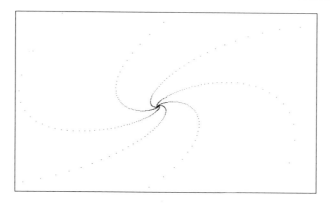

Figure 15.8. *Asymptotically stable fixed point of the delayed logistic map* (15.7) *at* $(1 - 1/\lambda,\ 1 - 1/\lambda)$ *for the parameter value* $\lambda = 1.98$. *This is a single orbit! The shadow of sixth root of unity is visible in the presence of six strands.*

We now turn our attention to the fine structure of the flow of a map near a hyperbolic fixed point. The counterpart of the results from differential equations that we presented in Chapter 9 can easily be formulated for maps when a map is differentiable with a differentiable inverse, that is, when it is a diffeomorphism. For the remainder of this section, let us assume that $\mathbf{f}$ is a diffeomorphism, thus excluding zero as an eigenvalue.

If a hyperbolic fixed point is asymptotically stable, then there is an open neighborhood of the fixed point in which all points tend to the fixed point under forward iterations; such a fixed point is often called a *sink*. When a fixed point is unstable, there are two qualitatively distinct cases. If both eigenvalues are greater than one in modulus, then there is an open neighborhood of the fixed point in which all points tend to the fixed point under backward iterations—a *source*. However, if one eigenvalue is smaller and the other is larger than one in absolute value, then in any open neighborhood of the fixed point, some points tend to the fixed point under forward iterations while others under backward iterations—a *saddle*.

The local fine structure of a saddle point of a diffeomorphism is analogous to that of a saddle point of a planar differential equation (as we described in Section 9.5).

Definition 15.18. *Let U be a neighborhood of a fixed point $\bar{x}$ of a diffeomorphism f defined in U. Then the local stable manifold $W^s(\bar{x}, U)$, and the local unstable manifold $W^u(\bar{x}, U)$ of $\bar{x}$ are defined, respectively, to be the following subsets of U:*

$$W^s(\bar{x}, U) \equiv \{\, x^0 \in U \,:\, f^n(x^0) \in U \text{ for all } n \geq 0,$$
$$\text{and } f^n(x^0) \to \bar{x} \text{ as } n \to +\infty \,\},$$

$$W^u(\bar{x}, U) \equiv \{\, x^0 \in U \,:\, f^{-n}(x^0) \in U \text{ for all } n \geq 0,$$
$$\text{and } f^{-n}(x^0) \to \bar{x} \text{ as } n \to +\infty \,\}.$$

The key geometric properties of these local invariant manifolds are the content of the theorem below, which we state without a proof.

Theorem 15.19. *(Stable and Unstable Manifolds) Let $f : \mathbb{R}^2 \to \mathbb{R}^2$ be a diffeomorphism with a hyperbolic saddle point $\bar{x}$, that is, the linearized map $Df(\bar{x})$ at the fixed point has nonzero eigenvalues $|\mu_1| < 1$ and $|\mu_2| > 1$. Then $W^s(\bar{x}, U)$ is a curve tangent at $\bar{x}$ to, and a graph over, the eigenspace corresponding to μ_1, while $W^u(\bar{x}, U)$ is a curve tangent at $\bar{x}$ to, and a graph over, the eigenspace corresponding to μ_2. These curves are as smooth as the map f.* $\Diamond$

If we do not confine our attention to a local neighborhood U of $\bar{x}$, then we are led to the global analogs of the two invariant sets:

Definition 15.20. *The global stable manifold $W^s(\bar{x})$, and the global unstable manifold $W^u(\bar{x})$ of an equilibrium point $\bar{x}$ are defined, respectively, to be the following sets:*

$$W^s(\bar{x}) \equiv \{x^0 \in \mathbb{R}^2 \,:\, f^n(x^0) \to \bar{x} \text{ as } n \to +\infty \,\},$$

$$W^u(\bar{x}) \equiv \{x^0 \in \mathbb{R}^2 \,:\, f^{-n}(x^0) \to \bar{x} \text{ as } n \to +\infty \,\}.$$

The global stable manifold $W^s(\bar{\mathbf{x}})$ and the global unstable manifold $W^u(\bar{\mathbf{x}})$ can be constructed from the corresponding local ones; to obtain $W^s(\bar{\mathbf{x}})$, let the points in $W^s(\bar{\mathbf{x}}, U)$ flow under backward iterates, and to obtain $W^u(\bar{\mathbf{x}})$, let the points in $W^u(\bar{\mathbf{x}}, U)$ flow under forward iterates:

$$W^s(\bar{\mathbf{x}}) \equiv \bigcup_{n \geq 0} \mathbf{f}^{-n}\big(W^s(\bar{\mathbf{x}}, U)\big),$$

$$W^u(\bar{\mathbf{x}}) \equiv \bigcup_{n \geq 0} \mathbf{f}^{n}\big(W^u(\bar{\mathbf{x}}, U)\big).$$

It is important to note that the behavior of global stable and unstable fixed points differs from the similar invariant manifolds of equilibria of differential equation. In particular, it is possible for the invariant manifolds of a fixed point to cross each other at some other point without coinciding exactly.

Definition 15.21. *Let $\bar{\mathbf{x}}$ be a saddle point of a diffeomorphism $\mathbf{f}$. A point $\mathbf{p}$ is called a homoclinic point for $\mathbf{f}$ if*

$$\mathbf{f}^n(\mathbf{p}) \to \bar{\mathbf{x}} \qquad \text{as } n \to +\infty \quad \text{and as } n \to -\infty.$$

The homoclinic point $\mathbf{p}$ is said to be transversal if the tangent vectors to $W^u(\bar{\mathbf{x}})$ and $W^s(\bar{\mathbf{x}})$ at $\mathbf{p}$ do not coincide.

The presence of transversal homoclinic points of planar maps is an unmistakable sign of dynamical complexity. It is, of course, rare that one can theoretically establish the existence of homoclinic points in specific equations; however, numerical evidence can at times be convincing, as in the case of one of the most famous planar maps.

Example 15.22. *The Hénon map:* Consider the quadratic map

$$\begin{pmatrix} x_1 \\ x_2 \end{pmatrix} \mapsto \begin{pmatrix} 1 + x_2 - ax_1^2 \\ bx_1 \end{pmatrix} \tag{15.9}$$

depending on two real parameters a and b. As indicated in the exercises, this map is a diffeomorphism when $b \neq 0$.

Following Hénon, let us set, for the moment,

$$a = 1.4, \qquad b = 0.3.$$

At these parameter values there are two fixed points and their coordinates are

$$\bar{x}_1 = \frac{-(1-b) \pm \sqrt{(1-b)^2 + 4a}}{2a}, \qquad \bar{x}_2 = b\bar{x}_1.$$

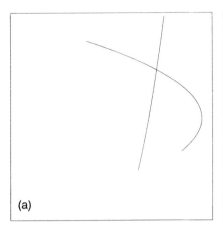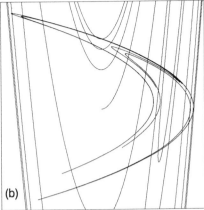

(a)   (b)

Figure 15.9. (a) *Local stable and unstable manifolds of a saddle point of the Hénon map; and* (b) *global stable and unstable manifolds of the same saddle point with many intersections—transversal homoclinic points.*

Linearization about the fixed points reveals that the fixed point

$$\bar{x}_1 = 0.63135448\ldots, \quad \bar{x}_2 = 0.18940534\ldots$$

is a saddle point with the eigenvalues

$$\mu_1 = 0.15594632\ldots, \quad \mu_2 = -1.92373886\ldots\,.$$

The slope of eigenvector of μ_1 is $1.92373886\ldots$ and the slope of eigenvector of μ_2 is $-0.15594632\ldots$.

For a crude approximation of the unstable manifold of the saddle point, we can take a sufficient number of initial points lying on the eigenvector of μ_2 and iterate forward. For the stable manifold, we iterate points on the eigenvector of μ_2 using the inverse of the map. The result of such an experiment, with further refinements, is depicted in Figure 15.9, where intersection points of the stable and the unstable manifolds are visible.

The global dynamics of the Hénon map are quite complicated and the mathematical details are still fragmentary for many parameter values, including the ones chosen by Hénon. However, we should note that for $a = 1.4$ and $b = 0.3$, Hénon discovered through numerical experiments that in a region of the plane almost all solutions eventually get attracted to a set—the Hénon attractor—that is neither a fixed point nor a periodic orbit. For a picture of this *strange attractor*, see Figure 15.10. It is interesting to observe that the attractor appears to be the global unstable manifold of the saddle point. We will provide further information on the Hénon map later in this chapter, including a novel reason why a numerical analyst might also be interested in this map. $\Diamond$

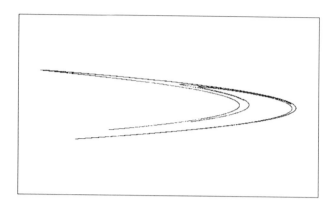

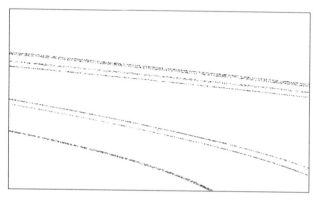

Figure 15.10. *The "strange attractor" of the Hénon map at the parameter values $a = 1.4$ and $b = 0.3$. The picture is obtained by iterating the initial value (0.63135448, 0.18940634) and plotting the iterates 1000–5000. The second picture is an enlargement of a region near the initial point.*

Exercises _____

15.6. *Linearization and nonhyperbolicity:* Find two planar maps with the following properties: both have a fixed point and the linearization at the fixed points are the same linear map but the fixed point of one of the maps is asymptotically stable while the fixed point of the other map is unstable.

15.7. *Nonhyperbolic fixed point of the delayed logistic map:* The delayed logistic map is stored in the library of PHASER as a planar map under the name *dellogis*. Using the machine, investigate the following questions:

   (a) At the parameter value $a = 1$, the only fixed point of the delayed logistic map is the origin. Since one of the eigenvalues is equal to 1, the fixed point is not hyperbolic and thus linearization yields no information

about its stability type. Is the origin asymptotically stable because of
the form of the nonlinearity?

(b) At the parameter value $a = 2$, the fixed point $(1/2, 1/2)$ is nonhyperbolic. Is it asymptotically stable?

(c) To anticipate what is to come, increase the parameter value very slowly past $a = 2$. Does the resulting bifurcation resemble anything familiar?

15.8. Is the delayed logistic map invertible on $\mathbb{R}^2$ for all parameter values? If not, determine the parameter values and a region of the plane on which the map is invertible.

15.9. *The inverse of Hénon:* Verify that, when $b \neq 0$, the quadratic map

$$\begin{pmatrix} x_1 \\ x_2 \end{pmatrix} \mapsto \begin{pmatrix} x_2/b \\ x_1 - 1 + ax_2^2/b^2 \end{pmatrix}$$

is indeed the inverse of the Hénon map.

15.10. *The Hénon attractor:* The Hénon map is stored in the library of PHASER under the name *henon*; while you are thinking about the problems below, you may wish to experiment on the machine as well.

(a) As Hénon did, set $a = 1.4$ and $b = 0.3$. Consider the quadrangle in the plane whose vertices are given by the four points $(-1.33, 0.42)$, $(1.32, 0.133)$, $(1.245, -0.14)$, and $(-1.06, -0.5)$. Show that under the Hénon map this quadrangle is mapped into itself; such a region is often said to be "trapping."

(b) Select an initial point in this quadrangle and plot its iterates 100–2000. Try this for several other initial vectors in the quadrangle. Do you get about the same picture?

15.11. *Hénon and logistic:* Show that using the linear change of variables $x_1 \mapsto x_1/a$ and $x_2 \mapsto bx_2/a$, the Hénon map (15.9) transforms to

$$\begin{pmatrix} x_1 \\ x_2 \end{pmatrix} \mapsto \begin{pmatrix} a + bx_2 - x_1^2 \\ x_1 \end{pmatrix}.$$

The interesting observation about the new form of the map is that when $b = 0$, the first component is the logistic map. Consequently, the Hénon map contains all the complexity of the logistic map, and more. You should experiment numerically on PHASER using very small nonzero values of b.

15.12. *The Lozi map:* This map is a piecewise linear version of the Hénon map obtained by replacing the square term with absolute value:

$$\begin{pmatrix} x_1 \\ x_2 \end{pmatrix} \mapsto \begin{pmatrix} 1 + x_2 - a|x_1| \\ bx_2 \end{pmatrix},$$

where a and b are real parameters. The map of Lozi is stored in the library of PHASER under the name *lozi*. Take $a = 1.7$ and $b = 0.5$; plot the iterates 1000–3000 of the initial vector $(0.11, 0.21)$. The resulting picture is the Lozi attractor. Try several other initial points. Can you determine a trapping region?

Notes: It is often the case that mathematical analysis tends to be easier for piecewise linear maps than smooth nonlinear ones. In fact, unlike the case of the Hénon map, it has been proved that the Lozi map has a hyperbolic strange attractor; see Lozi [1978] and Misiurewicz [1980].

15.13. *Liapunov functions for maps:* The stability type, more importantly, the basin of attraction, of a fixed point of a planar map can also be determined using an appropriate modification of Liapunov functions that we explored in Part II.

A real-valued function $V : \mathbb{R}^2 \to \mathbb{R}$ is called a *Liapunov function* centered about a fixed point $\bar{\mathbf{x}}$ of a planar map $\mathbf{f}$ if

(i) $V(\mathbf{x}) > 0$ for $\mathbf{x} \neq \bar{\mathbf{x}}$;

(ii) $V(\bar{\mathbf{x}}) = 0$;

(iii) $(V \circ \mathbf{f})(\mathbf{x}) \leq V(\mathbf{x})$.

Using the quadratic function $V(x_1, x_2) = x_1^2 + x_2^2$, analyze the stability and the domain of attraction of the fixed point $(0, 0)$ of the planar map

$$
\begin{pmatrix} x_1 \\ x_2 \end{pmatrix} \mapsto \begin{pmatrix} ax_2/(1 + x_1^2) \\ bx_1/(1 + x_2^2) \end{pmatrix},
$$

where a and b are two real parameters. You should consider several cases depending on the values of the parameters; $a^2 = b^2 = 1$ requires special care.

References: For more information on this example, Liapunov functions, and the Invariance Principle for maps at large, see the expository article by La Salle in Hale [1977], or his books (La Salle [1976 and 1987]).

15.3. Numerical Algorithms and Maps

We saw in Section 3.1 that the numerical approximation of solutions of an autonomous scalar differential equation using Euler's algorithm naturally gave rise to a scalar difference equation. The accuracy of the numerical approximation of the solutions could then be determined by analyzing the dynamics of the resulting scalar map depending on one parameter, the step size. In this section, we pursue this important idea further and explore the effects of several other numerical methods of approximation of solutions of differential equations. Even if you are not a great fan of numerical analysis, read on to see the incarnation of the Hénon map from the logistic equation.

As our first example, we will consider the ramifications of approximating solutions of a planar linear differential equation using Euler's algorithm. Before we embark on numerical analysis, however, let us recall our notational convention. In the numerical algorithms below, we will use a constant step size h. The approximate value of a vector solution of a planar differential equation $\mathbf{x}(t) = (x_1(t), x_2(t))$ at time $t = nh$, where n is an integer, will be denoted by $\mathbf{x^n} = (x_1^n, x_2^n)$.

Example 15.23. *Linear systems and Euler:* Consider the linear planar differential equation $\dot{\mathbf{x}} = \mathbf{A}\mathbf{x}$, where $\mathbf{A}$ is a 2×2 matrix. If we use Euler's method with step size h to approximate the derivative

$$\dot{\mathbf{x}}(nh) \approx \frac{\mathbf{x}^{n+1} - \mathbf{x}^n}{h}, \tag{15.10}$$

then the linear planar differential equation becomes the planar linear map

$$\mathbf{x}^{n+1} = (\mathbf{I} + h\mathbf{A})\mathbf{x}^n. \tag{15.11}$$

The origin is a fixed point of this linear planar map corresponding to the equilibrium point of the differential equation. Let us now analyze the dynamics of the linear planar map (15.11) as a function of the step size for various coefficient matrices in Jordan Normal Form.

(i) *Asymptotic stability preserved:* Consider the linear differential system with the coefficient matrix

$$\mathbf{A} = \begin{pmatrix} \lambda_1 & 0 \\ 0 & \lambda_2 \end{pmatrix},$$

where both λ_1 and λ_2 are negative so that the equilibrium point at the origin is asymptotically stable. The eigenvalues of $\mathbf{I} + h\mathbf{A}$ are $1 + h\lambda_1$ and $1 + h\lambda_2$, both of which have modulus less than 1 if the step size h is assumed to be sufficiently small and positive. In this case, the fixed point at the origin of the planar linear map (15.11) is asymptotically stable also.

(ii) *Saddle preserved:* If $\lambda_1 > 0$ and $\lambda_2 < 0$, then one of the eigenvalues of $\mathbf{I}+h\mathbf{A}$ must have modulus greater than 1 and the other less than 1 when h is small and positive. Consequently, the saddle character of the equilibrium is reflected in the phase portrait of the approximate linear map about the corresponding fixed point at the origin.

(iii) *Center destroyed:* Since nonhyperbolic equilibria are very sensitive to small perturbations, we have reason to fear that numerical approximations will destroy a center. To validate that such fears are indeed well-founded, let us consider the planar linear differential equations with the coefficient matrix

$$\mathbf{A} = \begin{pmatrix} 0 & 1 \\ -1 & 0 \end{pmatrix},$$

which has eigenvalues $\pm i$. The eigenvalues of $\mathbf{I} + h\mathbf{A}$ are $1 \pm ih$, both of which have modulus greater than 1. Consequently, the fixed point $\mathbf{0}$ of the linear map (15.11) is unstable (source) for every positive step size h. However, the equilibrium point at the origin is a center for the differential equation. The disturbing realization now is that no matter how small a step size we use, it is impossible to capture even the qualitative features of a center using Euler's algorithm; see Figure 15.11. ◇

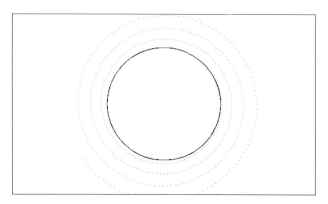

Figure 15.11. *A (nonhyperbolic) periodic orbit of a linear center, and its Euler approximation with step size 0.05.*

We now consider the consequences of using a "two-step" numerical algorithm to approximate the solutions of a scalar differential equation. Our example is, of course, the logistic equation so that we may compare new findings with the results in Section 3.1.

Example 15.24. *Central difference and the Logistic equation:* Let us consider the scalar logistic differential equation (3.2) at the parameter value $a = 1$,

$$\dot{y} = y(1 - y).\tag{15.12}$$

We saw there that Euler's algorithm yielded a reasonable result to capture the asymptotic stability of the equilibrium point at $\bar{y} = 1$. Here, we will study this equilibrium point by approximating the derivative $\dot{y}(t)$ at $t = nh$ with the *central difference*, or the *midpoint*, algorithm

$$\dot{y}(nh) \approx \frac{y_{n+1} - y_{n-1}}{2h}.\tag{15.13}$$

If we use the central difference (15.13) on the logistic equation (15.12), we obtain the second-order difference equation

$$y_{n+1} = y_{n-1} + 2hy_n(1 - y_n).$$

By setting $x_1^n = y_n$ and $x_2^n = y_{n-1}$, we can convert this second-order difference equation to the equivalent planar map

$$\begin{pmatrix} x_1 \\ x_2 \end{pmatrix} \mapsto \begin{pmatrix} 2hx_1(1 - x_1) + x_2 \\ x_1 \end{pmatrix}.\tag{15.14}$$

The equilibrium point $\bar{y} = 1$ of Eq. (15.12) corresponds to the fixed point of the map (15.14) at $\bar{\mathbf{x}} = (1, 1)$. To determine its stability type, we compute that the Jacobian matrix at this fixed point is

$$\begin{pmatrix} -2h & 1 \\ 1 & 0 \end{pmatrix}.$$

The eigenvalues of this matrix are

$$\mu_1 = -h - \sqrt{h^2 + 1}, \qquad \mu_2 = -h + \sqrt{h^2 + 1}.$$

When $h > 0$, one eigenvalue, μ_1, is smaller than -1 and the other is between 0 and 1. Consequently, the fixed point $(1, 1)$ is a saddle point of the map (15.14).

Now, the disappointing conclusion is that no matter how small a positive step size we use, it is impossible to capture the asymptotic stability of the equilibrium point $\bar{y} = 1$ with the central difference algorithm. $\Diamond$

We now come to our final example in this section; we trust you will find it exciting. The numerical approximation scheme we are about to introduce may not have a wide practical appeal from the viewpoint of numerical analysis, but the resulting planar map is certainly a famous one—the Hénon map.

Example 15.25. *The Logistic equation, a mixed difference scheme, and the Hénon map:* We saw above that while Euler's algorithm was capable of capturing the asymptotic stability of the fixed point $\bar{y} = 1$ of the logistic equation (15.12), the central difference algorithm failed hopelessly. To see how things might degrade from good to bad, let us now use a weighted "average" of the two methods to approximate $\dot{y}$ at $t = nh$:

$$\dot{y}(nh) \approx (1 - \lambda)\frac{y_{n+1} - y_{n-1}}{2h} + \lambda\frac{y_{n+1} - y_n}{h}, \qquad (15.15)$$

where $0 \leq \lambda \leq 1$. At $\lambda = 0$ this is reduced to the central difference (15.13), and at $\lambda = 1$ it becomes Euler's algorithm.

Using the mixed-difference algorithm (15.15) to approximate and then converting, as usual, the resulting second-order difference equation yields the following equivalent map:

$$\begin{pmatrix} z_1 \\ z_2 \end{pmatrix} \mapsto \begin{pmatrix} \frac{2}{1+\lambda}[(\lambda + h)z_1 - hz_1^2] + \frac{1-\lambda}{1+\lambda}z_2 \\ z_1 \end{pmatrix}. \qquad (15.16)$$

The unstable equilibrium point $\bar{y} = 0$ and the asymptotically stable equilibrium point $\bar{y} = 1$ of the logistic equation (15.12) correspond, respectively, to the fixed points $(0, 0)$ and $(1, 1)$ of this map. We will indicate the stability types and some of the bifurcations of these fixed points in the exercises. For now, let us manipulate the form of the map (15.16) a bit more.

If we make the affine transformation

$$z_1 = \frac{\lambda + h}{2h} \left[\frac{h - \lambda}{1 + \lambda} x_1 + 1 \right], \qquad z_2 = \frac{\lambda + h}{2h} \left[\frac{h - \lambda}{1 - \lambda} x_2 + 1 \right]$$

and put

$$a = \frac{h^2 - \lambda^2}{(1 + \lambda^2)^2}, \qquad b = \frac{1 - \lambda}{1 + \lambda},$$

we obtain

$$\begin{pmatrix} x_1 \\ x_2 \end{pmatrix} \mapsto \begin{pmatrix} 1 + x_2 - a x_1^2 \\ b x_1 \end{pmatrix}. \tag{15.17}$$

This, of course, is the map we have been seeking—the Hénon map. The parameter values $a = 1.4$ and $b = 0.3$ chosen by Hénon correspond approximately to $\lambda = 0.538$ and $h = 1.9$. $\Diamond$

After this interlude on numerical analysis, we now return to basics and investigate bifurcations of fixed points of planar maps.

Exercises _____ ♣ ♡ ♠ ◇

15.14. Suppose that $\mathbf{A}$ has an eigenvalue with positive real part. Show that $\mathbf{I} + h\mathbf{A}$ must have an eigenvalue with modulus greater than 1 if h is positive.

15.15. *Safe step size:* Consider the linear system of differential equations

$$\dot{\mathbf{x}} = \begin{pmatrix} -1.5 & 2 \\ 0 & -1 \end{pmatrix} \mathbf{x}.$$

What is the largest safe step size in the numerical integration of the system using Euler's algorithm?

15.16. *Central difference and hyperbolic equilibria:* Consider the linear scalar differential equation

$$\dot{y} = ay,$$

where a is a nonzero real number so that the origin is a hyperbolic equilibrium. Approximate the derivative of y with the central difference

$$\dot{y}(nh) \approx \frac{y_{n+1} - y_{n-1}}{2h},$$

and arrive at the linear planar map:

$$\mathbf{x}^{n+1} = \begin{pmatrix} 2a & 1 \\ 1 & 0 \end{pmatrix} \mathbf{x}^n.$$

Show that the origin is a saddle point of this map. It is evident from this example that using the central difference one cannot preserve the stability of a hyperbolic equilibrium point of a scalar differential equation for either $t \to +\infty$ or $t \to -\infty$.

15.17. *A center-preserving algorithm:* Here, we describe an algorithm that captures the qualitative feature of a linear center. Consider the second-order scalar differential equation

$$\ddot{y} + y = 0$$

whose phase portrait, when written as a first-order system, is a center.

(a) Show that approximating the second derivative of y by the *central difference*

$$\ddot{y}(t) \approx \frac{y(t+h) - 2y(t) + y(t-h)}{h^2},$$

turns the differential equation into the second-order difference equation

$$y_{n+1} - 2y_n + y_{n-1} + h^2 y_n = 0.$$

(b) Set $x_1^n = y_n$ and $x_2^n = y_{n-1}$, and convert the second-order difference equation to the equivalent planar linear map

$$\mathbf{x}^{n+1} = \begin{pmatrix} 2 - h^2 & -1 \\ 1 & 0 \end{pmatrix} \mathbf{x}^n.$$

(c) Compute that the eigenvalues of the coefficient matrix are

$$\lambda_1 = \tfrac{1}{2} \left[2 - h^2 + ih\sqrt{4 - h^2} \right], \qquad \lambda_2 = \bar{\lambda}_1,$$

and verify that both are complex and have unit moduli.

(d) Show that each orbit of the planar linear map lies on an ellipse and the ellipses are concentric. Determine the equations of the ellipses.

(e) We captured the qualitative character of the center, but not the quantitative one because the orbits of the differential equation lie on circles in the $(x, \dot{x})$-plane. Can you choose a better coordinate system for the corresponding map so as to capture the quantitative character of the center?

15.18. *A hyperbolic limit cycle and Euler:* Consider our old-time favorite example of a planar differential equation:

$$\dot{x}_1 = -x_2 + x_1(1 - x_1^2 - x_2^2)$$
$$\dot{x}_2 = x_1 + x_2(1 - x_1^2 - x_2^2),$$

possessing a hyperbolic orbitally asymptotically stable limit cycle at $x_1^2 + x_2^2 = 1$. Show that the planar map resulting from approximating the solutions of this system using Euler has an invariant circle of radius $r_h =$

$\sqrt{1 + (1 - \sqrt{1 - h^2})/h}$, where h is the step size. For $0 < h \le 1$, we have $r_h = 1 + O(h)$.

15.19. *Mixed difference and the fixed points:* Consider the planar map (15.16). Show that, for positive step size h, the fixed point $(0, 0)$ is a saddle if $0 < \lambda < 1$. The other fixed point $(1, 1)$ is a sink if the Euler component of the mixed difference scheme (15.15) is sufficiently dominant; more precisely, if $0 < h < 1$ and $\lambda > h/2$.

15.20. *Fast computation of elliptic integrals:* This perplexing numerical analysis problem concerns the computation of the following elliptic integral of the second kind:

$$I(a, b) = \int_0^{\pi/2} \frac{d\theta}{\sqrt{a^2 \cos^2 \theta + b^2 \sin^2 \theta}},$$

where a and b are some constants. To compute the numerical value of this integral for given values of a and b, consider the planar map

$$\begin{pmatrix} x_1 \\ x_2 \end{pmatrix} \mapsto \begin{pmatrix} \frac{1}{2}(x_1 + x_2) \\ \sqrt{x_1 x_2} \end{pmatrix}.$$

For the initial conditions $x_1^0 = a$ and $x_2^0 = b$, both components x_1 and x_2 of the orbit converge quadratically to the value of the integral $I(a, b)$. Perform such iterations on the machine.

Reference: This method is called an arithmetic-geometric mean iteration and was known to Gauss and Legendre. The map above is stored in the library of PHASER under the name *gauss*. If you are curious why this strange iteration scheme works, consult Borwein and Borwein [1984].

15.4. Saddle Node and Period Doubling

If a fixed point of a nonlinear map is nonhyperbolic—that is, if at least one eigenvalue of the linearization at the fixed point has modulus one—then the stability type of the fixed point cannot be determined from the linear approximation. Moreover, if the nonlinear map depends on parameters, then we expect the fixed point to undergo a bifurcation as the parameters are varied. In this section we consider several such bifurcations in the cases where the eigenvalues of the linearization are real. Bifurcations of nonhyperbolic fixed points with complex eigenvalues will be the subject of the next section.

If one eigenvalue of the linearization at a fixed point of a planar map is 1 while the other eigenvalue is different from 1 or -1, then such a fixed point typically undergoes a saddle-node bifurcation. In special situations, for example, in the presence of symmetry, other bifurcations such as pitchfork are also possible. In the case where one eigenvalue is -1 and the other is again different from ± 1, the typical bifurcation is period doubling. As in the case of bifurcations of equilibrium points of planar differential equations with one zero eigenvalue, which we have considered in great detail in Chapter 10, there are two possible ways to investigate the bifurcations of nonhyperbolic fixed points of planar maps with a 1 or -1 eigenvalue. One way is to use a center manifold for maps and reduce the problem to one dimension and then apply the results from Chapter 3 on bifurcations of scalar maps. The second way is to employ a variant of the bifurcation equation for finding the zeros of functions on $\mathbb{R}^2$ as developed in Chapter 10; in the case of maps, the appropriate functions are $\mathbf{f} - \mathbf{I}$ for saddle node, and $\mathbf{f}^2 - \mathbf{I}$ for period doubling.

For reasons similar to the ones we have explained at the end of Chapter 10, it is often advantageous to use the method of bifurcation function when analyzing specific maps. Let us now briefly summarize this procedure for maps depending on a parameter λ. We may assume that the origin is a fixed point for $\lambda = 0$ and that the linear part of the map at the fixed point is in Jordan Normal Form. Thus, it suffices to consider the map

$$\mathbf{x} \mapsto \begin{pmatrix} 1 & 0 \\ 0 & \mu \end{pmatrix} \mathbf{x} + \mathbf{g}(\lambda, \mathbf{x}), \tag{15.18}$$

where

$$\mu \neq 1, \quad \mathbf{g}(0, 0) = \mathbf{0}, \quad D_{\mathbf{x}}\mathbf{g}(0, 0) = \mathbf{0}.$$

The fixed points of the map (15.18) are the solutions of the pair of equations

$$g_1(\lambda, x_1, x_2) = 0,$$
$$(1 - \mu)x_2 - g_2(\lambda, x_1, x_2) = 0.$$

Since $\mu \neq 1$, we can apply the Implicit Function Theorem to obtain a solution $x_2 = \psi(\lambda, x_1)$, with $\psi(0, 0) = 0$ and $\partial\psi(0, 0)/\partial x_1 = 0$, of the second equation. Therefore, the fixed points of the map (15.18) are given by $(x_1, \psi(\lambda, x_1))$, where x_1 are the solutions of the *bifurcation equation*

$$G(\lambda, x_1) \equiv g_1\big(\lambda, x_1, \psi(\lambda, x_1)\big) = 0. \tag{15.19}$$

One can prove the following result which is comparable to Theorem 10.8 on the stability types of equilibrium points of a planar differential equation with a zero eigenvalue.

Theorem 15.26. *Suppose that* $\bar{\mathbf{x}} = \big(\bar{x}_1, \psi(\bar{\lambda}, \bar{x}_1)\big)$ *is a fixed point of Eq. (15.18) satisfying the bifurcation equation* $G(\bar{\lambda}, \bar{x}_1) = 0$. *Then*
  (i) *when* $\mu < 1$, *the fixed point* $\bar{\mathbf{x}}$ *is hyperbolic and asymptotically stable if* $\partial G(\bar{\lambda}, \bar{x}_1)/\partial x_1 < 0$, *and is a saddle if* $\partial G(\bar{\lambda}, \bar{x}_1)/\partial x_1 > 0$;
  (ii) *when* $\mu > 1$, *the fixed point* $\bar{\mathbf{x}}$ *is a saddle if* $\partial G(\bar{\lambda}, \bar{x}_1)/\partial x_1 < 0$, *and is an unstable node if* $\partial G(\bar{\lambda}, \bar{x}_1)/\partial x_1 > 0$. $\diamondsuit$

Using the product of two scalar maps, it is rather easy to construct examples of planar maps exhibiting certain elementary bifurcations of fixed points. Here is a nonproduct example to illustrate the utility of the theorem above.

Example 15.27. *A saddle-node bifurcation:* Consider the planar map

$$\begin{pmatrix} x_1 \\ x_2 \end{pmatrix} \mapsto \begin{pmatrix} \lambda + x_1 + \lambda x_2 + x_1^2 \\ 0.5 x_2 + \lambda x_1 + x_1^2 \end{pmatrix},$$

where λ is a scalar real parameter ranging in a small neighborhood of 0. Simple computations yield $x_2 = \psi(\lambda, x_1) = 2x_1^2 + 2\lambda x_1$, and the bifurcation equation (15.19) for the fixed points of the map becomes

$$\lambda + 2\lambda^2 x_1 + (1 + 2\lambda) x_1^2 = 0.$$

If $\lambda < 0$, there are two zeros of the bifurcation function. From part (i) of Theorem 15.26, the fixed point with the smaller x_1 coordinate is asymptotically stable, while the other fixed point is a saddle. When $\lambda = 0$, these two fixed points coalesce at the origin which is a nonhyperbolic fixed point. If $\lambda > 0$, there are no zeros of the bifurcation function and thus no fixed points either. This, of course, is a saddle-node bifurcation for the nonhyperbolic fixed point at the origin as the parameter λ passes through zero. $\diamond$

Example 15.28. *Saddle node in the Hénon map:* Let us consider the Hénon map from Section 15.2:

$$\begin{pmatrix} x_1 \\ x_2 \end{pmatrix} \mapsto \begin{pmatrix} 1 + x_2 - ax_1^2 \\ bx_1 \end{pmatrix}. \tag{15.20}$$

Here, we will fix $b = 0.3$ and locate a saddle-node bifurcation while we vary the parameter a. As we indicated before, the fixed points of the Hénon map are given by the solutions of $x_2 = bx_1$ and $ax_1^2 + (1 - b)x_1 - 1 = 0$:

$$\bar{x}_1 = \frac{-(1 - b) \pm \sqrt{(1 - b)^2 + 4a}}{2a}, \qquad \bar{x}_2 = b\bar{x}_1.$$

At $a = -(1 - b)^2/4$, there is a unique fixed point. Linearization at this fixed point shows that the fixed point is nonhyperbolic with 1 as an eigenvalue; the other eigenvalue is less than one in absolute value. Close to this bifurcation value, there are no fixed points when $a < -(1 - b)^2/4$, and two hyperbolic fixed points when $a > -(1 - b)^2/4$. One of these fixed points is unstable and the other stable. $\diamond$

We refrain from presenting a detailed statement of period-doubling bifurcation when one eigenvalue passes through -1. Instead, we will be content to demonstrate this important bifurcation in the Hénon map.

Example 15.29. *Period doubling in the Hénon map:* Here, we continue to work in the setting of the example above. If we proceed to increase the parameter a, the stable fixed point becomes nonhyperbolic at $a = 3(1 - b)^2/4$ when one of the eigenvalues becomes -1; the other eigenvalue remains less than one in absolute value.

When a is increased through $3(1 - b)^2/4$, the negative eigenvalue becomes larger than one in absolute value and the fixed point loses its stability. It is possible to show that the fixed point gives up its stability to an asymptotically stable periodic orbit of period 2. As the parameter a is increased further, the Hénon map undergoes successive period-doubling bifurcations; see Figure 15.12.

The successive bifurcation values of the parameter forms a geometric sequence, much like in the case of the period-doubling bifurcations of the logistic map as we described in Section 3.5; however, the ratio of the successive terms in the geometric sequence of the Hénon map turns out to be a different constant than that of the logistic map. ◊

Exercises ━━━━━━━━━━━━━━━━━━━━━━━━━━━━━

15.21. *A period-7 orbit for Hénon:* For the parameter values $a = 1.24$ and $b = 0.3$, the Hénon map (15.20) has an attracting periodic orbit of period 7. Find it on the computer by using appropriate initial data, for example, $\mathbf{x}^0 = (0, 0)$. Also, try other initial data. While leaving b fixed, increase a gradually until this periodic orbit doubles its period. Can you double the period again?
Hint: When plotting orbits, leave out the transients; that is, do not plot, say, the first 1000 iterates.

15.22. *A pitchfork bifurcation:* Consider the following planar map depending on a scalar parameter λ:

$$\begin{pmatrix} x_1 \\ x_2 \end{pmatrix} \mapsto \begin{pmatrix} x_2 \\ -0.5x_1 + \lambda x_2 - x_2^3 \end{pmatrix}.$$

(a) Notice that the origin is a fixed point for all values of λ. Show that the origin undergoes a pitchfork bifurcation near $\lambda = 3/2$.

(b) There are additional fixed points at $(\pm\sqrt{\lambda - 3/2}, \pm\sqrt{\lambda - 3/2})$. At $\lambda = 3$, compute that the eigenvalues at these fixed points are -1 and $-1/2$. Analyze the bifurcation as λ passes through 3. In your computations you should translate the fixed point to the origin and then put the linear part into Jordan Canonical Form.

15.23. *Bifurcation in mixed-difference algorithm:* Consider the map (15.16) from the previous section. Fix the step size h and, while varying λ, show the following:

(a) One of the fixed points of the map undergoes a pitchfork bifurcation at $\lambda = h/2$;

(b) determine the value of λ at which the fixed point $(1, 1)$ loses its stability to an asymptotically stable periodic orbit of period 2.

Figure 15.12. *Successive period-doubling bifurcations in the Hénon map. The parameter values are $b = 0.3$, and a is varied through 0.3, 0.7, 1.0, and 1.05. The initial point for all parameter values is $(0, 0)$. Transients (first several hundred iterates) have been discarded.*

15.24. *A two-parameter map:* Consider the following map depending on two real parameters, λ_1 and λ_2:

$$\begin{pmatrix} x_1 \\ x_2 \end{pmatrix} \mapsto \begin{pmatrix} \lambda_1 + x_1 + \lambda_2 x_2 + x_1^2 \\ 0.5 x_2 + \lambda_2 x_1 + x_1^2 \end{pmatrix}.$$

Analyze the bifurcation of fixed points near the origin for (λ_1, λ_2) small in norm. What is happening near the curve $\lambda_2^4 - \lambda_1 = 0$?
Hint: The bifurcation equation is $\lambda_1 + 2\lambda_2^2 x_1 + (1 + 2\lambda_2) x_1^2 = 0$.

15.5. Poincaré–Andronov–Hopf Bifurcation

In this section, we consider bifurcations of a fixed point of a planar map depending on one parameter in the case where the eigenvalues are complex conjugates and of unit moduli. As the eigenvalues move off the unit circle, the generic result is that there appears a closed invariant curve—all the iterates of any point on the curve remain on the curve—encircling the fixed point. Below, we give a precise statement of this bifurcation and analyze several illustrative examples.

We begin with a prototypical specific planar map exhibiting the birth of an invariant circle surrounding a fixed point.

Example 15.30. *The canonical example:* Consider the following map depending on a real scalar parameter λ:

$$\begin{pmatrix} x_1 \\ x_2 \end{pmatrix} \mapsto (\lambda - x_1^2 - x_2^2) \begin{pmatrix} \cos \omega & -\sin \omega \\ \sin \omega & \cos \omega \end{pmatrix} \begin{pmatrix} x_1 \\ x_2 \end{pmatrix}. \tag{15.21}$$

The origin is a fixed point of this map for all values of λ. The linear part of this map is the same as the linear map (15.4) that we examined in Example 15.11: as the parameter λ is increased through 1, the eigenvalues cross the unit circle from the inside to the outside, and thus the stability type of the origin changes from one of asymptotic stability to that of instability. To see what else might be happening near the origin, we need to investigate the effects of the nonlinear terms. For this purpose, it is convenient to transform the map (15.21) to polar coordinates to obtain

$$\begin{pmatrix} r \\ \theta \end{pmatrix} \mapsto \begin{pmatrix} \lambda r - r^3 \\ \theta + \omega \end{pmatrix}. \tag{15.22}$$

It is now evident that for $\lambda > 1$, the map (15.22) has an invariant circle of radius $r = \sqrt{\lambda - 1}$. Moreover, the omega-limit set of every positive orbit, except the origin, is contained in this circle. The flow on the invariant circle is simple rotation with rotation number ω. For the bifurcation diagram of Eq. (15.21), see Figure 15.13. ◊

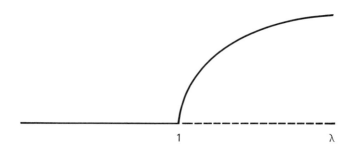

Figure 15.13. *The bifurcation diagram of the canonical example* (15.21). *As* λ *is increased pass 1, the asymptotically stable fixed point yields its stability to an invariant circle. The vertical variable is the radius of the invariant circle.*

The birth of an invariant circle in the example above is the "typical" local bifurcation near a fixed point as a pair of complex conjugate eigenvalues move off the unit circle. Indeed, a general result can be established by transforming a reasonably well-behaved nonlinear map locally to a "normal form" resembling the canonical example above. Here is a statement of the main theorem of this subject in all its fine details:

Theorem 15.31. (Poincaré–Andronov–Hopf Bifurcation for Maps) *Let*

$$\mathbf{F} : \mathbb{R} \times \mathbb{R}^2 \to \mathbb{R}^2; \qquad (\lambda, \mathbf{x}) \mapsto \mathbf{F}(\lambda, \mathbf{x})$$

be a C^4 map depending on a real parameter λ satisfying the following conditions:
 (i) $\mathbf{F}(\lambda, \mathbf{0}) = \mathbf{0}$ *for λ near some fixed λ_0;*
 (ii) $D\mathbf{F}(\lambda, \mathbf{0})$ *has two non-real eigenvalues $\mu(\lambda)$ and $\bar\mu(\lambda)$ for λ near λ_0 with $|\mu(\lambda_0)| = 1$;*
 (iii) $\frac{d}{d\lambda}|\mu(\lambda)| > 0$ *at $\lambda = \lambda_0$;*
 (iv) $\mu^k(\lambda_0) \neq 1$ *for $k = 1, 2, 3, 4$.*

Then there is a smooth λ-dependent change of coordinates bringing $\mathbf{F}$ into the form

$$\mathbf{F}(\lambda, \mathbf{x}) = \mathcal{F}(\lambda, \mathbf{x}) + O\left(\|\mathbf{x}\|^5\right)$$

and there are smooth functions $a(\lambda)$, $b(\lambda)$, and $\omega(\lambda)$ so that in polar coordinates the function $\mathcal{F}(\lambda, \mathbf{x})$ is given by

$$\begin{pmatrix} r \\ \theta \end{pmatrix} \mapsto \begin{pmatrix} |\mu(\lambda)|r - a(\lambda)r^3 \\ \theta + \omega(\lambda) + b(\lambda)r^2 \end{pmatrix}. \qquad (15.23)$$

If $a(\lambda_0) > 0$, then there is a neighborhood U of the origin and a $\delta > 0$ such that, for $|\lambda - \lambda_0| < \delta$ and $\mathbf{x}^0 \in U$, the ω-limit set of $\mathbf{x}^0$ is the origin *if* $\lambda < \lambda_0$ *and belongs to a closed invariant* C^1 *curve* $\Gamma(\lambda)$ *encircling the origin if* $\lambda > \lambda_0$. *Furthermore,* $\Gamma(\lambda_0) = \mathbf{0}$.

If $a(\lambda_0) < 0$, *then there is a neighborhood* U *of the origin and a* $\delta > 0$ *such that, for* $|\lambda - \lambda_0| < \delta$ *and* $\mathbf{x}^0 \in U$, *the* α-*limit set of* $\mathbf{x}^0$ *is the origin if* $\lambda > \lambda_0$ *and belongs to a closed invariant* C^1 *curve* $\Gamma(\lambda)$ *encircling the origin if* $\lambda < \lambda_0$. *Furthermore,* $\Gamma(\lambda_0) = \mathbf{0}$. $\diamond$

A proof of this theorem is beyond the intended scope of our book. However, let us examine the hypotheses of the theorem carefully so that we may apply it to specific maps. The first four hypotheses concern only the linear part of a map and they can easily be checked. The somewhat mysterious fourth condition excluding the first four roots of unity is included to arrive at the desired normal form, as we shall explain momentarily. The sign of the coefficient $a(\lambda_0)$ of the cubic term is what characterizes the bifurcation: when $a(\lambda_0) > 0$, the bifurcation is said to be *supercritical*; when $a(\lambda_0) < 0$ is *subcritical*; the case $a(\lambda_0) = 0$ cannot be determined from the cubic terms alone. The main practical difficulty in applications lies in the determination of this sign. Luckily, there is an explicit formula for this tedious task, which we now present.

To simplify the notation, let us begin by observing that we need to compute the value of the coefficient $a(\lambda)$ at only one fixed value of λ; namely, at the bifurcation value λ_0. Consequently, we omit λ_0 from our notation and consider a general map $\mathbf{f} : \mathbb{R}^2 \to \mathbb{R}^2$ that has a fixed point at the origin with complex eigenvalues $\mu = \alpha + i\beta$ and $\bar{\mu} = \alpha - i\beta$ satisfying $\alpha^2 + \beta^2 = 1$ and $\beta \neq 0$. By putting the linear part of such a map into Jordan Canonical Form, we may assume $\mathbf{f}$ to have the following form near the origin:

$$\mathbf{x} \mapsto \mathbf{f}(\mathbf{x}) = \begin{pmatrix} \alpha & -\beta \\ \beta & \alpha \end{pmatrix} \begin{pmatrix} x_1 \\ x_2 \end{pmatrix} + \begin{pmatrix} g_1(x_1, x_2) \\ g_2(x_1, x_2) \end{pmatrix}. \tag{15.24}$$

Then the magic coefficient a of the cubic term in Eq. (15.23) in polar coordinates is equal to

$$a = \mathrm{Re}\left[\frac{(1 - 2\mu)\bar{\mu}^2}{1 - \mu} \xi_{11}\xi_{20} \right] + \tfrac{1}{2}|\xi_{11}|^2 + |\xi_{02}|^2 - \mathrm{Re}\,(\bar{\mu}\xi_{21}), \tag{15.25}$$

where

$$\xi_{20} = \tfrac{1}{8}\Big\{ (g_1)_{x_1 x_1} - (g_1)_{x_2 x_2} + 2(g_2)_{x_1 x_2}$$
$$+ i\big[(g_2)_{x_1 x_1} - (g_2)_{x_2 x_2} - 2(g_1)_{x_1 x_2}\big]\Big\},$$

$$\xi_{11} = \tfrac{1}{4}\Big\{ (g_1)_{x_1 x_1} + (g_1)_{x_2 x_2}$$
$$+ i\big[(g_2)_{x_1 x_1} + (g_2)_{x_2 x_2}\big]\Big\},$$

$$\xi_{02} = \frac{1}{8}\Big\{(g_1)_{x_1x_1} - (g_1)_{x_2x_2} - 2(g_2)_{x_1x_2}$$
$$+ i\big[(g_2)_{x_1x_1} - (g_2)_{x_2x_2} + 2(g_1)_{x_1x_2}\big]\Big\},$$

and

$$\xi_{21} = \frac{1}{16}\Big\{(g_1)_{x_1x_1x_1} + (g_1)_{x_1x_2x_2} + (g_2)_{x_1x_1x_2} + (g_2)_{x_2x_2x_2}$$
$$+ i\big[(g_2)_{x_1x_1x_1} + (g_2)_{x_1x_2x_2} - (g_1)_{x_1x_1x_2} - (g_1)_{x_2x_2x_2}\big]\Big\}.$$

"Re" in formula (15.25) represents the real parts of those complex numbers, and all the partial derivatives are evaluated at the fixed point at the origin.

Admittedly, the formula above appears rather formidable, but rest assured that we will not attempt to derive it. Instead, let us demonstrate its utility on one of our earlier maps.

Example 15.32. *Delayed logistic map continued:* Here, we continue with the analysis in Example 15.17 near the parameter value $\lambda_0 = 2$ and establish, as consequence of the theorem above, that the fixed point $(1/2, 1/2)$ undergoes a supercritical Poincaré–Andronov–Hopf bifurcation.

From our earlier computations, it is evident that the hypotheses (i)–(iv) of Theorem 15.31 are satisfied. The only remaining hypothesis to verify is the computation of the coefficient $a(2)$ of the cubic term in normal form [Eq. (15.23)]. To accomplish this, we set $\lambda = 2$ and put the delayed logistic map

$$\begin{pmatrix} x_1 \\ x_2 \end{pmatrix} \mapsto \begin{pmatrix} x_2 \\ 2x_2(1 - x_1) \end{pmatrix} \tag{15.26}$$

into the form (15.24). For this purpose, we first translate the fixed point $(1/2, 1/2)$ to the origin by applying the transformation

$$\begin{pmatrix} x_1 \\ x_2 \end{pmatrix} \mapsto \begin{pmatrix} x_1 + \frac{1}{2} \\ x_2 + \frac{1}{2} \end{pmatrix}.$$

Next, we put the linear part of the map into Jordan Normal Form by making one further change of variables:

$$\begin{pmatrix} x_1 \\ x_2 \end{pmatrix} \mapsto \mathbf{P}\begin{pmatrix} x_1 \\ x_2 \end{pmatrix},$$

where

$$\mathbf{P} = \begin{pmatrix} 0 & 1 \\ \sqrt{3}/2 & 1/2 \end{pmatrix}, \qquad \mathbf{P}^{-1} = \begin{pmatrix} -1/\sqrt{3} & 2\sqrt{3} \\ 1 & 0 \end{pmatrix}.$$

The resulting map $\mathbf{x} \mapsto \mathbf{f}(\mathbf{x})$ is given by

$$\mathbf{f}\begin{pmatrix} x_1 \\ x_2 \end{pmatrix} = \begin{pmatrix} 1/2 & -\sqrt{3}/2 \\ \sqrt{3}/2 & 1/2 \end{pmatrix} \begin{pmatrix} x_1 \\ x_2 \end{pmatrix} + \begin{pmatrix} -2x_1 x_2 - 2x_2^2 \\ 0 \end{pmatrix}.$$

This map is now in the form (15.24) and we can use formula (15.25) with the nonlinear terms $g_1(x_1, x_2) = -2x_1 x_2 - 2x_2^2$ and $g_2(x_1, x_2) = 0$. A short computation yields

$$\xi_{20} = \tfrac{1}{2}(1 + i), \qquad \xi_{11} = -1, \qquad \xi_{02} = \tfrac{1}{2}(1 - i), \qquad \xi_{21} = 0,$$

and

$$a = \frac{-\sqrt{3} + 7}{4} > 0.$$

Consequently, the bifurcation is supercritical: for $\lambda > 2$ and $\lambda - 2$ sufficiently small, the delayed logistic map (15.26) has an attracting closed curve encircling the fixed point $(1 - 1/\lambda, \, 1 - 1/\lambda)$; see Figure 15.14. We should emphasize that the smooth invariant curve is guaranteed for values of the parameter that are sufficiently near the bifurcation value. What happens "far away" from the bifurcation value may be rather complicated and perplexing, as shown in Figure 15.15. ◊

We now once again return to the general setting of Theorem 15.31 and indicate how to arrive at the normal form (15.23). During our normal form computations, the necessity of the hypothesis (iv) will become self-evident. To simplify the notation, let us suppose that, by a λ-dependent transformation, the linear part of our map is already in normal form [Eq. (15.24)]. Of course, α, β, g_1, and g_2 would depend on λ; however, let us leave λ out of our formulas. With these simplifications, our difficult computational task is made feasible by introducing the complex change of variables

$$z = x_1 + ix_2, \qquad \bar{z} = x_1 - ix_2. \tag{15.27}$$

This complex coordinate system $(z, \bar{z})$ is called the *"zee-zeebar" coordinates*. It is comforting to notice that the change of variables is invertible with the inverse

$$x_1 = \frac{1}{2}(z + \bar{z}), \qquad x_2 = \frac{1}{2i}(z - \bar{z})$$

so that we may transform the results of our computations back to reality. To consolidate our notation, let

$$P^{-1} = \begin{pmatrix} 1 & i \\ 1 & -i \end{pmatrix}, \qquad P = \frac{1}{2i}\begin{pmatrix} i & i \\ 1 & -1 \end{pmatrix}$$

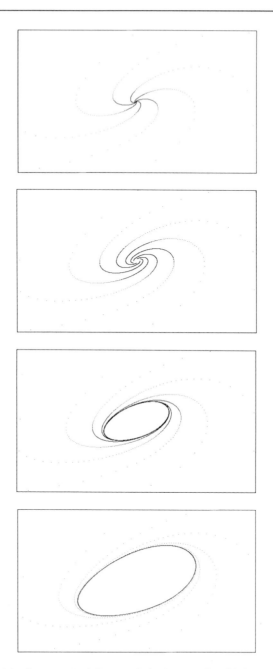

Figure 15.14. *Supercritical Poincaré–Andronov–Hopf bifurcation in the delayed logistic map* (15.26) *for* $\lambda = 1.990$, 2.000, 2.003, *and* 2.010.

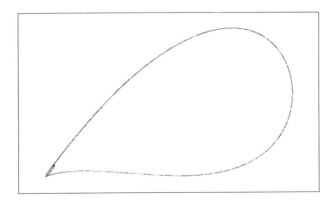

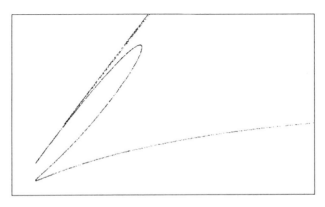

Figure 15.15. *Far away from the bifurcation value, at* $\lambda = 2.265$, *the invariant "circle" of the delayed logistic map* (15.6) *folds onto itself and becomes nonsmooth. For emphasis, a piece of the invariant curve near its tip is enlarged.*

so that the transformation and its inverse can be written as

$$\begin{pmatrix} z \\ \bar{z} \end{pmatrix} = P^{-1} \begin{pmatrix} x_1 \\ x_2 \end{pmatrix}, \qquad \begin{pmatrix} x_1 \\ x_2 \end{pmatrix} = P \begin{pmatrix} z \\ \bar{z} \end{pmatrix}.$$

In the $(z, \bar{z})$ coordinates, the map **f** in Eq. (15.24) transforms to

$$\begin{pmatrix} z \\ \bar{z} \end{pmatrix} \mapsto P^{-1} \mathbf{f} \left(P \begin{pmatrix} z \\ \bar{z} \end{pmatrix} \right) \equiv \begin{pmatrix} \psi(z, \bar{z}) \\ \overline{\psi}(z, \bar{z}) \end{pmatrix}.$$

Now, a short calculation yields that the function ψ has the form

$$\psi(z, \bar{z}) = \mu z + O\left(|z|^2\right);$$

the other function $\overline{\psi}$ is simply the conjugate of ψ and has the linear term $\overline{\mu}\overline{z}$. Thus it suffices to consider only one of the component functions, say, ψ, which is really a function of a scalar complex variable z. Indeed, this is the main advantage of working in complex coordinates: a real planar map becomes a scalar map of a complex variable.

Our next task is to simplify, or to eliminate, as many of the higher-order terms as possible in the Taylor series of ψ using successive complex transformations. The following lemma indicates the possible simplifications:

Lemma 15.33. *If $\mu^k \neq 1$ for $k = 1, 2, 3,$ or 4, then there is smooth change of variables bringing ψ into the form*

$$\psi(z, \overline{z}) = \mu z + \xi_{21} z^2 \overline{z} + \xi_{04} \overline{z}^4 + O\left(|z|^5\right), \qquad (15.28)$$

where ξ_{21} and ξ_{04} are complex numbers.

Proof. Let us begin by observing that all possible quadratic terms are absent in the normal form above. To show how to eliminate the quadratic terms, we consider a general Taylor series of the form

$$\psi(z, \overline{z}) = \mu z + \xi_{20} z^2 + \xi_{11} z\overline{z} + \xi_{02} \overline{z}^2 + O\left(|z|^3\right).$$

Now we try a change of variables of the form

$$Q(z, \overline{z}) = z + \gamma_{20} z^2 + \gamma_{11} z\overline{z} + \gamma_{02} \overline{z}^2$$

and determine appropriate coefficients to achieve our goal. Notice that near the origin this is an invertible transformation with inverse

$$Q^{-1}(z, \overline{z}) = z - \gamma_{20} z^2 - \gamma_{11} z\overline{z} - \gamma_{02} \overline{z}^2 + O\left(|z|^3\right).$$

A tedious computation shows that the transformed map $Q^{-1} \circ \psi \circ Q$ is given by

$$\begin{aligned}
Q^{-1} \circ \psi \circ Q(z, \overline{z}) = \mu z &+ \left(\xi_{20} - \mu\gamma_{20} + \mu^2\gamma_{20}\right) z^2 \\
&+ \left(\xi_{11} - \mu\gamma_{11} + \mu\overline{\mu}\gamma_{11}\right) z\overline{z} \\
&+ \left(\xi_{02} - \mu\gamma_{02} + \overline{\mu}^2\gamma_{02}\right) \overline{z}^2 + O\left(|z|^3\right).
\end{aligned}$$

We can remove all the quadratic terms by choosing

$$\gamma_{20} = \frac{\xi_{20}}{\mu(1 - \mu)}, \quad \gamma_{11} = \frac{\xi_{11}}{\mu(1 - \overline{\mu})}, \quad \gamma_{02} = \frac{\xi_{02}}{\mu - \overline{\mu}^2}.$$

These choices are possible because of our assumption that $\mu \neq 1$ and $\mu^3 \neq 1$, the latter one being necessary for the third choice.

The simplification of higher-order terms are accomplished in a similar manner. For the cubic terms, for instance, one considers a transformation of the form

$$Q(z, \bar{z}) = z + \sum_{p+q=3} \gamma_{pq} z^p \bar{z}^q,$$

which leaves the lower-order terms unchanged, and determines suitable coefficients where the remaining condition $\mu^4 \neq 1$ becomes necessary for success. We will spare you the tedious details. $\Diamond$

A routine computation shows that the complex normal form (15.28) yields the polar normal form (15.23) by letting $z = re^{-i\theta}$ and $\psi(z, \bar{z}) = \Re e^{-i\Theta}$. Now that we have indicated how to arrive at the normal form (15.23), you might get the impression that the rest of the proof of the Poincaré–Andronov–Hopf bifurcation theorem for maps should be easy. After all, the dynamics of the cubic part of the normal form in polar coordinates is easy to establish. Unfortunately, it is technically quite demanding to account for the effects of the higher-order terms.

In the paragraphs above we have explained the necessity of the somewhat mysterious requirement that the eigenvalues be different from the first four roots of unity. The question remains, however: what really happens if this condition is not met? The dynamics of such maps—*strong resonances*—can be exceedingly complicated and the answer is not yet completely known. We end this section with an example exhibiting a rather innocuous anomaly due to a strong resonance.

Example 15.34. *A strong resonance:* Consider the planar map depending on a parameter λ:

$$\begin{pmatrix} x_1 \\ x_2 \end{pmatrix} \mapsto \begin{pmatrix} x_2 \\ \lambda - x_1^2 \end{pmatrix}. \tag{15.29}$$

This map has a fixed point with coordinates

$$\bar{x}_1 = \bar{x}_2 = \tfrac{1}{2} \left(-1 + \sqrt{4\lambda + 1} \right).$$

When the parameter passes through $\lambda = 3/4$, it follows from linearization that this fixed point changes from a sink to a source while the eigenvalues leave the unit circle through $\pm i$. Consequently, the nonresonance condition in the Poincaré–Andronov–Hopf bifurcation theorem is not satisfied. Do we still get an invariant smooth closed curve encircling this fixed point as the eigenvalues cross the unit circle?

The answer to this question in the case of our example is delightfully simple. As λ increases through $\lambda = 3/4$, two orbits of period 4 bifurcate from the fixed point: a sink

$$(p_+, p_+) \to (p_+, p_-) \to (p_-, p_-) \to (p_-, p_+) \to (p_+, p_+)$$

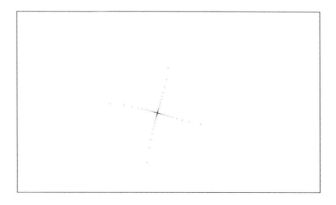

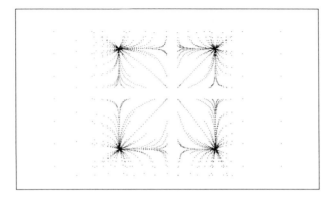

Figure 15.16. *A strong resonance in Eq.* (15.29) *as eigenvalues leave the unit circle at* $\pm i$. *The bifurcating invariant closed curve, the square, is not smooth, and the orbits on the invariant square are not simple rotations.*

and a saddle

$$(p_+, q) \to (q, p_-) \to (p_-, q) \to (q, p_+) \to (p_+, q),$$

where

$$p_\pm = \tfrac{1}{2}(1 \pm \sqrt{4\lambda - 3}), \qquad q = \tfrac{1}{2}\left(-1 + \sqrt{4\lambda + 1}\right).$$

The four horizontal and vertical line segments connecting the points of the periodic sink contain the saddle points, and form an invariant square which bifurcate from the fixed point. Thus, there is still an invariant closed curve—a square—encircling the fixed point, but the curve is no longer smooth; moreover, the dynamics on the square is not simply a rotation. This invariant square is illustrated in Figure 15.16. Further information on this system is contained in the exercises. ◊

Exercises

15.25. Verify that the canonical example (15.21) satisfies all the hypotheses of the Poincaré–Andronov–Hopf bifurcation theorem. Using the elaborate formula (15.25), revalidate the stability type of the invariant circle.

15.26. *A subcritical Poincaré–Andronov–Hopf bifurcation:* Analyze the bifurcations of the map

$$\begin{pmatrix} x_1 \\ x_2 \end{pmatrix} \mapsto \left(\lambda + x_1^2 + x_2^2\right) \begin{pmatrix} \cos\omega & -\sin\omega \\ \sin\omega & \cos\omega \end{pmatrix} \begin{pmatrix} x_1 \\ x_2 \end{pmatrix}.$$

Suggestions: Do this problem two ways. First, transform the map to polar coordinates and analyze its dynamics directly. Second, verify all the hypotheses of Theorem 15.31 and compute the sign of the cubic term in cartesian coordinates using formula (15.25).

15.27. *No Hopf in Hénon:* Show that the fixed points of the Hénon map (15.20) do not undergo a Poincaré–Andronov–Hopf bifurcation for any values of the parameters.

15.28. *On the invariant square:* Consider the map (15.29):

$$\begin{pmatrix} x_1 \\ x_2 \end{pmatrix} \mapsto \begin{pmatrix} x_2 \\ \lambda - x_1^2 \end{pmatrix}$$

from Example 15.34, where λ is a real parameter.
(a) Show that the second iterate of this map is the product system

$$\begin{pmatrix} x_1 \\ x_2 \end{pmatrix} \mapsto \begin{pmatrix} \lambda - x_1^2 \\ \lambda - x_2^2 \end{pmatrix}.$$

Up to a simple change of variables, each coordinate is really the logistic map. Determine now that each scalar map undergoes a period-doubling bifurcation at $\lambda = 3/4$.
(b) After translating the fixed point to the origin, show that at $\lambda = 3/4$, the map has the complex form

$$z \mapsto iz - \tfrac{3}{4}iz^2\bar{z} - \tfrac{1}{4}i\bar{z}^3 + O(|z|^5).$$

15.29. *A discrete predator-prey model:* Consider the following planar map depending on two parameters a and b:

$$\begin{pmatrix} x_1 \\ x_2 \end{pmatrix} \mapsto \begin{pmatrix} ax_1(1.0 - x_1) - x_1x_2 \\ (1/b)x_1x_2 \end{pmatrix}.$$

This map describes a discrete model of interactions between two species: a prey, x_1, and a predator, x_2. In the absence of the predator the population of the prey is governed by the logistic equation. Furthermore, it is assumed that each predator consumes a number of prey proportional to the abundance of that prey. For the growth rate of the predator, the assumption is that the number of offsprings produced by each predator is proportional to the number of preys it kills. For further information on this and other biological models, see Maynard-Smith [1968].

(a) Fix the parameter $b = 0.31$, and vary a, say from 2.0 to 3.6545, on the computer. Observe the birth of an attracting invariant closed curve through Poincaré–Andronov–Hopf bifurcation. Can you determine the precise value of the parameter a at which the bifurcation occurs?

(b) Observe that as the parameter a is increased by and by relatively far away from the bifurcation value the invariant closed curve becomes nonsmooth and eventually breaks up.

Help: This model is stored in the library of PHASER under the name *dispprey*. Be warned, however, that what happens to the invariant curve is rather complicated and the details are not easy to decipher by simple simulations on the machine.

15.30. *Symbolic computations:* Consider the delayed logistic map (15.26) and carry out all the necessary power series computations to arrive at the normal form (15.23) in polar coordinates.

Help: Computational demands of normal forms in bifurcation theory, especially in higher dimensions, can be formidable and it is best to resort to the machine. For such matters, see Hassard et al. [1980], and Rand and Armbruster [1987].

15.6. Area-preserving Maps

In this section, we consider a rather special class of maps: area-preserving maps. Despite their specialized nature, such maps play a prominent role in Hamiltonian mechanics. Indeed, several theorems we state below are some of the most spectacular results of this time-honored subject.

As the adjective suggests, an area-preserving map is a map that preserves the area of a planar region under the forward iterates of the map. Following our discussions in Section 8.5, we make the following definition:

Definition 15.35. *A map* $\mathbf{f} : \mathbb{R}^2 \to \mathbb{R}^2$ *is said to be area-preserving if*

$$\det D\mathbf{f}(\mathbf{x}) = 1 \tag{15.30}$$

at all points $\mathbf{x} \in \mathbb{R}^2$.

As our first example of area-preserving maps, let us reassemble the dynamics of linear area-preserving maps.

Example 15.36. *Linear area-preserving maps:* It is rather easy to determine if a planar linear map $\mathbf{x} \mapsto \mathbf{A}\mathbf{x}$ preserves area in terms of the

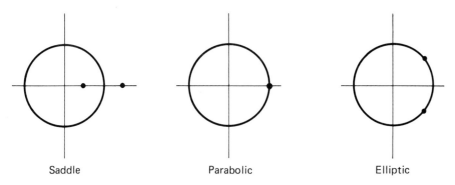

Saddle   Parabolic   Elliptic

Figure 15.17. *Three possible eigenvalue configurations of a linear area-preserving map relative to the unit circle on the complex plane.*

eigenvalues of $\mathbf{A}$. Indeed, $\mathbf{A}$ is area-preserving if the product of its eigenvalues μ_1 and μ_2 is 1:

$$\mu_1 \mu_2 = 1. \tag{15.31}$$

This observation follows, of course, from the fact that the derivative of a linear map is itself and that the determinant of a square matrix is the product of its eigenvalues.

The constraint (15.31) on the eigenvalues suggests three dynamically distinct possibilities to consider:

(i) *Saddle:* μ_1 and μ_2 are real, of the same sign, and $|\mu_1| < 1 < |\mu_2|$.

(ii) *Parabolic:* $\mu_1 = \mu_2 = 1$ or $\mu_1 = \mu_2 = -1$.

(iii) *Elliptic:* μ_1 and μ_2 are nonreal, $\mu_1 = \bar{\mu}_2$, and $|\mu_1| = |\mu_2| = 1$.

These three configurations of the eigenvalues with respect to the unit circle on the complex plane are depicted in Figure 15.17. Corresponding phase portraits can easily be inferred from Section 15.1. ◊

The main utility of the knowledge of linear systems is, of course, in deciphering the local dynamics of a fixed point of a nonlinear planar map through linearization. The saddle case (i) is hyperbolic and thus linearization suffices to conclude instability. The remaining two cases are nonhyperbolic and linearization is inconclusive. Let us choose to ignore the parabolic case since it is rather specialized. The elliptic case, however, is one of the most commonly encountered situations in conservative mechanical systems exhibiting some sort of stability. We now turn to a resolution of this very difficult case.

The stability of an elliptic fixed point of a nonlinear area-preserving map cannot be determined solely from linearization and the effects of the nonlinear terms in local dynamics must be accounted for. As you might suspect by now, this task is facilitated by simplifying the nonlinear terms through appropriate coordinate transformations—normal form. One possible reduction to a well-suited normal form is the content of the lemma below.

Lemma 15.37. (Birkhoff Normal Form) *Let* $\mathbf{f} : \mathbb{R}^2 \to \mathbb{R}^2$ *be an area-preserving* C^n *map with a fixed point at the origin whose complex eigenvalues* μ *and* $\bar{\mu}$ *are on the unit circle. If there is some integer* q *with*

$$4 \leq q \leq n+1$$

and the eigenvalues satisfy

$$\mu^k \neq 1 \quad \text{for} \quad k = 1, 2, \ldots, q,$$

then in a suitable complex coordinate system the map $\mathbf{f}$ *can be put into the normal form*

$$z \mapsto \mathbf{f}(z, \bar{z}) = \mu z \, e^{ia(z\bar{z})} + g(z, \bar{z}),$$

where

$$a(z\bar{z}) = a_1 |z|^2 + \cdots + a_s |z|^{2s}, \qquad s = \left[\frac{q}{2}\right] - 1,$$

is a real polynomial in $|z|^2$ *and the function* g *vanishes with its derivatives up to order* $q-1$ *at* $z = \bar{z} = 0$. *The square brackets denote the largest integer in* $q/2$. $\diamond$

Equipped with this normal form result, we are now ready to state our main stability theorem.

Theorem 15.38. (Stability of an Elliptic Fixed Point) *Let* $\mathbf{f}$ *be an area-preserving planar map with an elliptic fixed point at the origin satisfying the conditions in the lemma above. If the polynomial* $a\left(|z|^2\right)$ *does not vanish identically, then the origin is a stable fixed point.* $\diamond$

In the stability criterion above, the typical situation is that $q = 4$, $s = 1$, and $a_1 \neq 0$; it is interesting to notice that this is exactly the same nonresonance condition as in the Poincaré–Andronov–Hopf theorem of the previous section.

As we will not validate this stability theorem, let us explore its ramifications on a substantial example. In order not to burden you with extensive normal form computations, we will be content to explore the dynamics of the example below through numerical simulations.

Example 15.39. *Cremona map:* Consider the quadratic planar map

$$\begin{pmatrix} x_1 \\ x_2 \end{pmatrix} \mapsto \begin{pmatrix} x_1 \cos\lambda - [x_2 - x_1^2]\sin\lambda \\ x_1 \sin\lambda + [x_2 - x_1^2]\cos\lambda \end{pmatrix} \tag{15.32}$$

depending on a real parameter λ. It is easy to verify that this map is area-preserving with a fixed point at the origin.

A particularly interesting parameter value is $\lambda = 2\pi/3$. At this parameter value the origin is an elliptic fixed point in strong resonance: the eigenvalues are third roots of unity. In each frame of Figure 15.18, we have plotted several positive orbits near the origin for a different parameter value. First, the parameter λ is set to just a little less than $2\pi/3$; notice that the origin is a stable elliptic fixed point surrounded by closed invariant curves. At $\lambda = 2\pi/3$, the invariant curves shrink to the origin and the origin appears to be unstable. Finally, as λ is increased to a value that is a little larger than $2\pi/3$, the invariant curves are reborn and the origin is again stable.

Assuming the fact that the higher-order terms of the Birkhoff normal form are not identically zero, all these macroscopic observations are consistent with the theoretical results above. The microscopic details, however, are quite a bit subtler than they first appear, as we shall see momentarily. $\Diamond$

For technical reasons, we now restrict our attention to a special class of area-preserving maps defined in an annulus. More precisely, in polar coordinates consider an *annulus* defined by

$$a \leq r \leq b,$$

where $0 < a < b$, and an area-preserving map given by

$$\begin{pmatrix} r \\ \theta \end{pmatrix} \mapsto \begin{pmatrix} r \\ \theta + \gamma(r) \end{pmatrix}. \tag{15.33}$$

A map of the form (15.33) is called a *twist map* because such a map leaves each circle of constant radius invariant and the orbits are simple rotations on these circles. We will impose the further restriction

$$\frac{d\gamma}{dr} \neq 0$$

in the annulus so that the angle of rotation is not constant and depends on the radius.

From our investigations in Chapter 6 on circle maps, it is easy to describe the dynamics of a twist map (15.33): on each circle for which the rotation number is a rational number the orbits are periodic, otherwise they are dense on the circle. It is a remarkable fact that under small area-preserving perturbations some of these invariant circles persist.

Theorem 15.40. (Twist Theorem) *Consider in polar coordinates the following area-preserving perturbation of a twist map*

$$\begin{pmatrix} r \\ \theta \end{pmatrix} \mapsto \begin{pmatrix} r \\ \theta + \gamma(r) \end{pmatrix} + \varepsilon \mathbf{g}(\varepsilon, r, \theta) \tag{15.34}$$

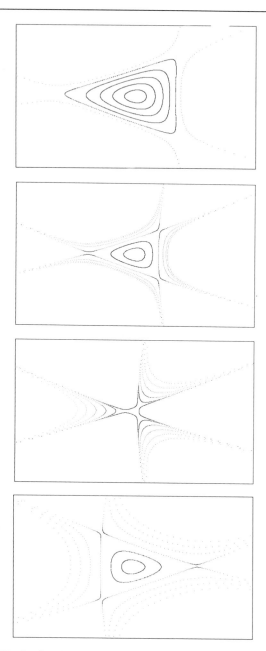

Figure 15.18. *In the vicinity of the origin of the Cremona map (15.32) as the parameter is increased by and by past* $\lambda = 2\pi/3$. *At* $\lambda = 2\pi/3$, *the origin is unstable.*

defined in an annulus $a \leq r \leq b$ such that $d\gamma/dr \neq 0$, the function $\mathbf{g}$ is C^5, and $|\varepsilon|$ is sufficiently small. Then, given any number ω between $\gamma(a)$ and $\gamma(b)$ incommensurable with 2π, and satisfying

$$\left| \frac{\omega}{2\pi} - \frac{p}{q} \right| \geq c|q|^{-5/2} \tag{15.35}$$

for all integers p and q, there exists a differentiable closed curve

$$\begin{aligned} r(\tau) &= G_1(\varepsilon, \tau) \\ \theta(\tau) &= \tau + G_2(\varepsilon, \tau) \end{aligned} \tag{15.36}$$

with G_1 and G_2 of period 2π in τ, which is invariant under the map (15.34). The positive orbits on the curve (15.36) are given by the simple rotation $\tau \mapsto \tau + \omega$. $\Diamond$

Verification of this theorem, which we will not attempt here, is exceedingly difficult. Some of its consequences, however, are easy to appreciate. The bizarre number theoretic condition (15.35) requires that ω must be an irrational number and must be badly approximated by the rationals. The geometric implication of this theorem is that such an invariant circle of the unperturbed twist map (15.33) survives small area-preserving perturbations, possibly deformed a bit. It is evident that there are infinitely many surviving concentric invariant closed curves, and that an orbit of the perturbed map remains between two such curves. From these observations, the stability result of Theorem 15.38 is now plausible by considering a small annulus about the origin.

What happens to the other invariant circles? Usually they break up, although the Twist Theorem says nothing about them. A disintegrating circle may give rise to periodic orbits and some of these periodic orbits can in turn be elliptic surrounded by invariant and disintegrating curves. This hierarchy can result in bewildering dynamical complexity, as seen in the Cremona map.

Example 15.41. *Cremona map continued:* For the purpose of illustrating the dynamical hierarchy described above, let us avoid strong resonances and consider the Cremona map (15.32) at the parameter value, say, $\cos \lambda = 0.24$. It is evident in Figure 15.19 that the origin is an elliptic fixed point surrounded by closed invariant curves. As we move outwards, these curves get distorted and finally break up. There appears an elliptic periodic orbit of period 5, the five small "islands," and an intermediate hyperbolic periodic orbit. Finally, we reach a "chaotic" region where there are no more visible surviving invariant curves.

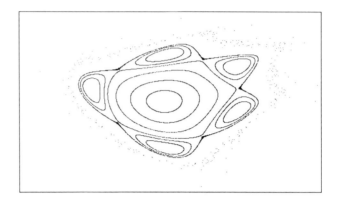

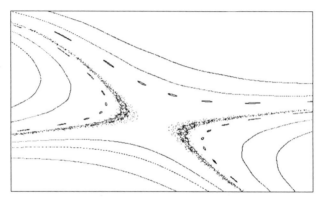

Figure 15.19. *Disintegration of invariant curves of the Cremona map away from the origin.*

If we examine the finer details, we expect to see a similar structure about each elliptic periodic point. This is substantiated in Figure 15.19 where we have enlarged a small region encompassing the far-right hyperbolic periodic point and parts of the neighboring two islands. About each new smaller island, the miniature replication of the larger structure continues.

The dynamical complexity described above is certainly in concert with the Twist Theorem. However, unlike the pictures, the theorem suggests the presence of the complex structure arbitrarily near the origin. This is not easily discernible from numerical simulations because very near the origin the effects of the nonlinear terms are almost negligible. Sufficiently far away from the origin the nonlinear terms act as a large enough perturbation to make their presence pronounced.

In a couple of pictures and a descriptive paragraph we have tried here to convey the incredible complexity of the global dynamics of the Cremona map, but the mathematical details are still fragmented. Nevertheless, you should feel the excitement on the computer by exploring the finer details for various parameter values. $\diamond$

As we have pointed out in the opening of the chapter, our findings about planar maps have ramifications for differential equations in the form of Poincaré maps. We now turn to an exploration of some of these in Part IV of our book.

Exercises ⎯⎯⎯⎯⎯⎯⎯⎯⎯⎯⎯⎯⎯⎯⎯⎯⎯⎯⎯⎯ ♣ ♡ ♠ ◇

15.31. *Fixed points of Cremona:* Show that the Cremona map (15.32) has another fixed point $(\bar{x}_1, \bar{x}_2)$, where

$$\bar{x}_1 = 2\tan(\lambda/2), \qquad \bar{x}_2 = 2\tan^2(\lambda/2).$$

Can you determine the stability type of this fixed point? This map also has many periodic orbits; up to period 4, they can be found explicitly; see Hénon [1969].
Aid: This map is stored in the library of PHASER under the name *cremona*.

15.32. *Inverse of Cremona:* Verify that the planar quadratic map

$$\begin{pmatrix} x_1 \\ x_2 \end{pmatrix} \mapsto \begin{pmatrix} x_1\cos\lambda + x_2\sin\lambda \\ -x_1\sin\lambda + x_2\cos\lambda + [x_1\cos\lambda + x_2\sin\lambda]^2 \end{pmatrix}$$

is the inverse of Cremona map (15.32). To follow full, not just positive, orbits of an invertible map on the computer an explicit expression for the inverse map is needed. You may wish to use the inverse map above while investigating the dynamics of the Cremona map.
Aid: This map is stored in the library of PHASER under the name *icremona*.

15.33. Consider the cubic planar mapping

$$\begin{pmatrix} x_1 \\ x_2 \end{pmatrix} \mapsto \begin{pmatrix} [x_1 + (x_2)^3]\cos\lambda - x_2\sin\lambda \\ [x_1 + (x_2)^3]\sin\lambda + x_2\cos\lambda \end{pmatrix}.$$

Verify that it is an area-preserving map. Observe on the computer that when $\sin 2\lambda \neq 1$ the origin is a stable fixed point.
Reference: A theoretical proof of the stability of the origin using the Twist Theorem, as well as a reason why this map is of any interest, is contained in Siegel and Moser [1971], p. 246.

15.34. *Area-preserving Hénon:* Consider once more the Hénon map

$$\begin{pmatrix} x_1 \\ x_2 \end{pmatrix} \mapsto \begin{pmatrix} 1 + x_2 - a(x_1)^2 \\ bx_1 \end{pmatrix}.$$

Fix $b = -1$ and observe that the map is area-preserving. Determine the ranges of the parameter a for which one of the fixed points is elliptic. Now explore, on the machine, when necessary, the dynamics in the proximity of the elliptic fixed point, looking for resonances, invariant curves, etc.

15.35. *A map on the torus:* Consider the map

$$\begin{pmatrix} x_1 \\ x_2 \end{pmatrix} \mapsto \begin{pmatrix} 2 & 1 \\ 1 & 1 \end{pmatrix} \begin{pmatrix} x_1 \\ x_2 \end{pmatrix} \pmod 1.$$

(a) Observe that the determinant of this map is 1; thus, it is a diffeomorphism and "area"-preserving.

(b) Show that $(0, 0)$ is the only fixed point by solving

$$2x_1 + x_2 = x_1 + m$$
$$x_1 + x_2 = x_2 + n,$$

where m and n are integers.

(c) Determine that the origin is a saddle, and compute its stable and unstable manifolds. You may want to do this on the plane and then identify the squares with integer coordinates. Do the stable and unstable manifolds intersect?

(d) Show that the point $(1/2, 1/2)$ is a periodic point of period 3. Find some other periodic points with different periods. In fact, show that its periodic points are dense in the unit square.

Notes: This famous map is stored in the library of PHASER under the name *anosov*. Also, see Arnold [1968 and 1983], Devaney [1986], and Smale [1966] for several important occasions in which *anosov* appears.

15.36. *Gingerman—a piecewise linear area-preserving map:*

$$\begin{pmatrix} x_1 \\ x_2 \end{pmatrix} \mapsto \begin{pmatrix} 1 - x_2 + |x_1| \\ x_1 \end{pmatrix}.$$

As we saw in the text, an elliptic fixed point of a sufficiently smooth area-preserving twist mapping is usually surrounded by infinitely many invariant circles; between these circles are annular zones of complexity. The piecewise linear map above exhibits such dynamics in the proximity of its elliptic fixed point at $(1, 1)$. Try various initial conditions.

Notes: This map is stored in the library of PHASER under the name *gingerman*. For further details, consult Devaney [1984].

15.37. *Mathieu's equation:* In light of your new knowledge of area-preserving maps, you may wish to return to Section 8.6 and reexamine the dynamics of the equation of Mathieu.

Note: Mathieu's equation is stored in the library of PHASER under the name *mathieu* in dimension three.

Bibliographical Notes _____ ⊚⌒⊚

Some of the specific maps described in the text have been the subject of many studies. The delayed logistic map is described in Maynard-Smith [1968] from a biological viewpoint. The finer details are elucidated in Pounder and Rogers [1980], and particularly in Aronson et al. [1980 and 1982]. The mathematical literature on the Hénon map is very extensive and still growing; while Hénon [1976] is, of course, the place to start, you should consult Benedicks and Carleson [1991] for an exciting new development. The Cremona map, suggested by Siegel and Moser [1971], is studied extensively by Hénon [1969].

The details of the local theory of hyperbolic fixed points, including the Hartman–Grobman theorem for maps (see Section 2.6), as well as stable and unstable manifolds are given in Hartman [1964], Irwin [1980], and Nitecki [1971]. The center manifold theorem for maps is described by Carr [1981] and Lanford [1973]. Poincaré [1892] remarked on the dynamical complexity associated with homoclinic points. Smale [1965 and 1967] and Šil'nikov [1967] showed the depth of this complexity; see also Newhouse's exposition in Guckenheimer et al. [1980]. A discussion of structural stability for diffeomorphisms, in the spirit of Chapter 13, is available in Palis and de Melo [1982].

In his review article, Ushiki [1986] describes some of the recent trends in numerical analysis of differential equations from the viewpoint of dynamical systems. Several representative references for "safe" approximations of, for example, hyperbolic periodic orbits, are Braun and Hershenov [1977], Beyn [1987], Eirola and Lorenz [1988], and Kloeden and Lorenz [1986]. The effects of floating point arithmetic in the Hénon map are addressed in Curry [1979] and Hammel et al. [1988]. Methods for computing invariant curves of maps are examined in van Veldhuisen [1988a].

The Poincaré–Andronov–Hopf bifurcation theorem for maps is of more recent origin than the one for differential equations. Indeed, contrary to the usual practice, it would be more appropriate to call it "invariant circle," or "Neimark–Sacker" bifurcation; see Neimark [1959], Sacker [1965], and Ruelle and Takens [1972]. The exposition in Lanford [1973] is particularly accessible and contains most of the necessary details; a general reference is Iooss [1979]. An account of the stability formula is given in Wan [1978]. Strong resonances were first investigated by Arnold [1983] and Takens [1974]; an exposition is contained in Whitley [1983]. The case of fourth roots of unity still remains unresolved.

The study of area-preserving maps was initiated by Poincaré and studied extensively by Birkhoff [1927]. The main reference for a proof of the Twist Theorem is Moser [1973]; see also Moser [1962 and 1967] and Siegel and Moser [1971]. More recent developments are described in a review by Moser [1986]. A comparable theorem was formulated for analytic Hamiltonians by Kolmogorov [1954] and proved by Arnold [1963]; expositions are in Arnold and Avez [1968] and Arnold [1978]. Consequently, these results are collectively called the KAM theory for **K**olmogorov, **A**rnold, and **M**oser. Long ago, Levi-Civita [1901] was concerned about strong resonances in mechanics; a newer account is Anosov and Katok [1972]. Even when an elliptic fixed point of an area-preserving map is stable so that there are invariant closed curves, the dynamics of the map between two such curves are usually quite complicated: there are homoclinic points, as shown in Zehnder [1973].

No exposition of planar maps would be complete without mentioning fractals; so we mention them. Some of the popular sources are Barnsley [1988] and Peitgen and Richter [1986].

$2\frac{1}{2}$+ D

16

Dimension
Two
and One Half

 In the final part of our book, we break the barrier of dimension two and venture into higher dimensions. Dynamical diversity in such dimensions is truly bewildering. Consequently, to bring our book to a conclusion in a finite number of pages, we attempt here to convey the current excitement of our subject with mere thumbnail sketches of several prominent examples. This chapter consists of abbreviated geometrical descriptions of two classical examples from the theory of forced oscillations: Van der Pol and Duffing. These nonautonomous planar systems contain a term that is a periodic function of time—hence the title of the chapter. Because of the time periodicity of the nonautonomous terms, the qualitative dynamics of these equations are studied most conveniently in the space $\mathbb{R}^2 \times S^1$. Indeed, since in this space both of these equations possess global Poincaré maps, the results from the previous chapter become the natural mathematical backdrop. In the chaotic behavior of Duffing's equation, the decisive role is played by transversal homoclinic points of its Poincaré map. We expound on this important connection by including a description of the dynamics of planar maps near such points.

16.1. Forced Van der Pol

As we have noted in Chapter 13, the oscillator of Van der Pol is structurally stable under small autonomous perturbations. Here, we sketch a geometric description of the changes in the structure of the solutions when the oscillator is subjected to nonautonomous periodic perturbations.

Let us consider the nonautonomous system

$$\dot{x}_1 = x_2$$
$$\dot{x}_2 = -x_1 + (1 - x_1^2)x_2 + \lambda f(t), \tag{16.1}$$

where f is a T-periodic function of the independent variable t, and λ is a real parameter. The term $\lambda f(t)$ is called the *forcing function*. When $\lambda = 0$, there is no forcing and the system (16.1) is the oscillator of Van der Pol.

As we pointed out in Chapter 4, it is necessary to examine the trajectories (x_1, x_2, t) of the nonautonomous system (16.1) in $\mathbb{R}^2 \times \mathbb{R}$ rather than the orbits in $\mathbb{R}^2$. Equivalently, we may consider the orbits of the three-dimensional autonomous system

$$\dot{x}_1 = x_2$$
$$\dot{x}_2 = -x_1 + (1 - x_1^2)x_2 + \lambda f(x_3) \tag{16.2}$$
$$\dot{x}_3 = 1.$$

In this setting, let us first re-examine the case of $\lambda = 0$: the limit cycle, the isolated periodic orbit, of the unforced oscillator of Van der Pol becomes a cylinder; that is, topologically it is homeomorphic to $S^1 \times \mathbb{R}$. This cylinder is an invariant manifold in the sense that any solution starting on the cylinder remains on it for all positive time. Moreover, this invariant cylinder attracts all nearby solutions. For $\lambda = 0$, the invariant cylinder is filled with a family of periodic solutions, as seen in Figure 16.1a. Of course, the cylinder under the projection $\mathbb{R}^2 \times \mathbb{R} \to \mathbb{R}^2$ simply becomes the limit cycle.

Now, let us consider the case of periodic forcing with small amplitude, that is, $|\lambda|$ small. In this case, there is still a cylinder in $\mathbb{R}^2 \times \mathbb{R}$ close to the invariant cylinder of the unforced oscillator. This new cylinder is again an invariant manifold of solutions of the forced equation (16.1) and attracts all nearby solutions. However, the flow on the invariant cylinder of the forced equation can be quite different from the one of the unforced oscillator. We have plotted in Figure 16.1b several solutions of the forced oscillator (16.1) with the forcing function $0.2 \cos 3t$.

Next, we would like to view the solutions of Eq. (16.1) in a different setting which offers several distinct advantages. Since the forcing function

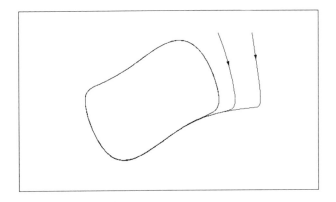

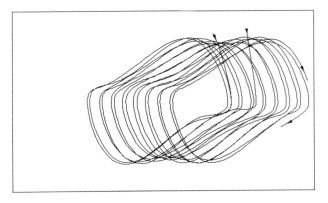

Figure 16.1. (a) *Solutions of the unforced oscillator of Van der Pol are attracted to an invariant cylinder in* $\mathbb{R}^2 \times \mathbb{R}$. (b) *The invariant cylinder persists under periodic forcing with small amplitude. The forcing term is* $0.2 \cos 3t$.

is periodic in time with period T, the solutions are translation invariant in time by a distance of T, as we have observed in Section 4.2. Consequently, by identifying the time dimension at $t = 0$ with $t = T$, we can view the solutions of Eq. (16.1) in $\mathbb{R}^2 \times S^1$. Equivalently, we may consider the orbits of the system

$$\dot{x}_1 = x_2$$
$$\dot{x}_2 = -x_1 + (1 - x_1^2)x_2 + \lambda f(x_3) \qquad (16.3)$$
$$\dot{x}_3 = 1 \quad \text{mod } T.$$

With the identification above, the invariant cylinder of the system (16.2) now becomes diffeomorphic to a torus: $S^1 \times S^1$. This torus is an invariant

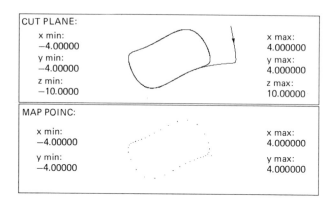

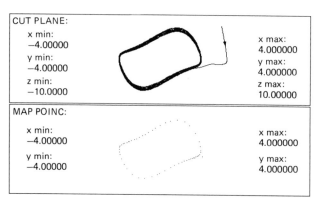

Figure 16.2. *An orbit and its Poincaré section of the oscillator of Van der Pol: (a) Unforced and (b) with forcing term* 0.3 cos 3t.

manifold of the forced oscillator (16.3) and attracts all nearby solutions. For $\lambda = 0$, the flow on this torus is qualitatively a parallel flow which we studied in Section 6.2. If $\lambda \neq 0$ with $|\lambda|$ is small, the torus still persists but the flow on it usually is not a parallel flow and may have orbitally asymptotically stable and unstable periodic orbits.

The qualitative dynamics of Eq. (16.3) that we have just described can be recast in terms of a Poincaré map; see Figure 16.2. To this end, we take the plane L consisting of the points $(x_1, x_2, 0)$ as our *cross section* which solutions pierce transversally in view of the last component of Eq. (16.3). The Poincaré map Π is then defined by following the solution through $(x_1, x_2) \in L$ for time T when the solution next intersects L, that is,

$$\Pi : L \to L; \qquad (x_1, x_2, 0) \mapsto (\varphi(T, x_1, x_2, 0), 0),$$

where $\varphi(T, x_1, x_2, 0)$ is the solution of Eq. (16.3) through $(x_1, x_2, 0)$ at $t = 0$. Now, the global dynamics of Eq. (16.3) have the following interpretation. For $|\lambda|$ small, the equilibrium point at the origin becomes an unstable fixed point of the Poincaré map Π. This fixed point is very near the origin. The iterates of all other points converge to an invariant closed curve encircling the fixed point; see Figure 16.2. This invariant curve is, of course, a cross section of the invariant torus. Consequently, the asymptotic dynamics of periodically forced oscillations with small amplitude forcing term can be captured in terms of the dynamics of a circle diffeomorphism that we have studied in Chapter 6. When $\lambda = 0$, the circle diffeomorphism is a simple rotation with no fixed points. For $|\lambda|$ small, however, it can have asymptotically stable and unstable fixed points.

   If the amplitude of periodic forcing is large, then the dynamics of the periodically forced Van der Pol's oscillator (16.1) gets rather complicated as the map on the invariant closed curve ceases to be a diffeomorphism. We refrain from delving further into these intricacies.

Exercises _____ ♣♡♠♢

16.1. *Forced Van der Pol on PHASER:* The oscillator of Van der Pol with the forcing term $\lambda \cos \omega t$ is stored in the 3D library of PHASER under the name *forcevdp*. Study large amplitude forcing and observe the dynamics on the invariant curve of the Poincaré map. Also, investigate the effect of the forcing frequency ω.

16.2. Forced Duffing

In this section, we outline the dynamics of a periodically forced conservative system, the equation of Duffing from Example 14.5. More specifically, we consider the system

$$\begin{aligned}
\dot{x}_1 &= x_2 \\
\dot{x}_2 &= x_1 - x_1^3 + \lambda f(t)
\end{aligned} \tag{16.4}$$

for parameter values $|\lambda|$ small, where the forcing function $f(t)$ is a periodic function with period T. We will examine the dynamics of Eq. (16.4) on $\mathbb{R}^2 \times S^1$, as we have done in the case of the periodically forced Van der Pol's equation in the preceding section. Equivalently, we will investigate the orbits of the system

$$\begin{aligned}
\dot{x}_1 &= x_2 \\
\dot{x}_2 &= x_1 - x_1^3 + \lambda f(x_3) \\
\dot{x}_3 &= 1 \qquad \mod T.
\end{aligned} \tag{16.5}$$

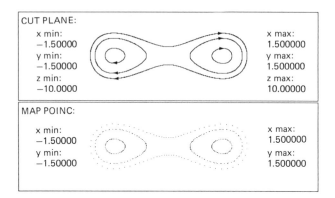

Figure 16.3. *Solutions and Poincaré map of Duffing's equation without forcing. The map is area-preserving.*

Let us begin (without force) by setting $\lambda = 0$. As depicted earlier in Figure 14.2, the planar phase portrait of the unforced Duffing consists of two centers, one at each equilibrium $(-1, 0)$ and $(1, 0)$, and a saddle point at $(0, 0)$ whose stable and unstable manifolds form a figure eight outside of which all orbits are again periodic. When viewed in $\mathbb{R}^2 \times S^1$, each periodic orbit of the unforced Duffing's equation becomes a torus and each equilibrium point corresponds to a periodic solution.

It is most convenient to study the dynamics of Eq. (16.4) in $\mathbb{R}^2 \times S^1$ with the aid of a Poincaré map. It is evident that for $\lambda = 0$, a Poincaré map of the unforced Duffing is area-preserving. In the terminology of Section 15.6, each center at $(-1, 0)$ and $(1, 0)$ corresponds to an elliptic fixed point of the Poincaré map surrounded by closed invariant curves which are the cross sections of the tori nearby; see Figure 16.3. Evidently, a Poincaré map is not only area-preserving but also a twist map in an annulus surrounding each elliptic fixed point.

Next, let us consider what may happen in the presence of periodic forcing with small amplitude. It is true that a Poincaré map is still an area-preserving map for $\lambda \neq 0$. Therefore, when $|\lambda|$ small, the celebrated Twist Theorem, Theorem 15.40, guarantees the existence of many closed invariant curves for the Poincaré map. Some of these invariant curves are visible in Figure 16.4. The dynamics between two closed invariant curves may be rather complicated. Rest assured, however, that all orbits are bounded for $|\lambda|$ small.

You have surely noticed that so far we have been avoiding the fate of the homoclinic loops under small perturbations. For $\lambda = 0$, the Poincaré map has a homoclinic fixed point at the origin, as in Definition 15.21. When $|\lambda| \neq 0$ but small, there is now a transversal homoclinic point in the vicinity

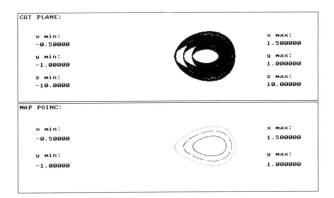

Figure 16.4. *Closed invariant curves surrounding an elliptic fixed point of a Poincaré map of the Duffing's equation with forcing* $0.2 \cos t$.

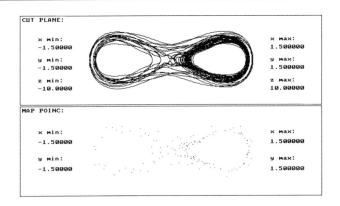

Figure 16.5. *An orbit of a Poincaré map of the periodically forced Duffing equation near the homoclinic orbit. The cross section of the orbit does not appear to lie on a closed invariant curve which indicates that the orbit is not periodic or quasiperiodic. The initial values are* $(0, 0)$ *and the forcing term is* $0.2 \cos t$, *as in the previous figure.*

of the origin and the dynamics near the homoclinic loops of the unforced equation becomes rather complicated, as illustrated in Figure 16.5. In order to describe this complexity in a bit more detail, we now turn to a general exposition of orbits near a transversal homoclinic point of planar maps.

Exercises ——————————————————————————— ♣ ♡ ♠ ◇

16.2. *Forced linear oscillator:* In the spirit of this section, investigate the dynamics of the periodically forced linear harmonic oscillator $\ddot{y} + y = \lambda \cos \omega t$ as a function of the forcing amplitude λ and the forcing frequency ω.

16.3. *Forced Duffing on PHASER:* The equation of Duffing with the periodic forcing term $\lambda \cos \omega t$ is stored in the 3D library of PHASER under the name *forceduf.* First, try to reproduce the figures in the text. Then, experiment with various values of the forcing amplitude and frequency. Also, try some initial data far away from the origin. Can you get any unbounded solutions even for relatively large forcing amplitudes?

16.3. Near a Transversal Homoclinic Point

Many of the chaotic motions that are observed in dynamical systems are intimately associated with the presence of transversal homoclinic points of maps. In this section, we provide specific information about the intricate dynamics near such points.

We begin by introducing an abstract dynamical system whose ubiquity near transversal homoclinic points will be the main assertion of this section. Let $A = \{0, 1\}$ be the set consisting of two elements 0 and 1; the set A is referred to as the *symbol set.* Let S be the set of all bi-infinite sequences of the form

$$s = \{ \ldots s_{-n} \ldots s_{-2} \, s_{-1} . s_0 \, s_1 \, s_2 \ldots s_n \ldots \}$$

whose entries are chosen from the symbol set A. The set S can be endowed with a metric to make it a topological space. Indeed, it can be verified that

$$d(s, \bar{s}) = \sum_{n=-\infty}^{+\infty} \frac{\delta_n}{2^{|n|}}, \quad \text{where } \delta_n = \begin{cases} 0 & \text{if } s_n = \bar{s}_n \\ 1 & \text{if } s_n \neq \bar{s}_n, \end{cases}$$

defines a distance between two bi-infinite sequences s and $\bar{s}$. In words, two sequences are "close" if they agree on a sufficiently long central block.

We next define a map $\sigma : S \to S$ by

$$\sigma(s)_n = s_{n+1}.$$

This map is a homeomorphism of S and it is called, appropriately, the *shift map* because it simply shifts the entries of a sequence by one place to the right.

For the convenience of labeling the orbits of σ, let us agree to use an overline to denote the repeating segment of a bi-infinite sequence; for example, $\{ \ldots 010101.01010101 \ldots \}$ is denoted by $\{ \overline{01.01} \}$. With this convention, it is evident that σ has two fixed points $\{ \overline{0.0} \}$ and $\{ \overline{1.1} \}$. Also, it is easy to see that σ has many periodic orbits; for example, here is a periodic orbit of period 2:

$$\{ \overline{01.01} \} \mapsto \{ \overline{10.10} \} \mapsto \{ \overline{01.01} \}.$$

Indeed, the shift map σ has an amazing variety of orbits, as listed in the following lemma.

Lemma 16.1. *The shift map on the space of bi-infinite sequences with two symbols has*

- *a countably infinite number of periodic orbits, including periodic orbits of arbitrarily high period;*
- *an uncountably infinite number of nonperiodic orbits, including countably many homoclinic and heteroclinic orbits;*
- *a dense orbit.* ◇

Now, we state the main result of this section which asserts that the dynamics in a neighborhood of a transversal homoclinic point is at least as complicated as that of the shift map.

Theorem 16.2. *Let* $\Pi : \mathbb{R}^2 \to \mathbb{R}^2$ *be a planar diffeomorphism with a transversal homoclinic point* **q**. *Then, in any neighborhood of* **q**, *the map* Π *has a hyperbolic invariant set on which the iterate* Π^n, *for some positive integer* n, *is topologically equivalent to the shift map on two symbols.* ◇

Since topological equivalence preserves qualitative features of orbits, on the invariant set, there are many fixed points, periodic, nonperiodic, homoclinic, and heteroclinic orbits. An equally important fact is the hyperbolicity of the invariant set. Although we have not defined hyperbolicity of a general invariant set, the implication is that the invariant set persists under small perturbation of the map.

The actual dynamics near a transversal homoclinic point is in fact more complicated than that of the shift map on two symbols. Further generalizations of shift maps on bi-infinite sequences with infinite number of symbols are needed to account for the full dynamics. Instead of pursuing such extensions, we now turn to an investigation of the creation of transversal homoclinic points in a specific system.

Exercises  ♣ ♡ ♠ ◇

16.4. *For the shift map* σ : Establish the following for the shift map on two symbols.

    (a) List all possible period 3 orbits.

    (b) When an irrational number in the interval $[0, 1]$ is expressed in base 2, it yields a non-repeating sequence of 0 and 1. Using the fact that the set of irrational numbers in the unit interval is uncountable, prove that σ has an uncountable number of nonperiodic orbits.

    (c) What are the stable and unstable sets of the fixed point $\{\overline{0.0}\}$?

    (d) Construct a homoclinic orbit to $\{\overline{0.0}\}$.

    (e) Construct a heteroclinic orbit to $\{\overline{0.0}\}$ and $\{\overline{01.01}\}$.

16.4. Forced and Damped Duffing

As we have just seen, the existence of a transversal homoclinic point of
a planar map leads to very complicated behavior of orbits nearby. Such
dynamical complexity is often dubbed as chaos. In this section, we numerically explore the onset of chaos in the balancing act of forcing and damping
of the equation of Duffing.

In a family of planar maps depending on parameters, the onset of chaos
typically occurs at the parameter values for which the stable and unstable
manifolds of a saddle point come into contact tangentially—homoclinic
tangency. As we have drawn schematically in Figure 16.6, there is no
homoclinic point before such a parameter value, and there are transversal
homoclinic points afterwards. This scenario for the creation of transversal
homoclinic points and the accompanying complicated dynamics can indeed
be established for the forced damped Duffing equation

$$\dot{x}_1 = x_2$$
$$\dot{x}_2 = x_1 - x_1^3 - \mu x_2 + \lambda f(t). \tag{16.6}$$

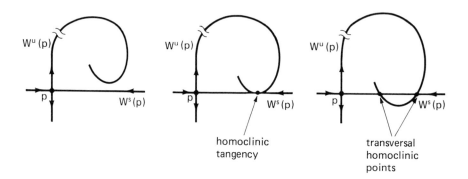

homoclinic
tangency

transversal
homoclinic
points

Figure 16.6. *Creation of transversal homoclinic points through a homoclinic tangency.*

Let us first quickly note that the dynamics of Eq. (16.6) with damping
but no forcing, that is, when $\mu > 0$ and $\lambda = 0$. In this case, the ω-limit set of
each orbit is one of the three equilibria $(-1, 0)$, $(0, 0)$, or $(1, 0)$. Following
our previous convention, let us now view the solutions of Eq. (16.6) as the
orbits of

$$\dot{x}_1 = x_2$$
$$\dot{x}_2 = x_1 - x_1^3 - \mu x_2 + \lambda f(x_3) \tag{16.7}$$
$$\dot{x}_3 = 1 \mod T$$

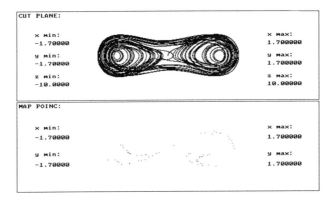

Figure 16.7. *Passage through homoclinic tangency in the damped forced Duffing equation. Damping is fixed at $\mu = 0.2$ and forcing is $\lambda = 0.2$ and $\lambda = 0.25$. The initial data for the computed orbit is* $(0, 0, 0)$.

in the space $\mathbb{R}^2 \times S^1$. Then, the equilibrium point $(0, 0)$ corresponds to a fixed point of saddle type of the Poincaré map while the other two equilibria are represented by asymptotically stable fixed points.

Now, let us fix the damping parameter μ at some positive value. If $|\lambda|$ is sufficiently small, then the dynamics of the Poincaré map remains the same and the ω-limit set of each orbit is one of the hyperbolic fixed points. As we increase the forcing parameter λ, however, there is a value of the parameter λ for which the stable and the unstable manifolds of the saddle point have a homoclinic tangency. When we increase the amplitude of the forcing term further, transversal homoclinic points are created and the associated complicated dynamics sets in. As depicted in Figure 16.7, this mode of transition to chaos in the forced damped Duffing equation (16.7) that we have just outlined is indeed visible in the numerically computed orbits and their Poincaré sections.

It is true that the forced damped Duffing has a hyperbolic invariant set with complicated dynamics as we have described in Theorem 16.2. However, the invariant set provided by this theorem is not an attracting set. Consequently, it is not certain that the computed trajectories typically turn out to be chaotic. Indeed, it is possible to observe complicated transient behavior which ultimately turns out to be asymptotically periodic motion. Presently, determination of the attracting invariant set of the damped forced Duffing remains an open problem.

Exercises _____

16.5. *Damped Duffing is dissipative:* Prove that the equation of Duffing is dissipative if there is positive damping but no forcing. Identify the global attractor.

16.6. *Duffing on PHASER:* The equation of Duffing with damping and periodic forcing term $\lambda \cos \omega t$ is stored in the 3D library of PHASER under the name *forceduf*. In Figure 16.7, we implied that for $\mu = 0.2$, there is a homoclinic tangency for some value of λ in the interval [2.0, 2.5]. Try to locate the value of λ for which the orbit through the origin first becomes complicated. Also, investigate the effect of the forcing frequency on the dynamics.

16.7. *Two attractors:* For the parameter values $\mu = 0.3$ and $\lambda = 0.15$, the forced damped Duffing has both a large stable periodic orbit and a more complicated attractor. Find them on PHASER. Try the initial data (1.8, 0.0, 0.0) for the periodic orbit.

Bibliographical Notes _____

There is an extensive theory of invariant manifolds of nonautonomous differential equations which are perturbations of autonomous equations with a "normally hyperbolic" set. The existence of invariant tori for the periodically forced Van der Pol was first proved by Krylov and Bogoliubov in 1938, an exposition of which is given in Bogoliubov and Mitropolski [1961] and Hale [1961]. Perhaps unaware of their work, Levinson [1950] obtained similar results. More recent work on such invariant manifolds can be found in, for example, Hale [1980], or Hirsch, Pugh, and Shub [1977].

Beginning with Van der Pol and Van der Mark [1927], several variations of the oscillator of Van der Pol with possibly large amplitude forcing have received much attention. It was noted in Cartwright and Littlewood [1945] that in some parameter ranges there were two periodic solutions with different periods. More importantly, they discovered a family of solutions with "random" behavior. Related sources on this subject are Levinson [1949], Levi [1981], and Guckenheimer and Holmes [1983].

The original reference to Duffing's Equation seems to be Duffing [1918]. The existence of the invariant curves of Duffing's equation with small amplitude forcing was established by Moser [1961] as one of the first illustrations of the famous KAM theory. The complicated behavior near the homoclinic orbit was considered by McGehee and Meyer [1974].

Poincaré was well aware of the complexity of dynamics associated with transversal homoclinic orbits. Birkhoff proved that there must be infinitely many periodic points nearby. Smale [1963 and 1967] gave the result in the text relating the dynamics to the shift map with a finite number of symbols. He also introduced the *horseshoe map* as an abstract prototype of the dynamics that should be prevalent near a transversal homoclinic orbit. A variant of this theory for an infinite number of symbols is in Moser [1973]. Sil'nikov [1967] classified the types of orbits that remain in a neighborhood of a transverse homoclinic orbit. A detailed exposition of related topics and some applications are given in Wiggins [1988].

Some of the deeper properties of homoclinic tangencies of planar diffeomorphisms are investigated from an abstract point of view by Newhouse [1979]; see also Gavrilov and Silnikov [1972 and 1973], Newhouse in Guckenheimer, Newhouse, and Moser [1982], Kan, Koçak, and Yorke [1990], and Robinson [1983]. A specific criterion for the existence of a transversal homoclinic orbit for the forced damped Duffing equation was given by Melnikov [1963]. A detailed discussion of the creation of a homoclinic tangency for the same equation is provided by Chow, Hale, and Mallet-Paret [1980]. Further generalizations of these methods are available in, for example, Palmer [1984] and the books of Guckenheimer and Holmes [1983] and Wiggins [1988].

17

Dimension Three

 In this chapter, we introduce four vector fields to illustrate selected highlights from three-dimensional dynamics and bifurcations. In the first example, a periodic orbit of a vector field in $\mathbb{R}^3$ yields its stability to another periodic orbit of approximately twice the period. In the second example, as a periodic orbit becomes unstable, an invariant torus appears nearby. Using an appropriate Poincaré map, these two local bifurcations are, respectively, the counterparts of the period-doubling and Poincaré–Andronov–Hopf bifurcations of a fixed point of a planar map from Chapter 15. The third example illustrates an important source of chaotic dynamics other than successive period-doubling bifurcations, a special type of homoclinic orbit—Sil'nikov orbit or saddle-focus. The final example features the Lorenz equations, the strange attractor most photographed.

17.1. Period Doubling

Dynamics of a differential equation in $\mathbb{R}^3$ near one of its periodic orbits can be reduced to the local dynamics of a planar diffeomorphism near a fixed or periodic orbit. In this section, we first briefly indicate how such a reduction is accomplished. Then, we illustrate the consequences for a periodic orbit of a differential equation in $\mathbb{R}^3$ when the corresponding fixed point of the map undergoes a period-doubling bifurcation as described in Section 15.4.

The planar reduction of dynamics near a periodic orbit in $\mathbb{R}^3$ is analogous to the construction of a Poincaré map as we have described in Section 12.2. Let Γ be a periodic orbit of a vector field in $\mathbb{R}^3$ passing through a point $\mathbf{p}$. Also, let L be a planar section through the point $\mathbf{p}$ and transverse to Γ. Then, the orbit through a point $\mathbf{x}$ in L which is sufficiently close to $\mathbf{p}$ will return to L. Therefore, we can define a planar map $\Pi : L \to L$ by letting the image of $\mathbf{x}$ be the first point at which the orbit through $\mathbf{x}$ pierces L in the same direction as Γ. The planar map Π is called the *Poincaré map* and it is a local diffeomorphism on L. The point $\mathbf{p}$ is a fixed point of Π and the dynamics of the vector field near Γ is determined from the properties of the Poincaré map near the fixed point $\mathbf{p}$.

Using a Poincaré map, the bifurcations of a nonhyperbolic fixed point of a planar map from Section 15.4 can now be interpreted as bifurcations of periodic orbits of differential equations in $\mathbb{R}^3$. Here are two such interpretations.

If one of the eigenvalues of the linearization $D\Pi(\mathbf{p})$ of the Poincaré map Π at the fixed point $\mathbf{p}$ is 1 and the other eigenvalue has absolute value different than 1, then under perturbation there can be a saddle-node bifurcation. The implication for the vector field is the coalescing and disappearance of two periodic orbits. This is similar to the bifurcation we have already encountered in Section 12.3 and thus we will not pursue it.

The second situation is more interesting as it cannot occur in planar vector fields. If one of the eigenvalues of the linearization of the Poincaré map $D\Pi(\mathbf{p})$ is -1 and the other has absolute value different than 1, then the nonhyperbolic fixed point $\mathbf{p}$ of Π usually undergoes a period-doubling bifurcation when the vector field is subjected to perturbations. The corresponding bifurcation in the flow of the differential equations in $\mathbb{R}^3$ is the appearance of a periodic orbit near Γ with a period approximately twice that of Γ. This important bifurcation is readily observable (see Figure 171.1) in the numerical simulations of the following three-dimensional vector field:

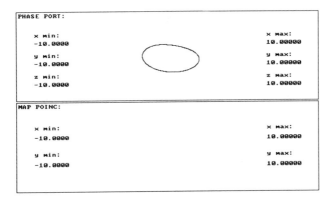

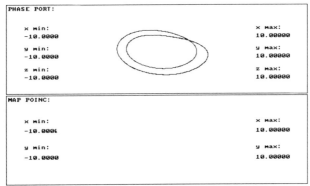

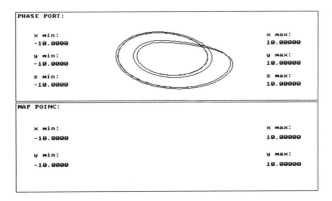

Figure 17.1. *Period doubling in three dimensions: Successive doubling of the period of an asymptotically stable periodic orbit (limit cycle) and a Poincaré map for Example 17.1. The parameter* λ *is varied through 2.2, 3.1, and 3.95. The initial point for all parameter values is* (2, 2, 0). *To discard the transients, the orbits are plotted for time 50 to 80.*

Example 17.1. *Period doubling in* $\mathbb{R}^3$: Consider the three-dimensional system of equations

$$\dot{x}_1 = -(x_2 + x_3)$$
$$\dot{x}_2 = x_1 + 0.2x_2 \qquad\qquad (17.1)$$
$$\dot{x}_3 = 0.2 + x_3(x_1 - \lambda)$$

depending on the scalar parameter λ. It is, of course, theoretically not possible to locate a periodic orbit, let alone a Poincaré map, of this system. Numerically, however, it is a routine matter to see that there is an orbitally asymptotically stable periodic orbit for $\lambda = 2.2$. For $\lambda = 3.1$, this periodic orbit becomes unstable and a new stable periodic orbit of twice the period appears. As the parameter λ is increased further, a periodic orbit of period 4 appears, and the process continues. In Figure 17.1, we have plotted the periodic orbit and successive period-doubling bifurcations. ◇

Exercises ———————————————————————— ♣♡♠◇

17.1. *On Example 17.1:* This system is from Rössler [1976] and stored in the library of PHASER under the name *rossler*. Investigate the fate of the periodic orbit as you gradually increase the parameter to, say, $\lambda = 5.7$. Observe, in particular, how successive period-doubling bifurcations give rise to a "strange attractor."

17.2. *Period doubling in Duffing:* Period doubling occurs in forced damped oscillators also. For the damping coefficient $\mu = 0.22$ and the forcing coefficient $\lambda = 0.3$, the equation of Duffing from Section 16.4 has a stable periodic orbit of period 3. On PHASER, use the initial data (1.4, 0) and discard transient to find the periodic orbit. Next, take $\mu = 0.222$ to double the period of the orbit.

17.2. Bifurcation to Invariant Torus

We saw in Section 15.5 that bifurcation of a nonhyperbolic fixed point of a planar map with complex eigenvalues of unit modulus is usually accompanied with an invariant closed curve surrounding the fixed point—the Poincaré–Andronov–Hopf bifurcation. When we consider a Poincaré map near a periodic orbit of a vector field in $\mathbb{R}^3$, the implications of this bifurcation is that the nonhyperbolic periodic orbit corresponding to such a fixed point of the Poincaré map is usually accompanied with an *invariant torus*. In this section, we exhibit the birth of an attracting invariant torus in a specific vector field in dimension three.

Example 17.2. *Birth of an invariant torus in* $\mathbb{R}^3$: Consider the following three-dimensional vector field:

$$\dot{x}_1 =(\lambda - b)x_1 - cx_2 + x_1\big[x_3 + d(1.0 - x_3^2)\big]$$
$$\dot{x}_2 =cx_1 + (\lambda - b)x_2 + x_2\big[x_3 + d(1.0 - x_3^2)\big] \qquad (17.2)$$
$$\dot{x}_3 =\lambda x_3 - (x_1^2 + x_2^2 + x_3^2),$$

where λ is a parameter, and b, c, and d are constants which we will fix below. These equations represent an unfolding of the equilibrium at the origin with a pair of purely imaginary and a zero eigenvalue. Here, we will not attempt to elucidate the dynamics of this system in its full generality; rather, we will be content to numerically illustrate only the birth of an invariant torus. For this purpose, we set $b = 3.0$, $c = 0.25$, and $d = 0.2$, while varying the parameter λ.

For $\lambda > 0$ and small, there is an asymptotically stable equilibrium point, with a positive x_3-coordinate near the origin. At approximately $\lambda \approx 1.68$, this equilibrium point becomes unstable and undergoes a Poincaré–Andronov–Hopf bifurcation. The resulting periodic orbit is hyperbolic and orbitally asymptotically stable.

At $\lambda = 2.0$, the periodic orbit is no longer hyperbolic, but is still orbitally asymptotically stable due to the nonlinear terms. For $\lambda > 2.0$, the periodic orbit becomes unstable and an attracting invariant torus appears near the periodic orbit. As the parameter is increased past 2.0, the invariant torus grows quite rapidly. This sequence of bifurcations is illustrated in Figure 17.2. $\Diamond$

Exercises _____ ♣ ♡ ♠ ◇

17.3. *On Example 17.2:* This vector field is from Langford [1985].
  (a) Identify the bifurcation of the origin at $\lambda = 0$.
  (b) Show that one of the equilibria resulting from the bifurcation above undergoes a Poincaré–Andronov–Hopf bifurcation at $\lambda \approx 1.68$.
  (c) Verify numerically that for $\lambda = 2.0$, the nonhyperbolic periodic orbit is still attracting.
  (d) Increase λ further and investigate the fate of the invariant torus.
  Help: Example 17.2 is stored in the library of PHASER under the name *zeroim*.

17.3. Sil'nikov Orbits

We saw in Sections 16.3 and 16.4 that chaotic dynamics occurs for a Poincaré map when the stable and unstable manifolds of a hyperbolic fixed point intersect transversally. It is also possible for chaotic behavior to be present in a flow in $\mathbb{R}^3$ from the mere existence of a homoclinic orbit of a special type—Sil'nikov orbit or saddle focus. In this section, we outline this important source of dynamical complications.

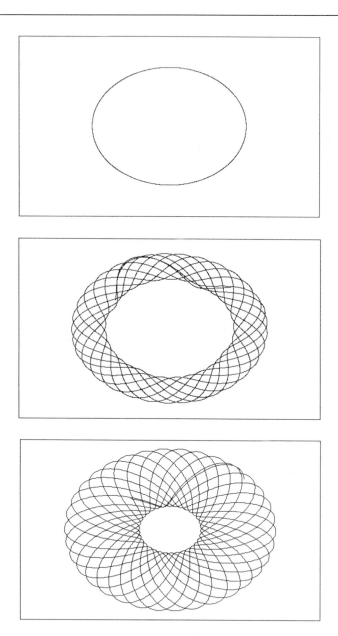

Figure 17.2. *Birth of an invariant torus:* Example 17.2 has an asymptotically stable limit cycle for $\lambda = 1.95$, which looses its stability to an attracting torus for $\lambda = 2.005$. The torus grows as the parameter is increased to $\lambda = 2.02$. The initial values for all pictures are (0.1, 0.1, 0.1).

Consider a system of autonomous differential equations in $\mathbb{R}^3$ possessing an equilibrium point **p** such that the matrix of the linearized equations at **p** has a real positive eigenvalue λ and a pair of complex eigenvalues $\alpha \mp i\beta$ with negative real parts α. Thus, the equilibrium point has a one-dimensional unstable manifold and a two-dimensional stable manifold. Moreover, let us impose the condition $|\alpha| < \lambda$ on the rates of contraction and expansion. Finally, let us suppose that there is a homoclinic orbit for **p**, that is, an orbit which tends to the equilibrium point **p** in both forward and reverse time. Then, a theorem of Sil'nikov asserts that every neighborhood of the homoclinic orbit contains a countably infinite number of unstable periodic orbits.

The special type of homoclinic orbit above is often referred to as a *Sil'nikov orbit,* and its presence implies chaotic dynamics. When the vector field is subjected to small perturbations, a Sil'nikov orbit is broken. To describe the resulting dynamics near the broken orbit, it is possible to construct a planar map and establish the presence of transversal homoclinic points for this map. This, of course, implies the continuing presence of chaos and it can be described in terms of a certain shift map on infinitely many symbols.

It is important to observe a fundamental difference between a homoclinic tangency of a Poincaré map and a Sil'nikov orbit of a flow in $\mathbb{R}^3$ as a source of chaotic dynamics. In a system containing parameters, as a parameter is varied, there is no transversal homoclinic point before a homoclinic tangency, but chaos associated with transversal homoclinic points afterwards. In contrast, chaotic behavior exists before, during, and after the creation of a Sil'nikov orbit.

Let us now examine a specific system of three ordinary differential equations possessing a Sil'nikov-type homoclinic orbit.

Example 17.3. *A Sil'nikov orbit:* Consider the following piecewise linear system:

$$
\begin{aligned}
\dot{x}_1 &= x_2 \\
\dot{x}_2 &= x_3 \\
\dot{x}_3 &= -x_2 - ax_3 + \begin{cases} 1.0 - bx_1 & \text{if } x > 0.0 \\ 1.0 + cx_1 & \text{if } x \le 0.0. \end{cases}
\end{aligned}
\tag{17.3}
$$

Here, we will set $a = 0.3375$ and $c = 0.633625$, while varying the remaining parameter b.

There is an equilibrium point $\mathbf{p} = (-1/c, 0, 0)$ and the eigenvalues of the matrix of linearization at **p** are approximately 0.4625 and $-0.4 \pm 1.1i$. Therefore, the first set of conditions for the theorem of Sil'nikov listed above are satisfied. The last hypothesis on the existence of a Sil'nikov-type homoclinic orbit is usually very difficult to establish. However, for some value of the parameter b near $b \approx 2.16$, it is possible to verify for our

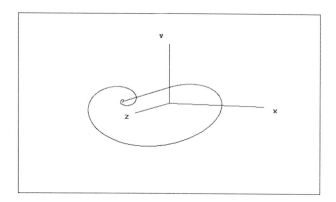

Figure 17.3. *A Sil'nikov type homoclinic orbit in Example 17.3. The initial values are* $(-1.576, 0.0, 0.0)$ *and* $b = 2.16$.

example the presence of a homoclinic orbit to **p** because of the piecewise linear character of the vector field (17.3). We have plotted a reasonably good numerical approximation to this Sil'nikov orbit in Figure 17.3; the one-dimensional unstable manifold of the equilibrium point corresponding to the positive real eigenvalue moves out and later spirals in along the two-dimensional stable manifold corresponding to the complex eigenvalues with negative real parts.

Many of the complicated orbits that arise after the breaking of the homoclinic orbit are unstable and thus not readily visible in numerical computations. In our example, however, there seems to be a rather intricate attracting set for parameter values relatively far from the value for which the homoclinic orbit of Eq. (17.3) is present. We refrain from further exploration of this system, but simply offer you the orbit in Figure 17.4 for the parameter value $b = 0.82$. $\Diamond$

Exercises _____ ♣ ♡ ♠ ◇

17.4. *On Example 17.3:* This example is stored in the library of PHASER under the name *silnikov*. Use it to recreate the figures in this section and for further exploration of the dynamics.

17.5. *Piecewise linear vs. smooth:* The statement of the theorem of Sil'nikov is really for smooth vector fields. With appropriate modifications, it is also applicable to the piecewise linear example in the text. Now, consider the following smooth version of our vector field:

$$\dot{x}_1 = x_2$$
$$\dot{x}_2 = x_3$$
$$\dot{x}_3 = -x_2 - ax_3 + bx_1(1.0 - x_1).$$

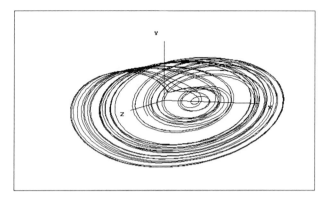

Figure 17.4. *A complicated orbit of Example 17.3 with the same initial data as in the previous figure but with $b = 0.82$.*

(a) Take $a = 0.4$ and $b = 1.6064$. Verify that the linearization of the equilibrium point at the origin satisfies the requirements on the eigenvalues.

(b) As is often the case with smooth vector fields, it does not appear to be possible to establish the existence of a homoclinic orbit to the origin. However, experiment numerically to convince yourself that there may be such a Sil'nikov orbit. Use initial data near the origin close to the unstable manifold.

(c) Study the parameter values $b = 1.0232$ and $b = 0.872$.

Assistance: These equations are stored in the library of PHASER under the name *silnikov2*. Also consult Arneodo, Coullet, and Tresser [1982].

17.4. The Lorenz Equations

The Lorenz equations are a quadratic system of autonomous differential equations in three dimensions modelling a three-mode approximation to the motion of a layer of fluid heated from below. When integrated numerically, they appear to posses extremely complicated solutions. Perhaps this bewildering complexity and the long standing interest in turbulence explain why they have caught the imagination of many people in both theoretical and applied dynamical systems. In this section, we simply introduce these famous equations.

Example 17.4. *The Lorenz equations:* The set of three ordinary differential equations of Lorenz are

$$\dot{x}_1 = s(-x_1 + x_2)$$
$$\dot{x}_2 = rx_1 - x_2 - x_1 x_3 \qquad (17.4)$$
$$\dot{x}_3 = -bx_3 + x_1 x_2,$$

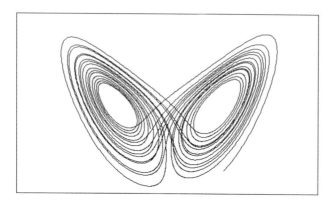

Figure 17.5. *A numerically computed nonperiodic orbit of the Lorenz equations projected onto the (x_1, x_3)-plane.*

where s, r, and b are three real positive parameters. Following Lorenz, we will set the parameter values to $s = 10.0$, $r = 28.0$, and $b = 8/3$. Representative dynamics and bifurcations for other parameter values are indicated in the exercises.

It is relatively easy to establish that the Lorenz equations (17.4) are dissipative. Therefore, it follows from Theorem 13.13 that they have a global attractor that is a compact, connected invariant set. The geometry of the attractor is exceedingly complicated, however, because there are no asymptotically stable equilibria or periodic orbits. Consequently, the attractor of Lorenz is dubbed as a *strange attractor*. All solutions approach the attractor quite rapidly; once on the attractor, most solutions seem to exhibit aperiodic and almost random behavior. Moreover, two solutions with close initial data display radically different dynamical behavior; a behavior called *sensitive dependence on initial data*.

We have illustrated these two key features, nonperiodic solutions and sensitive dependence on initial data, of the Lorenz equations in Figures 17.5 and 17.6. It appears that some of the most profound investigations of Lorenz were motivated by these two factors. You might like to recreate these numerical experiments using PHASER. Be forewarned, however, that many of the numerically observed properties of these innocuous looking equations still await a proof, despite the concerted efforts of many mathematicians. ◇

Exercises ————————————————————————————

17.6. *Lorenz is dissipative:* Using the quadratic Liapunov function

$$V(x_1, x_2, x_3) = rx_1^2 + \sigma x_2^2 + \sigma(x_3 - 2r)^2,$$

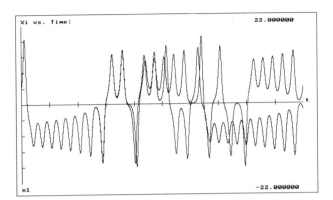

Figure 17.6. *Sensitive dependence on initial data in the solutions of the Lorenz equations. The x_1-coordinates of two solutions against time are plotted. Although their initial data are close, the two solutions behave quite differently.*

show that the Lorenz equations are dissipative for the parameter values set in the text.

Reference: Consult Appendix C of Sparrow [1982] for the details of the necessary computations and other possible Liapunov functions.

17.7. *Elementary bifurcations in Lorenz:* Fix $b = 8/3$ and $\sigma = 10.0$, and establish the following bifurcations as the parameter r is varied:

(a) For $0 < r < 1$, the origin is the global attractor.

(b) At $r = 1$, the origin undergoes a pitchfork bifurcation. To explain the presence of pitchfork, rather than saddle-node bifurcation, verify that the differential equations are invariant under the reflection symmetry $(x_1, x_2, x_3) \mapsto (-x_1, -x_2, x_3)$.

(c) For $r > 1$, many global things happen, but they are hard to identify theoretically or numerically. For example, at $r = 13.962\ldots$ one piece of the unstable manifold of the origin forms a homoclinic loop. Consequently, for values of r larger than this one, the dynamics get noticeably complicated.

(d) For $r = 24.74\ldots$, two of the equilibrium points undergo Poincaré–Andronov–Hopf bifurcation. It is subcritical, in case you look for it numerically.

17.8. *Period doubling in Lorenz:* Fix, as usual, $b = 8/3$ and $\sigma = 10$ for the following numerical experiments:

(a) For $r = 100.5$, use the initial data $(0.0, 5.0, 75.0)$ to locate an asymptotically stable periodic orbit. Watch its projection on the (x_1, x_3)-plane. Make sure to discard transients.

(b) Observe period doubling in the parameter range $99.524 < r < 100.795$. Try, for example, the initial data above with $r = 99.65$.

(c) Explore period-doubling bifurcations also in the parameter windows $145.0 < r < 166.0$ and for $r > 214.364$.

Help: The equations of Lorenz are stored in the library of PHASER under the name *lorenz.*

Bibliographical Notes

Example 17.1 displaying a sequence of period-doubling bifurcations is taken from Rössler [1976]. Period-doubling sequences occur frequently in systems that eventually become chaotic. In case you have missed them, see the exercises for the forced damped oscillator of Duffing in the previous chapter, or the Lorenz equations.

Example 17.2 exhibiting an invariant torus is from Langford [1985]. The unfolding of a purely imaginary and a zero eigenvalue is a codimension-two bifurcation. This important bifurcation, "interaction of periodic and steady state mode," is studied in Langford [1979]; see also Spirig [1983].

The dynamics near a homoclinic orbit in three dimensions as described in the text is due to Sil'nikov [1965]; generalizations to higher dimensions are in Sil'nikov [1970]. Example 17.3 is from Arneodo, Coullet, and Tresser [1982]. Recently, homoclinic orbits of ordinary differential equations have attracted much attention; several representative sources are Glendinning and Sparrow [1984], Guckenheimer and Holmes [1983], and especially Wiggins [1988]. Sil'nikov-type homoclinic orbits have been found in applications such as nerve axon equations; see, for example, Evans, Fenichel, and Feroe [1982], and Hastings [1982].

The Lorenz equations command vast mathematical and experimental literature. The original paper Lorenz [1963] is must reading. The book by Sparrow [1982] and the references therein should provide sufficient resources for further exploration.

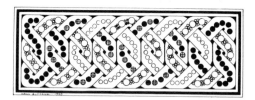

18

Dimension
Four

 This is the final chapter! At the same time, it is the beginning of a new geometric adventure into dimension four—the hyperspace. Arguably, the most natural differential equations residing in dimension four are the Hamiltonian systems with two degrees of freedom. Hence, we have chosen them as the subject of this chapter. Following a rapid introduction to the setting of Hamiltonian systems, we outline a topological program for the study of a small class of Hamiltonians—completely integrable systems—that can be analyzed successfully. From this contemporary viewpoint, we then study the flow of a pair of linear harmonic oscillators. Here, the term bifurcation gains yet another meaning in the context of level sets of the energy-momentum mapping. Our success with completely integrable systems is somewhat overshadowed by their rarity. Indeed, a satisfactory analysis of a general Hamiltonian system in four dimensions—unlike the case of the plane, one degree of freedom—is currently beyond reach. To hint at this complexity, we conclude the chapter with an example of a Hamiltonian that, in all likelihood, is nonintegrable.

18.1. Integrable Hamiltonians

We begin this section with a rapid description of a setting for Hamiltonian mechanics in four dimensions. Then, we implement a topological programme for the global analysis of particularly well-behaved Hamiltonians, the completely integrable systems, using a pair of linear harmonic oscillators.

A classical mechanical conservative system with a four-dimensional phase space can be characterized by its Hamiltonian function

$$H : \mathbb{R}^4 \to \mathbb{R}; \qquad (x_1, \, x_2, \, y_1, \, y_2) \mapsto H(x_1, \, x_2, \, y_1, \, y_2),$$

where x_i are the *position* and y_i are the generalized *momentum* variables. We will assume in the sequel that the Hamiltonian is at least a C^1 function. The time evolution of the system is then governed by the vector field X_H given by the following system of four first-order differential equations of Hamilton:

$$X_H \; : \; \begin{cases} \dot{x}_i = \dfrac{\partial H}{\partial y_i} \\[4mm] \dot{y}_i = -\dfrac{\partial H}{\partial x_i} \end{cases} \qquad \text{for} \;\; i = 1, \, 2. \qquad (18.1)$$

The next important concept of Hamiltonian mechanics is the notion of conservation.

Definition 18.1. *A real-valued C^1 function $L \; : \; \mathbb{R}^4 \to \mathbb{R}$ is called a conserved quantity or a first integral of H, or X_H, if L is constant on the orbits of X_H. Equivalently, L is a first integral of H if their Poisson bracket*

$$\{H, \, L\} = \sum_{i=1}^{2} \left(\frac{\partial H}{\partial y_i} \frac{\partial L}{\partial x_i} - \frac{\partial H}{\partial x_i} \frac{\partial L}{\partial y_i} \right)$$

vanishes identically for all $(\mathbf{x}, \, \mathbf{y}) \in \mathbb{R}^4$.

A trivial but important observation is that H is always a first integral of X_H. Consequently, an orbit of X_H cannot wander in $\mathbb{R}^4$. More precisely, if $H(\mathbf{x}^0, \, \mathbf{y}^0) = h$, then the orbit through $(\mathbf{x}^0, \, \mathbf{y}^0)$ lies on the level set $H^{-1}(h)$, which is called a *constant energy surface*. It usually is a three-dimensional surface (submanifold) in $\mathbb{R}^4$. This is of little comfort for any general observations, however, because of the immense variety of flows in three dimensions, as we have previously seen. Therefore, for the remainder of this section, we confine our attention to a special class of Hamiltonians.

Definition 18.2. *A Hamiltonian H is called completely integrable if it has another first integral L that is functionally independent from H; equivalently, $\{H, L\} = 0$ and the differentials of H and L are linearly independent except on a subset of $\mathbb{R}^4$ of measure zero.*

If L is a first integral of H and $L(\mathbf{x}^0, \mathbf{y}^0) = \ell$, then the orbit of X_H through $(\mathbf{x}^0, \mathbf{y}^0)$ lies on the level set $L^{-1}(\ell)$, which is called a *constant momentum surface*. Consequently, in the case of a completely integrable Hamiltonian, the orbit through $(\mathbf{x}^0, \mathbf{y}^0)$ lies on the intersection of two three-dimensional submanifolds $H^{-1}(h)$ and $L^{-1}(\ell)$. As a result, the dynamics of a completely integrable Hamiltonian H with an integral L is essentially determined by the vector-valued function

$$\mathcal{EM} : \mathbb{R}^4 \to \mathbb{R}^2,$$
$$(x_1, x_2, y_1, y_2) \mapsto \big(H(x_1, x_2, y_1, y_2), \, L(x_1, x_2, y_1, y_2)\big), \tag{18.2}$$

which is called the *energy-momentum mapping* of X_H.

The qualitative study of a completely integrable Hamiltonian system consists of a thorough analysis of its energy-momentum mapping. We begin the local analysis by identifying the topological type of the two-dimensional *energy-momentum surfaces* $\mathcal{EM}^{-1}(h, \ell)$ for all $(h, \ell) \in \mathbb{R}^2$, and determining the flow of X_H on them. For this purpose, we need to identify two types of values of the energy and momentum. A value $(h, \ell) \in \mathbb{R}^2$ is called a *regular value* of $\mathcal{EM}$ if the derivative of $\mathcal{EM}$ has maximum rank on all of $\mathcal{EM}^{-1}(h, \ell)$; otherwise it is a *critical value*. In the case of regular values, we have the following general result:

Theorem 18.3. *Let H be a completely integrable Hamiltonian and consider the (h, ℓ) level set $\mathcal{EM}^{-1}(h, \ell)$ of its energy-momentum mapping $\mathcal{EM}$. If (h, ℓ) is a regular value of $\mathcal{EM}$, then*

(i) *$\mathcal{EM}^{-1}(h, \ell)$ is a smooth two-dimensional surface that is invariant under the flow of X_H.*

(ii) *Connected pieces of $\mathcal{EM}^{-1}(h, \ell)$ are diffeomorphic to either the two-dimensional torus $T^2 = S^1 \times S^1$ if they are compact, or the cylinder $S^1 \times \mathbb{R}$ or $\mathbb{R}^2$ if they are noncompact.*

(iii) *The flow of X_H on each torus piece T^2 is a conditionally periodic motion, that is, it has constant velocity field in suitable toral coordinates.* $\diamondsuit$

To complete the qualitative analysis of the energy-momentum mapping we need to locate the critical values of $\mathcal{EM}$ and address the following global questions:

- What is the *bifurcation set* of $\mathcal{EM}$; that is, the set of $(h, \ell) \in \mathbb{R}^2$ at which the topological type of the energy-momentum level sets change?

- How do the energy-momentum surfaces $\mathcal{EM}^{-1}(h, \ell)$ fit together to "foliate" the constant energy surface $H^{-1}(h)$ for a fixed value of h?

Let us now carry out this ambitious topological program for the analysis of completely integrable systems on one of the simplest yet fundamental examples.

Example 18.4. *Pair of linear harmonic oscillators:* Consider the following family of Hamiltonian functions with $m > 0$ and $n > 0$:

$$H(x_1, x_2, y_1, y_2) = \tfrac{1}{2}m\left(x_1^2 + y_1^2\right) + \tfrac{1}{2}n\left(x_2^2 + y_2^2\right). \qquad (18.3)$$

This is the sum of the energy functions of two harmonic oscillators with frequencies m and n. The corresponding linear system has eigenvalues $\pm im$ and $\pm in$. It is easy to verify that the function

$$L(x_1, x_2, y_1, y_2) = \tfrac{1}{2}m\left(x_1^2 + y_1^2\right) - \tfrac{1}{2}n\left(x_2^2 + y_2^2\right)$$

is a first integral of the Hamiltonian, so that H is completely integrable.

We begin the topological analysis of the energy-momentum mapping $\mathcal{EM} = (H, L)$ of the harmonic oscillators by finding its critical points. A point $(\mathbf{x}, \mathbf{y}) \in \mathbb{R}^4$ is a critical point of $\mathcal{EM}$ if and only if the derivative $D\mathcal{EM}(\mathbf{x}, \mathbf{y})$ at the point $(\mathbf{x}, \mathbf{y})$ is not surjective. This happens when

(i) the point $(\mathbf{x}, \mathbf{y})$ is a critical point of H or L; that is, $DH(\mathbf{x}, \mathbf{y}) = \mathbf{0}$ or $DL(\mathbf{x}, \mathbf{y}) = \mathbf{0}$,
(ii) the point $(\mathbf{x}, \mathbf{y})$ is not a critical point of H but a critical point of $L|H^{-1}(h)$, where $H(\mathbf{x}, \mathbf{y}) = h$. This is equivalent to requiring that

$$DH(\mathbf{x}, \mathbf{y}) + \lambda DL(\mathbf{x}, \mathbf{y}) = \mathbf{0}$$

for some real number $\lambda \neq 0$, which is a Lagrange multiplier problem.

In case (i), the only critical point of H or L is the origin $(0, 0, 0, 0)$ and the corresponding critical value is 0. In case (ii), the solution of the Lagrange multiplier problem shows that, for $h > 0$, the function $L|H^{-1}(h)$ has two circular critical sets:

$$S_{h+}^1 = \left\{(x_1, 0, y_1, 0) \in \mathbb{R}^4 \mid \tfrac{1}{2}m(x_1^2 + y_1^2) = h\right\},$$

$$S_{h-}^1 = \left\{(0, x_2, 0, y_2) \in \mathbb{R}^4 \mid \tfrac{1}{2}n(x_2^2 + y_2^2) = h\right\},$$

with the corresponding critical values (h, h) and $(h, -h)$, respectively.

Next, we will show that a regular energy-momentum surface is a two-dimensional torus $T^2 = S^1 \times S^1$. A point (x_1, x_2, y_1, y_2) belongs to the level set $\mathcal{EM}^{-1}(h, \ell)$ if

$$h = \tfrac{1}{2}m\left(x_1^2 + y_1^2\right) + \tfrac{1}{2}n\left(x_2^2 + y_2^2\right), \quad \ell = \tfrac{1}{2}m\left(x_1^2 + y_1^2\right) - \tfrac{1}{2}n\left(x_2^2 + y_2^2\right),$$

which is equivalent to

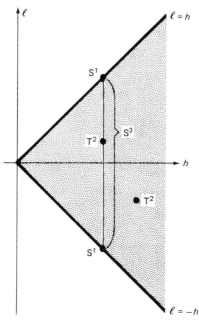

Figure 18.1. *Bifurcation diagram of harmonic oscillators: The range of the energy-momentum mapping is the shaded wedge. On the darkened line segments the preimage of a point, except the origin, is a critical circle. In the inside of the wedge away from the boundaries the preimage of a point is a two-dimensional torus. A constant energy surface is diffeomorphic to S^3, and it is foliated with two critical circles and a family of tori in between.*

$$x_1^2 + y_1^2 = \frac{h+\ell}{m}, \qquad x_2^2 + y_2^2 = \frac{h-\ell}{n}.$$

Thus, $\mathcal{EM}^{-1}(h, \ell)$ is the product of two circles if $h \neq \ell$ and $h \neq -\ell$.

With these bits of information, we now draw a bifurcation diagram for the level sets of the energy momentum mapping of the harmonic oscillators, as seen in Figure 18.1. The set of critical values of $\mathcal{EM}$ are the darkened half-lines, the range of the map is the shaded wedge, and the inside of the wedge away from the boundary contains the regular values. Unlike our earlier bifurcation diagrams, in the bifurcation diagram of $\mathcal{EM}$ we record not the dynamics of the flow of X_H but the changes in the topology of the level sets of $\mathcal{EM}$. As such, the bifurcation diagram is independent of the frequencies m and n.

The flow of X_H on the level sets of the energy-momentum mapping are easy to describe. The level set $\mathcal{EM}^{-1}(0, 0)$ consists of the origin, which is an equilibrium point. For $h > 0$, the circular critical level sets $\mathcal{EM}^{-1}(h, h)$ and $\mathcal{EM}^{-1}(h, -h)$ are always periodic orbits regardless of the frequencies. The flow on each regular toral level set depends strongly on m and n. If m and n are rationally related so that $m = (a/b)n$ for a/b a rational number

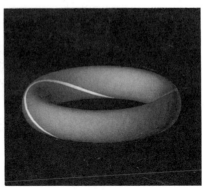

Figure 18.2. *Periodic orbits on a toroidal energy-momentum surface: If the frequencies m and n of oscillators are rationally related so that m = (a/b)n for a/b a rational number in lowest terms, then every orbit, except the critical circles, is a (a, b) toral knot. In these figures, (1, 1) and (2, 1) toral knots are shown.*

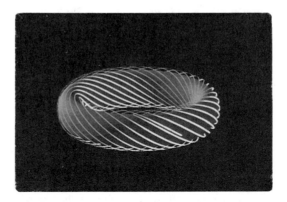

Figure 18.3. *An ergodic orbit on a toroidal energy-momentum surface: If the frequencies of oscillators are not rationally related, then each orbit is dense on the torus.*

in lowest terms, then the flow consists of parallel (a, b) toral knots, and every orbit is periodic, as illustrated in Figure 18.2. If m and n are not rationally related, then every orbit is dense on the torus; see Figure 18.3. This concludes our answer to the first global question we have posed above.

We now briefly turn to the second global question. A constant energy surface $H^{-1}(h)$ of the harmonic oscillators is diffeomorphic to the three-

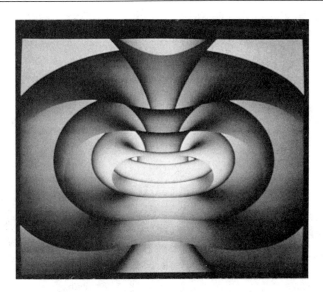

Figure 18.4. *A view of S^3: The three-sphere S^3 is foliated with two critical circles and a one-parameter family of tori in between. Under the four-dimensional stereographic projection, S^3 minus one point is mapped into all of $\mathbb{R}^3$. Here, a sliced view of the foliated S^3 under stereographic projection is shown. One critical circle is inside the smallest torus, while the other is a vertical line through the hole.*

sphere S^3. As indicated in the bifurcation diagram, Figure 18.1, such a constant energy surface is the union of the two critical circles S^1_{h+} and S^1_{h-}, and a one-parameter family of tori "in between" parametrized by the values of the first integral L. This foliation of S^3 is indeed intricate; the two critical circles are linked, and concentric tori enveloping the critical circles fill in the rest of S^3. This complicated four-dimensional geometry can be visualized in three dimensions with the aid of computer graphics, as shown in Figures 18.4 and 18.5.

It is noteworthy that although the solutions of the Hamilton's equations for the linear harmonic oscillators can be written down in terms of trigonometric functions, global considerations lead to deep topological questions. We refrain from presenting further topological details and simply invite you to ponder the graphics. $\diamondsuit$

Exercises _____ ♣♡♠♦

18.1. *Lissajous figures:* The projection of a solution of a pair of linear harmonic oscillators onto the (x_1, x_2)-plane consisting of the position variables is called a *Lissajous figure.* Draw some Lissajous figures for various values of the

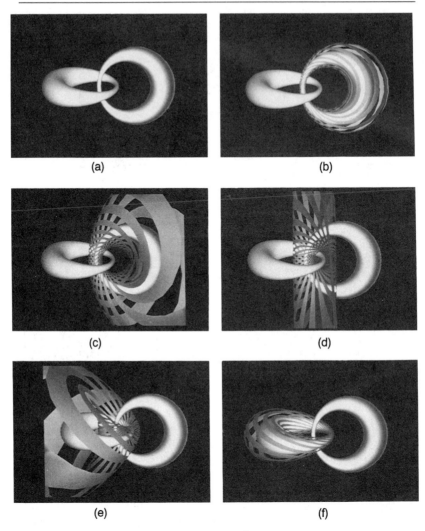

(a) (b)

(c) (d)

(e) (f)

Figure 18.5. *Another animated view of S^3 : The two tori are closely enveloping the two critical circles. A third torus is shown as it moves from one torus to the other. The third torus has been cut into bands to reveal its linkage with the other tori.*

frequencies m and n. For your convenience, the differential equations of linear harmonic oscillators are stored in the 4D library of PHASER under the name *harmoscil.*

18.2. *Opposite harmonic oscillators:* Consider the Hamiltonian function

$$H(x_1,\, x_2,\, y_1,\, y_2) = \tfrac{1}{2}m\big(x_1^2 + y_1^2\big) - \tfrac{1}{2}n\big(x_2^2 + y_2^2\big),$$

with $m > 0$ and $n > 0$. This is the energy function of two linear oscillators "running opposite in time."

(a) Verify that H is completely integrable with the first integral

$$L(x_1, x_2, y_1, y_2) = \tfrac{1}{2}m(x_1^2 + y_1^2) + \tfrac{1}{2}n(x_2^2 + y_2^2).$$

(b) Carry out the topological program for the energy-momentum mapping of this system.

(c) Establish that the origin is a stable equilibrium point of the Hamiltonian vector field X_H even though the Hamiltonian is not positive definite.

Reference: For this and the following problem, you may wish to consult Koçak et al. [1986].

18.3. *At Lagrange libration point L_4:* The quadratic part of the Hamiltonian function of the planar restricted three-body problem at one of the relative equilibrium points is the following function:

$$H(q_1, q_2, p_1, p_2) = \frac{\sqrt{2}}{2}(q_1 p_2 - q_2 p_1) + \tfrac{1}{2}(q_1^2 + q_2^2).$$

(a) Verify that the quadratic function

$$L(q_1, q_2, p_1, p_2) = \frac{\sqrt{2}}{2}(q_1 p_2 - q_2 p_1)$$

is a first integral; thus H is completely integrable.

(b) Write down the corresponding four-dimensional Hamiltonian differential equations; they are constant coefficient.

(c) Compute that the eigenvalues are $\pm\sqrt{2}/2i$.

(d) Show that even though all the eigenvalues are purely imaginary, the origin is an unstable equilibrium point of the linear system. You may need the explicit solution.

18.2. A Nonintegrable Hamiltonian

Although we have devoted a lengthy section to integrable Hamiltonians, they are quite rare, in a precise sense, in the set of all Hamiltonian systems. This was apparent to Poincaré in his work on celestial mechanics. Presently, striking examples of nonintegrable Hamiltonians are abundant due, in part, to computer experiments. In this section, we introduce one such influential example of a nonintegrable Hamiltonian from a model galaxy in an asymmetrical potential.

Example 18.5. *Hénon–Heiles Hamiltonian:* Consider the following cubic Hamiltonian function in $\mathbb{R}^4$:

$$H(x_1, x_2, y_1, y_2) = \tfrac{1}{2}(x_1^2 + x_2^2 + y_1^2 + y_2^2) + x_1^2 x_2 - \tfrac{1}{3}x_2^3. \qquad (18.4)$$

Let us begin with a relatively easy local question. Is the origin a stable equilibrium point of the Hamiltonian vector field X_H? The matrix of the linearized vector field at the origin has purely imaginary eigenvalues in $1:1$ resonance, and thus its stability type cannot be determined from linearization. However, since the quadratic part of the Hamiltonian function is positive definite, it is not difficult to establish that the origin is stable.

The important question regarding the global dynamics of the Hamiltonian (18.4) is the existence or the lack thereof of an additional first integral. To gather experimental evidence towards a resolution of this question, we compute an orbit numerically and manipulate it graphically. For a crude idea, we first project the solution into three dimensions, for example, the (x_1, x_2, x_4)-space. This projection can be rather complicated. To gain further geometric insight, we then plot the points of intersections of the computed orbit with a two-dimensional plane, say the (x_2, x_4)-plane.

Since the Hamiltonian itself is conserved, a computed solution, barring numerical inaccuracies, usually lies on a three-dimensional constant energy surface in $\mathbb{R}^4$. If there happens to be another first integral in addition to the Hamiltonian, then the solution will be confined to a two-dimensional submanifold such as a torus or a cylinder. If, however, there is no additional first integral, then the solution will typically wander in this three-dimensional constant energy surface. These two cases can be distinguished most easily on the planar section. In the presence of an additional first integral, the points of intersection on the plane will yield curves; otherwise, they will be a sprinkling of dots not lying on any discernible curve.

We have plotted in Figure 18.6 a sequence of numerically computed orbits and their planar sections. In this sequence, we have selected the initial data with increasing energy values. The results are rather revealing. For small energy values, there appears to be closed curves on the planar section. As the value of the Hamiltonian gets larger, the curves begin to disintegrate, which is a strong indication that the Hénon–Heiles Hamiltonian is not completely integrable.

A cursory explanation of the disintegration of the invariant tori for increasing energy values is the following. For small energy values, we may view the Hénon–Heiles Hamiltonian (18.4) as a small perturbation of the completely integrable Hamiltonian (18.3) of a pair of harmonic oscillators because the quadratic terms dominate the cubic ones. Now the presence of closed curves on the planar section for small energy values can be inferred from the Twist Theorem by noting that on the planar section near the origin we have an area-preserving perturbation of a twist map. Consequently, for larger energy values the perturbations become stronger leading to disintegration of more of the invariant curves.

Despite the strong numerical evidence above, the important question still remains: Does the Hamiltonian (18.4) of Hénon–Heiles have a first integral that is functionally independent from the Hamiltonian? The answer

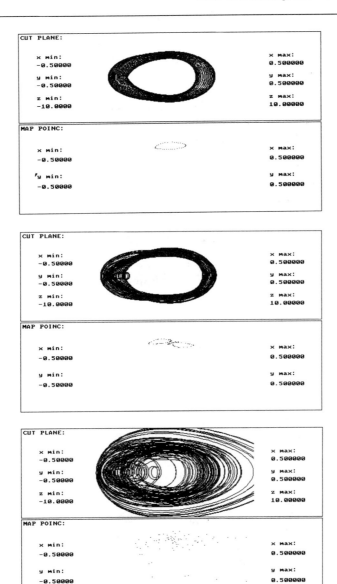

Figure 18.6. *Three numerically computed orbits and their planar sections of the Hénon–Heiles Hamiltonian with increasing energy. The initial data for the first orbit is* $x_1 = 0.34$, $x_2 = 0.25$, $y_1 = 0.1$, *and* $y_2 = 0.1$. *For the other two orbits,* x_1 *is increased to 0.375 and 0.383. The orbits are projected into the* (x_2, y_2, x_1)-*space and the section plane is* $x_1 = 0$.

is not yet entirely satisfactory. It recently has been proved that the Hénon–Heiles Hamiltonian has no *analytic* first integral; however, the nonexistence of a *differentiable* first integral has not yet been established. ◇

Exercises _____ ♣ ♡ ♠ ◇

18.4. *Consult Hénon–Heiles:* There are some famous pictures in the original article by Hénon and Heiles [1964] which you should try to duplicate on the computer. How did they pick so many initial conditions on a given constant energy surface? Do not be fooled by pictures so readily. When inspected in detail, some of those nice looking invariant curves may not be curves at all. Substantiate this remark on the machine using extreme enlargements.

18.5. *Generalized Hénon–Heiles:* Consider the Hamiltonian

$$H(x_1, x_2, y_1, y_2) = \tfrac{1}{2}(ax_1^2 + bx_2^2 + y_1^2 + y_2^2) + x_1^2 x_2 + \tfrac{1}{3}cx_2^3,$$

where a, b, and c are three real parameters. Notice that when $a = b = 1$ and $c = -1$, this is the Hénon–Heiles Hamiltonian. It is known that the Hamiltonian above is completely integrable for (only?) the following sets of values of the parameters:
  (i) $a = b$ and $c = 1$;
  (ii) a and b arbitrary, $c = 6$;
  (iii) $b = 16a$ and $c = 16$.
The corresponding Hamiltonian vector field of the Hamiltonian function above is stored in the library of PHASER under the name *henheile*. Using a variety of initial values, substantiate this assertion.

18.6. *The last problem:* We hope that you have enjoyed the classical designs sprinkled throughout the book. Reexamine them with a mathematician's eye.

Bibliographical Notes _____

A general account of Hamiltonian mechanics, including complete integrability and proof of Theorem 18.3, is given by Abraham and Marsden [1978] and Arnold [1978]. Generic properties of Hamiltonian systems are investigated by Robinson [1970]; nongenericity of complete integrability is discussed by Marcus and Meyer [1974].

Topology of the energy-momentum mappings of linear Hamiltonian systems in four dimensions are studied further using computer graphics by Koçak et al. [1986]; see also Cushman [1974]. Additional illustrations and mathematical details of the foliation of S^3, which is related to the famous *Hopf map*, are in Koçak and Laidlaw [1987]. Global analysis of a nonlinear completely integrable example, the spherical pendulum, is in Cushman [1983].

The original reference to the Hénon–Heiles Hamiltonian is Hénon and Heiles [1964]. A survey on this famous Hamiltonian is Churchill, Pecelli, and Rod [1979]; its analytic nonintegrability was established by Ito [1985] and Churchill and Rod [1988]. General results on the nonexistence of analytic first integrals are in Ziglin [1982 and 1983]; see also the survey by Kozlov [1987]. The question of C^k vs. analytic complete integrability of certain Hamiltonians is investigated by Oliva and Castilla [1989] and Gorni and Zampieri [1989].

Celestial mechanics occupies a special place among Hamiltonians; the classic source is Siegel and Moser [1971]. Moser [1973] describes some of the more recent developments such as random motions, homoclinic points, and analytic nonintegrability. For lighter reading, try Pollard [1966] and Milnor [1983].

A delightful survey of some of the influential numerical experiments in Hamiltonian mechanics, including the planar restricted three-body problem, is in Hénon [1983]. Many of the examples in this article are stored in the library of PHASER, if you wish to experiment. All-purpose numerical algorithms can fail miserably for Hamiltonian systems; see, for example, Hockett [1990] for spurious chaos in the Eulerian numerics of the central force problem, which is a completely integrable system. The development of special purpose symplectic numerical integration algorithms appears to be promising, as exposed by Channell and Scovel [1990].

Farewell

It has indeed been a long and occasionally arduous journey through dynamics together; thank you for your perseverance. We do hope that we have not made the struggle inherent to learning any more painful than necessary. To echo the opening paragraph of our book, we trust that you have found many of the ideas, and especially the examples, informative and enjoyable. In any case, we would like to hear from you with your suggestions and criticisms so that we can make our book a better one in the future.

Despite the innumerable pages we have already filled, we have been unable to mention many exciting new developments in dynamical systems. With the presumption of joint enthusiasm, we would like to offer in these closing sentences an inevitably biased selection from the current directions of research.

The chaotic behavior that we briefly encountered in PART IV is more ubiquitous in applications than previously realized. Unfortunately, a detailed understanding of all aspects of a chaotic flow is nearly impossible, and most current efforts are directed towards identification and computation of distinguishing characteristics in specific systems. Dimension of an attractor, Liapunov exponents, and entropy are some of the most significant such entities.

One can often construct an invariant measure so as to identify those parts of the space where the flow of a dynamical system is most likely to be for a long time. Construction of such measures and the study of the ensuing benefits are the subject of ergodic theory.

Complicated dynamics arise through bifurcations, and the more degenerate the bifurcation, the more complex the resulting dynamics. Ideally, one attempts to find a bifurcation point—an organizing center—with the property that all possible dynamics are encountered in a neighborhood of this point. A search for these organizing centers is underway through the exploitation of symmetries and singular perturbations.

In applications, many phenomena are modelled in the framework of infinite dimensional dynamical systems: partial differential equations, functional equations, etc. Adaptation of ideas from finite dimensional dynamics to this setting promises profound insights into fundamental processes such as turbulence in fluid flow, pattern formation in biological and ecological systems, and phase transitions in condensed matter.

Perhaps we will have the opportunity to explore some of these exciting topics together in a future volume. Perhaps. For now, FAREWELL.

Appendix

 For your convenience, we collect here a bit of notation as well as statements of several fundamental theorems from analysis and differential equations that we had occasion to use in our book. We have also cited references containing the proofs of these results. Let us begin with precise definitions of O and o notations which we have utilized rather liberally throughout the text; for further information, see Hardy [1952].

Little Big Oooh. *Let f and g be two given functions. We say that*

$$f(x) = O\big(g(x)\big) \quad \text{as} \quad x \to 0$$

if there are constants $\alpha > 0$ and $A > 0$ such that $|f(x)| \leq A|g(x)|$ for $|x| < \alpha$. We say that

$$f(x) = o\big(g(x)\big) \quad \text{as} \quad x \to 0$$

if, for any $\varepsilon > 0$, there is a $\delta > 0$ such that $|f(x)| \leq \varepsilon|g(x)|$ for $|x| < \delta$.

Here are some examples of these notations:

$\sin x = O(x), \quad \cos x = O(1), \quad x^{-1} + x^{-3} = O(x^{-3})$ as $x \to 0$;
$\sin x = o(1), \quad 1 - \cos x = o(x), \quad e^{-1/|x|} = o(x^n)$ as $x \to 0$.

We next state a series of theorems from analysis. Until further notice, the proofs of these results can be found in, for example, Smith [1983]. We commence with the Intermediate-value Theorem for real-valued functions of a real variable.

Intermediate-value Theorem. *If the function $f : [a, b] \to \mathbb{R}$ is continuous and $f(a) < 0 < f(b)$, then there exists a point $c \in (a, b)$ so that $f(c) = 0$.*

The next two statements are the Mean-value Theorem of a real-valued function of a real or vector variable. The gradient is denoted by ∇.

Scalar Mean-value Theorem. *Let* $f : [a, b] \to \mathbb{R}$ *be continuous, and differentiable on the open interval* (a, b). *Then for any two points* x_1 *and* x_2 *on the interval* $[a, b]$, *there is a point* $\bar{x}$ *between them such that*

$$f(x_2) - f(x_1) = (x_2 - x_1)f'(\bar{x}).$$

Mean-value Theorem. *Suppose that* U *is an open subset in* $\mathbb{R}^n$, *and that* $\mathbf{x}^1$ *and* $\mathbf{x}^2$ *are two points of* U *such that* U *contains the line segment* L *from* $\mathbf{x}^1$ *to* $\mathbf{x}^2$. *If* $f : U \to \mathbb{R}$ *is a real-valued function defined on the set* U, *then*

$$f(\mathbf{x}^2) - f(\mathbf{x}^1) = \nabla f(\bar{\mathbf{x}}) \cdot (\mathbf{x}^2 - \mathbf{x}^1)$$

for some point $\bar{\mathbf{x}}$ *on the line segment* L.

The most useful tool in local analysis is approximation of functions with polynomials. The next two statements are the scalar and vector versions of Taylor's Theorem regarding such approximations.

Scalar Taylor's Theorem. *Suppose that* $f : I \to \mathbb{R}$ *defined on an open interval* I *is a* C^{m+1} *function. If* a *and* x *are any two points in the interval* I, *then there is a point* ξ *between them such that*

$$f(x) = \sum_{k=0}^{m} \frac{1}{k!} f^{(k)}(a) (x-a)^k + \frac{1}{(m+1)!} f^{(m+1)}(\xi) (x-a)^{m+1},$$

where $f^{(k)}$ *denotes the* kth *derivative of the function* f.

For the general version of Taylor's Theorem, we introduce a bit of notation. Let $\mathbf{i} = (i_1, i_2, \ldots, i_n)$ be an n-vector with nonnegative integer components. The norm of $\mathbf{i}$ is $|\mathbf{i}| = i_1 + i_2 + \cdots + i_n$. For $\mathbf{x} \in \mathbb{R}^n$, let $\mathbf{x}^{\mathbf{i}}$ be the product $\mathbf{x}^{\mathbf{i}} = x_1^{i_1} x_2^{i_2} \cdots x_n^{i_n}$. Finally, if $f : \mathbb{R}^n \to \mathbb{R}$ has $|\mathbf{i}|$ derivatives, let

$$D_{\mathbf{i}} f(\mathbf{x}) = \frac{\partial^{|\mathbf{i}|}}{\partial x_1^{i_1} \cdots \partial x_n^{i_n}} f(\mathbf{x}) \, .$$

Taylor's Theorem. *If* $f : \mathbb{R}^n \to \mathbb{R}$ *is a* C^{m+1} *function on an open set containing the line segment from points* $\mathbf{a}$ *to* $\mathbf{x}$, *then there is a point* ξ *on this line segment such that*

$$f(\mathbf{x}) = \sum_{|\mathbf{i}| \leq m} \frac{1}{|\mathbf{i}|!} D_{\mathbf{i}} f(\mathbf{a}) (\mathbf{x} - \mathbf{a})^{\mathbf{i}} + \frac{1}{(m+1)!} \sum_{|\mathbf{i}| = m+1} D_{\mathbf{i}} f(\xi) (\mathbf{x} - \mathbf{a})^{\mathbf{i}}.$$

The next two statements are about the derivatives of the composition of two functions, both in the scalar and vector cases.

Scalar Chain Rule. *Let* $f : \mathbb{R} \to \mathbb{R}$ *and* $g : \mathbb{R} \to \mathbb{R}$ *be two real-valued functions of a real variable. If* f *is differentiable at point* a *and* g *is differentiable at* $f(a)$, *then the composite function* $(g \circ f)(x) \equiv g(f(x))$ *is differentiable at* a, *and*

$$(g \circ f)'(a) = g'(f(a)) \, f'(a).$$

Chain Rule. *Let* $\mathbf{f} : \mathbb{R}^k \to \mathbb{R}^m$ *and* $\mathbf{g} : \mathbb{R}^m \to \mathbb{R}^n$ *such that* $\mathbf{f}$ *is differentiable at* $\mathbf{x}$ *and* $\mathbf{g}$ *is differentiable at* $\mathbf{f}(\mathbf{x})$. *Then the composite function* $\mathbf{g} \circ \mathbf{f} : \mathbb{R}^k \to \mathbb{R}^n$ *is differentiable at* $\mathbf{x}$, *and*

$$D(\mathbf{g} \circ \mathbf{f})(\mathbf{x}) = D\mathbf{g}(\mathbf{f}(\mathbf{x})) \circ D\mathbf{f}(\mathbf{x}).$$

The second most important tool in local analysis is the Inverse and Implicit Function Theorems.

Inverse Function Theorem. *Let* U *be an open set in* $\mathbb{R}^n$ *and let* $\mathbf{f} : U \to \mathbb{R}^n$ *be a* C^k *function with* $k \geq 1$. *If a point* $\bar{\mathbf{x}} \in U$ *is such that the* $n \times n$ *matrix* $D\mathbf{f}(\bar{\mathbf{x}})$ *is invertible, then there is an open neighborhood* V *of* $\bar{\mathbf{x}}$ *in* U *such that* $\mathbf{f} : V \to \mathbf{f}(V)$ *is invertible with a* C^k *inverse.*

Implicit Function Theorem. *Let* U *be an open set in* $\mathbb{R}^m \times \mathbb{R}^n$ *and let* $\mathbf{f} : U \to \mathbb{R}^n$ *be a* C^k *function with* $k \geq 1$. *Consider a point* $(\bar{\mathbf{x}}, \bar{\mathbf{y}}) \in U$, *where* $\bar{\mathbf{x}} \in \mathbb{R}^m$ *and* $\bar{\mathbf{y}} \in \mathbb{R}^n$, *with* $\mathbf{f}(\bar{\mathbf{x}}, \bar{\mathbf{y}}) = \mathbf{c}$. *If the* $n \times n$ *matrix* $D_{\mathbf{y}}\mathbf{f}(\bar{\mathbf{x}}, \bar{\mathbf{y}})$ *of partial derivatives is invertible, then there are open sets* $V_m \subset \mathbb{R}^m$ *and* $V_n \subset \mathbb{R}^n$ *with* $(\bar{\mathbf{x}}, \bar{\mathbf{y}}) \in V_m \times V_n \subset U$ *and a unique* C^k *function* $\psi : V_m \to V_n$ *such that* $\mathbf{f}(\mathbf{x}, \psi(\mathbf{x})) = \mathbf{c}$ *for all* $\mathbf{x} \in V_m$. *Moreover,* $\mathbf{f}(\mathbf{x}, \mathbf{y}) \neq \mathbf{c}$ *if* $(\mathbf{x}, \mathbf{y}) \in V_m \times V_n$ *and* $\mathbf{y} \neq \psi(\mathbf{x})$. *The derivative of the function* ψ *is given by the formula*

$$D\psi(\mathbf{x}) = - \left[D_{\mathbf{y}}\mathbf{f}(\mathbf{x}, \psi(\mathbf{x})) \right]^{-1} D_{\mathbf{x}}\mathbf{f}(\mathbf{x}, \psi(\mathbf{x})).$$

A useful corollary of the Implicit Function Theorem is the following geometric fact:

Submanifold Theorem. *Let* U *be an open set in* $\mathbb{R}^n$ *and let* $\mathbf{f} : U \to \mathbb{R}^p$ *be a differentiable function such that* $D\mathbf{f}(\mathbf{x})$ *has rank* p *whenever* $\mathbf{f}(\mathbf{x}) = \mathbf{0}$. *Then* $\mathbf{f}^{-1}(\mathbf{0})$ *is an* $(n - p)$-*dimensional manifold in* $\mathbb{R}^n$.

In case you are wondering what the third most important result of local analysis might be, some contend that it is the Malgrange Preparation Theorem; see, for example, Golubitsky and Guillemin [1973]. This theorem is useful in bifurcation theory, although we did not refer to it; consult Chow and Hale [1982]. Two results to which we did refer in the text are the Lemma of Morse and the Theorem of Sard; for proofs, see Milnor [1963 and 1965].

Morse Lemma. Let $f : \mathbb{R}^n \to \mathbb{R}$ be a sufficiently differentiable function. If $\bar{x}$ is a nondegenerate critical point of f, that is, $Df(\bar{x}) = 0$ and the Hessian matrix $\left(\partial^2 f(\bar{x})/\partial x_i \partial x_j\right)$ is nonsingular, then there is a local coordinate system $(y_1, \ldots, y_n)$ in a neighborhood U of $\bar{x}$, with $y_i(\bar{x}) = 0$ for all i, such that

$$f(\mathbf{y}) = f(\bar{x}) - y_1^2 - \cdots - y_k^2 + y_{k+1}^2 + \cdots + y_n^2$$

for all $\mathbf{y} \in U$. The integer k is the number of negative eigenvalues of the Hessian matrix.

Sard's Theorem. Let U be an open set in $\mathbb{R}^n$ and consider a sufficiently differentiable function $\mathbf{f} : U \to \mathbb{R}^p$. Let C be the set of critical points of $\mathbf{f}$, that is, the set of all $\mathbf{x} \in U$ with rank $D\mathbf{f}(\mathbf{x}) < p$. Then $\mathbf{f}(C)$ has measure zero in $\mathbb{R}^p$.

Finally, we come to the generalizations of both the first theorem of our book and the subject at large: the existence, continuation, uniqueness, and dependence on initial data and parameters of an initial-value problem.

We proceed with a notational interlude. Let U be an open set in $\mathbb{R} \times \mathbb{R}^n$ and

$$\mathbf{f} : U \to \mathbb{R}^n; \qquad (t, \mathbf{x}) \mapsto \mathbf{f}(t, \mathbf{x}).$$

Now, suppose that $(t_0, \mathbf{x}^0) \in U$ and consider the initial-value problem

$$\dot{\mathbf{x}} = \mathbf{f}(t, \mathbf{x}), \qquad \mathbf{x}(t_0) = \mathbf{x}^0. \qquad \text{(A1)}$$

A function $\varphi(t, t_0, \mathbf{x}^0)$ is said to be a solution of the initial-value problem (A1) in an interval I containing t_0 if φ is a C^1 function of t on I and satisfies the initial data and the differential equation (A1) for each $t \in I$. If φ is a solution of Eq. (A1) on I, then a function Ψ is said to be a *continuation* of the solution φ if Ψ is a solution of Eq. (A1) on a larger open interval containing I and $\Psi(t, t_0, \mathbf{x}^0) = \varphi(t, t_0, \mathbf{x}^0)$ for $t \in I$. An interval I is said to be a *maximal interval of existence* if φ has no continuation to a larger interval.

Existence, Uniqueness, and Smoothness. Let $\mathbf{f} : U \to \mathbb{R}^n$, where U is an open set in $\mathbb{R} \times \mathbb{R}^n$ and consider the initial-value problem

$$\dot{\mathbf{x}} = \mathbf{f}(t, \mathbf{x}), \qquad \mathbf{x}(t_0) = \mathbf{x}^0.$$

(i) If $\mathbf{f} \in C^0(U, \mathbb{R}^n)$, then there exists a solution $\varphi(t, t_0, \mathbf{x}^0)$ of the initial-value problem defined for all t on a maximal interval of existence $(\alpha_{t_0, \mathbf{x}^0}, \beta_{t_0, \mathbf{x}^0})$.

(ii) For any closed bounded set $W \subset U$, there is $\delta > 0$, depending on W, such that $(t, \varphi(t, t_0, \mathbf{x}^0)) \notin W$ for $t \notin (\alpha_{t_0, \mathbf{x}^0} + \delta, \beta_{t_0, \mathbf{x}^0} - \delta)$.

(iii) If $\mathbf{f} \in C^k(U, \mathbb{R}^n)$, with $k \geq 1$, then there exists a unique solution $\varphi(t, t_0, \mathbf{x}^0)$ of the initial-value problem defined on a maximal interval of existence; moreover, φ is C^k in $(t, t_0, \mathbf{x}^0)$.

Dependence on Parameters. *Let U be an open set in $\mathbb{R} \times \mathbb{R}^n$, and λ be a vector parameter in an open subset Λ of $\mathbb{R}^m$. If $\mathbf{f} \in C^k(\Lambda \times U, \mathbb{R}^n)$, with $k \geq 1$, then the solution $\varphi(\lambda, t, t_0, \mathbf{x}^0)$ of the initial-value problem*

$$\dot{\mathbf{x}} = \mathbf{f}(\lambda, t, \mathbf{x}), \qquad \mathbf{x}(t_0) = \mathbf{x}^0$$

is a C^k function of $(\lambda, t, t_0, \mathbf{x}^0)$.

The proofs of these two theorems are, among others, in Coddington and Levinson [1955], Hale [1980], Hartman [1964], and Robbin [1968].

References

ABRAHAM, R. and MARSDEN, J. [1978]. *Foundations of Mechanics, Second Edition.* Benjamin/Cummings: Reading, Massachusetts.

ABRAHAM, R. and SHAW, C. [1982]. *Dynamics, the Geometry of Behavior, I–IV.* Aerial Press: P.O. Box 1360, Santa Cruz, California 95061.

ANDRONOV, A. [1929]. "Application of Poincaré's theorem on 'bifurcation points' and 'change in stability' to simple autooscillatory systems," *C. R. Acad. Sci. (Paris),* **189**, 559–561.

ANDRONOV, A., LEONTOVICH, E.A., GORDON, I.I., and MAIER, A.G. [1973]. *Theory of Bifurcations of Dynamic Systems on a Plane.* Wiley: New York, New York.

ANDRONOV, A. and PONTRJAGIN, L. [1937]. "Systemes grossiers," *Dokl. Akad. Nauk., SSSR,* **14**, 247–251.

ANDRONOV, A., VITT, A., and XHAIKIN, S.E. [1966]. *Theory of Oscillators.* Pergamon Press: New York, New York.

ANOSOV, D.V. [1967]. "Geodesic flows and closed Riemannian manifolds with negative curvature," *Proc. Steclov Inst. Math.,* **90**.

ANOSOV, D.V. and KATOK, A.B. [1972]. "New examples in smooth ergodic theory, Ergodic diffeomorphisms," *Trans. Mosc. Math. Soc., Am. Math. Soc.,* **23**, 1–32.

ARNEODO, A., COULLET, P., and TRESSER, C. [1981]. "Possible new strange attractor with spiral structure," *Commun. Math. Phys.,* **79**, 573–579.

[1982]. "Oscillations with chaotic behavior: An illustration of a theorem of Shilnikov," *J. Stat. Phys.,* **27**, 171–182.

ARNOLD, V.I. [1963]. " Proof of a theorem of A.N. Kolmogorov on the invariance of quasi-periodic motions under small perturbations of the Hamiltonian," *Russian Math. Surveys,* **18**, 9–36.

[1965]. "Small denominators, I: Mappings of the circumference onto itself," *AMS Trans. Ser. 2,* **46**, 213–284.

[1971]. "Matrices depending on parameters," *Russian Math. Surveys,* **26**, 29–43.

[1972]. "Lectures on bifurcations in versal families," *Russian Math. Surveys*, **27**, 54–119.

[1973]. *Ordinary Differential Equations.* M.I.T. Press: Cambridge, Massachusetts.

[1978]. *Mathematical Methods of Classical Mechanics.* Springer-Verlag, New York, New York, Heidelberg, Berlin.

[1983]. *Geometrical Methods in the Theory of Ordinary Differential Equations,* Grundlehren der mathematischen Wissenschaften, **250**. Springer-Verlag: New York, New York, Heidelberg, Berlin.

[1984]. *Catastrophe Theory.* Springer-Verlag: New York, New York, Heidelberg, Berlin.

ARNOLD, V.I. and AVEZ , A. [1968]. *Ergodic Problems of Classical Mechanics.* Benjamin: New York, New York, Amsterdam.

ARONSON, D.G., CHORY, M.A., HALL, G.R., and McGEHEE, R. [1980]. "A discrete dynamical system with subtly wild behavior," in *New Approaches to Nonlinear Problems in Dynamics*, Holmes, P. (Ed.), 339–359. SIAM Publications: Philadelphia, Pennsylvania.

[1982]. "Bifurcations from an invariant circle for two-parameter families of maps of the plane: a computer assisted study," *Commun. Math. Phys.*, **83**, 303–354.

AULBACH, B. [1984]. *Continuous and Discrete Dynamical Systems near Manifolds of Equilibria.* Lect. Notes in Math., Vol. 1094. Springer-Verlag: New York, New York, Heidelberg, Berlin.

BAK, P. [1986]. "The Devil's staircase," *Physics Today*, **39** (12), 38–45.

BAMÓN, R. [1987]. "Quadratic vector fields in the plane have a finite number of limit cycles," *Publications Mathématiques, I.H.E.S.*, **64**, 111–142.

BARNSLEY, M. [1988]. *Fractals Everywhere.* Academic Press: San Diego.

BENDIXSON, I. [1901]. "Sur les courbes définies par des équations differéntiélles," *Acta. Math.*, **24**, 1–88.

BENEDICKS, M. and CARLESON, L. [1991]. "The dynamics of the Hénon map," *Ann. Math.*, **133**, 73–169.

BEYN, W.J. [1987]. "On invariant closed curves for one-step methods," *Numerische Mathematik*, **51**, 103–122.

BIBIKOV, Y.N. [1979]. *Local Theory of Nonlinear Analytic Ordinary Differential Equations.* Lecture Notes in Math., Vol. 702. Springer-Verlag: New York, New York, Heidelberg.

BIRKHOFF, G. [1927]. *Dynamical Systems*, Amer. Math. Soc. Colloq. Publ., **9**. American Mathematical Society: Providence, Rhode Island.

BOGDANOV, R.I. [1981]. "Versal deformation of a singularity of a vector field on the plane in the case of zero eigenvalues," *Sel. Math. Sov.*, **1**, 389–421.

BOGOLIUBOV, N. and MITROPOLSKI, Y.A. [1961]. *Asymptotic Methods in the Theory of Nonlinear Oscillations.* Gordon and Breach: New York, New York.

BORWEIN, J.M. and BORWEIN, P.B. [1984]. "The arithmetic-geometric mean and fast computation of elementary functions," *SIAM Rev.*, **26**, 351–366.

BOWEN, R. [1975]. "ω-limit sets for Axiom A diffeomorphisms," *J. Differential Equations*, **18**.

BRAUN, M. [1983]. *Differential Equations and Their Applications, Third Edition,* Applied Mathematical Sciences, **15**. Springer-Verlag: New York, New York, Heidelberg, Berlin.

BRAUN, M. and HERSHENOV, J. [1977]. "Periodic solutions of finite difference equations," *Quart. Appl. Math.*, **35**, 139–147.

BROWDER, F. [1976]. Editor; *Mathematical Developments Arising from Hilbert Problems, Proceedings of Symposia in Pure Mathematics, XXVII.* Amer. Math. Soc.: Providence, Rhode Island.

CARR, J. [1981]. *Applications of Center Manifold Theory,* Applied Mathematical Sciences, **45**. Springer-Verlag: New York, New York, Heidelberg, Berlin.

CARTWRIGHT, M.L. and LITTLEWOOD, J.E. [1945]. "On nonlinear differential equations of the second order: I. The equation $\ddot{y} - k(1-y^2)\dot{y} + y = b\lambda k \cos(\lambda t + \alpha)$, k large," *J. London Math. Soc.*, **20**, 180–189.

CESARI, L. [1963]. *Asymptotic Behavior and Stability Problems, Second Edition.* Springer-Verlag: New York, New York, Heidelberg, Berlin.

ČETAEV, N.G. [1934]. "Un théorème sur l'instabilité," *Dokl. Akad. Nauk SSSR,* **2**, 529–534.

CHANNEL, P.J. and SCOVEL, C. [1990]. "Symplectic integration of Hamiltonian systems," *Nonlinearity,* **3**, 231–259.

CHERRY, T. [1938]. "Analytic quasiperiodic curves of discontinuous type on a torus," *Proc. London Math. Soc.,* **44**.

CHICONE, C. and TIAN, J.H. [1982]. "On general properties of quadratic systems," *Amer. Math. Monthly,* **89**, 167–179.

CHOW, S.N. and HALE, J.K. [1982]. *Methods of Bifurcation Theory.* Springer-Verlag: New York, New York, Heidelberg, Berlin.

CHOW, S.N., HALE, J., and MALLET-PARET, J. [1975]. "Applications of generic bifurcations, I," *Arch. Rat. Mech. Anal.,* **59**, 159–188.
[1976]. "Applications of generic bifurcations, II," *Arch. Rat. Mech. Anal.,* **62**, 209–236.
[1980]. "An example of bifurcation to homoclinic orbits," *J. Diff. Eqns.,* **37**, 351–373.

CHOW, S.N. and MALLET-PARET, J. [1977]. "Integral averaging and bifurcation," *J. Diff. Eqns.,* **26**, 112–159.

CHURCHILL, R.C. and ROD, D.L. [1988]. "Geometrical aspects of Ziglin's nonintegrability theorem for complex Hamiltonian systems," *J. Diff. Eqns.,* **76**, 91–114.

CODDINGTON, E. and LEVINSON, N. [1955]. *Theory of Ordinary Differential Equations.* McGraw-Hill: New York, New York.

COLLET, P. and ECKMANN, J.-P. [1980]. *Iterated Maps on the Interval as a Dynamical System.* Birkhäuser: Boston, Massachusetts.

CONLEY, C. [1978]. *Isolated Invariant Sets and the Morse Index,* CBMS **38**. Amer. Math. Soc.: Providence, Rhode Island.

CONTE, S.D. and DeBOOR, C. [1972]. *Elementary Numerical Analysis, Second Edition.* McGraw–Hill: New York, New York.

COPPEL, W.A. [1965]. *Stability and Asymptotic Behavior of Differential Equations.* Heath Mathematical Monographs.

CURRY, J.H. [1979]. "On the Hénon transformation," *Commun. Math. Phys.,* **68**, 129–140.

CUSHMAN, R. [1974]. "The momentum mapping of the harmonic oscillator," *Sympos. Math.,* **14**, 323–342.

[1983]. "Geometry of the energy-momentum mapping of the spherical pendulum," *CWI Newslett.*, **1**, 4–18.

D'ANCONA, U. [1954]. *The Struggle for Existence*. Brill: Leiden.

de MOTTONI, P. and SCHIAFFINO, A. [1981]. "Bifurcation results for a class of periodic quasilinear parabolic equations," *Math. Meth. Appl. Sci.*, **3**, 11–20.

DENJOY, A. [1932]. "Sur les courbes définies par les équations différentielles à la surface du tore," *J. Math. Pures Appl.*, **17**, 333–375.

DESOER, C. and KUH, E. [1969]. *Basic Circuit Theory*. McGraw-Hill: New York, New York.

DEVANEY, R.L. [1978]. "Transversal homoclinic orbits in an integrable system," *Am. J. Math.*, **100**, 631–642.

[1984]. "A piecewise linear model for the zones of instability of an area-preserving map," *Physica 10D*, 387–393.

[1986]. *An Introduction to Chaotic Dynamical Systems*. Benjamin/Cummings: Menlo Park, California, Reading, Massachusetts.

DIENER, M. and REEB, G. [1986]. "Champs polynomiaux: Nouvelles trajectoires remarquables," *Bull. Soc. Math. Belgique*, **38**, 131–150.

DOEDEL, E. [1986]. *AUTO: Software for Continuation and Bifurcation Problems in Ordinary Differential Equations*. CIT Press: Pasadena, California.

DOEDEL, E. and FRIEDMAN, M. [1989]. "Numerical computation of heteroclinic orbits," *J. Comp. Appl. Math.*, **25**, 1–16.

DUFFING, G. [1918]. *Erzwungene Schwingungen bei Veränderlicher Eigenfrequenz*. Braunschweig.

DULAC, M.H. [1923]. "Sur les cycles limites," *Bull. Soc. Math. France*, **51**, 45–188.

ECALLE, J., MARTINET, J., MOUSSU, R., and RAMIS, J.-P. [1987]. "Non-accumulation des cycles-limites," *C.R. Acad. Sci. Paris*, *Tom 304, Series I*, **13**, p. 375, and **14**, p. 431.

ECKMANN, J.-P. and RUELLE, D. [1985]. "Ergodic theory of chaos and strange attractors," *Rev. Modern Phys.*, **57**, 617–656.

EIROLA, T. and NEVANLINNA, O. [1988]. "What do multistep methods approximate?," *Numer. Math.*, **53**, 559–569.

EVANS, J.W., FENICHEL, N., and FEROE, J.A. [1982]. "Double impulse solutions in nerve axon equations," *SIAM J. Appl. Math.*, **42**, 219–234.

FEIGENBAUM, M.J. [1978]. "Quantitative universality for a class of nonlinear transformations," *J. Stat. Phys.*, **19**, 25–52.

[1980]. "Universal behavior in nonlinear systems," *Los Alamos Sci.*, **1**, 4–27.

FIFE, P. [1979]. *Mathematical Aspects of Reaction Diffusion Systems*, Lecture Notes in Mathematics, Vol. 28. Springer-Verlag: New York, New York, Heidelberg, Berlin.

FINK, A.M. [1974]. *Almost Periodic Differential Equations*, Lecture Notes in Mathematics, Vol. 377. Springer-Verlag: New York, New York, Heidelberg, Berlin.

FITZHUGH, R. [1961]. "Impulses and physiological states in theoretical models of nerve membrane," *Biophys. J.*, **1**, 445-466.

FRANKE, J. and SELGRADE, J. [1979]. "A computer method for verification of asymptotically stable periodic orbits," *SIAM J. Math. Anal.*, **10**, 614–628.

GAVRILOV, N.K. and SIL'NIKOV, L.P. [1972]. "On three-dimensional dynamical systems close to systems with a structurally unstable homoclinic curve, I," *Mat. Sbornik*, **17**, 467–485.

[1973]. *Ibid.* II, **19**, 139–156.

GLENDINNING, P. and SPARROW, C. [1984]. "Local and global behavior near homoclinic orbits," *J. Stat. Phys.*, **35**, 645–696.

GOLUB, G. and WILKINSON, J. [1976]. "Ill-conditioned eigensystems and the computation of Jordan canonical form," *SIAM Rev.*, **18**, 578–619.

GOLUBITSKY, M. and GUILLEMIN, V. [1973]. *Stable Mappings and their Singularities.* Springer-Verlag: New York, New York, Heidelberg, Berlin.

GOLUBITSKY, M. and SCHAEFFER, D.G. [1985]. *Singularities and Groups in Bifurcation Theory, Vol. I,* Applied Mathematical Sciences, **51**. Springer-Verlag: New York, New York, Heidelberg, Berlin.

GOLUBITSKY, M., STEWART, I., and SCHAEFFER, D.G. [1988]. *Singularities and Groups in Bifurcation Theory, Vol. II,* Applied Mathematical Sciences, **51**. Springer-Verlag: New York, New York, Heidelberg, Berlin.

GORNI, G. and ZAMPIERI, G. [1990]. "Complete integrability for Hamiltonian systems with a cone potential," *J. Diff. Eqns.*, **85**, 302–337.

GROBMAN, D. [1959]. "Homeomorphisms of systems of differential equations," *Dokl. Akad. Nauk SSSR*, **129**, 880–881.

GUCKENHEIMER, J. and HOLMES, P. [1983]. *Nonlinear Oscillations, Dynamical Systems, and Bifurcations of Vector Fields,* Applied Mathematical Sciences, **42**. Springer-Verlag: New York, New York, Heidelberg, Berlin.

GUCKENHEIMER, J., MOSER, J., and NEWHOUSE, S. [1980]. *Dynamical Systems.* Progress in Mathematics, No. 8. Birkhäuser: Boston, Massachusetts.

HALE, J.K. [1961]. "Integral manifolds of perturbed differential systems," *Ann. Math.*, **73**, 496–531.

[1963]. *Oscillations in Nonlinear Systems.* McGraw-Hill: New York, New York, Toronto, London.

[1977]. Editor, *Studies in Ordinary Differential Equations,* MAA Studies in Mathematics, **14**.

[1977a]. *Theory of Functional Differential Equations.* Springer-Verlag: New York, New York, Heidelberg, Berlin.

[1980]. *Ordinary Differential Equations, Second Edition.* Robert E. Krieger Publishing Company: Huntington, New York.

[1981]. *Topics in Dynamic Bifurcation Theory,"* NFS-CBMS Lectures, **47**, Am. Math. Soc., Providence, Rhode Island.

[1984]. "Introduction to dynamic bifurcation," in Lecture Notes in Mathematics, Vol. 1057. Springer-Verlag: New York, New York, Heidelberg, Berlin.

HALE, J.K. and LA SALLE, J. [1963]. " Differential equations: Linearity vs. Nonlinearity," *SIAM Rev.*, **5**, 249–272.

HALE, J.K. and MASSATT, P. [1982]. "Asymptotic behavior of gradient-like systems," in *Dynamical Systems II* (Eds. Bednarek, A. and Cesari, L.), 85–101. Academic Press: New York, New York, London.

HAMMEL, S.M., YORKE, J.A., and GREBOGI, C. [1987]. "Do numerical orbits of chaotic dynamical processes represent true orbits?," *J. Complexity*, **3**, 136–145.

[1988]. "Numerical orbits of chaotic dynamical processes represent true orbits," *Bull. Amer. Math. Soc.*, **19**, 465–469.

HARDY, G.H. [1952]. *A Course in Pure Mathematics*. Cambridge University Press: Cambridge.

HARTMAN, P. [1963]. "On the local linearization of differential equations," *Proc. Am. Math. Soc.*, **14**, 568–573.

[1964]. *Ordinary Differential Equations*. Wiley: New York, New York.

HASSARD, B.D., KAZARINOFF, N.D., and WAN, Y.-H. [1980]. *Theory and Applications of the Hopf Bifurcation*. Cambridge University Press: Cambridge.

HASSARD, B.D. and WAN, Y.-H. [1978]. "Bifurcation formulae derived from center manifold theory," *J. Math. Anal. Appl.*, **63**(1), 297–312.

HASTINGS, S. [1982]. "Single and multiple pulse waves for the Fitzhugh–Nagumo equations," *SIAM J. Appl. Math.*, **42**, 247–260.

HÉNON, M. [1969]. "Numerical study of quadratic area-preserving mappings," *Quart. Appl. Math.*, **27**, 291–312.

[1976]. "A two-dimensional mapping with a strange attractor," *Commun. Math. Phys.*, **50**, 69–77.

[1983]. "Numerical exploration of Hamiltonian systems," in *Les Houches, Session XXXVI, 1981 – Chaotic Behavior of Deterministic Systems* (Eds. Helleman, R. and Stora, R.). North-Holland: Amsterdam.

HÉNON, M. and HEILES, C. [1964]. "The applicability of the third integral of motion: Some numerical experiments," *Astron. J.*, **69**, 73–79.

HENRY, D. [1981]. *Geometric Theory of Semilinear Parabolic Equations*. Lecture Notes in Mathematics, Vol. 840. Springer-Verlag: New York, New York, Heidelberg, Berlin.

HIRSCH, M.W. [1976]. *Differential Topology*. Springer-Verlag: New York, New York, Heidelberg, Berlin.

[1984]. "The dynamical systems approach to differential equations," *Bull. Am. Math. Soc.*, **11**, 1–63.

HIRSCH, M.W., PUGH, C., and SHUB, M. [1977]. *Invariant Manifolds*. Lecture Notes in Mathematics, Vol. 583. Springer-Verlag: New York, New York, Heidelberg, Berlin.

HIRSCH, M.W. and SMALE, S. [1974].*Differential Equations, Dynamical Systems, and Linear Algebra*. Academic Press: New York, New York, London.

HOCKETT, K. [1990]. "Chaotic numerics from an integrable Hamiltonian system," *Proc. Am. Math. Soc.*, **108**, 271–281.

HOLMES, P.J. and RAND, D. [1978]. "Bifurcations of the forced Van der Pol oscillator," *Quart. Appl. Math.*, **35**, 495–509.

HOPF, E. [1943]. "Abzweigung einer periodischen lösung von einer stationären lösung eines differentialsystems," *Ber. Verh. Sächs, Acad. Wiss. Leipzig Math. Phys.*, **95**, 3–22.

IL'YASHENKO, Y. S. [1990]. "Finiteness theorems for limit cycles," *Uspekhi Mat. Nauk*, **45**, 143–200.

IOOSS, G. [1979]. *Bifurcation of Maps and Applications*. North-Holland: Amsterdam.

IRWIN, M.C. [1980]. *Smooth Dynamical Systems*. Academic Press: New York, New York.

ITO, H. [1985]. "Non-integrability of Hénon–Heiles system and theorem of Ziglin," *Kodai Math. J.*, **8**, 120–138.

JOHNSON, R.A. and SELL, G.R. [1981]. "Smoothness of spectral subbundles and reducibility of quasiperiodic linear differential systems," *J. Diff. Eqns.*, **41**, 262–288.

JONES, D.S. and SLEEMAN, B.D. [1983]. *Differential Equations of Mathematical Biology*. George Allen and Unwin: London.

KAN, I., KOÇAK, H., and YORKE, J. [1990]. "Antimonotonicity: Concurrent creation and annihilation of periodic orbits," *Ann. Math.*, to appear.

KELLEY, A. [1967]. "The stable, center stable, center, center unstable, and unstable manifolds," *J. Diff. Eqns.*, **3**, 546–570.

KIELHOFER, H. [1979]. "Hopf bifurcation at multiple eigenvalues," *Arch. Rational Mech. Anal.*, **69**, 53–83.

KIRCHGRABER, U. and STIEFEL, E. [1978]. *Methoden der Analytischen Sörungrechnung und ihre Anwendungen*. Teubner: Stutgart.

KLOEDEN, P.E. and LORENZ, J. [1986]. "Stable attracting sets in dynamical systems and in their one-step discretizations," *SIAM J. Numer. Anal.*, **23**, 986–995.

KOÇAK, H. [1984]. "Normal forms and versal deformations of linear Hamiltonian systems," *J. Diff. Eqns.*, **51**, 359–407.

[1989]. *Differential and Difference Equations through Computer Experiments, Second Edition*, with diskettes containing *PHASER: An Animator/Simulator for Dynamical Systems for I.B.M. Personal Computers*. Springer-Verlag: New York, New York, Heidelberg, Berlin.

KOÇAK, H., BISSHOPP, F., BANCHOFF, T., and LAIDLAW, D. [1986]. "Topology and mechanics with computer graphics: Linear Hamiltonian systems in four dimensions," *Advances in Applied Mathematics*, **7**, 282–308.

KOÇAK, H. and LAIDLAW, D. [1987]. "Computer graphics and the geometry of S^3," *The Mathematical Intelligencer*, **9**, 8–10.

KOLMOGOROV, A.N. [1954]. "La théorie générale des systèmes dynamique et méchanique classique," *Proc. Int. Congr. Math., Amsterdam*. 315–333. English translation in Abraham and Marsden [1978].

KOLMOGOROV, A.N., PETROVSKII, I.G., and PUSKINOV, N.S. [1937]. "A study of the equation of diffusion with increase in the quantity of matter, and its application to a biological problem," *Bjul. Moskovskovo Gos. Univ.*, **17**, 1–72.

KOZLOV, V.V. [1987]. "Phenomena of nonintegrability in Hamiltonian systems," in *Proc. International Congress of Mathematicians*, 1161–1170. Am. Math. Soc.: Providence, Rhode Island.

KRASOVSKII, N.N. [1963]. *Stability of Motion*. Stanford Univ. Press: Palo Alto, California.

KUBICEK, M. and MAREK, M. [1986]. *Computational Methods in Bifurcation Theory and Dissipative Structures*, Springer Series Comp. Physics. Springer-Verlag: New York, New York, Heidelberg, Berlin.

KUIPER, N.H. [1975]. "The topology of the solutions of a linear differential equation on $\mathbb{R}^n$," in *Manifolds–Tokyo 1973*, 195–203. University of Tokyo Press: Tokyo.

KUIPER, N.H. and ROBBIN, J. [1973]. "Topological classification of linear endomorphisms," *Invent. Math.*, **19**, 83–106.

LADIS, N.N. [1973]. "Topological equivalence of linear flows," *Differential Equations*, **9**, 938–947.

LANFORD, O. [1973]. "Bifurcations of periodic solutions into invariant tori: The work of Ruelle and Takens," in *Nonlinear Problems in the Physical Sciences and Biology*, Lecture Notes in Mathematics, Vol. 322, 159–192. Springer-Verlag: New York, New York, Heidelberg, Berlin.

[1982]. "A computer-assisted proof of the Feigenbaum conjectures," *Bull. Am. Math. Soc.*, **6**, 427–434.

[1987]. "Computer-assisted proofs in analysis," in *Proc. International Congress of Mathematicians*, 1385–1394. Am. Math. Soc.: Providence, Rhode Island.

LANGFORD, W. [1979]. "Periodic and steady mode interactions lead to tori," *SIAM J. Appl. Math.*, **37**, 22–48.

[1985]. "Unfolding of degenerate bifurcations," in *Chaos, Fractals, and Dynamics* (Eds. Fisher, P. and Smith, W.), 87–103. Marcel Dekker: New York, New York, Basel.

LA SALLE, J. [1960]. "The extent of asymptotic stability," *Prod. Natl. Acad. Sci. U.S.A.*, **46**, 363–365.

[1976]. *The Stability of Dynamical Systems*, Regional Conference Series in Applied Mathematics, **25**. SIAM: Philadelphia, Pennsylvania.

[1987]. *The Stability and Control of Discrete Processes*, Applied Mathematical Sciences, **62**. Springer-Verlag: New York, New York, Heidelberg, Berlin.

LA SALLE, J. and LEFSCHETZ, S. [1961]. *Stability by Liapunov's Direct Method, With Applications*. Academic Press: New York, New York.

LEFEVER, F. and NICHOLIS, G. [1971]. "Chemical instabilities and sustained oscillations," *J. Theoretical Biology*, **30**, 267–284.

LEFSCHETZ, S. [1965]. *Stability of Control Systems*. Academic Press: New York, New York.

LEVI, M. [1981]. "Qualitative analysis of the periodically forced relaxation oscillations," *Mem. AMS*, **214**, 1-147.

LEVI-CIVITA, T. [1901]. "Sopra alcuni criteri di instabilita," *Ann. Mat. Pura Appl.* **3**(5), 221–307.

LEVINSON, N. [1949]. "A second-order differential equation with singular solutions," *Ann. Math.*, **50**, 127–153.

[1950]. "Small periodic perturbations of an autonomous system with a stable orbit," *Ann. Math.*, **52**, 727–738.

LEVINSON, N. and SMITH, O.K. [1942]. "A general equation for relaxation oscillations," *Duke Math. J.*, **9**, 382–403.

LI, T.Y. and YORKE, J. [1975]. "Period three implies chaos," *American Math. Monthly*, **82**, 985–992.

[1979]. "Path following approaches for solving nonlinear equations: Homotopy, continuous Newton, and projection," in Lecture Notes in Mathematics, Vol. 370, 257–264. Springer-Verlag: New York, New York, Heidelberg, Berlin.

LIBCHABER, A. and MAUER, J. [1982]. "A Rayleigh–Bernard experiment: Helium in a small box," in *Nonlinear Phenomena at Phase Transitions and In-*

stabilities (Ed. Riste, T.), 259–286. Plenum Publication Corp.: New York, New York.

LLOYD, N.G. [1972]. "The number of periodic solutions of the equation $\dot{z} = z^N + p_1(t)z^{N-1} + \cdots + p_N(t)$," *Proc. London Math. Soc.*, **27**(3), 667–700.

LORENZ, E. [1963]. "Deterministic nonperiodic flow," *J. Atmospheric Sci.*, **20**, 130–141.

LOZI, R. [1978]. "Un attracteur etrange(?) du type attracteur de Hénon," *J. Phys.*, **39**, 9–10.

LYAPUNOV, A. [1947]. *Problème Général de la Stabilité de Mouvement*, Annals of Math. Studies, **17**. Princeton University Press: Princeton, New Jersey. (Translation of the Russian edition originally published in 1892 by the Mathematics Society of Kharkov.)

MALLET-PARET, J. and SELL, G. [1989]. "Inertial manifolds for reaction-diffusion equations in higher space dimension," *J. Amer. Math. Soc.*, **1**, 805–866.

MAGNUS, W. and WINKLER, S. [1979]. *Hill's Equation*. Dover: New York, New York.

MARCUS, L. and MEYER, K. [1974]. "Generic Hamiltonian systems are neither integrable nor ergodic," *Mem. Am. Math. Soc.*, **144**.

MARCUS, L. and YAMABE, H. [1960]. "Global stability criteria for differential systems," *Osaka Math. J.* **12**, 305–317.

MARSDEN, J. and McCRACKEN, M. [1976]. *Hopf Bifurcation and Its Applications*, Applied Mathematical Sciences, **19**. Springer-Verlag, New York, New York, Heidelberg, Berlin.

MAY, R. [1973]. *Stability and Complexity in Model Ecosystems*. Princeton University Press: Princeton, New Jersey.

MAYNARD SMITH, J. [1968]. *Mathematical Ideas in Biology*. Cambridge University Press: London, New York, New York.

[1974]. *Models in Ecology*. Cambridge University Press: London, New York, New York.

MEL'NIKOV, V.K. [1963]. "On the stability of the center for time periodic perturbations," *Trans. Moscow Math.*, **12**, 1–57.

McGEHEE, R. and MEYER, K. [1974]. "Homoclinic points of area-preserving diffeomorphisms," *Amer. J. Math.*, **96**, 409–421.

McLACHLAN, N.W. [1947]. *Theory and Application of Mathieu Functions*. The Clarendon Press: Oxford.

MILNOR, J. [1963]. *Morse Theory*. Princeton University Press: Princeton, New Jersey.

[1965]. *Topology from the Differentiable Viewpoint*. The University Press of Virginia, Charlottesville.

[1983]. "On the geometry of the Kepler problem," *Amer. Math. Monthly*, **90**, 353–364.

MISUREWICZ, M. [1980]. "Strange attractors for the Lozi mapping," in *Nonlinear Dynamics*, Annals of the New York Academy of Sciences, **357**, 348–358.

MOSER, J. [1962]. "On invariant curves of area-preserving mappings of an annulus," *Nachr. Akad. Wiss., Göttingen Math. Phys.*, **K1**, 1–20.

[1968]. "Lectures on Hamiltonian Systems," *Memoirs A.M.S.*, **81**.

[1973]. *Stable and Random Motions in Dynamical Systems.* Princeton University Press: Princeton, New Jersey.

[1986]. "Recent developments in the theory of Hamiltonian systems," *SIAM Review*, **28**, 459–486.

NEIMARK, J.I. [1959]. "On some cases of periodic motions depending on parameters," *Dokl. Acad. Nauk SSSR*, **129**, 736–739.

NEWHOUSE, S. [1979]. "The abundance of wild hyperbolic sets and non-smooth stable sets," *Publ. I.H.E.S.*, **50**, 101–151.

NEWHOUSE, S., PALIS, J., and TAKENS, F. [1976]. "Stable arcs of diffeomorphisms," *Bull. Amer. Math. Soc.*, **82**, 491–502.

NEWMAN, M.H.A. [1953]. *Topology of Plane Sets.* Cambridge University Press: London, New York, New York.

NITECKI, Z. [1971]. *Differentiable Dynamics.* M.I.T. Press: Cambridge, Massachusetts.

OBI, C. [1954]. "Researches on the equation $\ddot{x} + (\varepsilon_1 + \varepsilon_2 x)\dot{x} + x + \xi x^3 = 0$," *Proc. Cambridge Philos. Soc.*, **50**, 26–32.

OLIVA, W.M. and CASTILLA, M.S. [1989]. "On a class of C^∞-integrable Hamiltonian systems," *Proc. Royal Soc. Edinburgh*, **113**, 293–314.

PALIS, J. and de MELO, W. [1982]. *Geometric Theory of Dynamical Systems: An Introduction.* Springer-Verlag: New York, New York, Heidelberg, Berlin.

PALIS, J. and SMALE, S. [1970]. "Structural stability theorems," in *Global Analysis* (Eds. Chern, S.S. and Smale, S.), Proceedings of Symposia in Pure Mathematics, Vol. XIV, 223–232. American Mathematical Society: Providence, Rhode Island.

PALMER, K. [1984]. "Exponential dichotomies and transversal homoclinic points," *J. Differential Equations*, **55**, 225–256.

PEITGEN, H.O. and RICHTER, P.H. [1986]. *The Beauty of Fractals: Images of Complex Dynamical Systems.* Springer-Verlag: New York, New York, Heidelberg, Berlin.

PEITGEN, H.O. and PRÜFER, M. [1979]. "The Leray-Schauder continuation method in a constructive element in the numerical study of nonlinear eigenvalue and bifurcation problems," *Lect. Notes in Math.*, **730**, 326–409.

PEIXOTO, M.M. [1959]. "On structural stability," *Ann. Math.*, **69**(2), 199-222.

[1962]. "Structural stability on two dimensional manifolds," *Topology*, **1**, 101–120.

[1973]. Editor, *Dynamical Systems.* Academic Press: New York, New York, London.

PIELOU, E.C. [1969]. *An Introduction to Mathematical Ecology.* McGraw-Hill: New York, New York.

PLISS, V.A. [1966]. *Nonlocal Problems in the Theory of Oscillations.* Academic Press: New York, New York.

POINCARÉ, H. [1881]. "Memoire sur les curves définies par une équation differentielle," *J. Math. Pures Appl.*, **7**(3), 375–422.

[1899]. *Les Méthodes Nouevelles de la Méchanique Céleste.* Gauthier-Villar: Paris.

POLLARD, H. [1966]. *Mathematical Introduction to Celestial Mechanics.* Prentice-Hall: Englewood Cliffs, New Jersey.

POSTON, T. and STEWART, I. [1978]. *Catastrophe Theory and its Applications.* Pitman: San Francisco.

POUNDER, J.R. and ROGERS, T.D. [1980]. "The geometry of chaos: Dynamics of a nonlinear second-order difference equation," *Bull. Math. Biol.,* **42**(4), 551–597.

RAND, D. [1988]. "Universality and renormalization in dynamical systems," in *New Directions in Dynamical Systems* (Eds. Bedford, T. and Swift, J.W.). Cambridge University Press: Cambridge.

RAND, R.H. and ARMBRUSTER, D. [1987]. *Perturbation Methods, Bifurcation Theory and Computer Algebra.* Applied Mathematical Sciences, **65**. Springer-Verlag: New York, New York, Heidelberg, Berlin.

RAYLEIGH, L. [1883]. "On maintained vibrations," *Philos. Magazine,* **15**, 229.

ROBBIN, J. [1968]. "On the existence theorem for differential equations," *Proc. Am. Math. Soc.,* **19**, 1005–1006.

[1972]. "Topological conjugacy and structural stability for discrete dynamical systems," *Bull. Amer. Math. Soc.,* **78**, 923–952.

ROBINSON, C. [1970]. "Generic properties of conservative systems, I, II," *Am. J. Math.,* **92**, 562–603 and 897–906.

[1983]. "Bifurcation to infinitely many sinks," *Commun. Math. Phys.,* **90**, 433–459.

ROGERS, T.D. and WHITLEY, D.C. [1983]. "Chaos in the cubic mapping," *Math. Modelling,* **4**, 9–25.

RÖSSLER, O.E. [1976]. "An equation for continuous chaos," *Phys. Lett.,* **A57**, 397–398.

RUELLE, D. [1980]. "Strange attractors," *Math. Intelligencer,* **2**, 126–132.

RUELLE, D. and TAKENS, F. [1971]. "On the nature of turbulence," *Commun. Math. Phys.,* **20**, 167–192.

[1971a]. "Note concerning our paper 'On the nature of turbulence'," *Commun. Math. Phys.,* **23**, 343–344.

RYBAKOWSKI, K. [1987]. *The Homotopy Index and Partial Differential Equations.* Springer-Verlag: New York, New York, Heidelberg, Berlin.

SAARI, D.G. and URENKO, J.B. [1984]. "Newton's method, circle maps, and chaotic motion," *Am. Math. Monthly,* **91**, 3–18.

SACKER, R.J. [1965]. "A new approach to the perturbation theory of invariant surfaces," *Commun. Pure Appl. Math.,* **18**, 717–732.

SANCHEZ, D.A. [1982]. "Periodic environments, harvesting, and a Riccati equation," in *Nonlinear Phenomena in Mathematical Sciences,* 883–886. Academic Press: New York, New York, London.

SANDERS, J. and VERLHULST, F. [1987]. *Averaging Methods in Nonlinear Dynamical Systems.* Springer-Verlag: New York, New York, Heidelberg, Berlin.

SCHMIDT, D.S. [1978]. "Hopf's bifurcation theorem and the center theorem of Liapunov with resonant cases," *J. Math. Anal. Appl.,* **63**, 354–370.

SCHMIDT, E. [1908]. "Zur Theorie der linearen und nichtlinearen Integralgleichungen," *Math. Ann.,* **65**, 370–379.

SCHWARTZ, A.J. [1963]. "A generalization of the Poincaré–Bendixson theorem to closed two-dimensional manifolds," *Amer. J. Math.,* **85**, 453–458. Errata, *ibid,* **85**, 753.

SELL, G. [1971]. *Topological Dynamics and Ordinary Differential Equations.* Van Nostrand Reinhold: London.

SHARKOVSKII, A.N. [1964]. "Coexistence of cycles of a continuous map of a line into itself," *Ukr. Math. Z.,* **16**, 61–71.

SHENKER, S.J. [1982]. "Scaling behavior in a map of a circle onto itself: Empirical results," *Physica* **5D**, 405–411.

SHI, S.-L. [1980]. "A concrete example of the existence of four limit cycles for plane quadratic systems," *Sci. Acta,* **23**, 153–158.

SIEGEL, C.L. and MOSER, J.K. [1971]. *Lectures on Celestial Mechanics.* Springer-Verlag: New York, New York, Heidelberg, Berlin.

ŠIL'NIKOV, L.P. [1965]. "The case of the existence of a denumerable set of periodic motions," *Sov. Math. Dokl.,* **6**, 163–166.

[1967]. "On a Poincaré–Birkhoff problem," *Math. USSR-Sbornik,* **3**, 353–371.

[1970]. "A contribution to the problem of the structure of an extended neighborhood of a rough equilibrium state with saddle-focus type," *Math. USSR-Sbornik,* **10**, 91–102.

SINGER, D. [1978]. "Stable orbits and bifurcations of maps of the interval," *SIAM J. Appl. Math.,* **35**, 260–268.

SMALE, S. [1961]. "On gradient dynamical systems," *Ann. Math.,* **74**, 199–206.

[1961a]. "Generalized Poincaré's conjecture in dimensions greater than four," *Ann. Math.,* **74**, 391–406.

[1963]. "Diffeomorphisms with many periodic points," in *Differential and Combinatorial Topology* (Ed. Cairns, S.S.), 63–80. Princeton University Press: Princeton, New Jersey.

[1966]. "Structurally stable systems are not dense," *Amer. J. Math.,* **88**, 491–496.

[1967]. "Differentiable dynamical systems," *Bull. Amer. Math. Soc.,* **73**, 747–817.

[1972]. "On the mathematical foundations of electrical circuit theory," *J. Differential Geom.,* **7**, 193–210.

[1976]. "A convergent process of price adjustment and global Newton's method," *J. Math. Econ.,* **3**, 1–14.

[1981]. "The fundamental theorem of algebra and complexity theory," *Bull. Amer. Math. Soc.,* **4**, 1–36.

[1987]. "Algorithms for solving equations," in *Proc. International Congress of Mathematicians,* 172–195. Am. Math. Soc.: Providence, Rhode Island.

SMITH, K.T. [1983]. *Primer of Modern Analysis.* Springer-Verlag: New York, New York, Heidelberg, Berlin.

SOTOMAYOR, J. [1974]. "Generic one-parameter families of vector fields on two-dimensional manifolds," *Publ. Math. IHES,* **43**, 5–46.

[1984]. *Introduccion al Estudio de las Bifurcationes de los Sistemas Dinamocos.* Universidad Simon Bolivar, Fodo Editorial: Caracas.

[1985]. "Stable planar polynomial vector fields," *Rev. Mat. Iberoamericana,* **1**, 15–23.

SOTOMAYOR, J. and PATERLINI, R. [1987]. "Bifurcations of polynomial vector fields in the plane," in *Oscillations, Bifurcations, and Chaos* (Toronto, Ont., 1986), 665–685, *CMS Conf. Proc.,* **8**. Amer. Math. Soc.: Providence, Rhode Island.

SPARROW, C. [1982]. *The Lorenz Equations: Bifurcations, Chaos and Strange Attractors*, Applied Mathematical Sciences, **41**. Springer-Verlag: New York, New York, Heidelberg, Berlin.

SPIRIG, F. [1983]. "Sequence of bifurcations in a three-dimensional system near a critical point," *J. Appl. Math. Phys. (ZAMP)*, **34**, 259–276.

STEFAN, P. [1977]. "A theorem of Sarkovskii on the existence of periodic orbits of continuous endomorphisms of the real line," *Commun. Math. Phys.*, **54**, 237–248.

STOKER, J.J. [1950]. *Nonlinear Vibrations*. Interscience: New York, New York.

TAKENS, F. [1971]. "A C^1 counterexample to Moser's twist theorem," *Indag. Math.*, **33**, 379–386.

[1974a]. "Singularities of vector fileds," *Publ. Math. IHES*, **43**, 47–100.

[1974b]. "Forced oscillations and bifurcations," *Commun. Math. Inst., Rijksuniversiteit Utrecht*, **3**, 1–59.

[1987]. "Homoclinic bifurcations," in *Proc. International Congress of Mathematicians*, 1229–1236. Am. Math. Soc.: Providence, Rhode Island.

TEMAM, R. [1988]. *Infinite Dimensional Dynamical Systems in Mechanics and Physics*. Springer-Verlag: New York, New York, Heidelberg, Berlin.

THOM, R. [1975]. *Structural Stability and Morphogenesis*. Addison Wesley: New York, New York.

THOMPSON, J.M.T. and HUNT, G.W. [1973]. *A General Theory of Elastic Stability*. Wiley: New York, New York.

TODA, M. [1981]. *Theory of Nonlinear Lattices*. Springer-Verlag: New York, New York, Berlin, Heidelberg.

ULAM, S.M. and von NEUMANN, J. [1947]. "On combinations of stochastic and deterministic processes," *Bull. Amer. Math. Soc.*, **53**, 1120.

USHIKI, S. [1986]. "Chaotic phenomena and fractal objects in numerical analysis," in *Patterns and Waves*, 221–258. North Holland: Amsterdam.

VAN DER POL, B. [1927]. "Forced oscillations in a circuit with nonlinear resistance," *London, Edinburg and Dublin Philos. Mag.*, **3**, 65–80.

VAN DER POL, B. and VAN DER MARK, J. [1927]. "Frequency demultiplication," *Nature*, **120**, 363–364.

VAN VELDHUIZEN, M. [1988]. "On the numerical approximation of the rotation number," *J. Computational and Applied Math.*, **21**, 203–212.

[1988a]. "Convergence results for invariant curve algorithms," *Math. Computation*, **51**, 677–697.

WAN, Y.H. [1978]. "Computations of the stability condition for the Hopf bifurcation of diffeomorphisms on $\mathbb{R}^2$," *SIAM J. Appl. Math.*, **34**, 167–175.

WHITLEY, D.C. [1983]. "Discrete dynamical systems in dimensions one and two," *Bull. London Math. Soc.*, **15**, 177–217.

WHITNEY, H. [1955]. "On singularities of mappings of Euclidean spaces: I. Mappings of the plane into the plane," *Ann. Math.*, **62**, 374–410.

WIGGINS, S. [1988]. *Global Bifurcations and Chaos*. Springer-Verlag: New York, New York, Berlin, Heidelberg.

WILLEMS, J.C. [1980]. "Topological classification and structural stability of linear systems," *J. Diff. Eqns.*, **35**, 306–318.

Y E YAN-QIAN and Others [1986]. *Theory of Limit Cycles.* Translations of Mathematical Monographs, **66**. Amer. Math. Soc.: Providence, Rhode Island.

YORKE, J. [1989]. "Dynamics." Software for dynamical systems. University of Maryland.

YOSHIZAWA, T. [1966]. *Stability Theory of Liapunov's Second Method.* Math. Soc. Japan: Tokyo.

[1974]. *Stability Theory and the Existence of Periodic and Almost Periodic Solutions.* Springer-Verlag: New York, New York, Heidelberg, Berlin.

Z EEMAN, C. [1977]. *Catastrophe Theory: Selected Papers, 1972–1977.* Addison-Wesley: Reading, Massachusetts.

ZEHNDER, E. [1973]. "Homoclinic points near elliptic fixed points," *Commun. Pure Appl. Math.,* **26**, 131–182.

ZIGLIN, S.L. [1982]. "Branching of solutions and nonexistence of first integrals in Hamiltonian mechanics, I," *Functional Anal. Appl.,* **16**, 181–189.

[1983]. "Branching of solutions and nonexistence of first integrals in Hamiltonian mechanics, II," *Functional Anal. Appl.,* **17**, 6–17.

Index

To enhance the utility of the index, we have followed certain typesetting conventions regarding the page numbers. An <u>underlined</u> page number indicates a main source of information such as a definition or a theorem. The suffix "r" refers to a reference in the mathematical literature. An *italicized* page number points to an example. Underlined *italicized* page numbers are reserved for the three hundred and fourteen figures.

Texts in Applied Mathematics

(continued from page ii)

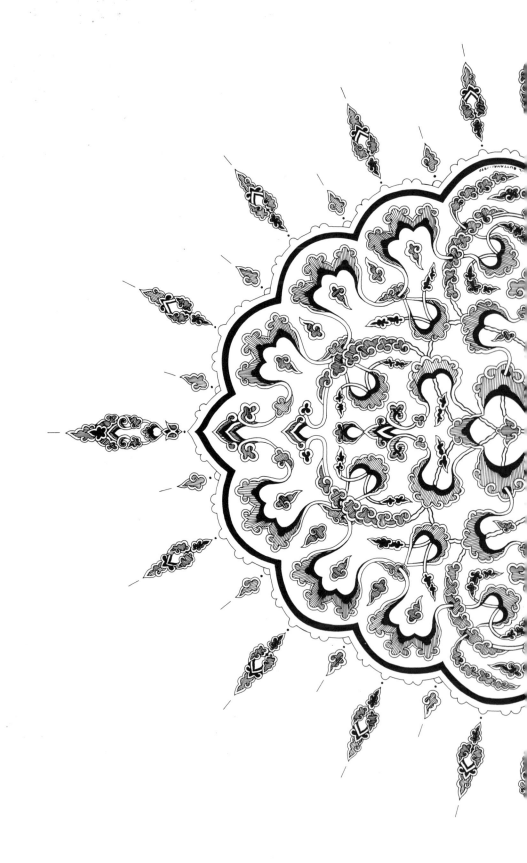